A GUIDE TO SOIL MECHANICS

Also from The Macmillan Press

An Introduction to Engineering Fluid Mechanics, Second Edition
J. A. Fox

Civil Engineering Materials
edited by N. Jackson

Reinforced Concrete Design
W. H. Mosley and J. H. Bungey

Highway Traffic Analysis and Design
R. J. Salter

Structural Theory and Analysis
J. D. Todd

Surveying for Engineers
J. Uren and W. F. Price

Engineering Hydrology, Second Edition
E. M. Wilson

Essential Solid Mechanics, theory, worked examples and problems
B. W. Young

A Guide to
Soil Mechanics

MALCOLM BOLTON

Department of Civil and Structural Engineering,
UMIST

First published 1979 by
THE MACMILLAN PRESS LTD
London and Basingstoke
Associated companies in Delhi Dublin
Hong Kong Johannesburg Lagos Melbourne
New York Singapore and Tokyo

Typeset in 10/11 IBM Press Roman by
Styleset Limited, Salisbury, Wilts.
and Printed in Great Britain by
J. W. Arrowsmith Ltd, Bristol

British Library Cataloguing in Publication Data

Bolton, Malcolm
A guide to soil mechanics.
1. Soil mechanics
I. Title
624'. 1513 TA710

ISBN 0–333–18931–0
ISBN 0–333–18932–0 Pbk

Contents

Preface

Aim

This book covers all the soil mechanics and foundation engineering topics that are commonly included in civil engineering degree courses, and provides a number of springboards into related advanced studies. All the fundamental concepts are carefully nurtured from seed, although a knowledge equivalent to a study of A-level mathematics and physics is expected. The book is intended principally to satisfy the needs of school-leavers turned student civil engineers, but it should also prove useful to those practising engineers who would like to come to some more educated opinion on the relevance of this relatively new discipline.

I sensed that the prevailing attitude amongst student engineers was the same as that among qualified engineers: utilitarianism. So I have tried to write a useful book, which I take to mean that it should be read with interest, understood, and put to work. My chief ally in this three-pronged attack has been the notion of 'the engineer', whom I have invoked at frequent intervals, and whose need to solve problems, create designs and take decisions forms the backdrop to the theory. Whereas scientists need only to understand, engineers must also act. Although the engineer's creative role should not diminish his need to be able to argue scientifically, it should remind him of his responsibility to his client and to his fellow man when the argument has become sterile and the time for action has come. My references to the work and attitudes of professional civil engineers are sincerely felt but are also intended to be provocative: I hope that those who seek a more classical relationship with a pure body of knowledge will not be too irritated.

Format

My requirement that the book should be put to work by readers who intend to solve problems, take decisions and create designs, has led me to adopt an unconventional format. The first chapter consists of three case histories, which serve the triple function of demonstrating the role of the engineer as a troubleshooter, introducing some ideas which can be used to clarify diverse soil problems, and making a number of basic definitions. Chapters 2, 3, 4 and 5 then offer a basic familiarisation with the more distinctive modes of soil behaviour. Each mode is carefully presented as a mental 'model' built of interrelated concepts, which is meant to generate predictions of one aspect of the behaviour of engineering constructions. For example, soil is considered as though it were: 'sandstone', a rigid porous skeleton through which water can pass, chapter 3; 'sand', an assembly of rough particles, chapter 4; 'plasticene', a cohesive mass of constant volume, chapter 5. By the close of chapter 5 the reader has been introduced to such problems as the pumping of excavations in sand beneath the water table, the stabilisation of landslides by drainage, and the collapse of foundations on clay at constant volume. He has also become aware of the physical significance of water, water pressure, effective stress, seepage, friction, dilation, and cohesion, without having been forced to employ any analysis which would be beyond the sixth-former. Chapters 6, 7 and 8 then develop a number of fundamental models of stress and strain, making reference to problems of foundation settlement, the transient flow of pore water in compressible soils and its implication for the designers and constructors of works in soil, and the interpretation of the triaxial compression test with regard to useful soil parameters. Chapter 9 presents models covering the collapse of a wide range of soil constructions, contrasting the design methods which must be used for brittle and plastic materials, and generating notions of the meaning of safety and safety factors. Chapter 10 capitalises on the models that have already been introduced in order to make clear what is the nature of design as an activity. Aspects of the design of cuttings, compacted soils, embankments, trenches, foundations and retaining walls are individually discussed in such a fashion as to clarify the relative roles of experience, judgement, soil mechanics, decision-making, construction techniques and craftsmanship, while enlarging chiefly on the role of soil mechanics. Like the rest of the book, the prime objective has been to stimulate thought and to coach-in useful mental skills, rather than to describe detailed provisions of the codes of practice which are, in any event, in a state of flux.

Application

I have attempted to organise the material in a flexible fashion so that it might suit a variety of teaching styles and course programmes. My own preference will be to use the first five short chapters as a one-year introduction to soil mechanics, the central chapters as a basis for a more detailed review of soil behaviour at an intermediate stage, and to ask the graduation class to steer a course through chapters 9 and 10 in parallel with their design and project work. The later sections in each of chapters 6, 7 and 8 may be too advanced for most undergraduate courses, though they may be useful in postgraduate courses.

This is a book to be used. Follow the worked examples in detail, and attempt to repeat them on your own. Tackle the problems at the ends of the chapters, and discuss them with your colleagues and tutors. The book is finely subdivided into 'models', which makes it relatively easy to alter the order or to make additions or deletions: use notes in the margin to indicate where your lecturer agrees and disagrees about their validity or usefulness. Above all try to *listen* to your lecturer. Do not try to write your own book: you have paid for this one already, so decide that *this* is your compendium of what is known and spend the rest of your time trying to master skills. Try comparing this book critically with another after every identifiable topic: you will be surprised how much more enjoyable it is to try to criticise rather than merely absorb. Start with the case studies in chapter 1, and re-read them from time to time throughout your course of study. Do not worry if a few details escape you on a first reading: it's always the last reading that counts.

Acknowledgements

This book has been written on the assumption that ideas rather than authorities, codes or formulae are the backbone of successful engineering. I might not have given the centre of the engineering stage to creative thought had I learnt my soil mechanics other than from Professor Andrew Schofield. And I might not have supposed that such a standpoint was viable for the average undergraduate or graduate engineer had I not been associated with the students of the Civil Engineering Department at U.M.I.S.T. who eventually convinced me that formal analytical logic should be wholly subservient to creative and practical endeavour. This they did partly by the most intelligent being unable to learn any analysis which was introduced in an abstract fashion, and partly by demonstrating that even a mediocre student could not only use his intuition to take sensible decisions in the absence of formal analysis but could also make giant strides in his formal learning when he was given an 'incidental' personal task such as a design. I owe a great deal, also, to the attitudes of Mr. I. L. Whyte, Dr. F. T. Howell and my other staff colleagues at U.M.I.S.T., and was fortunate to have easy access to the challenging work being pursued in the University of Manchester by Professor P. W. Rowe. My thanks for great practical assistance go to Mrs. D. Pollard who typed the manuscript. On a more personal level, I would not have completed this work without the practical and emotional support of my wife Kate, and would have been more frequently discouraged without our daughter Vicki's continuous reminder that life is for living.

Malcolm Bolton

1

Introduction

1.1 The Case of the Collapsed Boilerhouse

1.1.1 The Client's Troubles

In 1971 a light engineering firm was taken over, and the new owners spent over £100 000 on expansion and modernisation. Included in this total was £40 000 for a new boilerhouse, which supplied steam under pressure to a plastics moulding shed, in addition to its main role of heating the whole factory. Only a matter of weeks after going into full production, the boilerhouse was giving trouble. First windows began to shatter, then cracks appeared in the concrete floor. The local builder who had organised the whole modernisation plan was called back, but he seemed unable to effect any substantial improvement. The first threats of industrial action came when two men were scalded by a steam burst on the fitting which gathered the pipes from the boiler before they were taken through the boilerhouse wall to the plastics shed. By January 1972 it was clear that the building would have to be closed. The walls had suffered substantial inward rotation, the roof leaked, and the boiler itself was subsiding on a badly crazed floor slab. There being no heating and no steam for the moulding process, those men who were not laid off walked out. The owners, despairing of their builder's remedial works, and aware that their loan for modernisation was only repayable if full production could be restored soon, approached a consulting civil engineer for advice.

1.1.2 The Engineer's Investigation

A preliminary site investigation revealed that the foundation of the brick structure housing the boiler had settled and tilted. When earth was removed from around the outside of the walls in two or three places, substantial cracks could be seen in the concrete raft which supported the walls and the boiler structure inside the building. The engineer noted that the raft did not appear to be unduly thin; indeed, the steel reinforcement, which he could just discern in one region of very badly broken concrete, looked rather heavier than he had been expecting. Inside the building it was impossible to elicit much from a visual inspection since the floor had been covered recently with a layer of asphalt. It was clear from the broken pipe brackets, however, that the boiler had settled by at least 0.1 m. In walking back to the managing director's office, the engineer observed that the other new buildings were showing no signs of distress.

It was not long before the team of contractors used by the engineer for rush jobs such as this were reporting that they had cut through the foundation slab at the place near the door which the engineer had indicated. The closely spaced steel reinforcing mesh had been burned away from the roughly chiseled hole. By the light of the flashlamp the assembled group could clearly see the culprit, directly under the concrete slab: nothing! Where there should have been compacted rubble there was a gap so deep that it was only just possible to confirm that the rubble did indeed exist, somewhere below. As the engineer ruefully withdrew his hand he took back with him another clue in the form of a blistered finger. The ground was very hot indeed; a thermometer installed later registered 100 °C in the rubble, while the concrete raft was only warm to the touch.

In order to complete his report, it was necessary only for the engineer to ask a drilling company to sink a few small boreholes through the raft and rubble, and also in the general vicinity of the boilerhouse, away from the influence of the heat. Clayey dust overlying very compact dried clay up to 2 m in depth below the boiler confirmed his opinion. The builder had established the raft on compacted rubble overlying clay. The final configuration of the foundation slab can be seen in section in figure 1.1, deduced from the inspection pits and

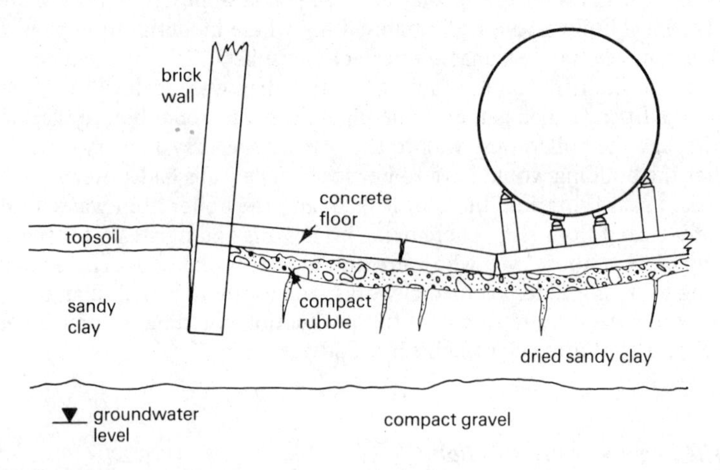

Figure 1.1 *Exaggerated sketch of boilerhouse problem*

borehole information. No doubt the clay stratum had initially been strong enough to support the raft, walls and boiler. Certainly the slab was more than equal to its task of spreading the foundation loads to the soil, had the soil remained in contact with it. Unfortunately, as the clay dried out, it shrank away from the raft that it was supposed to be supporting and, although the soil was becoming stronger and stronger, it was also completely losing its function. In the words of the engineer, there was an 'incompatibility' between the clay and its hot load. The builder had fallen into the trap of suiting materials to only one aspect of their working environment, in this case forgetting that clay soils settle owing to drying as readily as they compress under load.

Back in his office, the engineer began to prepare his report. In order to substantiate his findings in a fashion that would stand up in court if the builder refused to pay compensation for the damage, the engineer wanted to show that the drying of the clay layer 2 m thick could account for the measured ground settlement of up to 0.2 m. He therefore sent samples of the soil to a soil testing laboratory for moisture content determinations.

1.1.3 Definitions

In the field of soil mechanics there are many parameters, which are used to define the relative proportions of solids (usually silica or alumino-silicates, commonly as sands and clay minerals), liquids (usually fresh or saline water) and gases (usually air, or methane from rotted vegetation). In particular, these four are fundamental

$$G_s = \text{specific gravity of solids}$$

$$m = \text{moisture content} = \frac{\text{weight of water in a sample}}{\text{weight of dried solids in a sample}}$$

$$e = \text{void ratio} = \frac{\text{volume of voids in a sample}}{\text{volume of dried solids in the sample}}$$

$$S = \text{saturation} = \frac{\text{volume of voids which are filled with water}}{\text{total volume of voids}}$$

A useful way of visualising these, in order to relate them, is to draw elements of soil with the three phases separated, so that the air bubbles are imagined collected into one large bubble, contained within water in one large void, and the individual soil grains welded into a chunk of solid rock: this is done in table 1.1. From this it can be seen that the volume ratio and weight ratio definitions run in parallel, and that we can write

$$m = \frac{eS}{G_s} \tag{1.1}$$

1.1.4 The Laboratory Report

It is always easier to report on changes of weight rather than volume. The standard procedure for natural moisture content determination is to weigh a small sample of the soil in its natural state, then dry it in an oven at 105 °C for 24 hours to evaporate the free water, then re-weigh. The laboratory carried out this procedure on the soil sample which the engineer recovered some distance away from the boilerhouse.

Weight of container = 55.12 g

Weight of container + moist soil = 117.85 g

Weight of container + dry soil = 108.06 g

Therefore

$$\text{moisture content} = \frac{117.85 - 108.06}{108.06 - 55.12}$$

$$= 18.5 \text{ per cent}$$

The same test carried out on the soil from underneath the boilerhouse showed that there was still water trapped there, the moisture content being 3.1 per cent.

TABLE 1.1
Proportions of Soil Constituents

Constituents	Volume ratio definitions (i)	Density ratio definitions (ii)	Weight ratio deduced (iii) = (i) x (ii)	Weight ratio definitions (iv)
Air voids	e { $e(1-S)$	0	0	0
Water voids	eS	1	eS	m
Rock	1	G_s	G_s	1

1.1.5 The Engineer's Calculation

In order to arrive at volume changes, according to equation 1.1 one needs to be able to estimate both S and G_s. Standard procedures exist for the determination of these quantities, but the engineer was prepared to guess both. He knew that most natural clays are fully saturated ($S = 1.00$) and that well-compacted, man-made clays rarely have a saturation of less than 0.80. He also knew that the range of specific gravities met with in soil minerals is quite limited, say 2.60 to 2.75 in 99 per cent of cases. He therefore opted to take $S = 1.00$ and $G_s = 2.70$ to deduce the original void ratio of the clay as

$$e = \frac{2.70 \times 0.185}{1.00} = 0.50$$

It was clear, therefore, that roughly one part in three by volume of the original clay material (voids 0.50, solids 1.00, total 1.50) was water. This could be visualised as 0.67 m of water within the 2 m total thickness of clay.

In order to complete the calculation it would be necessary to find the void ratio of the clay under the boilerhouse. If the same assumptions were used as before ($S = 1.00$, $G_s = 2.70$) we would obtain

$$e = \frac{2.70 \times 0.031}{1.00} = 0.08$$

This is not likely. As clay is dried, air enters the fabric and the saturation eventually falls to zero. If a bucket exactly full of saturated sand is dried, the surface of the sand will hardly settle at all as the saturation ratio changes from 1 to 0: the void ratio hardly changes as the air replaces the water. When a bucket full of clay is dried, however, the very strong capillary action of water caught between the tiny clay particles has a marked effect. At first, the surface of the soil retreats as water is driven from the clay. The void ratio is reduced, but the soil is still saturated. Air cannot enter the clay until the air bubbles at the surface overcome the tendency for 'capillary rise': this only happens when the water has a strong enough suction (at a pressure below atmospheric pressure). As the suction due to surface tension in the water is increased by drying, the clay particles are forced together: the same effect can be seen in a much milder form with damp sand. Now clay is, in addition to being fine grained, quite compressible. The extra compression due to suction causes the clay matrix to contract. The result is that the clay matrix contracts as water is removed by drying, while the whole mass of soil remains saturated. Only when the drying is fierce enough to create the necessary suction can air enter the clay: when this happens the void ratio then remains almost constant while air replaces the pore water as it does with sand.

It is rare for the minimum void ratio of clay after drying to be above 0.45 or below 0.35, and so the engineer was prepared to assume that 0.40 was the final value for the soil underneath the boilerhouse. He could then estimate that

$$\frac{\text{final volume of clay}}{\text{original volume of clay}} = \frac{1 + 0.40}{1 + 0.50} = 0.93$$

Assuming that there was no lateral movement (a poor assumption considering the cracks in the clay, but good enough considering the general deterioration in stringency)

$$\frac{\text{final thickness of clay layer}}{\text{original thickness of clay layer}} = 0.93$$

Taking 2 m as the original thickness we obtain 1.86 m as the final thickness, implying 0.14 m settlement of the clay surface.

The engineer was fairly pleased with the correspondence he had achieved. He saved samples of the soil that he had retrieved in case his arguments were challenged. He was aware that a much more careful analysis was possible if it should be necessary, but considering the urgent need for progress, he submitted his report without further delay.

1.1.6 The Engineer's Succinct Conclusion

The boilerhouse was supported on a slab, which was designed on the assumption that it would be supported across its whole span by the underlying soil. The soil was a very firm sandy clay, which could lose nearly 10 per cent of its volume on being dried. The heat from the badly ventilated boiler was sufficient to dry the underlying clay, which subsequently shrank away from the raft by up to 0.2 m. The raft then broke, being unable to span the void, causing great structural deformation and malfunction.

The most efficient remedial action would be to jack the slab back into its original position, and pump concrete into the void underneath, to relevel the boiler on adjustable supports, to make the windows and roof safe, and to provide ventilation. This would bring the factory quickly back into production and would even be a satisfactory medium-term solution since only 3 per cent water remained in the underlying clay. The engineer did not doubt that a new boilerhouse with improved ventilation should be designed and built near the old one as quickly as possible, so that the eyesore could be demolished.

1.1.7 Questions for Tutorial Discussion

(1) How should the new boilerhouse be founded?

(2) Why was there still 3 per cent water in the soil under the boiler, although it had been there for 6 months at over 100 °C?

(3) Would compact sand under the slab have had the same effect as the compact clay, and, if not, why not?

(4) How much compensation should the original builder pay – for the continual maintenance, loss of production, remedial measures described above, the new structure, the consultant's fee?

1.2 The Case of the Eroded Dam

1.2.1 The Problem

Water supply in the vast 'undeveloped' plains of south-east Africa is very much a local affair, with village chiefs vying for state agriculture funds to provide reliable sources of drinking water for cattle and villagers. One popular solution in Swaziland is to construct small earth dams across the gullies, which otherwise run heavy with murky flood waters in the wet season, only to dry out immediately when the rains in the highlands cease. If the situation of the dam is favourable, it is able to retain sufficient water to last through harsh evaporation during the dry spell. Unfortunately, the ideal situation is rarely found, and some compromise must be sought among the relevant criteria, which include siting within walking distance of the village, the suitability of local earth for the construction, the possibility of the provision of a safe overflow, and the shape and volume of the storage provided. Although the designs were simple they were often quite effective, and accidents were rare. One particular incident, however, came to the attention of an engineer, newly graduated, who was employed as assistant resident engineer on a road construction project, which was designed by the British firm of consulting engineers that employed him, constructed by the Swaziland government and funded by the World Bank.

Early rains brought catastrophe to one of the villages near his road construction site, when the local dam failed, drowning two children and a number of cows, and causing a great deal of damage to the newly rebuilt school. The young engineer decided to investigate.

1.2.2 The Investigation

He began with a visit to the site, but could not refrain from feelings of awe on his way up to the reservoir as he encountered the deep ravine cut through the firm sandy clay by the flood waters. The embankment itself was, at 9 m high, rather larger than any he had encountered previously, but nothing about the design immediately appeared odd. The soil, exposed in cross-section where a block 3 m wide had been completely washed out, seemed well compacted, and was almost indistinguishable from the natural clayey foundation material. After filling two small polythene bags with samples from the bank and the natural ground, the engineer took the rare opportunity of measuring the cross-section of the bank adjoining the missing section, although he was rather concerned about the possibility of earth slips into the gap during his survey.

Enquiries revealed that the pond was only two-thirds full just before the failure, and that it had not dried out at all since the construction 3 years previously. Although the spillway looked rather small, and poorly constructed, failure could not be attributed in this case to overtopping or bad spillway detailing since the bank failure was many metres away. Clearly the earth embankment itself was at fault, but was the problem one of shear failure, dry-weather cracking followed by flood, or of erosion due to high water tables, or foundation instability, or some other cause?

The engineer could not discern any sign of the dry cracking that has been the downfall of countless embankments constructed with plastic clays, which shrink as they dry out in summer, so that the first winter floods find a direct route through the bank, which can be eroded away in a matter of hours. In fact he was so impressed with the vegetation cover, which was dense and lush around the base, that his mind wandered directly to the possible failure of the bank due to the unsuitability of the soil itself to withstand seepage. Perhaps the finer fractions had been washed out, leaving an unstable honeycomb structure through which major erosion channels could form. He left orders for two workmen to clear a patch of vegetation at the base of the bank near where it had failed, so that he could later look for tell-tale 'runs' of fine material.

Back in his rudimentary laboratory – a wooden shed set aside from the road construction quarters – he decided to run a few simple tests on the soil he had brought back. The first was a grading analysis, with a standard set of sieves. By washing a known mass of soil down through a nest of sieves with the coarsest at the top and the finest at the bottom, he arrived at a grading curve of 'percentage passing' against sieve size. This cumulative frequency distribution of particle

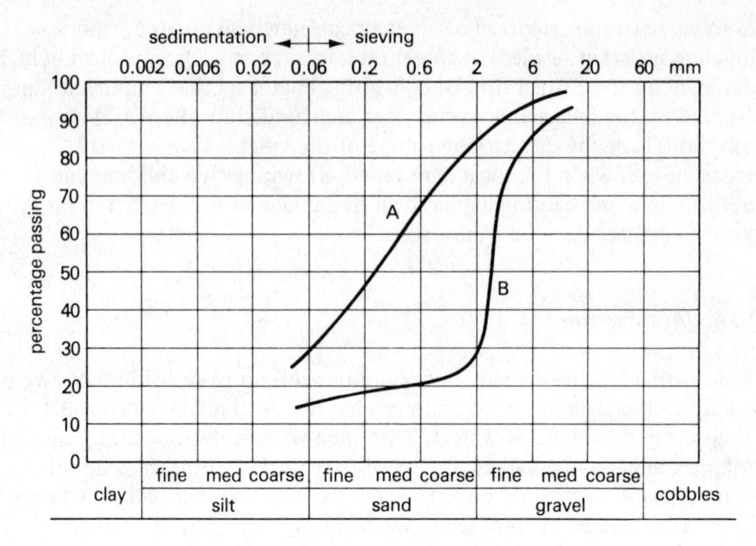

Figure 1.2 *Particle size chart*

sizes, shown in figure 1.2, is a most useful classification of soils, and BS 1377 describes the precise routines of mechanical analysis, which can be carried out on either an oven-dried or a wet mixture. Wet deposits on the sieves must be dried out in an oven before they are weighed, so that the final analysis can refer accurately to percentages by weight of the solid constituents of the soil. Where they contain a substantial proportion of clay, soils are normally washed through the sieves, as in this example, since oven-dried clay can set into hard lumps which would be retained on the sieves, whereas the clay particles in the natural soil might be smaller than 10^{-3} mm, and act quite individually.

From the grading analysis, curve A, the engineer was able to describe the soil as well graded with 30 per cent fine-grained constituents. If he had found the badly graded curve B, he would have been concerned lest the fine particles wash through the voids between the 2 mm gravel. The lack of material, in curve B, between 0.1 mm and 2 mm makes this structural instability quite likely. The ideal well-graded material possesses fine particles, which can just fill the voids left between the medium particles, which can just fill the voids between the coarse particles. The whole structure is then interlocking, and is easily compacted to high densities, and then remains stable when water runs through it.

The mechanical properties of soil are largely determined by the finest 20 per cent, and so it was necessary for the engineer to submit further samples of natural soil to tests more suited to clays and silts which all pass the finest sieve. He chose to determine the Atterberg limits of the soil, which refer to the moisture content of the soil as it changes from crumbly to plastic (plastic limit), and as it changes from plastic to liquid (liquid limit). BS 1377 describes the arbitrary dividing lines that have been chosen to embody these two changes of character, which are the limits of what is known as the plastic zone of behaviour. The plastic limit is defined as that moisture content at which the soil can just be rolled out into 3 mm diameter threads without crumbling. The liquid limit is defined as that moisture content at which a standard V-groove cut in the soil just closes when it is shaken in a standard manner, in a standard cup, by letting it fall a standard distance of 5 mm a standard number of 25 times. Both tests are effectively strength tests, although they are highly empirical, and the ratio of strengths that they measure is roughly 100:1. The plasticity index, which is the difference between the liquid limit and the plastic limit, is therefore the amount of water that must be added to the soil to reduce its strength a hundredfold, taking the soil from one extreme of plastic clay behaviour to the other.

Since the equipment needed for these Atterberg limit tests is minimal, it has been thought expedient to relate them to other soil properties, quite empirically, so that they may be used as guidelines in just such a case as the one being described. Therefore, by finding that

(1) liquid limit = 41 per cent

(2) plastic limit = 19 per cent

(3) plasticity index = 41 – 19 = 22 per cent

the engineer was able to use Casagrande's extended classification system, figure 1.3, to further describe his soil as a clay of medium plasticity.

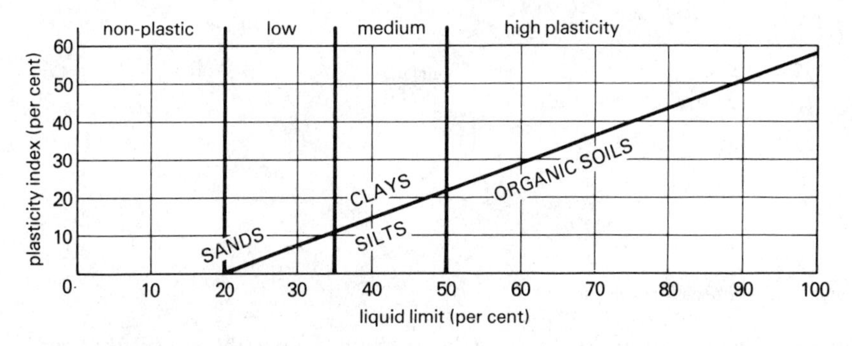

Figure 1.3 *Soil classification system*

Materials with low plasticity do not compress unduly under load, nor do they shrink excessively when dried, so that once they have acquired a firm consistency, near their plastic limit, they are quite stable. A check showed that the natural water content of the clay, both in the foundation to the bank and in the bank itself, was 20 per cent. Problems of settlement, shrinkage, and softening were therefore all ruled out at this stage as unlikely causes of instability.

So far the engineer had deduced merely negative indications concerning the cause of the failure. The soil was an ideal material for earth dam construction, being well graded and clayey enough to guarantee a low permeability. It had been well compacted, and was just 1 per cent wetter than its plastic limit – an ideal water content, which should have guaranteed maximum stability. Perhaps it had failed simply because the bank was steeper than friction would allow.

The profile of the bank is shown in figure 1.4, with a possible failure mechanism in which a toe failure was triggered by seepage and a progressive 'back-sapping' was then responsible for the final 'wash-out'.

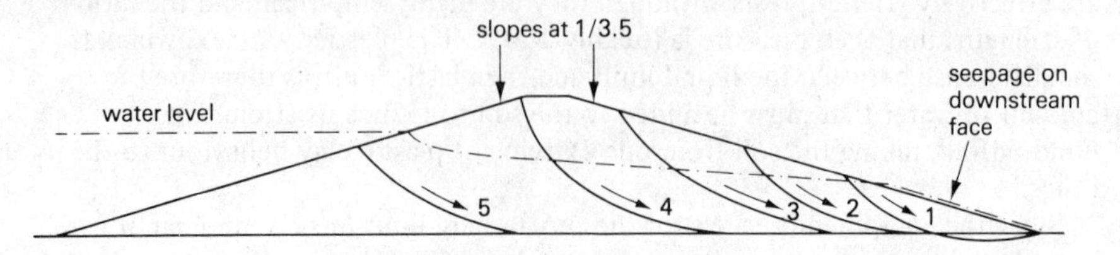

Figure 1.4 *Hypothetical failure sequence*

Although the more precise methods of slope stability analysis are beyond simple description at this stage in the book, they were also beyond the engineer with the bank failure, since he did not have the time to execute them. Instead he resorted to a simple model of behaviour which was intuitively instructive rather than authoritative. Consider the failure of a long slope of soil, by slipping on a plane parallel to the slope surface, and in particular the stability of a 'slice' of the soil, as shown in figure 1.5. Allow the side forces X to cancel, so that the

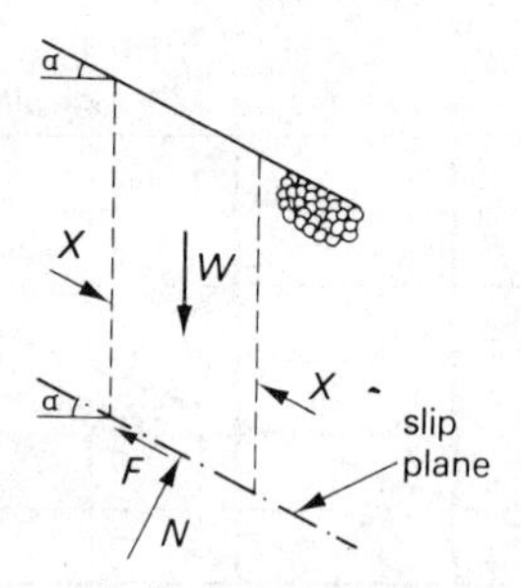

Figure 1.5 *Forces on soil slice*

forces W (the weight of the slice), F (the shear force along the slip surface) and N (the normal force) must balance if the 'slice' is to be in equilibrium. That is to say, the triangle of forces W, F, N, must close as shown in figure 1.6. It is 200 years since Coulomb suggested that soil could be ascribed a maximum frictional strength $F = \mu N$, corresponding to $F = \tan \phi \times N$, where ϕ is the angle of friction of the soil. In this case it is clear that the maximum angle of the slope is simply ϕ, beyond which Coulomb's friction law would predict failure by sliding, according to figure 1.6. Although Coulomb's law proved useful in the design of soil constructions, it was pointed out by Terzaghi in 1925 that the ϕ value described above would not be a material constant unless the law were rewritten as $F = \tan \phi \times (N - U)$, where U is caused by the water pressure in the soil voids, which of course acts perpendicularly to any surface that we consider. Soil found in nature is nearly always completely saturated with water, and that water is usually under pressure, corresponding to its depth below what might be called the water table. The water table can be simply located by driving a standpipe into the ground and observing the level to which water eventually rises in it.

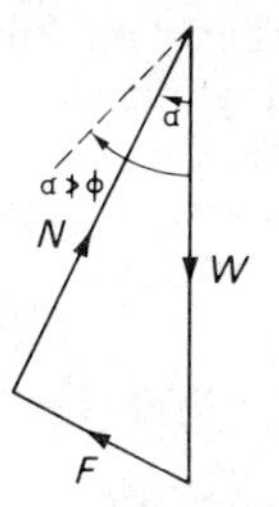

Figure 1.6 *Triangle of forces*

Since saturated soil normally weighs in the region of 20 kN/m^3, and water in the region of 10 kN/m^3, it would be fair to say that roughly half the weight of the land is supported by the soil grains and half by water, if the water table happens to coincide with ground level. This means that only half the weight of the soil is concerned in generating friction, while the whole weight must be supported.

The engineer was aware that the standpipe levels in the soil at the toe of the embankment would be close to ground level, so that the triangle of forces would appear, in the case of limiting friction, as shown in figure 1.7. In this case, if

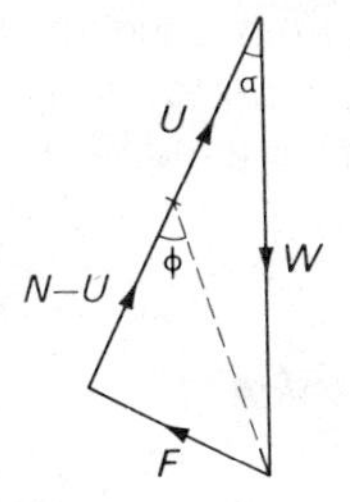

Figure 1.7 *Subdivision into water pressures U, granular forces N – U, F implies onset of slipping*

$N \approx 2U$, then $N - U \approx U$ so that a crude approximation would be $\alpha \approx \phi/2$. A stable slope, streaming with water, therefore, could only stand at roughly one-half its friction angle. Using his experience that pure clay soils were least frictional ($\phi \approx 20°$), silts more so ($\phi \approx 30°$) and sands and gravels most frictional ($\phi \approx 40°$), the engineer was able to guess a friction angle of 30° or so for his particular soil. Half this angle, 15°, would give a slope tangent of 1/3.5. Although there was some correlation there, it gave the engineer little satisfaction. He could not believe that so marginal a result could be used to predict a sudden catastrophic embankment failure. Surely the progress of back-sapping here would take months to produce significant effects? Moreover, if a gravity-generated mass slide had occurred, it must surely have involved more than a 3 m wide section of the bank. Since he did not have on hand the equipment that would allow him to estimate more accurately either the water pressures in the bank on a plane of sliding, or the friction angle, the engineer simply grouped loss of frictional support along with the previous failure mechanisms as rather unlikely. This left him without a failure mechanism.

By now it was late afternoon, and although his interest in the case had grown throughout the day he felt rather nonplussed at the inability of his European soil mechanics to deal with an African dam. Since he knew that his responsibilities on the road project would shortly have to claim his full attention, he decided to take a last look at the site of the dam. He arrived at the bank as the sun was going down, hoping to see new clues in the patch of ground at the base of the downstream toe of the bank which had, by then, been cleared of vegetation. However, there were no 'runs' of clay stains, nor was there any other indication that the bank was anything less than solid. As the sun set the engineer surveyed the lonely scene.

To any other human eyes, at that time, the ears of the rabbit which popped over the artificial horizon of the dam crest would have indicated simply the presence of other life, but as the animal sensed the surprised twist of the engineer's head and darted out of sight into the foliage close to the wash-out, it had already been assigned the role of 'failure mechanism'. It took only a matter of minutes to discover the entrance to a burrow, and a matter of hours to elucidate, in the beer hall down in the village, that the solitary family of rabbits which had inhabited the dam for over 12 months had been the popular villains of the tragedy from the moment it had occurred. They had naturally chosen a particularly moist site for their burrow, where excavation would be easiest, and where their extensive nest could be constructed without obstacle, and under the cover of rich and nutritious vegetation. Once the water had eroded a channel to meet a burrow it would take only minutes for the flow to enlarge the diameter, and for the whole bank section to wash out. The engineer resolved to spend more time drinking beer when he investigated his next collapse.

1.2.3 Questions for Tutorial Discussion

(1) Why are the engineering properties of soils largely determined by their finest 20 per cent fraction?

(2) Why should soils of low plasticity index be less subject to compression, shrinkage and swelling?

(3) Calculate the maximum and minimum possible void ratios of a perfectly uniform soil, that is, an assemblage of spheres of equal diameter. If a 1 m deep stratum of such a soil were as loose as possible, and then it was vibrated until it was as dense as possible, how deep would it then be?

(4) If the loosest possible assemblage described in (3) above were to be locked together by a family of identical smaller spheres, what would be their relative diameter and what proportion by weight would be required?

(5) What were the main strong and weak points of the argument which led to the discounting of a slope failure due to loss of frictional support?

(6) Assuming that the soil in the embankment was completely stable, so that water could not disturb even the fine particles, describe, with sketches, a possible mechanism for the connection of a rabbit burrow extending only up to the dam centre line, with the impounded water.

1.3 The Case of the Unhappy Home Owner

1.3.1 The Problem

John Doe was delighted with his £10 000 bungalow when he moved into it, with his wife and children, in September 1973. The price was comparatively low because the builder had acquired the land very cheaply, and he was keen to have the early houses on the estate occupied so as to attract further buyers. For a further concession John agreed to the bungalow being used as a show house. As time passed the family found a few of the usual faults, warping door frames, a leaking sewer pipe and cracking plaster: the builder rectified all of them. When the crisis came, it came suddenly. During the second week of March 1974 a crack appeared which seemed to divide the bungalow into two parts, as depicted in figure 1.8. It quickly widened to a maximum of 40 mm at roof level, tapering to 0 mm at ground level. John Doe immediately brought the builder to inspect the damage: thereafter work ceased on the estate and John was forced to consult his solicitors. On their advice he consulted a specialist foundation engineer.

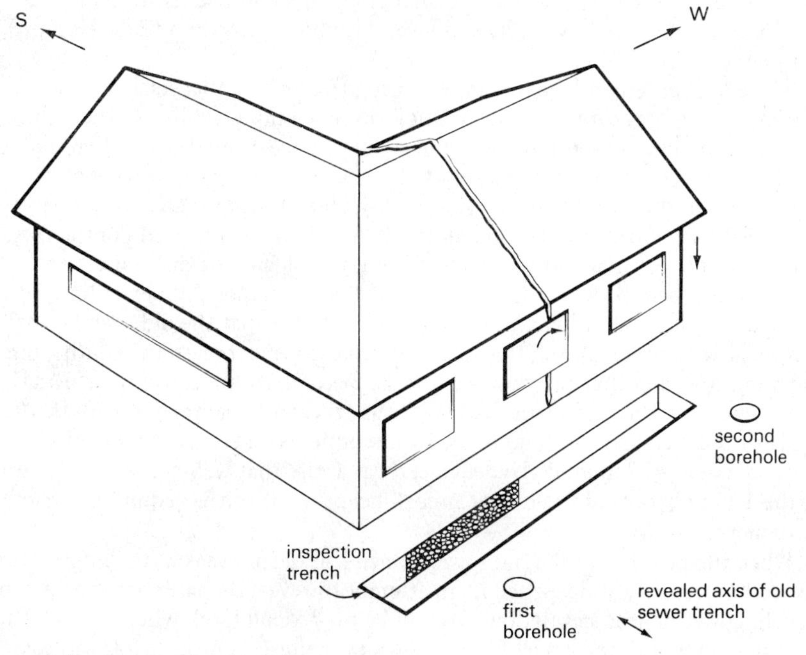

Figure 1.8 *The problem bungalow*

1.3.2 The Site Investigation

The engineer made an inspection of the property on 1 April and he could hardly resist a smile when he saw the show-house sign in front of the stricken bungalow.

He could measure a crack width of only 30 mm, and he observed that the hardboard sheet which had replaced the broken window had buckled outwards as though the crack had closed a little. The engineer speculated a little on whether this might mean that the west wing had begun to bounce back, or whether the south wing had begun to tilt in unison. Whichever was the case, an inspection of the foundations revealed a crack in the concrete raft, on which the bungalow was evidently built, at the junction of the two wings, and a 15 mm settlement of the west wing relative to the south. There was no doubt about the costliness of the damage.

The engineer automatically made a note that there were no large trees near the west wing, and that none of the neighbouring houses showed signs of distress. The housing estate was fairly near the sea coast, and the engineer made the presumption that the builder was obliged to use concrete rafts because the superficial soils were soft. On his way back to the office he called into the local library, and unearthed all the old maps of the district which might cover his client's site. He was immediately rewarded by a town plan which showed that the bungalow was built directly over the line of a sewage outfall pipe! The local authority said that the pipe had been disused since a sewage treatment works had been built in 1969, and that the builder had been instructed to remove it and to fill in the trench when he was given planning permission to develop the housing estate.

The engineer felt that he was making progress. Differential settlement within a building is often due to variations in the compressibility of the ground: soft spots are usually more flexible and spongy. Very often disturbed ground, or fill, is weak relative to the parent body. Perhaps the dipping west wing of the bungalow was built on the old sewer line, which had perhaps been improperly backfilled with loose soil. The engineer asked a small local firm of contractors, who were able to carry out site investigation, to come on to site and dig an inspection trench, shown in figure 1.8, to intersect the possible sewer line. Soon after the mechanical digger started work he could see that the old sewer trench was filled with beach gravel, that the trench passed underneath the south wing of the bungalow, and that the west wing was apparently founded on the natural clayey ground. The surface of the gravel was repeatedly hammered with the flat blade of the machine and it could be seen that no extra compaction of the material could be achieved. The engineer was aware that well-compacted gravel, as this had now proved to be, was almost incapable of further compression and settlement.

When the position of the old sewer trench had been revealed, the intersecting inspection trench was deepened in an attempt to reveal the underlying soils. The trenching went ahead easily down to 1 m below ground level, when groundwater was encountered in the gravel fill, as shown in figure 1.9. Although it was easy to continue excavating the natural soft blue clays on either side of the sewer trench, it proved difficult to remove the gravel fill from beneath the water table owing to the uprush of water which attended every attempt to create a cavity. The engineer decided that it would be necessary to establish the depth of the sewer trench by boring a hole through it. In the meantime he asked the contractors to deepen one small section of the trench in the natural ground down to a depth of 2.5 m, which was the maximum reach of the shovel. He did

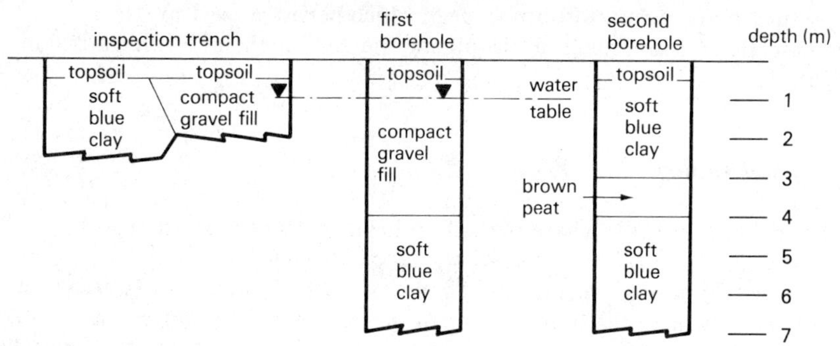

Figure 1.9 *Site investigation details*

not dare enter the pit thus formed for fear of a collapse, and was content to observe that the soft blue clay seemed to continue down to the base. The engineer decided that two boreholes, one to prove the depth of the sewer trench, and the other to go down at least 7 m into the natural ground, should complete his initial investigations.

The contractors used a simple tripod rig, figure 1.10, with a shell to remove

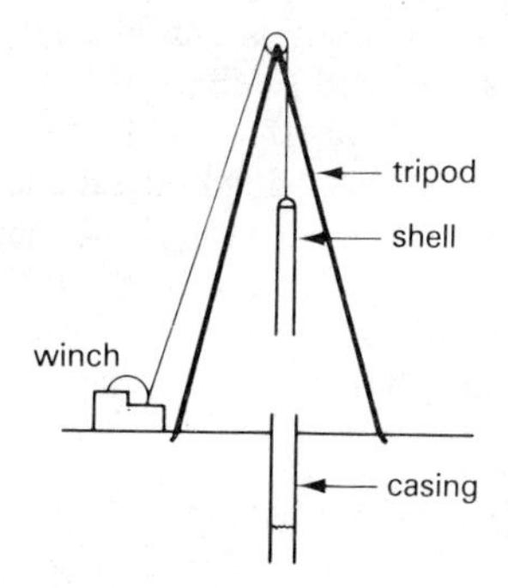

Figure 1.10 *Tripod rig*

the soil, which was prevented from washing into the hole by a steel casing which had previously been driven. The rig was initially positioned over the proven line of the old sewer trench. Each time the shell was brought to the surface to be emptied the engineer checked the soil chippings so that he could establish when the hole had penetrated through the gravel fill. This eventually happened at a depth of 4 m. From this point down to 7 m the engineer asked the contractors to drive 100 mm diameter sample tubes one after another. When these were extruded and laid side by side a continuous core of the natural soils could thereby be inspected. In order to prevent the samples from drying before they could be extruded, he asked that the two exposed faces be waxed, and that blank ends be screwed into place. This routine was repeated in the second borehole, which was sited away from the old sewer trench.

When the samples were extruded in the laboratory a dramatic change in soil properties could be seen at a depth of 3 m in the second bore. Here the soft blue clay turned into a very soft brown peat, which persisted for 1 m. The cross-section arising from the trial pit and the two boreholes is summarised in figure 1.9.

1.3.3 Soil Testing

The engineer decided to have the soil testing laboratory estimate the compressibility of the brown peat. The sample from a depth of 3 m was shaped into a disc 75 mm in diameter and 50 mm thick. This proved very difficult to carry out, owing to the fibrous fabric of the peat, but the engineer was prepared to accept the rough approximation that the test would offer him. The disc of peat was confined in a steel tube in such a way that pressure could be applied to the top surface while water could drain out of the soil as it would out of a sponge as the soil skeleton consolidated. After the sample had been submerged in water to stop it drying, an initial pressure was applied to it which was intended to duplicate the weight of earth that was compressing the soil sample in its natural position. To estimate this initial pressure, the engineer applied the most significant principle in the mechanics of soil, the effective stress concept. Most soils are composed of soil particles and voids filled with water: the deformation of the soil body is caused solely by rearrangements of the soil particles, and these are related solely to the stresses acting through the soil particles, which are known as the effective stresses.

Considering any plane section through a homogeneous soil, the stress components perpendicular and parallel to the plane may be σ and τ respectively. If the water pressure in the voids at that position is u then the effective stresses acting on the soil skeleton are σ' and τ, where $\sigma' = \sigma - u$, as depicted in figure 1.11. In the case of pure compression no shear stress τ exists and so the

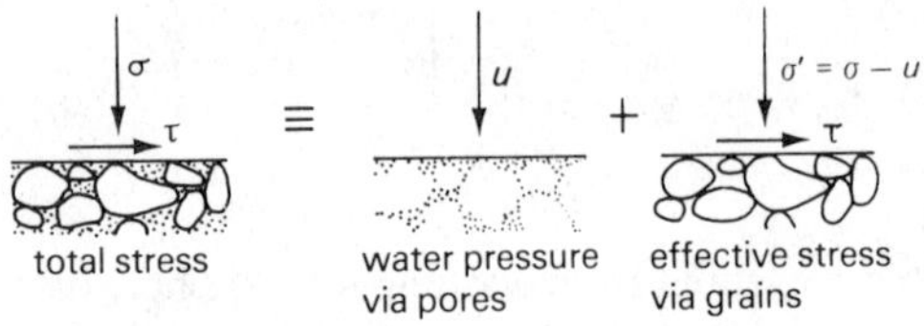

Figure 1.11 *Effective stress*

calculation of the initial effective vertical stress in the peat runs thus

depth of sample = 3 m

density of overlying clay = 20 kN/m^3

total vertical stress = 60 kN/m^2

head of water above sample = 2 m

density of water = 10 kN/m^3

water pressure = 20 kN/m^2

therefore

$$\text{effective vertical stress} = 60 - 20 = 40 \text{ kN/m}^2$$

This was accordingly the vertical pressure that was chosen as the initial state in the laboratory. The water pressure in the peat sample was allowed to be zero in the laboratory for convenience: this did not matter since only the effective soil stresses can cause deformations. Time was allowed for the sample to come into equilibrium in its initial condition.

The engineer then chose an arbitrary stress increase of 20 kN/m^2 so that he could measure the resulting compression of the disc of soil. When the extra load was imposed the sample quickly began to compress as water was squeezed out: the rate of compression slowed down until it was imperceptible after 15 minutes had elapsed and a compression of 5 mm had been observed, which was 10 per cent of the original thickness. It follows that if the sample was representative of the rest of the peat, then the whole 1 m thick layer would have compressed by 100 mm under the same increase in effective stress.

1.3.4 The Explanation

The engineer now had some idea of the geometry of the bungalow and its supporting soils, and a calculation for the flexibility of the brown peat, which was the weakest link in the mechanical chain that should have propped up the bungalow. The west wing was founded over the peat and had settled, whereas the south wing, which was resting on the highly competent gravels in the old sewer trench that had replaced the peat, had evidently remained in position. A number of questions remained, however. Why was the damage delayed until 6 months after completion of the bungalow? How had other buildings escaped damage, and would they remain safe in future? Was the builder negligent?

The clue that allowed the engineer to resolve these problems was the presence in the mystery of the dimension of time. Very often, in his experience, when the passage of time was an important parameter in a soils problem, then the agency at work was the drainage of water. The soil skeleton itself usually reacts very quickly to changes in effective stress; the most significant time-related phenomenon in soil mechanics is the slow seepage of water through the pores between the soil grains, which is due to changes in water pressure. The engineer therefore made the hypothesis that he should look for a change in water pressure shortly before John Doe found the damage to his bungalow.

A short excursion around the housing estate, together with one or two casual conversations, revealed the sequence of events. John Doe's bungalow was near the main road and on slightly higher ground. Even here the natural groundwater level was at a depth of 1 m, but it was not necessary to drain the ground in order to construct the foundation slab at a depth of 0.8 m. On the lower ground, however, the builder encountered the peaty and sandy remains of an old river bed in the sides of his foundation excavations, and groundwater flooded in at ground level. At the beginning of March 1974, therefore, when he started work on the foundations to six new houses, he was forced to hire pumping equipment and make arrangements to connect his land drains to the municipal sewer.

Having already been responsible for removing the old outfall sewer pipe and replacing it with well-compacted beach gravel, he had taken the opportunity of using a perforated pipe, running along the trench at a depth of 2 m below the lowest ground, as the spine of his land-drainage system. When the builder put the system into operation by pumping water out of it into the sewer, the groundwater table on site began to drop by 2 m. The water pressure in the gravel trench dropped almost immediately since the water was able to escape very quickly. More slowly, the sands, peats and clays that bordered on the trench also began to respond to the reduction in water pressure.

By chance, John Doe's bungalow was the only building straddling the new drain. The water pressure in the gravel trench supporting the south wing fell by 20 kN/m^2, corresponding to the 2 m head reduction. Since the total weight of the ground and bungalow remained almost the same this meant that the vertical effective soil stress in the gravel must have risen by 20 kN/m^2 in compensation, in order to preserve the rule $\sigma = \sigma' + u$. The gravel happened to be almost incompressible, so that no settlement occurred above it even though the stress on the soil skeleton increased. But when the water in the peat layer began to drain into the trench, the stress on the peat skeleton also began to rise by 20 kN/m^2 to compensate for the water pressure reduction. And, as the engineer had already determined from the laboratory results, the effect of this might be to cause a 100 mm compression of the initial 1 m thickness.

Although the reinforced concrete raft made a brave attempt to allow the west wing to cantilever out above the subsiding ground, it was not designed to withstand such enormous stresses, and it cracked. The west wing then pivoted about the crack, allowing the occupants to know of its distress. Almost immediately, John Doe had reported the damage to the builder, who had equally quickly retired home in a deep depression, after turning off his pumps. This allowed the groundwater table to recover slowly, which in turn allowed the west wing to spring back a little. Unfortunately soil is not elastic, and does not generally recover its original thickness when a stress is applied and then removed: this is particularly so in soft ground. It was clear that the shock wave of water pressure reduction had not been allowed to travel far enough along the peat bed to influence other houses. These would be safe only if they happened not to straddle a mixture of peat and other soils, or if no further attempts at groundwater lowering were ever made.

In the circumstances the foundation engineer had a great deal of sympathy for the builder, who seemed to have made a good job of the design and construction of the houses. In almost all circumstances John Doe would have found his bungalow very satisfactory. Since he was aware of the likely problem of differential settlement of houses constructed partly over peat the builder had adopted the very robust foundation system of a reinforced concrete raft established 0.8 m below ground. On the other hand differential settlement was not inevitable. A quick calculation showed that the entire mass of the bungalow with its likely contents was 110 tonnes, which matches almost exactly the mass of soil removed in the excavation for the foundation, being 84 m^2 x 0.8 m x 1.8 tonnes/m^3, which is 121 tonnes. By compensating for the weight of the structure in this way, the underlying soils are not compressed any further than they were before building started. The only deformation encountered as the bungalow went up must have been due to the peat being pushed back down after its previous small rebound due to the excavation above. This deformation probably caused the teething troubles that were referred to the builder before the crisis.

With such a design the bungalow would have been perfectly safe on the gravel or on the peat, or on any mixture of the two, except in the case of a draw-down in the groundwater level. The engineer felt that he could suggest ways in which the remaining houses could be built without lowering the water table, and he decided to approach his client's solicitor with a view to negotiating an agreement with the builder which might short-circuit the time-consuming process of making a claim under the 10-year guarantee of the NHBRC.

1.3.5 Questions for Discussion

(1) Why did the engineer check that there were no large trees in the neighbourhood of the bungalow?

(2) What might the settlement of the bungalow have been if it had been constructed entirely over the peat layer, and on the surface of the ground so that it was without the compensating buoyancy of the foundation excavation?

(3) The engineer later suggested to the builder that he should compact 1.5 m of granular fill over the low-lying ground before he started building operations. What problems would this solve, and create, for the builder?

(4) What is the unit weight γ of a dense quartz gravel ($G_s = 2.65$) at a void ratio of 0.5, both dry and saturated?

(5) It is often necessary to assume that soil properties are constant from point to point. Where did this assumption prove useful and where might it have proved misleading? What sort of expertise would be needed to choose the frequency and depth of boreholes on a site investigation?

(6) This case-history demonstrates the importance of groundwater-level fluctuations on the compression of soil. Does a variation in the level of a lake affect the compression of the mud at the bottom of it?

2

Constitution of Soil

2.1 Origins

Soils and rocks are the naturally occurring fabric of our planet. The individual crystal grains within a beach pebble may have witnessed each of the countless cycles in the 4 500 000 million year drama which has been the unfolding of Earth's history: cycles of mountain building, followed by erosion, transportation of the rock fragments by wind and water, sedimentation, cementation, compression and subsequent upthrust in a later collision of land-masses. Alternatively, the pebble may have hardened very recently from a molten state, having been flung into the arena during a recent volcanic eruption. Geologists recognise three main classes of rock: sedimentary, igneous and metamorphic. Sedimentary rocks are laid down 'cold' over millions of years of deposition, while igneous rocks owe their formation to relatively sudden outbursts or intrusions of liquefied rock. Metamorphic rocks are those whose nature has been altered as a result of the local application of pressure or heat.

Soil mechanics concerns itself with sedimentary materials, in particular with those that are not yet cemented, nor very greatly compressed. Soil to the engineer implies any uncemented and badly fitting accumulation of mineral fragments of various size, shape and composition with an appreciable percentage of interconnecting voids. The parallel discipline of rock mechanics, on the other hand, is concerned with intact blocks of a cemented, compressed or crystalline nature, separated by joints and fissures: the blocks themselves may or may not be porous, but the mechanics of the whole assemblage is dominated by the disposition of the joints in the same way as in a child's tower made of wooden blocks. This engineering classification does not exactly follow the 'drift'/'solid' subdivision of the geologist. Certainly, all 'drift' material is 'soil' to the engineer: but so are a number of the geologist's 'solid' rocks. Up to 100 m of shelly sands were laid down about 2 million years ago (Pliocene times) in East Anglia: they remain uncemented and are therefore soils, but they still lie where they were deposited and are therefore not 'drift' but 'solid'. The same can be said for many clays laid down over an even longer time scale. Clays are usually compressed into shales, rather than being cemented, and this compression requires burial under thousands of metres of overburden for millions of years. Ancient clays which are still soils to the civil engineer include London clay (Eocene: 50 million years), Gault clay (Cretaceous: 100 million years) and Oxford clay (Jurassic: 150 million years). Although this textbook will focus your attention on the mechanics of soil, much will also be relevant to a study of rock. The reverse also applies: you would benefit greatly from a parallel study of engineering geology, or by reading a textbook such as that by Blyth and de Freitas (1974). Only a civil engineer who is prepared to acquire geological and scientific concepts in addition to applied mechanics will be of value when considering earthquakes, volcanoes, coastal erosion, regional subsidence, dissolution cavities, mining effects, floods or landslides.

A study of geological processes also helps the engineer to understand the nature and likely distribution of a soil. The soil grains may have been split away from the parent rock by mechanical abrasion or the thermal shock of successive cycles of heating and cooling. An example of a combination of thermal and mechanical action is the frost damage that occurs when water freezes and expands in joints and fissures. Alternatively, the rock may have weathered chemically over a long period, with unstable minerals breaking down in the presence of water and oxygen. Having split away, the mineral grains are transported by wind, river or glacier, during which time they are further reduced in size and rounded. Finally, the grains are sorted as they are deposited. Most soils were transported by rivers and deposited in lakes and seas. The sorting of the grains in this case is due to the steady decline in the speed of the river currents, which results in larger fragments being deposited first and the smallest grains last, the extreme zones being geographically separated by many kilometres. Good geological interpretation can therefore be of the greatest value when a site investigation is being planned or conducted.

2.2 Minerals

There is an immense variety of minerals in the Earth's crust. They are made up of combinations of the 92 naturally occurring elements. Notwithstanding this, eleven oxides account for roughly 99 per cent by weight of the rocks in the Earth's crust, namely those in table 2.1. Even these few oxides, combined together, and usually with silica, are responsible for hundreds of chemically distinct minerals, but the overriding importance of silicates and alumino-silicates will be clear.

The three most significant engineering properties of a mineral grain are its scratch hardness H (measured on an arbitrary ten-point scale), its specific gravity G_s and its coefficient of friction. The coefficient of friction between interlocked soil particles is the only significant source of stability for earth construction: without it soil would behave rather like a heavy liquid. The specific gravity of the soil particles is important because Earth's gravity is the only agency which creates stress in ordinary soil construction: if all other parameters remain constant, the lateral pressure of earth against a retaining wall is simply proportional to the density of the soil grains. The hardness of a soil grain is allied to its crushing strength and is an important determinant of its likely size, the soil grain having been jostled and scratched by its neighbours for millions of years.

TABLE 2.1
Commonest Constituents of Earth's Crust

Element	Oxide	Percentage
Silicon	SiO_2	59.1
Aluminium	Al_2O_3	15.2
Calcium	CaO	5.1
Sodium	Na_2O	3.7
Iron	FeO	3.7
Iron	Fe_2O_3	3.1
Magnesium	MgO	3.5
Potassium	K_2O	3.1
Hydrogen	H_2O	1.3
Titanium	TiO_2	1.0
Phosphorus	P_2O_5	0.3
	Total:	99

Source: Blyth and de Freitas (1974)

These properties can be used to distinguish four broad groups of silicate minerals, which account for the overwhelming majority of soil constituents.

The first group consists of quartz (silica, H7, G_s 2.66) and feldspars (alumino-silicates with potassium, sodium or calcium, H6, G_s 2.58) which possess a strong three-dimensional atomic framework, responsible for their hardness. The very substantial resistance of quartz to abrasion leads to its widespread occurrence in gravels and sands: it is very rarely found in a finely ground condition. The second group consists of silicates, which possess a less substantial 'chain' lattice, but which nevertheless are quite stable mechanically and chemically. This includes olivine, enstatite, augite and hornblende (silicates with magnesium, iron, calcium or aluminium, H6, G_s 3.3). The third group contains the much softer and flaky micas (alumino-silicates with potassium and sometimes magnesium and iron, H2, G_s 2.9), which are chemically stable but easily ground down.

The fourth group consists of the clay minerals, which possess a weak and friable 'sheet' lattice structure that is made spongy by its ready acquisition of water between the 'sheets'. The minerals are kaolin and illite, which are alumino-silicates, and montmorillonite, which additionally uses the bases sodium, potassium and calcium. Their hardness is so small as to be unmeasurable, and they are so fragile that they are rarely found in single crystals above 1 micron in width (10^{-3} mm). Their 'sheet' structure means that a naturally occurring clay grain is likely to be much thinner than it is wide, perhaps only 10^{-5} mm. The specific gravity G_s of clay minerals is in the region of 2.6, similar to that of the feldspars from which kaolin, for example, is the chemically weathered residue.

The clay minerals can be chemically sensitive: the adding of calcium to a 'heavy' clay soil has long been known to offer the hope of reducing the compressibility, and therefore the swelling and softening of the soil, making it easier to work for the farmer.

The friction resistance of a soil is so dependent, as we shall see, on the density, arrangement, and surface roughness of the individual grains, that differences in the fundamental crystal-to-crystal coefficient of friction do not have such a large effect on the overall value. Nevertheless, a soil composed entirely of the soft and flaky clay minerals may possess only one-third the internal friction of a good quartz sand. It is within these bounds that engineers seek, by methods of soil testing described later, to measure the appropriate angle of internal friction of a soil sample. Mechanical testing, rather than visual or chemical analysis, is the engineer's method for assessing the strength of soil.

The model that an engineer carries around in his head concerning the mineralogy of soil can, therefore, be quite simple. All common soil grains have a specific gravity in the range 2.6 to 2.8. Large grains are likely to be strong, and indeed likely to be quartz. Smaller grains are likely to be weaker. Clays, made up of grains so small that an optical microscope cannot fully resolve them, are particularly soft and spongy. All soils possess friction, but clays are less frictional than quartz sands.

Much is made, in some textbooks, of special inter-particle forces that are said to exist between clay particles in close proximity. Authors point out that if clay flakes are only 20 atoms thick, more than 10 per cent of atoms within a block of clay soil would be on the surface of a particle. This, quite reasonably, leads them to expect that clay soils might behave quite differently from other soils. And so they do: they are particularly compressible. As we shall see, they may have a very small Young's modulus. Unfortunately, some writers leave in the minds of their readers the notion that clay is held together by special 'true cohesion' in addition to any friction that might exist. The cohesion is supposed to be due to the special inter-particle forces. This is now known to be misleading. Even if the cohesion is present, it is often devalued by a miriad of fissures. Only a tiny minority of clays seem to possess this extra quality of cohesion or cementation to an important degree. A rather greater proportion of sands are at least slightly cemented.

The student in search of a mental picture of the mechanism by which soii stands in slopes, or bears his weight, need look no further than friction between the particles, whether sand or clay, together with some mechanical interlocking.

2.3 Particle Size

As I have already remarked, the engineer's inclination is to take direct measurements of important properties rather than to attempt a pseudo-scientific analysis in terms of 'fundamental' parameters. This explains his liking for the direct measurement of soil particle sizes. From these he will get an insight into the likely origins and usefulness of the soil. He will also get some notion of the likely mineralogy of the soil in terms of its 'clay' content by assuming that small particles are 'clay' particles. If natural agencies have been able to whittle a grain down to a very small size, the assumption that it is a soft clay mineral seems reasonable to the engineer. Although this attempted identification of mineralogy with particle size is a little irrational and confusing, it appears to be sufficiently accurate to be useful. You should remember that the presence of clay minerals in a given proportion does not lead to a definite calculation, merely to the triggering of certain alarm-bells: the hunter doesn't need a tuning fork to detect the precise frequency of the call of a wild goose. The names given to particle sizes appeared in figure 1.2, being gravel, sand, silt and clay.

Particles larger than coarse silt (0.06 mm) can be sieved quite effectively. Soil containing a known weight of solids is washed through a nest of sieves with the mesh size reducing progressively, and the proportions by weight of the solid fractions retained on each sieve can be determined by weighing after drying. The conventional curve of 'percentage passing' against mesh aperture is illustrated in figure 1.2 for two particular soils.

Particle sizes smaller than the sands are best analysed by sedimentation: Stokes's law allows a correspondence to be made between the terminal velocity of heavy particles in water of known viscosity and their diameter. Just as with the sieve method, the shape of the particles affects the results somewhat, but not so much as to invalidate the method. The remaining descriptions of shape and roughness are best made after examination of the soil under a microscope or magnifying glass. The standard routines for soil size analysis and description appear in BS 1377.

The first piece of information that the engineer abstracts from a grading curve is the proportion of silts and clays. Fine-grained soils have very small pores: this, above all else, makes a clay look and behave in a peculiar way. Surface tension effects are inversely proportional to the size of the soil pores, so that the capillary rise in a 10^{-4} mm clay should theoretically be hundreds of metres. Capillary rise is caused by suction. This suction pulls the soil particles hard together, and the friction at their points of contact makes the clay feel tough or tenacious. Just as dramatic is the effect of small pore size on the speed of drainage of a clayey soil: clays are often said to be impermeable, so slowly does water percolate through them. A velocity of seepage of 10^{-10} m/s or 3 mm/year would not be considered unusually slow in clay, whereas 10^{-2} m/s might be achieved in a clean sand, and, as I have already mentioned, the engineer would suspect that a soil with a high clay content would be subject to large volume changes due to water migration.

You must appreciate that a soil with 25 per cent clay-sized particles and the rest sand could well be a 'clay' to the layman, and might also be usefully described as 'clay' by an engineer. Take a typical example of a soil, with 25 per cent by mass of clay-sized particles and 75 per cent by mass of 1 mm spherical sand particles, and a typical overall void ratio of 0.50. If the density of the two types of soil grain is the same, then their volumes will likewise be in the proportion 1 clay: 3 sand. Moreover, the definition of void ratio means that the volumes of solids to voids will be in the proportion 1:0.5. The overall volumetric proportion will therefore be

$$\begin{array}{lll} \text{solid} & 1 & \begin{cases} \text{sand} & 0.75 \\ \text{clay} & 0.25 \end{cases} \\ \text{void} & 0.5 & \end{array}$$

This means that if the clay particles were removed, leaving voids, the sand particles would have a void ratio of 0.75/0.75, which is 1.00. A moment's thought will show you that equal-sized balls which are only just touching each other each occupy $(4/3)\pi r^3$ of a cubic space $(2r)^3$ and therefore have a void ratio of 0.9. The sand particles must be imagined, therefore, suspended in a matrix of clay and hardly in contact with each other. It is the finest 25 per cent of any soil that determines its mechanical behaviour, and perhaps the finest 10 per cent that dictates the size of its pores.

The engineer will secondly take an interest in the shape of the grading curve throughout its entire length. He will be most satisfied to see a smooth curve. A wide horizontal step would indicate a size gap, a grade of particles missing from the mix. A high vertical step would mean the presence of a large fraction of particles of one size. The ideal well-graded soil possesses small particles to fit into the voids between medium particles, which in their turn lock together the largest grains. Without this spread of particle sizes a soil is likely to be difficult to compact, subject to a reduction in volume due to vibration, and subject to washing-out of its finest fraction. The engineer would only want to see a uniform particle size if he were intending to use the soil in a drainage scheme, or if he could use it as an aggregate in concrete. Further information on the criteria that are commonly chosen from the grading curve to depict the usefulness of a soil for various purposes can be found in Terzaghi and Peck (1967).

2.4 Plasticity

A soil containing silt and clay particles is to be suspected of excessive compressibility: large volume change may accompany load or temperature changes. For this reason it is very common to subject these soils to an empirical assessment of their likely capacity for volume change. Two tests, known as the Atterberg limit tests, arbitrarily define the plastic limit at which soil can just be rolled into 3 mm diameter threads without crumbling, and the liquid limit at which a 10 mm deep V-groove made in a pool of soil contained in a steel cup just closes when the cup is struck 25 times in a standard fashion. As a sample of the soil is wetted to carry it from the plastic limit towards its liquid limit it is said to cover its whole plastic range: first of all it can be rolled into very fine threads, then it becomes sticky, and finally it achieves the consistency of whipped cream. The moisture contents at the extreme points are simply called the plastic limit (PL) and the liquid limit (LL): the range of moisture contents between the two limits is called the plasticity index (PI). The tests were first mentioned in section 1.2.

The value of the tests lies in the fact that they are rough strength indicators: the shear strength of a soil at its plastic limit may be roughly 200 kN/m^2, but only 2 kN/m^2 at its liquid limit. The plasticity index, therefore, is the increment of water necessary to reduce the strength of a soil roughly a hundredfold. It can be shown that the reverse argument also applies. If the stress on a saturated soil is increased under a foundation, for example, then the ensuing compression, which drives water out of the soil and causes the foundation to sink, is proportional in magnitude to the plasticity index of the soil. Using the language of mechanics: the Young's moduli of soils are inversely proportional to their plasticity indices, other variables being excluded. The correspondence is only mediocre, partly due to the variable nature of the Atterberg limit tests themselves and partly due to the masking effect of the other variables mentioned above, which are the subject of later chapters. Nevertheless, the Atterberg limit tests have been found by many engineers to provide, by empirical formulae and correlations, intriguingly good predictions of the stiffness moduli of fine-grained soils.

It is necessary to adopt a cautious mental attitude to the assimilation of these plasticity tests. They depend for their success on the moisture content, and therefore the void ratio, of a soil being constant during the test. The void ratio of sand is not easily controlled: if you try to mix one part of sand with two parts of water in order to obtain a very loose sand, the sand will fall through the water in a matter of seconds. Clay, on the other hand, can preserve almost any given void ratio and strength for many minutes or hours, even on a small scale. Clay dust mixed to a slurry in water will also fall through the water and settle as a mud, but it will take hours to do it (some particles may actually be so fine as to be colloidal: random molecular vibrations keep them in suspension). The particle size affects the rate of change of soil properties: once again we see that sand and clay place themselves in different time categories. Sand will not stand still while we 'ask' it how strong it feels, while clay will: sand is said to be granular, while clay is called cohesive. You must try to guard against imagining that clays have special cohesive forces like magnets: the Atterberg tests are yet another temptation to jump to this erroneous early impression.

2.5 Questions for Discussion

(1) What are the three most important non-silicate minerals found in Great Britain?

(2) A family of soils of similar mineralogy has a variable clay content (C per cent): how might the plasticity indices (PI per cent) of the soils vary?

(3) A soil sample is carefully mixed, and three moisture content samples have an average of 14 per cent. 400 g of the natural wet soil are washed through a nest of sieves. After drying in an oven, the masses retained on the sieves are

mesh (mm)	2.4	1.2	0.6	0.3	0.15	0.075
mass (g)	54.2	12.5	46.5	83.8	77.0	41.6

Draw the grading curve, and describe the likely nature of the soil. List the differences in technique between the test described in BS 1377 and that described above, and remark on the merits of each of them.

(4) A silty clay encountered in a deep excavation is found to have a moisture content of 20 per cent. Guess its void ratio and bulk density, remarking on your assumption.

3

Groundwater

3.1 Water in the Ground

Water exists in the ground above sea level because the Sun evaporates an estimated 5×10^{13} tonnes of water every year, 20 per cent of which falls back over the *land* as snow and rain. If gravity were the only source of power on our planet, then the spherical surface of the open seas would be matched by inland seas, lakes and invisible groundwaters saturating the soil and rock, with spherical surfaces of identical radius. The Sun's influence raises freshwater high on to the land; it must then find its way by gravity back down to the sea in rivers and glaciers, and by underground seepage of that fraction of the water which percolates into the pores of the soil and rock. The rainfall in Great Britain is sufficient to ensure that a continuous body of groundwater can be struck almost anywhere on digging only a few metres below ground level. To this water we owe our existence; indeed, it might be argued that the civil engineer also owes to it his livelihood, for it furnishes him with most of his work. This is certainly true for the soil mechanics specialist: almost every significant failure, the majority of contractual disputes and the overwhelming majority of design problems are due to the presence of water in the soil.

As we shall see, the mechanically damaging aspect of the presence of water in the ground is the pressure in the water. It is vital to understand that water would be almost insignificant were it not for the ubiquitous pressures within it. Water does not generally 'lubricate' soil constituents to any marked degree: it has little effect on the coefficient of effective friction of a soil. Nor has it much effect on the weight of a soil body. Pure water has a density ρ_w of 1000 kg/m^3, which at the surface of this planet confers on it a unit weight γ_w of 9810 N/m^3 or, more conveniently, 9.8 kN/m^3. Sea water on the other hand has a density of roughly 1030 kg/m^3 with a corresponding unit weight of 10.1 kN/m^3. In this book I shall use the symbols and values, ρ_w = 1000 kg/m^3 = 1 tonne/m^3, γ_w = 10 kN/m^3, so as to simplify arithmetic without sacrificing much accuracy. As we have seen in chapter 2, most common rocks have a specific gravity G_s in the range 2.6 to 2.8, so that solid rock can be expected to have a density of roughly 2700 kg/m^3 and a unit weight of roughly 27 kN/m^3. Porous rock (or soil) clearly does not weigh as much as this. Consider a typical soil with a void ratio (see table 1.1) of 0.50. If the soil is perfectly dry, and the voids are full of air, then the density of the soil is 2700 x 1/(1 + 0.5) or 1800 kg/m^3. If the same soil is saturated with water its density is (2700 x 1 + 1000 x 0.5)/(1 + 0.5) or 2130 kg/m^3. The unit weight of this soil with a constant void ratio of 0.5 cannot escape beyond the bounds 18 to 21 kN/m^3. I shall very often assume that the unit weight of soil is 20 kN/m^3, saturated or not. I hope you will appreciate the advantages and dangers of this step. By proclaiming that water weighs 10 kN/m^3 and that soil usually weighs 20 kN/m^3 we leave ourselves unencumbered to deal with the enormously complicated issue of water pressure. We also leave ourselves vulnerable to being caught out by exceptional soils: a sensible precaution in practice is to weigh the materials on hand.

3.2 Hydrostatics

The logical starting point for a study of water in the ground is a revision of hydrostatics, which is concerned solely with a body of water at rest. Consider a column of water with vertical, parallel sides and a horizontal base $Z_1 Z_2$. Let it have a plane free surface $S_1 S_2$ above which exists an atmospheric pressure a_s, as shown in figure 3.1. By assuming that the water is an ideal fluid, which can

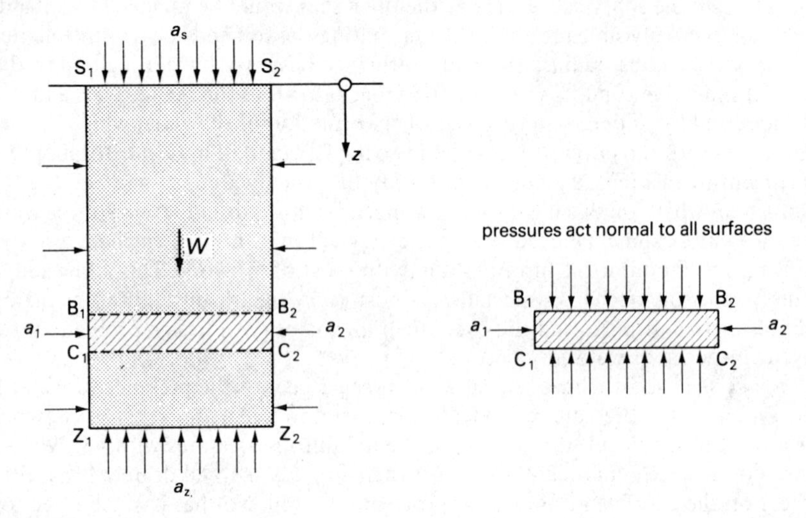

Figure 3.1 *Water at rest*

transmit no tangential or shear forces, but only pressures normal to any object, and applying Newton's laws for equilibrium of the static column, we can determine the water pressure distribution easily.

Firstly, by considering the horizontal equilibrium at a level disc $B_1 B_2 C_1 C_2$ of water at any depth, we can see that horizontal movement would take place unless $a_1 = a_2$. This shows us that the pressure in an ideal fluid at rest must be the same at all points on the same level. Similar argument shows that the free surface $S_1 S_2$ must be horizontal if it is not to be moving.

Secondly, the vertical equilibrium of the column can be guaranteed only when

$$a_s \times \text{column area} + W = a_z \times \text{column area}$$

since there can be no contribution from side forces in the ideal fluid. We can express the weight as

$$W = \text{column area} \times z \times \rho_w \times g$$

where g is the local gravitational acceleration, say 9.81 m/s^2, and ρ_w is the mass density of water, 1000 kg/m^3 or 1 tonne/m^3, so that

$$a_s + z\rho_w g = a_z \tag{3.1}$$

Engineers make two changes which alter the appearance of this simple relation: firstly they replace $\rho_w g$ by a single symbol, γ_w, the unit weight of water, roughly 9.8 kN/m^3 at the Earth's surface; secondly they replace the absolute pressures a used in the derivation above by gauge pressures u, that is, pressures above atmospheric.

On the level surface of the fluid, the gauge pressure is zero, that is

$$u = 0$$

but, for a point at any depth z below the surface, equation 3.1 is rewritten as

$$u_z = a_z - a_s = z\rho_w g = \gamma_w z$$

or simply

$$u = \gamma_w z \tag{3.2}$$

The use of gauge pressures will greatly simplify calculations, since the almost-constant atmospheric pressure invariably permeates every void in the soil and its effect is therefore nil. As an example, the upholstery trade would not be affected by a doubling of the atmospheric pressure since the extra air pressure which tended to compress the springs in the mattress even before the customer lay down would be exactly balanced by extra support from the selfsame pressure, which would have found its way into the mattress to support the load even before the springs were compressed.

This section has demonstrated that water 'finds its own level', and that the gauge pressure in static water is given by $u = \gamma_w z$ at a depth z beneath the water surface, which is at atmospheric pressure. This is as true when the water is saturating a stratum of soil or rock as it was when the water was in a lake, since it will always be possible to arrive at any point in the static groundwater under the water surface by a series of vertical and horizontal steps through water alone, each of which must conform to the same rules laid down above. Remember, however, that it only refers to water at rest in a rigid soil; moving water possesses viscous friction, which alters the simple equilibrium equations, and nonrigid soil would allow ground pressures to be transmitted to the water, which would then have to flow out of the compressed ground as bath water flows out of a sponge. The first of these cases, seepage, is precisely the future study of this chapter; the second is called transient flow and will be analysed in detail in chapter 7.

One last aspect of hydrostatics deserves mention, namely buoyancy. Soil particles that are completely immersed in water have an upward force acting on them which is exactly equal to the weight of water they displace, by the principle of Archimedes. Since most soil minerals have a specific gravity G_s in the region 2.6 to 2.8, the particles will not actually float, but will only become lighter, having a submerged effective specific gravity of 1.6 to 1.8. As we shall see in chapter 4, this loss of effective soil weight brings with it a loss in the frictional force between soil particles, which is very significant because friction is the principal source of strength and cohesion in soils.

The most powerful way of dealing with this buoyancy, however, is to account for water pressures from first principles rather than by using Archimedes' principle on individual soil particles. Suppose that you require to know the vertical effective stress at the point P depicted in figure 3.2 which is at a depth z

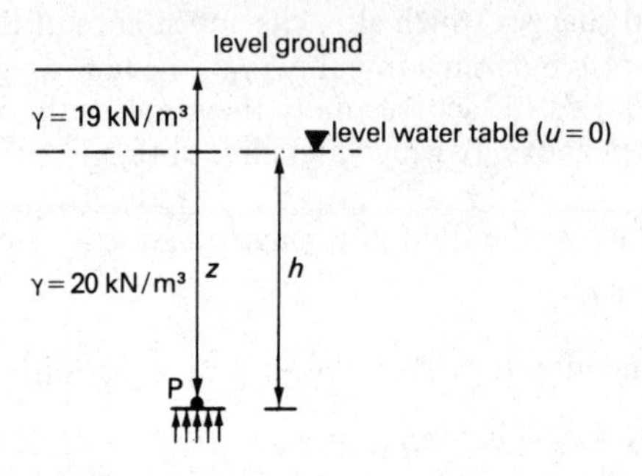

Figure 3.2 *Ground and water at rest*

beneath ground level and h beneath the water table under a large flat field. The most likely source of information about soil densities will come from the weighing of borehole samples of known volumes, so let us suppose that these weighings indicate unit weights of 19 kN/m^3 in the moist soil above the water table and 20 kN/m^3 in the saturated soil beneath the water table. The effective vertical stress at P is found by deducting the water pressure from the total vertical stress (see figure 1.11). We will make the calculations using gauge pressures relative to atmospheric pressure. The water pressure at P is simply

$$u = \gamma_w h$$

But what is the total soil pressure which the water is tending to buoy up? We can consider a column of wet soil, just as we considered a column of water in figure 3.1, in order to find the total vertical stress at its base. Whereas we obtained $u = \gamma_w z$ for water we will obtain $\sigma_v = \gamma z$ for the heavier soil. The only significant doubt arises from possible friction on the sides $S_1 Z_1$ and $S_2 Z_2$: might not the neighbouring soil columns support our central column by offering it upward friction at its edges? Since in statics every action must be balanced by a reaction, this would mean that the neighbouring columns would suffer a downdrag. And since there is no difference between column $S_1 Z_1 S_2 Z_2$ and its neighbours, it is unreasonable to suppose that its side forces act in opposite directions: there can be no side friction, therefore!

In figure 3.2, the soil is not of uniform density, so that at point P we have

$$\sigma_v = 19(z - h) + 20h$$

We have now almost completed our inquiry; by accounting for the unit weight of the soil with its water we have found the total vertical stress. We have previously found the water pressure, so that by subtraction we deduce that the vertical effective stress

$$\sigma'_v = \sigma_v - u = 19(z - h) + 20h - 10h$$

Although it is tempting to regroup the effective stress into two components due to a unit weight of 19 kN/m^3 above the water table and an 'effective' unit weight $20 - 10 = 10$ kN/m^3 where the soil is submerged, I appeal to you not to do it! It could prove your undoing on some future occasion when seepage of groundwater will destroy the simple hydrostatic relationships. Let a water pressure be a gauge water pressure, and a total stress be a gauge stress representing the whole superficial thrust of wet soil on a unit area. The effective stress, of vital significance in settlement and friction calculations, is then the difference between the two.

3.3 Suction

Water pressures can be smaller than the pressure of neighbouring atmospheric air. They can even be smaller by a margin that is greater than one atmospheric pressure. In other words not only can gauge water pressures be negative, but also absolute water pressures can be negative: water can carry tension. The source of this surprising ability is 'surface tension'. Although the physics of surface tension is derived from intermolecular attractions, it transpires that a simple model can be used to make predictions. The liquid/gas surface is notionally replaced by a thin membrane, which carries an omni-directional tension. Wherever the membrane is cut, the surface tension T acts perpendicularly to the wound: its magnitude for water at 10 °C is 7×10^{-5} kN/m so that the force tending to rip open a 1 m long crack in the surface of a lake of water is 7×10^{-5} kN, which is roughly the weight of a plain biscuit. Compared with the weight of a swimmer, for example, the force due to surface tension is negligible. It only becomes important when considering the behaviour of small volumes of water, when the weight of the water is reduced by perhaps the third power of the scale while the perimeter (governing the surface tension) is only scaled down linearly. This is typified by 'capillarity' or the ability of water to rise against gravity in a fine tube (see figure 3.3).

The air at points A and C is at atmospheric pressure, or zero gauge pressure. The surface of the water in the region of C is plane, so that the gauge pressure u in water at D must also be zero if it is to be stationary. If the gauge pressure at D is zero, then so must be that at E. Now hydrostatics on the column of water BE dictates that the gauge water pressure at B must be less than that at E by an amount $h_c\gamma_w$, in other words $u = -h_c\gamma_w$ at B. Although this relative suction acting over the area $\pi d^2/4$ is tending to draw down the hemispherical meniscus A/B, we observe that upward surface tension T acting on the perimeter πd of the meniscus is just able to hold it in equilibrium. Resolving vertically for the meniscus

$$T\pi d - h_c \gamma_w \frac{\pi d^2}{4} = 0$$

$$h_c = \frac{4T}{\gamma_w d} \qquad (3.3)$$

Making the substitutions $T \approx 7 \times 10^{-5}$ kN/m at 10 °C and $\gamma_w = 10$ kN/m^3 we obtain $h_c \approx 3 \times 10^{-5}/d$ metres where both h_c and d are in metres. So when $d = 1$ mm (perhaps representing the effective void size in a coarse sand) $h_c = 30$ mm; but when $d = 0.001$ mm (representing a uniform fine silt) $h_c = 30$ m: if a pure clay were thought to have effective void sizes of 10^{-5} mm then its capillary rise becomes almost unimaginably large. Of course, the imaginative leap between capillary tubes and the pattern of connected voids in a soil leaves much to be desired. Perhaps fortunately, the need to specialise this simple model is eroded away by three inescapable complications which assail soil in the field: these are the downward percolation of rainwater, the evaporation of groundwater into a relatively dry atmosphere, and the very slow rate of flow of water in fine-grained soils.

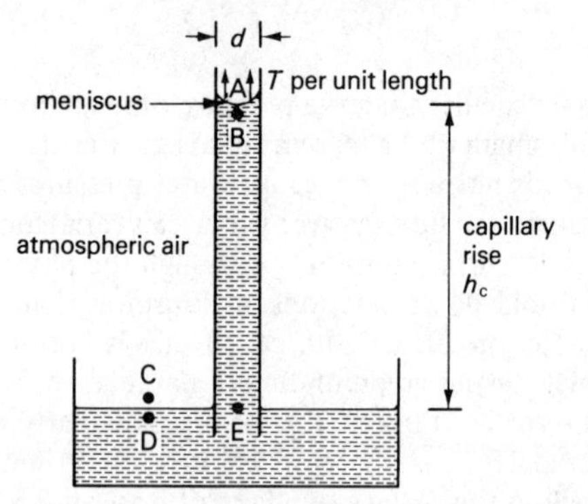

Figure 3.3 *A column of capillary water in equilibrium (not to scale)*

Most British soils have been deposited initially under water in a more or less saturated state. After the overlying water was removed, the surface of the soil may have been eroded: it would certainly have been subjected to the sun and rain, and would have suffered countless seasonal cycles of wetting and drying. This left the soil with a groundwater table at some depth, above which there may be a fully saturated capillary zone, capped by a very complex zone of partial saturation: this situation is depicted in figure 3.4. In a simple quasi-static situation the water table, which is the imaginary surface along which the pressure of the water in the voids is atmospheric, can be observed in a standpipe if it is left to fill for a sufficiently long period. However, when a pit is dug, a zone of soil above the water table will be found to be saturated with water. The depth of this capillary zone (as drawn in figure 3.4) may be of the same order as calculated in equation 3.3: in the case of fine-grained soils, however, it may be very much smaller. The theoretical height of capillary rise in a pure clay would exceed 100 m, and yet a water table in British clay which was as deep as this would be quite extraordinary. In a clay, therefore, the possible maximum capillary zone simply does not have room to develop. Moreover, most natural 'clays' are actually composed of alternating layers of clay particles and much larger silt particles with much larger pores which block capillary rise. In the static saturated zone, both above and below the water table, the gauge water pressure u in the voids must simply conform to equation 3.2, the straight line of which is depicted in figure 3.4, passing through zero at the level of the water

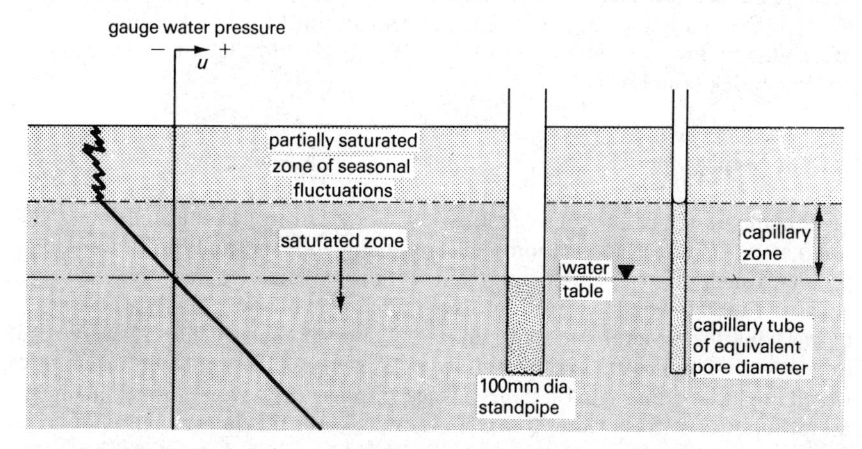

Figure 3.4 *A simple case of suction in sandy ground*

table. Above the saturated capillary zone there will be a partially saturated zone whose surface will be dried in summer and wetted in winter. The pore-air and pore-water pressures in this zone will be quite variable and unpredictable: this in turn makes effective stress analysis for strength or settlement very difficult. Engineers normally dig their foundations at or below a depth of 1 m. By doing this they hope to avoid the worst of the fluctuating zone.

Before we leave the topic of suction we should consider the enormously severe effects of dry air when it is allowed to 'steal' water from the soil. If a block of soil is exposed on a table in a room where the air is not saturated, only a relative suction in the water can prevent its total evaporation. The suction in the pore water can only be provided by surface tension, and the required magnitude is related to the relative humidity RH of the air

$$u = -150\,000\,(1 - \text{RH})\ \text{kN/m}^2 \tag{3.4}$$

In a centrally heated living room in winter RH may be 0.7, while in a cold unheated attic RH may be 0.9. Even if the exceptionally damp conditions relating to RH = 0.99 are assumed, you will see that the potential suction is $-1500\ \text{kN/m}^2$ (15 atmospheres, or 150 m of water). While soil in the ground may be protected from such severe drying by vegetation, or at worst by a thin dry crust, soil on the laboratory bench will have no such protection. It will remain saturated at first as its water pressure drops to the full extent of its capillary suction. Air will then enter the soil as the pore water continues to evaporate and as the menisci are forced to retreat into the smallest and tightest corners. Only when the suction of the dry air is matched by the suction of the tiniest capillaries will the air/water surface attain an equilibrium. If that equilibrium in a block of clay were to mean a suction of more than 1500 kN/m^2, then you will readily appreciate that the soil would have the consistency of a brick. At the moment of air entry, when $u = -4T/d$, it is certainly feasible to apply effective stress analysis. No total stress exists on the exposed soil block; the water pressure is $-4T/d$, therefore the effective soil stress must be $+4T/d$ in order to cancel out. The soil particles are effectively 'glued' together by surface tension: the menisci are almost as good as cement until the block is broken, after which point the water never finds its way back into the critical locations, and the block crumbles.

3.4 The Seepage Model

This model of behaviour considers the flow of water through the voids in a rigid soil matrix. Its direct application is exceptionally wide, not only in dam stability problems such as the example quoted in section 1.2, but also in every construction job where excavations have to be made below groundwater level, in every marine structure, every drainage or water supply scheme and every road. In very many applications its use simply rests in the prediction of flow quantities, which might be either troublesome leakage through a reservoir embankment, or useful supplies of fresh water to a well. In other cases the determination of forces, or water pressures, acting on and within an envisaged structure (earth, concrete, steel, etc.) will be the objective. The engineer will have in his mind, in these cases, that the flow of water may cause a ground movement. He will nevertheless apply the seepage model, which deals with flow through a rigid matrix, and deduce from the model the forces acting on that rigid matrix, following this by an analysis to determine whether or not the matrix can really remain rigid under these applied loads. This consecutive application of models that have specific limitations is typical of the engineer's method of establishing approximate solutions to problems that are essentially impossible. It makes him what he is, practical and flexible, but also vulnerable to appalling miscalculations when he unknowingly applies a model that contains a conventional assumption, such as the 'rigid skeleton of soil particles' in the seepage example, to a situation in which that assumption does not hold.

In 1856 the Frenchman, Henri Darcy, proposed the following simple flow relation

$$q_x = A_x k_x i_x \tag{3.5}$$

where

$$i_x = -\frac{\partial \bar{h}}{\partial x}$$

The meaning of these terms will be made clear below and demonstrated in figure 3.5. The quantity of water flowing perpendicular to an area A_x in unit time is given the symbol q_x. According to Darcy this flow quantity is proportional to the area A_x, a soil and water constant k_x called the permeability,

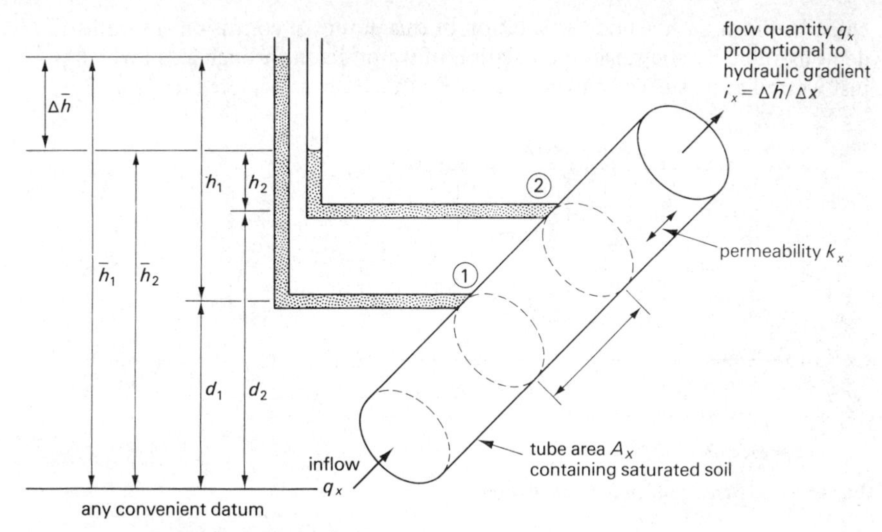

Figure 3.5 *Definitions of flow parameters*

and a measure of pressure gradient i_x, which is called the hydraulic gradient. In order to define the hydraulic gradient it is necessary to imagine standpipes connected to the points in question, with vertical heights of rise $\bar{h}$ above an arbitrary datum level. The hydraulic gradient is then the rate of loss of head $\bar{h}$ along the flow path $\Delta\bar{h}/\Delta x$ in the example in figure 3.5. According to Darcy the flow rate would remain the same whatever the position or orientation of the tube of soil, if the head drop $\Delta\bar{h}$ remained constant.

It should be clear that we have now defined two sets of useful water pressure parameters. The first set consists of gauge water pressures u, which give rise to standpipe water rises $h = u/\gamma_w$ above the point concerned; these are particularly useful in problems of statics. The second set is typified by standpipe rises $\bar{h}$ above a constant arbitrary datum, and they can be referred to as 'excess' pore-water pressures $\bar{u} = \gamma_w\bar{h}$; they are particularly useful in problems concerning the flow of water. In order to translate between an excess head, such as $\bar{h}_1$ in figure 3.5, and the actual pressure head h_1, it is necessary to know only the position of the point in question above the datum that was used to arrive at the excess head, that is, $h_1 = \bar{h}_1 - d_1$.

In many ways it is inconvenient that two parallel systems of water pressure measurement have to be used and related. Unfortunately, this situation is outside the scope of my authority. Water simply does not flow from high pressure to low pressure, otherwise a beaker of water would be in a constant turmoil as the high-pressure water at the bottom attempted to flow upwards towards the surface. Flow depends on relative heads, not on gauge pressures. On the other hand, the solutions to problems involving the equilibrium of soil or other structures will depend on actual water pressures, irrespective of whether the pressures have been expressed as heads above sea level in metres, or heads above the centre of the Earth in kilometres. I suggest that you attempt to remember what a real gauge water pressure is, and what an excess head of water represents, *independently*, and to translate between the two when necessary on an *ad hoc* basis. Both the source and the solution of this potential confusion are well demonstrated by the steady percolation of water down through a soil which is just wetted at its surface and well drained at its base, as depicted in figure 3.6.

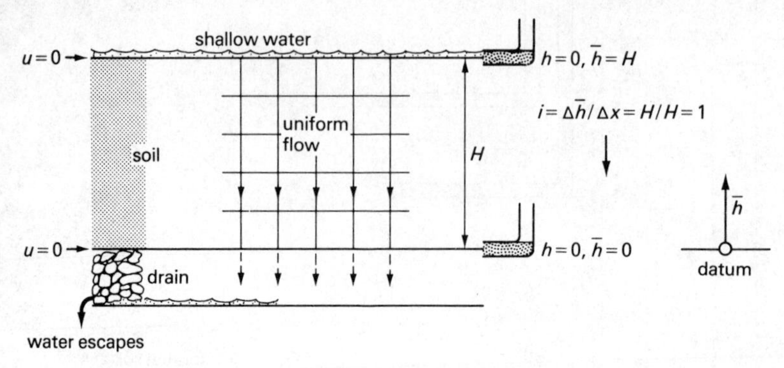

Figure 3.6 *Steady saturated percolation*

There is clearly zero gauge pore pressure in the soil at both its top and bottom surfaces. Only the sketching of standpipes makes clear that the excess head at any point in the soil is simply equal to its elevation above the base, so that the downward hydraulic gradient is unity while zero pore pressures persist through the whole layer. This elementary flownet is important because it demonstrates that a relatively permeable blanket drain laid beneath fill will be sufficient to prevent the build-up of pore-water pressure, even when the surface of the fill is flooded.

In modern work, the form of Darcy's equation is changed slightly to

$$v = ki \tag{3.6}$$

where all are assumed to refer to a single direction, and v is defined as q/A and referred to as the superficial seepage velocity. It is superficial in the sense that its definition implies that, over the whole superficial cross-sectional area A, the water can be thought of as flowing with velocity v. It should be recognised that this in no way represents the true average water velocity v_w, since water cannot flow through that part of the superficial area that is taken up with soil solids. Using the definition of void ratio given in section 1.1.3, we can say that

$$v_w \times e = v(1 + e)$$

or

$$v_w = \frac{v(1 + e)}{e} \tag{3.7}$$

so that in a soil with $e = 0.5$, $v_w = 3v$. Furthermore, the true velocities will depend on the shape of the tortuous connections between voids, and so they will be rather widely distributed about the value derived above. Fortunately the engineer usually requires simple estimates of total flow q in conditions where he knows areas and heads, and where he can estimate permeability k: he can therefore use equation 3.5. Only rarely, for example if he wants to compute the speed of spreading of a water pollutant, would he have to estimate the void ratio of the soil in order to arrive at a rough guess for v_w after he had calculated Darcy's v as before.

Although Darcy's law has proved very useful in the solution of engineering problems, it contains within it one of the most variable, unreliable, and unmeasurable, 'constants' ever imagined, permeability k. Part of the trouble rests in the extraordinarily wide distribution of permeabilities found in natural soils and rocks, as shown in table 3.1. Varying over perhaps thirteen orders of

TABLE 3.1

Permeability k(m/s)

10^{-1}	10^{-2}	10^{-3}	10^{-4}	10^{-5}	10^{-6}	10^{-7}	10^{-8}	10^{-9}	10^{-10}	10^{-11}	10^{-12}
pebble beds		sands			silts				clays		
	gravels			cemented sandstones							
						fissured mudstones					
					jointed, fissured granite						intact granite

magnitude, permeability covers the same range of magnitudes of velocity as are covered in the comparison between the sizes of an orange and the solar system. Furthermore, permeability is far from being constant even for one particular soil. Since it represents the ease with which viscous water can flow through the soil pores, it reduces as the soil is compressed, is affected by the chemistry of the soil/water interface and (due to viscosity variations) is quite dependent on temperature. Terzaghi and Peck (1967) include a number of suggested relationships. Engineers are usually pleased to be able to estimate the permeability of a stratum of soil to within a factor of 3 bearing in mind the enormous difficulty in recovering undisturbed laboratory samples or executing well-controlled field tests. The measurement of permeability is discussed later in section 3.8. Allen Hazen correlated the permeability k (m/s) with the size of the sieve which just allowed the first 10 per cent by weight of the soil to pass (referred to as D_{10} mm) in the case of a number of sands

$$k = 0.01 D_{10}^2 \tag{3.8}$$

in which the units must be as stated.

3.5 Two-dimensional Flow

3.5.1 The Extended Model

Most flow situations are three dimensional, such as the flow of water into a well, or into a square basement excavation. We therefore require an extension of Darcy's law from the simple one-dimensional statement in section 3.4. The first step along this way is to solve two-dimensional flow. We are then able to handle such problems as the flow of water into a long trench excavation or the leakage from a long canal or through a long dam. Here we draw a plane section of the flow as indicated in figure 3.7 and assume that there is no flow along the axis of the construction. These plane flow visualisations are responsible for the majority of flow predictions in civil engineering.

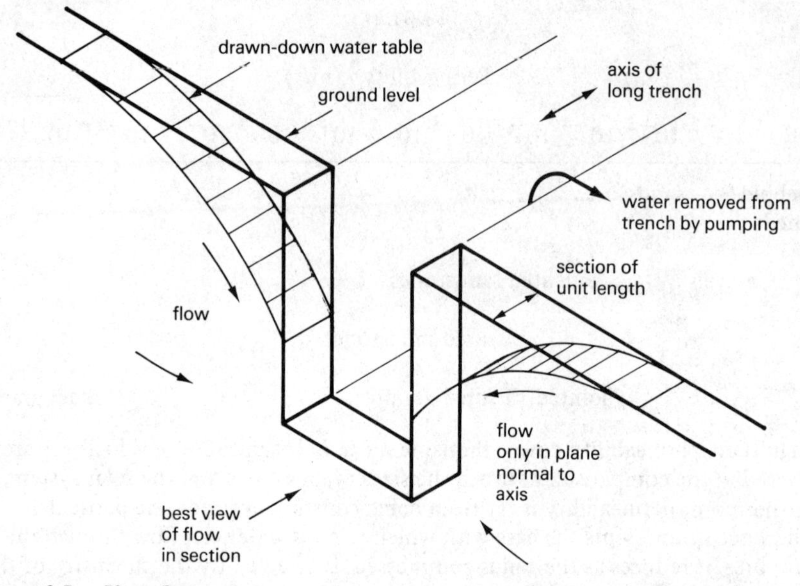

Figure 3.7 *Plane flow regime*

We now consider plane flow, and in particular the trajectories of imaginary packets of water, which we call flowlines. We also invoke the concept of equipotentials, which are lines joining points in the ground which share the same excess head over datum. Figure 3.8 shows one line of each sort. We now consider

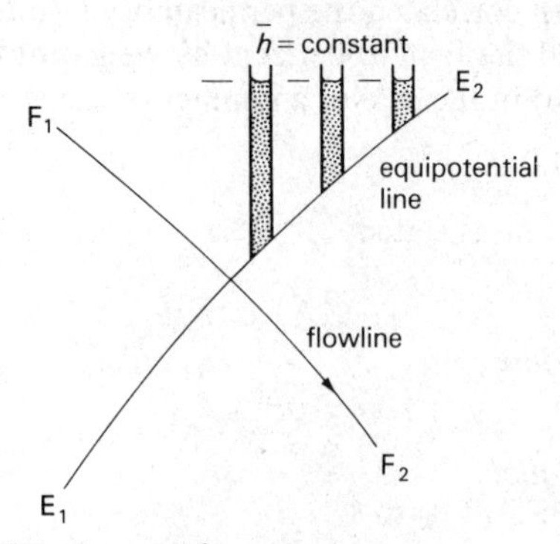

Figure 3.8 *Flowline and equipotential*

that permeability k is independent of position or flow direction. By applying Darcy's law along the equipotential

$$v_E = ki_E$$

Since there is no change of head along an equipotential we know that $i_E = 0$, so that there is no component of water velocity along an equipotential. Equipotentials and flowlines must, therefore, cross at right-angles.

The interlinking pattern of flowlines and equipotentials is called a flownet, and a typical part of such a net is shown in figure 3.9. Adjacent flowlines must

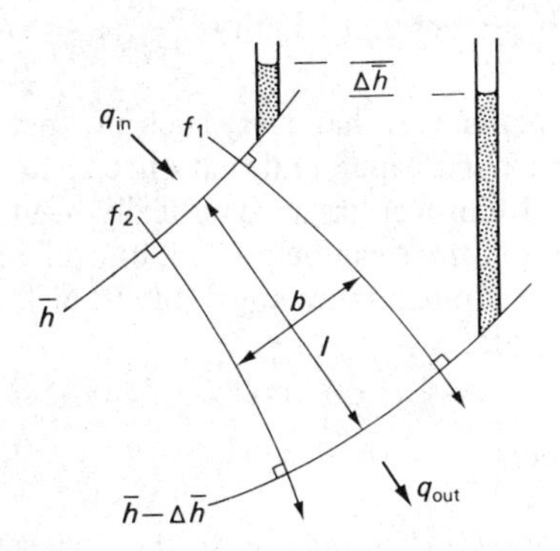

Figure 3.9 *Flow element*

never cross, since two packets of water cannot share the same volume in space. The quantity of water flowing between any two flowlines must, therefore, remain constant, q between f_1 and f_2, for example. Applying Darcy's law to a small part of the flow plane, then, which is so small that the flow unit can be considered rectangular $(b \times l)$

$$q = (1 \times b)k\frac{\Delta\bar{h}}{l} \text{ per channel per unit length}$$

where the area of flow is considered as one unit of length into the paper, multiplied by the average separation of the flowlines b. That is

$$q = k\Delta\bar{h}\frac{b}{l} \text{ per channel per unit length}$$

If the spacing of flowlines is such that all the flow elements are 'squares', $b = l$, then

$$q = k\Delta\bar{h} \text{ per channel per unit length} \tag{3.9}$$

So consider that water is flowing from one 'source' at constant excess head through a body of soil with set boundaries into a 'sink' at a lower constant excess head, so that each packet of water loses the same excess head H. Imagine that a flownet of slightly distorted squares of various sizes can be drawn, using N_F flow channels and N_H equal intervals of excess head, then by using equation 3.9

$$q = k\frac{H}{N_H} \text{ per channel per unit length}$$

so

$$Q = k\frac{H}{N_H}N_F \text{ total flow per unit length}$$

or

$$Q = kH\frac{N_F}{N_H} \text{ per unit length} \tag{3.10}$$

What is most surprising, perhaps, is that every such flownet can be arrived at by trial and error with a pencil and paper, and that once an arrangement of squares has been produced the right answer has automatically been obtained. The uniqueness of these flow solutions can be proved, but only after much advanced mathematical argument. It is much more enjoyable to pick up a pencil and paper and practise one's skill.

3.5.2 A Confined Flownet

An example with fixed and clear boundaries to the flow is shown in figure 3.10.

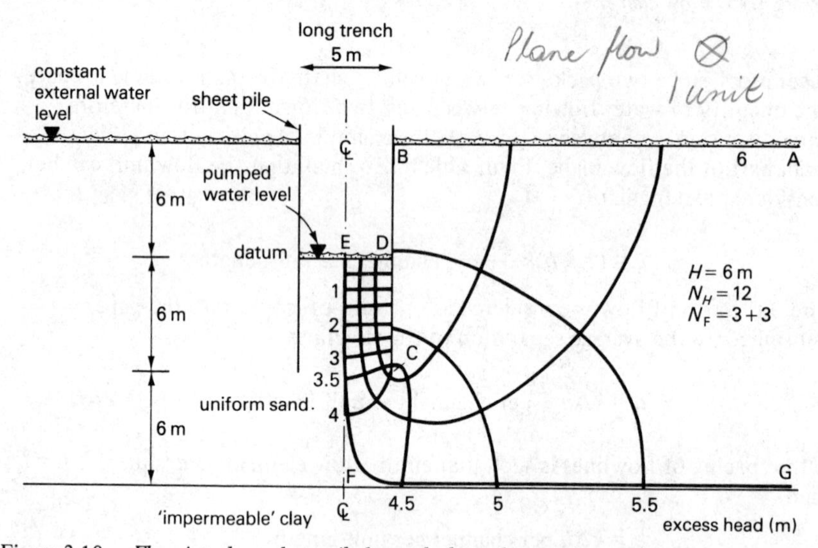

Figure 3.10 *Flow into long sheet-piled trench through estuarine sand*

The role of the boundaries in setting out the solution on the right lines cannot be overemphasised. The route to a trial-and-error solution of the problem in figure 3.10 is strewn with tricks and, in my case, guided by the memory of many similar attempts. The following points should be noted, however.

(1) There is symmetry about line EF, which means that FE must be a

flowline, since if a flowline crossed FE from right to left then one must cross from left to right at the same point, due to symmetry, and flowlines cannot cross. Only half the flownet need be drawn, therefore.

(2) Line GF must be a flowline, as must BC and CD, since flow cannot take place through these boundaries, which are impermeable relative to the sand.

(3) Line AB must be an equipotential since standpipes arranged along it would all contain water to the same head, in fact to the level of AB itself.

(4) Line DE is similarly an equipotential whose head happens to coincide with itself.

(5) Line DE has been taken as datum, above which all excess heads will be measured. AB, then, is the 6 m equipotential.

(6) Three flow channels were decided on, and sketched in, starting from the quite uniform EDC region.

(7) Starting at ED, equipotentials were sketched in so as to form squares, remembering that boundaries DC, EF and FG were flowlines. Both equipotentials and flowlines were then extended to cover the whole region with square figures of growing size.

(8) A rubber can be used to eliminate mistakes, which in this case centred around C and G, and a fair flownet is produced within about half an hour. Practice would reduce this considerably. After finishing, each equipotential can be labelled with its particular excess head over datum.

In this example we find $N_H = 12, N_F = 3 + 3$ (both sides count) and $H = 6$ m so that

$$Q = k6\frac{6}{12} = 3k \text{ m}^3\text{/s per m length}$$

If the contractor knew that the permeability of the sand concerned was roughly 3×10^{-4} m/s, and that he wanted 100 m of trench open at any one time, then he could estimate his required pumping capacity at

$$3 \times 3 \times 10^{-4} \times 100 = 0.09 \text{ m}^3\text{/s}$$

or roughly 5 tonnes per minute. This very large flow of water would be costly to pump, of course, and on seeing the relevant hire charges, the contractor may decide to reduce his pumping rate by working with only 5 m of trench open at once. He should not forget that, while 5 m of trench appears to give him simply 5 per cent of the flow of the 100 m trench, his flow problem is no longer plane but truly three dimensional. Visualising the hole in the ground as cubical, one can approximate the flow rate into two components, 5 per cent of 0.09 m^3/s from the north–south axis and 5 per cent of 0.09 m^3/s from the east–west axis.

It is instructive to speculate on the consequence of underpumping. Imagine, for example, that the contractor with the 100 m trench ordered a 0.09 m^3/s pump, but then encountered sand with a higher permeability than he had bargained for, let us say 9×10^{-4} m/s. Equation 3.10 then predicts

$$0.09 = 9 \times 10^{-4} \times H' \times \frac{N_F}{N_H} \times 100$$

where N_F/N_H is 1/2 as before, *since the flownet as such remains the same shape,* the geometry of the flownet depending only on the geometry of the flow boundaries. This means that a new head drop $H' = 2$ m is required which, of course, means that the excess heads of the twelve equipotentials are now at 1/6 m intervals instead of 1/2 m intervals. So the result of a threefold error in pumping requirements was simply a reduction in head by a factor of 3. The excavation could only be pumped down to a depth of 2 m below normal groundwater level.

But what if, on the contrary, the rate of abstraction by the pumps was greater than the rate at which the river and the sea could replenish it? The water table would then be drawn below the ground surface, creating an unconfined flow problem.

3.5.3 An Unconfined Flownet

Although the group of problems outlined above is important, the general flow of water through soil does not take place within given boundaries from one 'drowned' source to another 'drowned' sink. Problems such as the flow of groundwater to a river, the leakage from a canal, and the seepage through an earth dam, have one air/water boundary and the flownets are said to be unconfined since it is not immediately obvious where the top flowline should be drawn. Look, for example, at the flownet in figure 3.11, which represents the

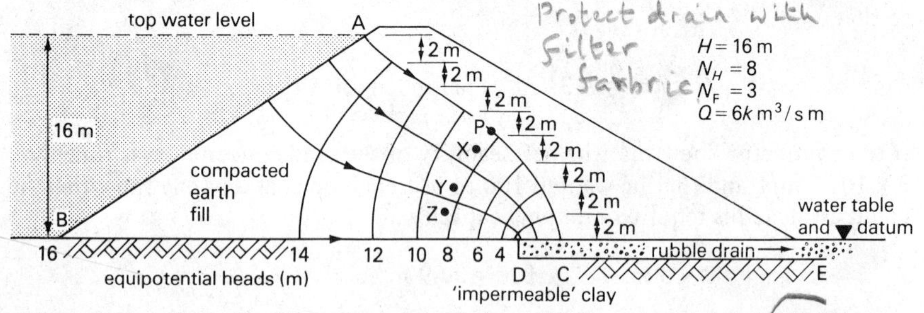

Figure 3.11 *Unconfined flow through earth dam with downstream drainage blanket*

flow of stored water through an earth bank with a downstream rubble drain DE laid down before the general fill, precisely to collect any leakage. Surely the top flowline AC could have been drawn in any position, and the space ACDB filled with 'squares'? This is so, but as you will see a further condition has been placed on the top flowline, in that equipotentials have been forced to start from it at equal intervals of vertical height. This mirrors the condition that the water pressure along the top flowline shall be atmospheric, as demonstrated below.

At any point P on the top flowline (often called the phreatic surface) the gauge water pressure must be zero if the capillary zone is ignored. Let its equipotential excess head be $\bar{h}$, and its elevation above the chosen datum be h'. If water in a standpipe placed at P would stand to a height $\bar{h}$ above datum, then it would stand to a height $(\bar{h} - h')$ above point P. But the gauge pressure at P is exactly zero, therefore there must be no head of water above P, that is, $\bar{h} = h'$. This proves that if contours of excess head $\bar{h}$ are to be drawn at equal intervals to form a set of equipotentials, they must intersect the top flowline within the dam at the same equal intervals of vertical height. In fact the excess head of any point on the top flowline is always equal to its height above the chosen datum. With this extra condition on the phreatic surface only one correct set of flowsquares can be drawn. At first you may find difficulty in drawing the phreatic line in the correct place to ensure the unique solution. Casagrande suggested an approximate method for this particular case on the assumption that the phreatic surface is parabolic; his method is described in Taylor (1948), chapter 9.

Once the flownet is completed, the equipotentials should again be labelled with the value of their excess heads above datum, so that at any future time the water pressure at any point within the earth mass can be declared. For example, the excess heads at the points X, Y and Z are all roughly 8.5 m above datum. Since the points themselves are 7 m, 4 m and 2 m above datum respectively, this means that their water pressure heads (above themselves) are 1.5 m, 4.5 m and 6.5 m. These latter heads imply physical water pressures at X, Y and Z of 15 kN/m^2, 44 kN/m^2 and 63 kN/m^2 respectively. It is essential to master the difference between a real water pressure (and its related head above the point of measurement) and an excess water head, which is a height above some arbitrary datum.

3.6 Seepage Forces

Engineers are continually required to assess the equilibrium of soil structures such as the sheet piled trench (figure 3.10) and the earth dam (figure 3.11). In such cases the engineer can attempt to identify the particular mass of wet soil that may be at risk, and then resolve the forces that act at the boundaries of the mass, and determine whether they can exactly balance its weight. It will always be possible, therefore, to solve such problems of equilibrium by accounting for the effect of gravity on a mass of wet soil, in addition to soil pressures and water pressures on its boundaries. Consider, for example, the 'quicksand' condition

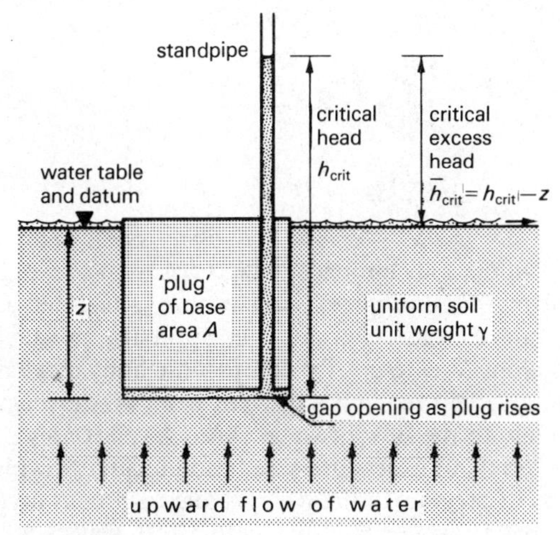

Figure 3.12 *Artesian quicksand condition*

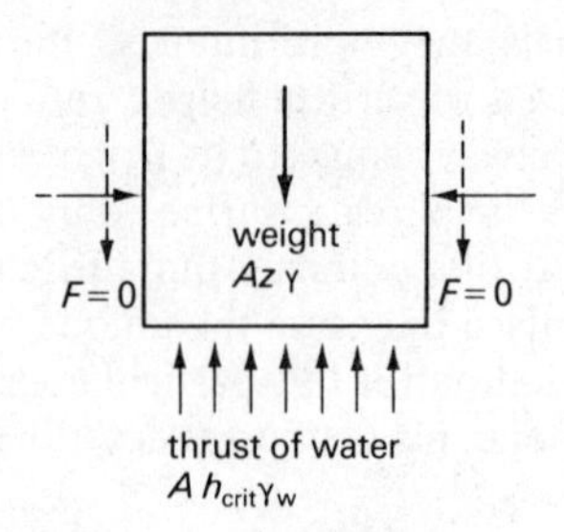

Figure 3.13 *Forces on plug assuming no side friction*

drawn in figure 3.12, where a 'plug' of soil is just being jacked upwards by the critical water pressure head h_{crit}. The forces on the 'plug' of saturated soil, which is to be tested for equilibrium, appear in figure 3.13. Resolving vertically

$$h_{crit}\gamma_w A = Az\gamma$$

which gives

$$h_{crit} = z\frac{\gamma}{\gamma_w} \tag{3.11}$$

In such a case the possible heaving failure would be blamed on the extraordinary artesian water pressure conditions, with standpipe levels rising above ground level. The engineer would frequently express the result in terms of the critical upward hydraulic gradient existing in the soil. Now

$$i = \text{rate of drop of standpipe level}$$

$$= \frac{(h - z) - 0}{z}$$

$$= \frac{h}{z} - 1$$

So

$$i_{crit} = \frac{h_{crit}}{z} - 1$$

or

$$i_{crit} = \frac{\gamma - \gamma_w}{\gamma_w} \tag{3.12}$$

Although the 'whole soil' approach, which was used above, gave a perfectly satisfactory solution to the problem, some engineers feel that a more substantial intuition can be gained from a 'soil skeleton' approach, in which the forces on the soil skeleton alone, due to gravity, hydrostatic buoyancy, and seepage are calculated separately. This requires a piece of work on the magnitude of force corresponding solely to the flow of water through soil. Darcy's law $v = ki$ predicts the rate of loss of excess pressure head in water flowing with a certain velocity. This loss of pressure in the water is due to viscous drag between the soil particles and the passing water. But drag works both ways, and the result of the

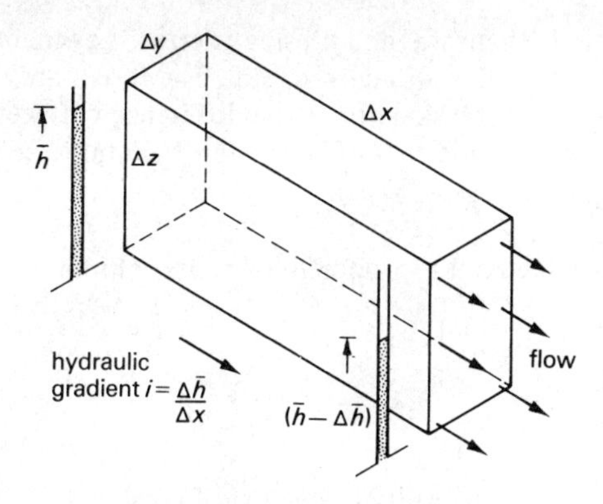

Figure 3.14

flow on the soil skeleton is to pluck it forward. Consider the force due to seepage alone on the soil skeleton sketched in figure 3.14. The water pressure drops by an amount $\gamma_w \Delta \bar{h}$, so that the restraining force on the water which instantaneously occupies the pores within the element (Δx, Δy, Δz) is $\gamma_w \Delta \bar{h} \Delta y \Delta z$, which can be expressed as $\gamma_w (\Delta \bar{h}/\Delta x) \Delta x \Delta y \Delta z$, which is simply $\gamma_w i$ per unit volume of soil. Since 'reaction' must be a consequence of 'action', there must be an identical forward drag on the soil particles within the element, of $\gamma_w i$ per unit volume of soil. This body force is rather similar to the action of gravity on all materials, which has the magnitude γ per unit volume. It requires a good deal of effort to imagine this internal body force, existing simply because

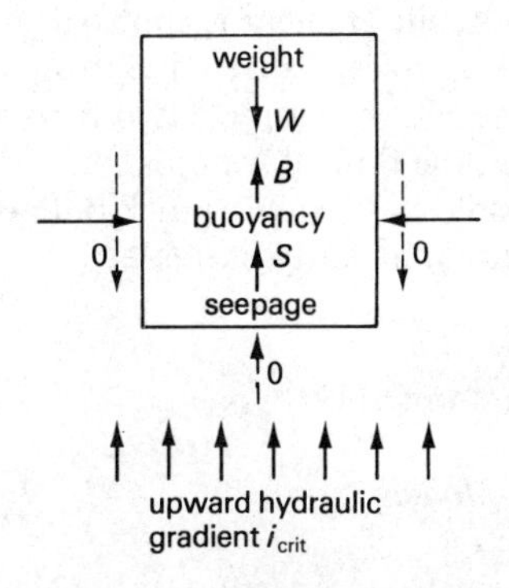

Figure 3.15 *Critical forces on soil skeleton assuming no external granular support*

water pressures are changing due to viscous friction, and some care to remember that, if soil is to be treated as a uniform block containing *both* soil and water, the internal action and reaction cancel out.

It is now straightforward to calculate, once again, the quicksand condition, this time considering *only the soil skeleton*. Let a region, volume V, of soil begin to float upwards due to hydraulic gradient i_{crit}, so that there are no supporting forces from the soil particles beneath, and no side friction. Let there be volume V_w of water contained within the volume V of soil. The forces are as shown in figure 3.15. The weight of the soil skeleton is the difference between the weight of the composite saturated soil and that of the water it contains, so that

$$W = \gamma V - \gamma_w V_w$$

The buoyancy of the soil skeleton is proportional to its volume

$$B = \gamma_w (V - V_w)$$

Finally, the seepage force is

$$S = \gamma_w i_{crit} V$$

After the resolution of forces vertically in the critical condition of limiting equilibrium

$$W = B + S$$

we obtain

$$i_{crit} = (\gamma V - \gamma_w V_w - \gamma_w V + \gamma_w V_w)/\gamma_w V$$

so that equation 3.12 is recovered afresh.

Since saturated soil often weighs in the region of 20 kN/m^3, and water in the region of 10 kN/m^3, the critical upward hydraulic gradient is in the region of unity. It will be the job of any engineer responsible for an excavation to ensure that upward hydraulic gradients are much less than unity. In our example in figure 3.10 we can scale off the diagram to see that the flow squares under the floor of the excavation are roughly 1.2 m long, and represent 0.5 m intervals of head, so that the upward hydraulic gradient is approximately 0.4. The quicksand condition, as it is met in coastal areas, is often due to an upward flow of sea water when the incoming tide is temporarily halted in its advance by a natural sand-bar, and especially when the flow of water is concentrated into one location by the presence of impermeable clay banks or rock outcrops. Fortunately such circumstances are rare and also inherently unstable.

3.7 Seepage through Non-uniform Soil

3.7.1 Application of Simple Models

I reproduce in figure 3.16 a single cross-section of the Mersey flood plain to the south of Manchester. No doubt a physicist or mathematician would throw his hands up in horror at such a variation in materials, and request a more fundamental (that is, a simpler) problem. The engineer must be somewhat more robust: knowing the inherent variability of ground he will restrict his development of those models which feature constant ground properties to roughly the level we have now arrived at, add to these a few conceptions about flow in nonuniform ground, and carry just these few simple weapons into the battle his client has asked him to fight. In the example cited above engineers had

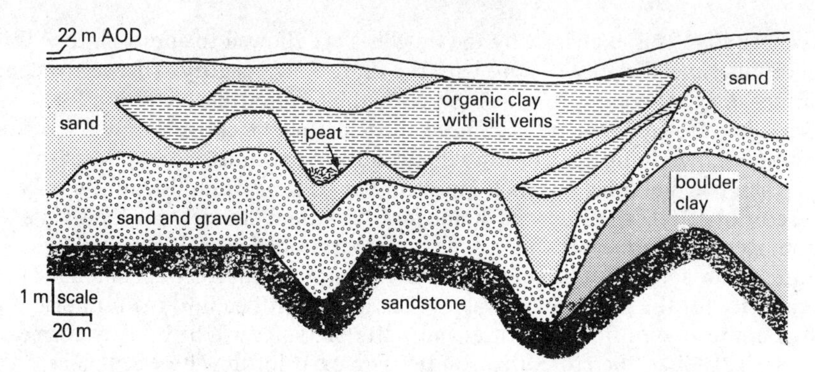

Figure 3.16 *One section through the Mersey flood plain near Manchester (taken from Rowe, 1972)*

to build a road embankment over the flood plain, and chose to excavate a huge 40 m deep 'borrow-pit' down to the boulder clays and Keuper marls, with which they constructed the embankment. Pumps in the lowest part of the pit were used to keep the excavation from filling with water.

What useful seepage models can be constructed for variable ground? We have already used the simplest model, that of 'permeable' ground meeting 'impermeable', with flow occurring in the permeable ground only parallel to the boundary. Such a model can be used freely when the 'impermeable' ground has a permeability that is less than one-hundredth of the permeability of the 'permeable', so that silty clay is impermeable relative to sand, and unfissured granite is impermeable relative to silt.

Entirely equivalent is the case when a certain block of soil can be considered so permeable that it is infinitely permeable relative to the remainder. This assumption was used to good effect in figure 3.11 with regard to the drainage blanket DE. No head losses were allowed in the blanket, so that the excess head was constant along its length, and flowlines were made to meet it at right-angles. Figure 3.17 shows how the earth dam would behave if the drainage layer were

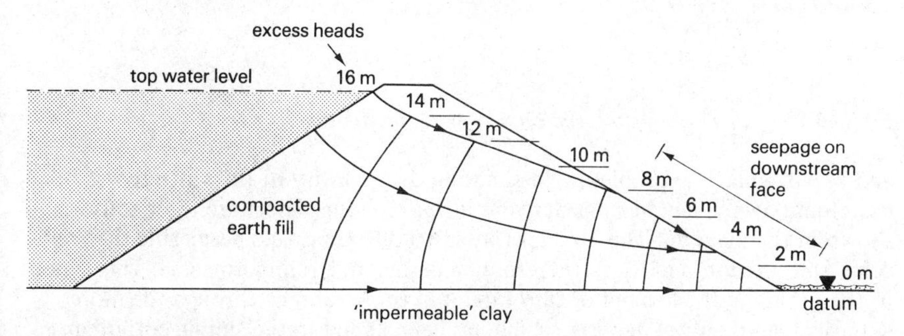

Figure 3.17 *Homogeneous earth dam: dangerous downstream seepage*

omitted. Here the flowlines emerge into the atmosphere at the downstream toe and seem to merge together. The flow down the slope in the zone of seepage is another example of flow in a material (air) which has an effectively infinite permeability. This explains why the flowlines are allowed to approach to within an infinitesimal distance to cope with the finite flow quantity. The wetted dam surface must, of course, continue to satisfy the top flowline condition for excess head to equal the position head above datum. Once again only the unique solution shown in figure 3.17 can satisfy all the conditions.

When two adjoining materials are such that their permeabilities differ only by a factor of 10, then neither the 'homogeneous ground' model, nor the simple 'homogeneous permeable–homogeneous impermeable' model, can give the correct flow solution. Used consecutively, however, with a certain amount of sensitivity for the permeability values that are substituted into the formulae, the total impression of flow quantities and water pressures will be broadly correct. Notwithstanding, therefore, that solutions do exist for flow between plane regions of different permeability (Taylor, 1948, paragraph 9.13), I shall not derive them.

Instead I will show you the power of the engineer's simple models. Consider again the sheet-piled cofferdam, shown in new circumstances in figure 3.18*a*,

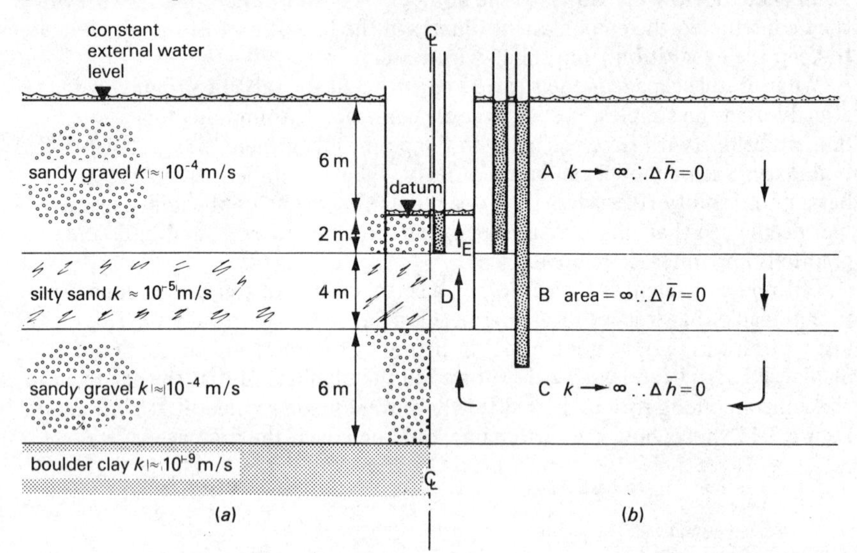

Figure 3.18 *(a) Two-soil problem; (b) pessimistic idealisation*

and in particular the problem of estimating the quantity of flow into the cofferdam, which can be considered as infinitely long. There are three soils shown in the diagram. The boulder clay is 10 000 times less permeable than any other soil present, and it can therefore be designated as impermeable. This leaves us the problem of the band of silty sand 4 m thick running through the more extensive sandy gravel deposit, which has been added to the simple cofferdam problem drawn in figure 3.10. In order to reach a solution a number of simple models will be attempted.

(i) Idealise the soil to be homogeneous throughout. The solution is already known (figure 3.10) and

$$q = kH\frac{N_F}{N_H} = k6\frac{6}{12} = 3k \text{ m}^3\text{/s per m length}$$

Various estimates can now be obtained which correspond to various permeability values.

(a) If all the soil were silty sand of $k = 10^{-5}$ m/s

$$q = 3 \times 10^{-5} \text{ m}^3\text{/s/m}$$

(b) If all the soil were sandy gravel of $k = 10^{-4}$ m/s

$$q = 3 \times 10^{-4} \text{ m}^3\text{/s/m}$$

(2) Idealise the soil by declaring that the sandy gravel is so much more permeable than the silty sand that it can be said to offer no resistance to flow at all. The flow problem is now different: see figure 3.18*b*. No head loss occurs in region A since k is set to infinity. No head loss occurs in region B since the hydraulic gradient through zone B need only be infinitesimal to generate any finite flow quantity due to its infinite extent. No head loss occurs in region C since this is, once again, frictionless, as is zone E. The only flow condition, therefore, is on the flow through zone D. Applying Darcy's law, $Q = Aki$, with $A = 5$ m^3/m length of excavation, $k = 10^{-5}$ m/s, and

$$i = (6 - 0)/4 = 1.5$$

we obtain

$$Q = 7.5 \times 10^{-5} \text{ m}^3\text{/s per m length}$$

The relative importance of predictions (1a), (1b) and (2) is clear. (1a) produces an optimistically small estimate, while (1b) is pessimistically high, and (2) is also pessimistic since it allows no break at all on the progress of the water due to the sandy gravel. We have therefore sandwiched the true flow

$$7.5 \times 10^{-5} \text{ m}^3\text{/s/m} > Q_{\text{true}} > 3 \times 10^{-5} \text{ m}^3\text{/s/m}$$

which leaves the contractor who is responsible for ordering the requisite pumps little choice but to provide at least 7.5×10^{-5} m^3/s/m (especially since he knows that the smallest pump could cope easily with the predicted 1 litre per second per 15 m of open trench).

I hope that this has demonstrated the need for, and application of, simple models in a complex engineering situation. I should not be surprised if you begin to be shocked by the apparent lack of formal mathematical solutions here, but I would admit to a certain disappointment if you did not concede that the approximate calculations which I have introduced have involved a fairly high degree of intelligent appreciation, and their own particular brand of flair.

3.7.2 Layer Systems

Since most soils and rocks have originally been laid down under water in discrete layers, and continue to display this organised layered construction even through the most severe ground movements, engineers have developed simple one-dimensional flow rules for seepage parallel and perpendicular to layers of soil, each with its own constant permeability.

In figure 3.19 the water is flowing parallel to the 'bedding' of the soil, rather

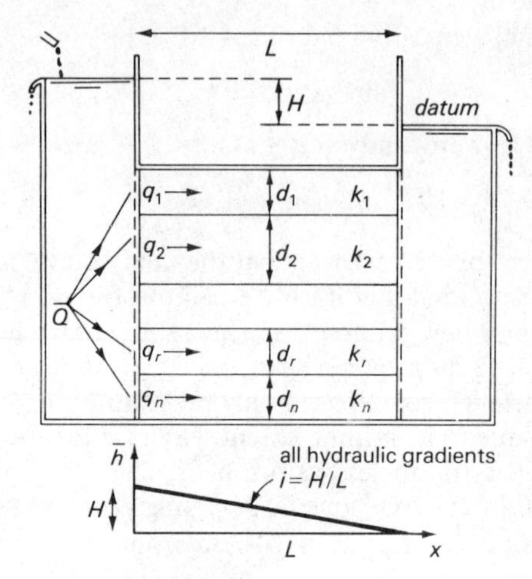

Figure 3.19 *Parallel flow per unit length*

as water within sandstone strata flows horizontally towards a well from which water is being pumped. We are interested in calculating the total quantity Q, which must be the sum of all individual contributions $q_1 + q_2 + \ldots + q_r$, that is

$$Q = \sum_{r=1}^{n} q_r$$

The flow in each stratum is easily estimated since the hydraulic gradient in each is the same

$$i_1 = i_2 \cdots = i_r = i_n = H/L$$

so

$$q_1 = d_1 k_1 H/L$$

$$q_2 = d_2 k_2 H/L$$

$$\vdots$$

$$q_r = d_r k_r H/L$$

etc.

The total flow Q is therefore given by

$$Q = \sum_{r=1}^{n} d_r k_r H/L = \frac{H}{L} \sum_{r=1}^{n} d_r k_r \tag{3.13}$$

It will be easier to remember this rule if we deduce the relevant average permeability k_a, which must satisfy

$$Q = Dk_aH/L \tag{3.14}$$

where D is the total depth of the soils, that is

$$D = \sum_{r=1}^{n} d_r$$

Substituting equation 3.14 into equation 3.13 we obtain

$$k_a = \frac{\sum_{r=1}^{n} d_r k_r}{D} \equiv \frac{\sum_{r=1}^{n} d_r k_r}{\sum_{r=1}^{n} d_r} \tag{3.15}$$

which is the arithmetic mean permeability.

In figure 3.20 the flow is perpendicular to the bedding plane of the soils,

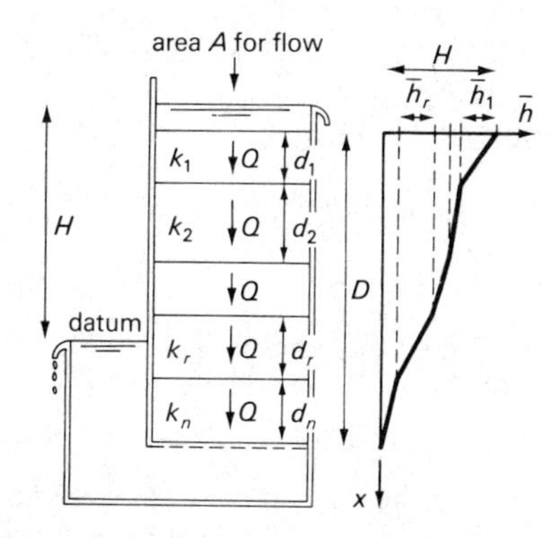

Figure 3.20 *Series flow through area A*

rather as rainfall might percolate into a system of sandstone strata to replenish the water which has been abstracted for drinking. Here the flow Q passes through each layer, and the head drops across the layers are different but must add up to H. We can write

$$q_1 = q_2 = \cdots = q_n \cdots = Q$$

But

$$i_1 = \frac{\bar{h}_1}{d_1}, i_2 = \frac{\bar{h}_2}{d_2}, \cdots, i_r = \frac{\bar{h}_r}{d_r}, \cdots, i_n = \frac{\bar{h}_n}{d_n}$$

and

$$q_1 = Ak_1i_1, q_2 = Ak_2i_2, \cdots, q_n = Ak_ni_n$$

so

$$\bar{h}_1 = \frac{Qd_1}{Ak_1}, \bar{h}_2 = \frac{Qd_2}{Ak_2}, \ldots, \bar{h}_n = \frac{Qd_n}{Ak_n}$$

but

$$\bar{h}_1 + \bar{h}_2 + \cdots + \bar{h}_r + \cdots + \bar{h}_n = H$$

Therefore

$$\frac{Q}{A} \sum_{r=1}^{n} d_r/k_r = H$$

or

$$Q = \frac{AH}{\sum_{r=1}^{n} d_r/k_r} \tag{3.16}$$

Using the notion of an average permeability k_b which satisfies

$$Q = Ak_b \frac{H}{D} \tag{3.17}$$

we can see that the equation for k_b must be

$$k_b = \frac{D}{\sum_{r=1}^{n} d_r/k_r} \equiv \frac{\sum_{r=1}^{n} d_r}{\sum_{r=1}^{n} d_r/k_r} \tag{3.18}$$

which is the harmonic mean permeability.

The perfect analogy between the flow of water in soil and the flow of electricity in conductors is quite apparent here, but it can lead to errors unless it is appreciated that permeability of soil is analogous to conductivity of electrical circuits, that is, to the inverse of Ohm's resistance: $k \equiv 1/R$. For this reason the harmonic and arithmetic equivalent averages are reversed with regard to series flow and parallel flow.

3.7.3 Anisotropic Model

Many soils have anisotropic permeation properties, due in large measure to their deposition under water in horizontal layers of different grain size. Since the arithmetic mean of any group of numbers is always greater than the harmonic mean, we can deduce from the foregoing analysis on the permeabilities of layered soils that the average horizontal permeability of such soils is larger than the vertical permeability, assuming that other factors do not intrude. Let us take as an example a silty clay with sandy silt partings an average of 3 mm thick at roughly 30 mm intervals: such a soil might have been laid down in a glacial lake which was for many years covered in ice during the winters, thus enabling the finest clay particles to settle in quiet water, but which suffered an inrush of sandladen melt water at every spring thaw. Let us impute permeabilities of 10^{-7} m/s to the silty clay and 10^{-5} m/s to the sandy silt. Taking as representative 27 mm of the clay together with one 3 mm silt band, we can use equation 3.15 for the average horizontal permeability

$$k_{\text{h}} = \frac{27 \times 10^{-7} + 3 \times 10^{-5}}{30} = 1.09 \times 10^{-6} \text{ m/s}$$

and equation 3.18 for the average vertical permeability

$$k_{\text{v}} = \frac{30}{27 \times 10^{7} + 3 \times 10^{5}} = 1.11 \times 10^{-7} \text{ m/s}$$

On viewing from a distance a soil with this fabric, therefore, one would see a homogeneous (same from point to point) material which was anisotropic (varying in direction) with regard to its permeability.

It is difficult to see immediately how to solve plane flow problems in cases where $k_{\text{h}} \neq k_{\text{v}}$: certainly the simple logic used in section 3.5.1 becomes inapplicable, and it becomes necessary to develop a more fundamental mathematical model for two-dimensional flow. Such a model can be generated by writing differential equations to govern the flow through an element δx, δy of soil in the plane of flow (see figure 3.21).

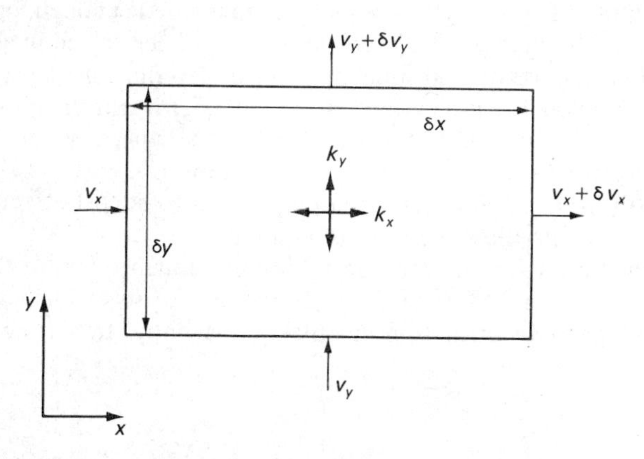

Figure 3.21 *Plane cartesian flow element*

Applying Darcy's law in the x-direction

$$v_x = k_x i_x = -\frac{k_x}{\gamma_{\text{w}}} \frac{\partial \bar{u}}{\partial x} \tag{3.19}$$

Applying Darcy's law in the y-direction

$$v_y = k_y i_y = -\frac{k_y}{\gamma_{\text{w}}} \frac{\partial \bar{u}}{\partial y} \tag{3.20}$$

Also, mass must be conserved, and by declaring that soil particles and water are incompressible then volume must be conserved, so that the flow rate into the element must equal the flow rate out, that is

$$v_x \delta y + v_y \delta x = (v_x + \delta v_x)\delta y + (v_y + \delta v_y)\delta x$$

or in the limit

$$\partial v_x / \partial x + \partial v_y / \partial y = 0 \tag{3.21}$$

Now by substituting 3.19 and 3.20 in 3.21, we obtain

$$k_x \frac{\partial^2 \bar{u}}{\partial x^2} + k_y \frac{\partial^2 \bar{u}}{\partial y^2} = 0 \tag{3.22}$$

When the material is isotropic then $k_x = k_y$, so that equation 3.22 reduces to Laplace's equation

$$\frac{\partial^2 \bar{u}}{\partial x^2} + \frac{\partial^2 \bar{u}}{\partial y^2} = 0 \tag{3.23}$$

Laplace's equation is known to cover, in addition to the flow of water through permeable media, the conduction of heat or electricity, problems in electrostatics, and gravitation. The equation simply puts conditions on the rate of change of pressure gradient from point to point in the plane: the difficulty in arriving at a meaningful solution is simply the difficulty of satisfying the boundary conditions. Although the equation looks simple enough, only a handful of useful boundary value problems have yet been solved in algebraic form, mostly by the use of conformal mapping as described in Harr (1962). The sketched flownet which was described in section 3.5.1 is still the most useful method available to a practising engineer of solving a Laplace equation with given boundary conditions. The use of electrical flow analogues or of automatic numerical analysis on a digital computer only begins to be cost-effective when a large number of similar flow solutions is required.

Whereas the flow solution for an isotropic soil is independent of the actual permeability, as is seen from equation 3.23, the flow in an anisotropic soil does depend on the ratio of vertical to horizontal permeability. Rewriting equation 3.22 as

$$\frac{k_x}{k_y} \frac{\partial^2 \bar{u}}{\partial x^2} + \frac{\partial^2 \bar{u}}{\partial y^2} = 0 \tag{3.24}$$

we can only achieve a simple Laplacian such as

$$\frac{\partial^2 \bar{u}}{\partial x'^2} + \frac{\partial^2 u}{\partial y^2} = 0 \tag{3.25}$$

if we can make a transformation of distances measured in the x-direction so that $x' = \alpha x$ where the constant α can be determined by equating 3.24 and 3.25, that is

$$\frac{k_x}{k_y} \frac{\partial^2 \bar{u}}{\partial x^2} = \frac{\partial^2 \bar{u}}{\partial (\alpha x)^2} = \frac{1}{\alpha^2} \frac{\partial^2 \bar{u}}{\partial x^2}$$

or

$$\alpha = \sqrt{(k_y/k_x)} \tag{3.26}$$

In anisotropic soil $k_h \neq k_v$, therefore the solution is obtained by drawing the flow plane at normal scale and then redrawing it with the same vertical scale but with the horizontal scale shortened by the factor $\sqrt{(k_v/k_h)}$, which in our numerical example is $\sqrt{(1/10)}$ or 0.32, on to which a normal flownet is drawn. Finally, the flownet is scaled back on to the natural scale drawing.

Although the water pressures within the soil are now known, there is some final confusion concerning the prediction of flow rates using $Q = kHN_F/N_H$, since the value of the appropriate permeability k has to be decided on. This can be resolved by solving a problem which can be answered either directly or by transforming the section, and equating the results obtained. Let k_t be the appropriate permeability coefficient to be used with the transformed section, and consider the case of uniform vertical flow, shown in figure 3.22. Equating

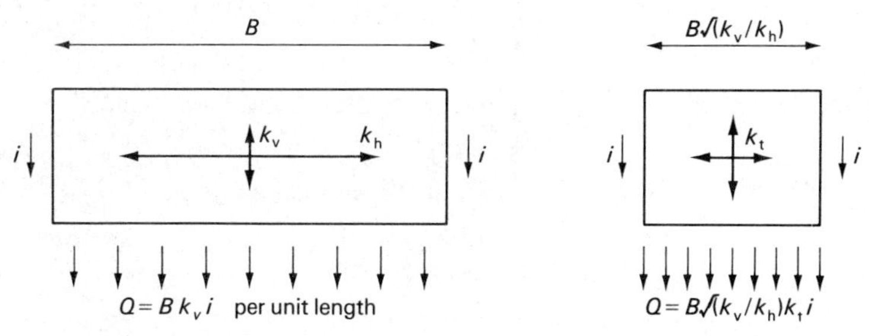

Figure 3.22 *Transformed permeability*

the flows we see that $k_v = k_t\sqrt{(k_v/k_h)}$ so that the appropriate permeability to use with the transformed section is $\sqrt{(k_v/k_h)}$, which certainly appears to possess the required degree of symmetry.

3.8 Measurement of Permeability

3.8.1 Laboratory Determinations

The most commonly used apparatus for determining the permeability k of a sand sample is the constant head apparatus shown in figure 3.23. By ensuring an approximately constant flow downwards through the sample, and by taking a measurement of the flow rate and excess head difference $\bar{h}$ between the tapping points, the permeability is found by using Darcy's law directly

$$q = Ak\bar{h}/L$$

$$k = \frac{q}{A}\frac{L}{\bar{h}} \tag{3.27}$$

The soil loaded into the confining tube will not be identical to the soil as it was in the ground. It is still almost impossible to recover undisturbed samples of sand from site. An attempt must therefore be made to compact the soil to a reasonable density, and pass through it deaired and distilled water in order to prevent air bubbles from partially blocking the pores of the soil, which certainly would occur if mains water were used. A discussion concerned with how to normalise these modelling attempts appears in Terzaghi and Peck (1967).

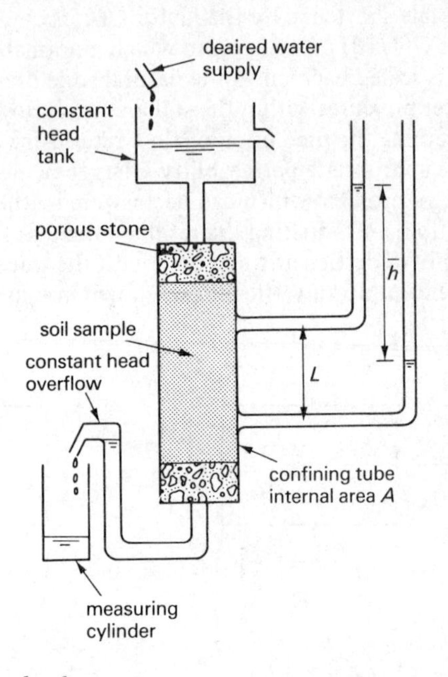

Figure 3.23 *Constant head permeameter*

It is instructive to substitute typical values into equation 3.27. If the tube is 10 cm in diameter then its area A is 0.0079 m^2, and if a hydraulic gradient of 0.5 is used, that is, L = 20 cm, $\bar{h}$ = 10 cm, then

$$q = 0.004k \text{ m}^3/\text{s}$$

If $k = 10^{-2}$ m/s, then $q = 4 \times 10^{-5}$ m^3/s = 40 cm^3/s, but if $k = 10^{-6}$ m/s, then q = 0.04 cm^3/s or roughly 140 cm^3/h. Now it may be possible to make such a determination over the period of an hour, but if the permeability of the soil is much less than 10^{-6} m/s, a silt with 10^{-8} m/s perhaps, then evaporation could well anihilate the very meagre flow rate of 1.4 cm^3/h unless extraordinary precautions were taken.

In such circumstances a falling-head permeameter might be used, as shown in figure 3.24, in which the excess head is allowed to fall from $\bar{h}_1$ to $\bar{h}_2$ as the water from the small bore tube of area A_2 flows through the soil sample in the larger tube of area A_1.

Since

$$q = A_1 \frac{k\bar{h}}{L} \text{ in tube 1}$$

and

$$q = -A_2 \frac{d\bar{h}}{dt} \text{ in tube 2}$$

we obtain

$$\frac{1}{\bar{h}} \frac{d\bar{h}}{dt} = -\frac{A_1}{A_2} \frac{k}{L}$$

so that

$$\bar{h} = \bar{h}_1 \left(1 - \exp\left(-\frac{A_1}{A_2} \frac{kt}{L}\right)\right)$$

and if T is the time elapsed as the head falls from $\bar{h}_1$ to $\bar{h}_2$ then

$$k = \frac{LA_2}{TA_1} \ln (1 - \bar{h}_2/\bar{h}_1) \tag{3.28}$$

using ln for a logarithm with the exponential base. This apparatus effectively solves the problem of measuring small flow rates, since the feeding tube area A_2 can be made as small as may be necessary, and evaporation losses are more easily eliminated.

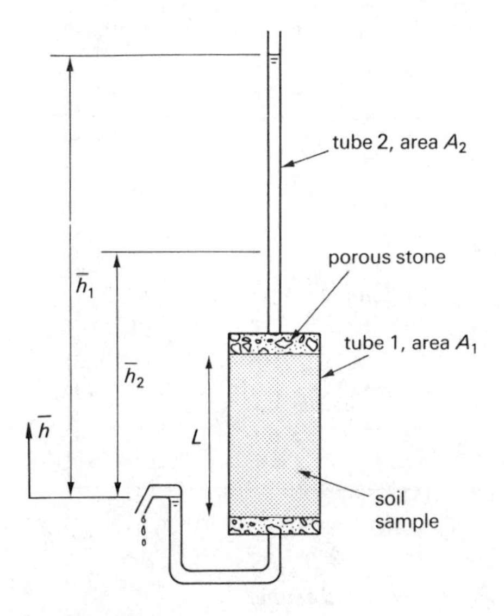

Figure 3.24 *Falling head permeameter*

Neither permeameter, however, solves the problem of soil sample disturbance, and laboratory values of permeability should be treated with caution, especially when the soil concerned may have possessed a layered fabric, as described in section 3.7. Since some experienced soils engineers have been heard to declare that they have never yet met a silt or clay that was *free* from such a fabric, it may be wise to anticipate very substantial errors however carefully the laboratory experiment was conducted. Even when samples are prepared by coring horizontally into an undisturbed sample of clayey soil that has been recovered carefully from a borehole as it was being augered downwards, it is exceptionally difficult to preserve the permeation fabric in such a way that k_h can be reliably estimated. The smearing with clay of the more granular layers is almost inevitable, so that laboratory permeability measurements are often underestimates, sometimes by factors of 10 or more, of field permeabilities in silt and clay. The permeability of fine-grained soil can also be overestimated if it is allowed to swell during the test.

3.8.2 Field Tests

Since soil is so variable, so difficult to sample, and so tricky to handle in the laboratory, the determination of permeability by testing *in situ* has become more popular. This usually involves pumping water in, or out, of boreholes in which a certain length has been left uncased so that water can migrate only through a known area of its perimeter. These tests depend on derivatives of a well-pumping formula established by J. Dupuit in 1863. To take a simple case, we shall establish the flownet for the case of the 'confined aquifer' shown in figure 3.25.

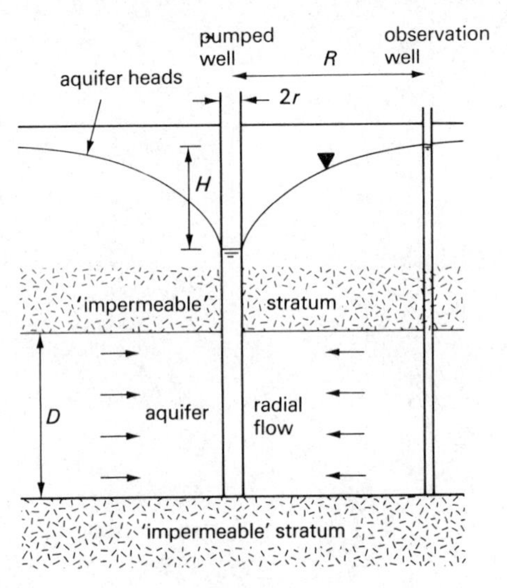

Figure 3.25 *Pumping test in confined aquifer*

Consider a disc of soil, height D, with an excess head H driving water towards the borehole, radius r, from an outer zone of radius R. The flownet on plan is shown in figure 3.26. This problem, although easily drawn by hand, can just as easily be calculated exactly. At any radius x, the perimeter is $2\pi x$ and so the flow area is $2\pi xD$. The rate of flow Q, which is simply the field pumping rate, must then be such that

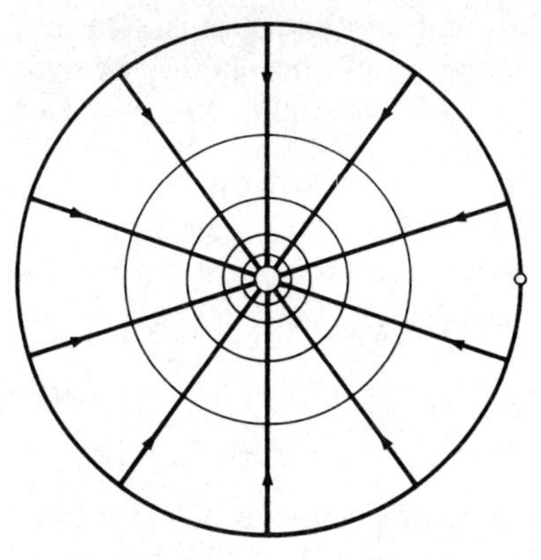

Figure 3.26 *Flow into well: plan view*

$$Q = Q_x = 2\pi x D k(\mathrm{d}\bar{h}/\mathrm{d}x)$$

so

$$\frac{\mathrm{d}\bar{h}}{\mathrm{d}x} = \frac{Q}{2\pi Dk}\frac{1}{x}$$

and

$$\int_0^H \mathrm{d}\bar{h} = \frac{Q}{2\pi Dk}\int_r^R \frac{1}{x}\,\mathrm{d}x$$

so

$$H = \frac{Q}{2\pi Dk}\ln(R/r)$$

or

$$k = \frac{Q}{2\pi DH}\ln(R/r) \tag{3.29}$$

While D and Q are relatively easy to obtain, the excess head difference H existing between radii r and R may not be so readily obtained. Let us suppose that in pumping 50 000 l/h from a 0.5 m diameter well, 100 m deep, it is possible to draw down the water level in the borehole by 12 m. Then

$$k = \frac{50\ 000 \times 10^{-3}}{3600 \times 2\pi \times 100 \times 12}\ln(R/0.25)$$

or

$$k = 1.84 \times 10^{-6}\ln(R/0.25)\ \mathrm{m/s}$$

Roughly $k = 4.25 \times 10^{-6}\log_{10}(R/0.25)$ m/s when the natural logarithms are replaced by logarithms to the base 10. The question is this: at which radius R can it be said that the pre-existing groundwater level is not disturbed? If other wells in the vicinity are within this 'radius of influence' then they will exhibit a draw-down due to pumping from the main well. The excess head difference H will then be the difference in draw-down in the two wells and the radius R the distance between them. But let us suppose that an existing well 1 km from the test borehole does not appear to be drawn down, even after many days of test pumping. It must be outside the zone of influence of the draw-down. But is R equal to 1000 m, 100 m or 10 m?

Examine the implications

$$R = 1000 \text{ m}, k = 4.25 \times 10^{-6} \log_{10} 4000 = 1.5 \times 10^{-5} \text{ m/s}$$

$$R = 100 \text{ m}, k = 4.25 \times 10^{-6} \log_{10} 400 = 1.1 \times 10^{-5} \text{ m/s}$$

$$R = 10 \text{ m}, k = 4.25 \times 10^{-6} \log_{10} 40 = 0.7 \times 10^{-5} \text{ m/s}$$

The accurate determination of R does not, therefore, matter a great deal. Whichever is the truth in the example quoted above, a water supply engineer would be quite happy that he was standing on a useful granular aquifer.

One significant point remains. Look at the hydraulic gradients surrounding the well

$$i_r = \left(\frac{\mathrm{d}\bar{h}}{\mathrm{d}x}\right)_{x=r} = \frac{Q}{2\pi Dkr} = \frac{H}{r \ln(R/r)} \tag{3.30}$$

Taking the above numerical example with R = 100 m to give us an order of size, then $i_r = 8$. The soil or rock surrounding the well is being pushed into it with a body force eight times greater than that caused by gravity (see section 3.6), and so the open section of the well would certainly have to be lined with a permeable but strong lining, otherwise it would collapse. It may be easier to reverse the flow and pump into the well; the analysis holds exactly but with a reversal of flow. The soil or rock is then forced away from the well with the same huge compressive force. Even this is clearly not good enough: fragile or compressible soils would be wholly damaged and disturbed. It is also quite possible for large-scale hydraulic fracturing to occur so that the high pressure water initiates and then jacks open an extended fissure system in the once intact soil. In all cases the hydraulic gradient should be kept as small as possible, and certainly no greater than 2. In general a pumping-out test in which the draw-down head is between one or two borehole diameters, and in which a carefully designed porous stone piezometer is used to collect the incoming water, can give reasonable results.

3.9 Summary

A model for the saturated flow of incompressible water through incompressible soil has been developed. Its cornerstone is the Darcy flow rule that flow velocity is proportional to the hydraulic gradient and is in the direction of the greatest possible hydraulic gradient. All flow phenomena are then related to excess water-pressure heads, which are the stable height of rise of water in standpipes measured *not above the base of the pipe* but above a chosen horizontal datum plane.

Darcy's rule leads directly to the approximate solution of two-dimensional flow problems by the free-hand sketching of flow nets composed throughout of distorted squares of various sizes. It can equally well lead to a Laplacian differential equation, but whereas any boundary condition can be met fairly easily by sketching, the analytical approach involves techniques outside the scope of this book. The flow solutions developed for isotropic, homogeneous soil are only dependent on the geometry and nature of the boundaries. All internal water pressures, seepage forces and flow paths are independent of the actual permeability constant in these cases. This is fortunate since permeability varies widely for soils, and is extraordinarily difficult to measure within a factor of 2. Flow quantities are directly proportional to the permeability, however, and in cases where the prediction of flow quantity is vital, the most carefully controlled slow pumping tests must be performed.

Since soil is intrinsically variable, the seepage model was extended to deal in a robust manner with ground variations from point to point, and in particular with the stratified condition. When soil stratification is on a micro or small scale, the engineer can step back and choose to see a homogeneous, but anisotropic, material with a greater permeability along the bedding planes than across them. He can apply the normal flow solutions only after transforming the flow cross section by shortening the axis along which flow is easiest.

Mention has been made of situations in which seepage forces $i\gamma_w$, which may be larger than gravitational forces and which are similar in nature, cause the soil skeleton to be less than rigid (heave, quicksand, fast well-pumping). Although this implies that the whole seepage model, based as it is on conservation of flow volume, can become completely inapplicable, we have been content to apply the model, find out whether any gross ground movements are to be expected, and report accordingly. Later in the book, chapter 7, we shall develop a more complex flow model in which any small compression of the soil matrix can be taken into account.

3.10 Problems

(1) Prove Archimedes' principle, using equation 3.2.

(2) A level stratum of silt with a permeability k is of depth D and is underlain by stiff clay. Supposing that the rainfall is sufficient just to keep its top surface wet, establish some concepts concerning land drainage by attempting the following. Draw a flow net for a cross section through the silt where a single porous pipe drain of diameter $0.05D$ is laid with its axis at a depth of $0.8D$, supposing that it is capable of removing the inflowing water without itself becoming choked. Show that the rate of flow into the drain is roughly kD per unit length. Show that the water pressure in the silt directly above the drain never exceeds $0.2D\gamma_w$. If the drain was suddenly blocked, show that the water pressure at mid-depth in the silt at a distance $0.5D$ laterally from the line of the drain would quickly increase by roughly 75 per cent. Expand your intuition by considering the effects on maximum flow quantities and pore-water pressures of doubling the diameter of the drain. Consult your tutor concerning the capacity of pipe drains laid at various gradients. Consider the criteria that would influence

the general choice of spacing between parallel pipe drains which were

(a) intended to prevent ponding of surface water or

(b) intended to prevent any pore-water pressure in the ground above from exceeding one-half its just-submerged hydrostatic value or

(c) intended to draw the water table down to at least one-half the depth of the drains.

(3) The contractor responsible for the construction of the earth dam shown in figure 3.11 allowed the earth fill to segregate into thin, fine and coarse layers. This gave a horizontal permeability of 10^{-6} m/s and a vertical permeability of 10^{-7} m/s. Redraw the flownet, recalculating the water pressures at points X, Y and Z, and comment on the difference between these and the original values. What is the rate of leakage, and by how much has this changed from the designed value if the intended isotropic permeability was 10^{-7} m/s?

Approximate answers: Excess heads, 10.5, 9, 7 m; pore pressures, 35, 50, 50 kN/m^2; leakage, 4×10^{-6} m^3/ s/m *cf.* 6×10^{-7} m^3/s/m.

(4) A constant-head permeameter was observed to have an internal diameter of 100 mm and a vertical distance of 200 mm between the standpipe tapping points. A mass of 4.16 kg of a dry quartz sand ($G_s = 2.65$) was compacted into the permeameter, and the experimenter noticed that the completed length of the sample was then 300 mm, extending 50 mm beyond either tapping. The experimenter arranged for the water to rise through the permeameter, so that it was free to rearrange the particles. He set the excess head between the tappings to various values, and at each one recorded the volume of water flowing through the tube in 1 min.

$\Delta\bar{h}$(mm)	40	80	120	160
Q(cm^3/min)	50	95	140	190

As he began to increase the flow rate again he noticed that the sand on the surface of the sample began to dance around at a head of roughly 180 mm and a flow rate of 230 cm^3/min. Beyond this, plugs of sand began rising bodily up the tube, only collapsing back as trapped water was able to escape at the top of the tube. When he had fully observed this effect the experimenter noticed that the height of the sample was now just stable at 340 mm, while the excess head difference had dropped to 120 mm and the flow rate had increased to 280 cm^3/min.

(a) find the initial void ratio of the sand

(b) show that Darcy's model roughly fits the initial flow data, and find the initial permeability constant k

(c) compare the initial experimental critical hydraulic gradient with a theoretical value and attempt to explain the difference

(d) fully explain and identify the final group of observations.

Answers (a) 0.50 (b) 5×10^{-4} m/s (c) 0.90 *cf.* 0.76, friction after compaction (d) $e \rightarrow 0.70$, $k \rightarrow 10^{-3}$ m/s, $i_{crit} \rightarrow 0.60$ *cf.* 0.56, after liquefaction.

4

Friction

4.1 Coulomb's Model

While the French Revolution brought the eighteenth century to a troubled close, Charles Augustin de Coulomb published in Paris his famous papers on friction, and in particular his treatment of the friction of soil, which allowed him to establish rules for the design of earth fortifications. Some of his work has recently been translated by Heyman (1972). Coulomb was drawing on the work of Amontons in the late seventeenth century, and indeed on that of Leonardo da Vinci in the sixteenth. Both men had formulated 'laws' for the frictional behaviour of materials, but because it was Coulomb who described useful design methods based on them, it is quite appropriate that twentieth century soils engineers should place his name against the friction model that is still central to most calculations of soil mechanics.

The Coulomb model describes the sliding of one solid object over another. You may imagine, as in figure 4.1, a wooden block sliding on a table, a single

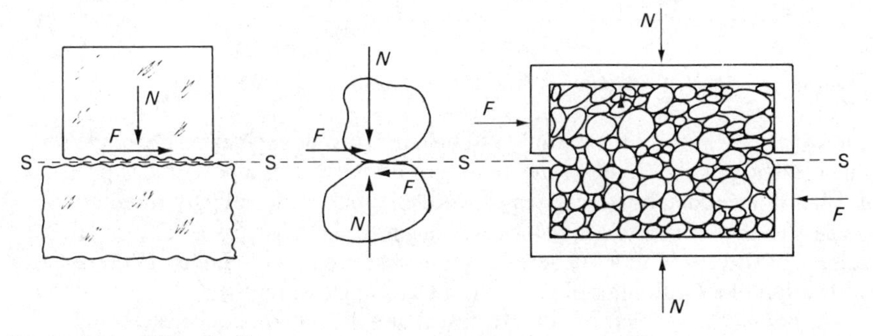

Figure 4.1 *Sliding bodies*

quartz grain sliding against another, or a whole accumulation of soil grains sliding bodily over a similar block of soil grains. The approximate surface of potential sliding (SS) must be known, ignoring the irregularities in the wooden surfaces or the interlocking of the soil grains for this purpose. The total force between the two 'bodies' must now be resolved into two components, N perpendicular to the surface of sliding and F parallel to it. The Coulomb friction model then predicts

(1) the bodies just begin to slide over each other on the slip surface when the ratio of F/N has increased to a certain limiting value μ; in other words the limiting frictional force

$$F_{max} = \mu N$$

where μ is the coefficient of friction of the sliding surfaces

(2) the coefficient μ does not depend at all on the superficial area of contact of the bodies.

Engineers prefer to use the definition

$$\mu = \tan \phi$$

in order to arrive at a definition of the angle of friction ϕ of a soil: figure 4.2

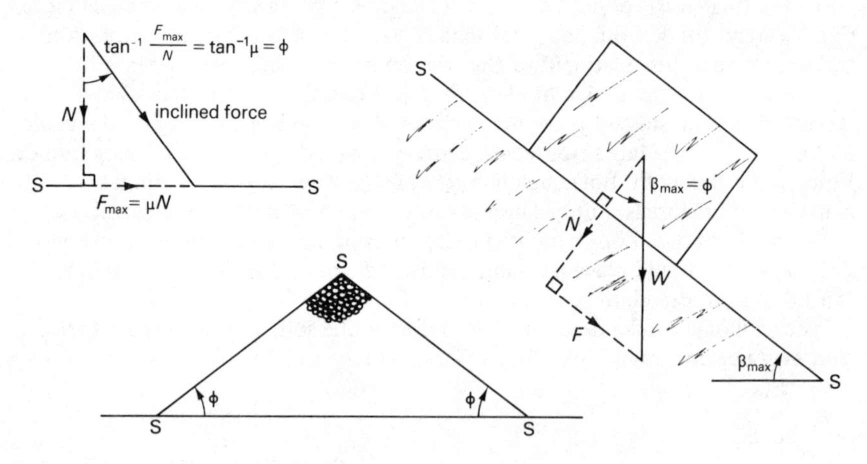

Figure 4.2 *Angle of friction with respect to sliding on surfaces SS*

indicates why angles of friction are so beloved of soils engineers. The angle ϕ is the greatest possible angle of inclination of a force across a potential slip surface, so that if a wooden board carrying a wooden block is increasingly tilted at an angle β to the horizontal, the block will begin to slip when $\beta = \phi$. The same argument will apply to a dry scree slope of shattered rock fragments: the angle of friction of a Coulomb soil is identical to its angle of repose.

Although the calculations do not reveal any distinction between a wooden block sliding on an inclined plane and an avalanche of soil taking part in a landslide, an important difference arises in their application. There is no doubt about the plane on which to measure μ or ϕ for a sliding wooden block, but an infinite number of possible rupture surfaces could be invoked in the design of soil slope. Only when a failure has occurred will it be known whether the appropriate surface was chosen. The prediction of the location of rupture surfaces in soil elements and in whole soil structures is still a matter of research and controversy, as is the most appropriate test for frictional properties. Some of these arguments, together with the conventional practices, will be exposed in later chapters.

4.2 Bowden and Tabor's Model

Recent work on the physical origins of friction, brought together in the books of Bowden and Tabor (1964), offers some explanation of the phenomenon of friction. Knowledge of the likely physics of a situation is usually of great help to the engineer, and the descriptions of friction on the microscopic scale are no exception. The real contact area A_c of two adjacent bodies is very much smaller than the whole superficial area A. Most surfaces are 'rough' on the atomic scale, and only come into intimate contact at very occasional points where the 'peaks' or asperities on the surfaces allow. It is important to get the scale right here: a surface that is fairly smooth to the touch must not have asperities higher than 0.01 mm, a surface that is optically 'flat' must not have asperities deeper than the wavelength of violet light, 0.0004 mm, while a surface that is atomically flat so that neighbouring surfaces have to be in overall intimate contact should have no asperities deeper than the molecular radius of the material, perhaps 0.000 02 mm. It is interesting to imagine the effect of placing two clean atomically flat surfaces face to face: would they not weld themselves permanently together?

Bowden and Tabor argue that the source of frictional resistance is indeed the welding together of molecules at the true points of contact of the two materials. They go on to derive Coulomb's friction law. The principal plank in their argument is that the true area of contact A_c is proportional to the normal force N. This follows from two assumptions: that the initial area of contact is infinitesimal and that there is an upper limit to the normal contact stress σ_c at an asperity which can be found at large scale simply by forcing a hard indenter into the surface. This indentation hardness can be correlated with the scratch hardness H mentioned in section 2.2; for example the indentation hardness of a tool steel may be 4000 MN/m^2, while its scratch hardness rates as H5.5. Now consider the placing of a 1 kg hammer head gently down on an anvil, before it is slid along the metal face. As it is slowly brought down, the weight of the hammer will begin to bear on a few asperities of infinitesimal area: these asperities will be flattened since the local normal stress σ_c will attempt unsuccessfully to exceed the plastic limit of 4000 MN/m^2. The asperities will only stop flattening when the load has been spread over an area A_c such that

$$N = \sigma_c A_c$$

In our example

$$A_c = \frac{1 \times 9.8}{4000 \times 10^6} \text{ m}^2$$

so that the approximate true contact area is 1/400 mm^2 or roughly 2 per cent of the area of this full stop. It is only over this tiny fraction of the hammer face that the anvil can come into intimate contact with it. The ensuing assumption is that the two contacting surfaces are effectively welded or locked at their points of contact, and that the tangential force required to break the lock is simply the shear strength of the junction, τ_c (perhaps 800 MN/m^2 for tool steel) multiplied by the contact area

$$F = \tau_c A_c$$

so that

$$\frac{F}{N} = \frac{\tau_c}{\sigma_c} = \mu = 0.2 \text{ in this case}$$

The coefficient of friction therefore shows itself to be the shear strength of the junction divided by the indentation strength of the surface. Materials that conform to pure plasticity theory have a definite relationship between their shear strength and indentation strength, and they therefore have a constant coefficient of friction. Bowden and Tabor point out that it is surprising to find that a wide variety of materials, including rock crystals which appear to fail in a brittle rather than a plastic fashion when they are crushed in the laboratory, also conform fairly well to their plastic model. They presume that the overall confining pressure in the vicinity of the points of contact is sufficient to suppress any tendency to fracture in a brittle fashion.

The significance of the presence of lubrication in the form of films of liquid on the surfaces of the two sliding solids is now easily understood. If the liquid is able to coat the surfaces and is not easily displaced by the approaching asperities, it will be able to reduce the shear strength of the junctions. It may then occur to you to enquire whether the properties of soils differ from wet to dry. Fortunately it happens that the presence of a minute amount of water has much the same effect as complete saturation. In other words a fairly constant angle of friction will be observed for a normal sand, whether it appears dry, damp or wet. Only if extraordinary precautions are taken to exclude water vapour and other impurities, such as by performing the friction test under a vacuum, will the 'pure' coefficient of friction be measured: this of course, will have no engineering application.

4.3 Terzaghi's Effective Stress

Although the overt presence of water does not affect the coefficient of friction between soil grains very greatly, the pressures that may exist in the water are vital. Consider the sliding of two rough blocks with pressurised water between

them, as in figure 4.3. The thrust of the water U will be normal to the potential slip surface and will tend to push the two sides apart as the total normal force N tends to clamp them together. The effective normal force $N' = N - U$ is the force which must be transmitted through the points of contact so that

$$F_{max} = \mu N' = \mu(N - U)$$

This is more usually written per unit area of surface

$$\frac{F_{max}}{A} = \mu\left(\frac{N}{A} - \frac{U}{A}\right)$$

or

$$\tau_{max} = \mu\left(\sigma - \frac{U}{A}\right)$$

where τ and σ are the shear stress and the total normal stress acting across the potential slip surface. Now the upthrust $U = (A - A_c)u$ where u is the water pressure between the blocks and A_c is the true solid/solid contact area. We have

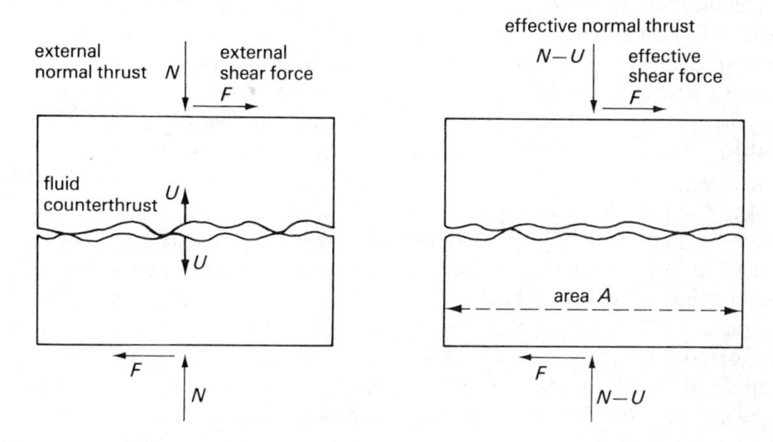

Figure 4.3 *Dealing with fluid pressure on sliding surfaces*

already seen that $A_c \ll A$ in the case of a small block of tool steel sliding under its own weight. It is necessary to ask whether $A_c \ll A$ in the case of soil. The indentation hardness of quartz is 7000 MN/m^2, and of feldspar 5000 MN/m^2, both greater than tool steel. The greatest normal stresses exerted in common soil mechanics might be 1 MN/m^2, corresponding to roughly 50 m of overburden above the point in question. This means that less than 1/5000 part of the superficial area of a typical sand is actually responsible for transmitting the effective intergranular stresses. Only in the case of very heavily stressed clay minerals, or organic soils which contain very soft fragments, might the engineer anticipate that A_c could become a significant proportion of A. Discounting these for the present, $A_c/A \to 0$, so that

$$\frac{U}{A} = \frac{(A - A_c)}{A} u \to u$$

which means that

$$\tau_{max} = \mu(\sigma - u)$$

Karl Terzaghi, from the 1920s onward, pioneered this simple approach; he defined the effective stress normal to a plane as

$$\sigma' = \sigma - u$$

and showed that only the effective stresses σ' and τ, acting through the soil grains, could be responsible for deformations of the soil skeleton. Using this definition

$$\tau_{max} = \mu\sigma'$$

and, as we have already seen, engineers prefer to use the angle of friction so that we write

$$\tau_{max} = \tan\phi'\sigma' \qquad (4.1)$$

in which the dash over the ϕ reminds the user that effective stresses are to be used. This dash convention will be used from now onwards to discriminate between quantities that refer to total stresses σ and τ and those that involve effective stresses σ' and τ'. You will recall that we investigated the sinking bungalow of section 1.3 in terms of effective stresses and that, since water carries no shear stresses, $\tau' = \tau$, so that the dash on the shear stress becomes redundant.

Terzaghi's revision of Coulomb's friction model to deal with pore fluid pressures appears very simple, but it would be unfair not to point out some problems. Terzaghi (1943) rightly proclaims the relevance of pore fluid pressures to the frictional strength of soils, and his method is found to be very accurate when both the friction angle ϕ' and the pore pressure u are known with some accuracy. The central problem for the engineer lies in the prediction of pore fluid pressures. If you are asked to determine the factor of safety of an existing earth slope, then you may be able to measure the water pressures at a few points in the soil directly. In general you would then proceed to link the water pressures in a flownet so that it would be possible to estimate the water pressure at any point. But what pore fluid pressure would you assign to wet soils above the water table? How would substantial air pockets in the groundwater be treated? And how, as a designer, would you cope with the problem of predicting the worst groundwater regime in the next 50 years? Would you be able to collect the necessary meteorological and hydrological data for the whole region in which your client's small problem is located?

Finally, and most significantly, there are further difficulties in the prediction of water pressures in fine-grained soils due to their impermeability. A standpipe in sand might display an equilibrium column of water in a matter of minutes: a standpipe in homogeneous clay might take months or years to fill. It is necessary to develop special pore pressure instruments (piezometers) for use with clays, which do not demand much water movement before the pressure is indicated. A saturated porous stone tip connected by means of a water-filled small-bore tube to a sensitive diaphragm can respond to pore pressures in clay within seconds, since only the water necessary to deflect the diaphragm need flow from the clay. The deflection of the diaphragm can be converted into an electrical signal by the use of electric resistance strain gauges.

The very slow movement of water in highly compressible clays is also responsible for their very slow accommodation to changes in normal stress. If you attempt to slide your feet over a patch of mud and over a stiff clay of the same mineral composition, you will report that the angle of friction of the former is much less than that of the latter: and in effective stress terms you will be wrong. The friction angle of the two soils will be almost identical, the difference will lie in the high water pressure induced in the pores of the mud by your foot, and the low water pressure you induced in the stiff clay. Much effort will be devoted in later chapters to this unfortunate chain of logic. A load is to be placed on clay and the strength must be assessed; the strength depends on the water pressure in the clay; the water pressure in the clay depends not only on the original groundwater table but also on the load that is to be applied. It is easy to see that stepping on a rubber hot-water bottle causes the water pressure to rise if the stopper is almost leakproof (*cf.* clay), whereas the load is almost immediately carried by the solid rubber skeleton if the stopper is left out (*cf.* sand) because the water simply runs out of the way of the load. It is sufficient here to sound the warning that water pressures in clay change when the total stresses change. It is imperative either to measure water pressures or to allow the clay to come slowly into equilibrium, if effective stress analysis is to be applied.

4.4 The Shear Box Test

The direct shear box test affords the most simple method of determining the angle of friction ϕ' on a predetermined rupture plane. The principle is simply to compress a sample of soil contained in a split brass box under a given vertical load N, and then to shear the two halves of the box while measuring the development of the shear force F: see figure 4.4. The conventional apparatus is

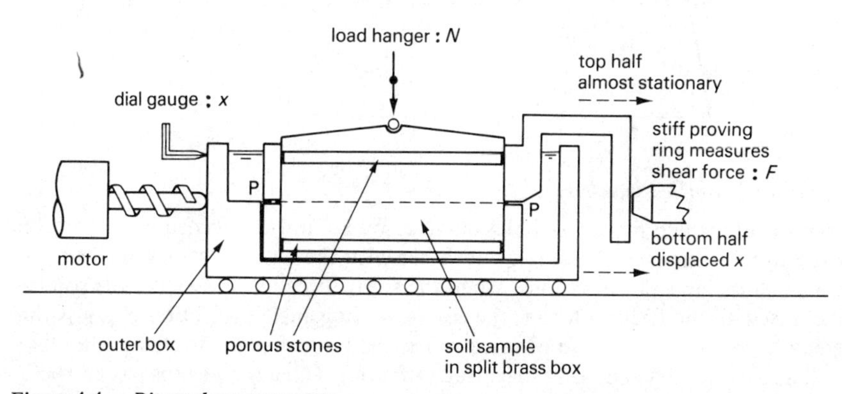

Figure 4.4 *Direct shear apparatus*

based on a soil sample 60 mm square on plan and 20 mm deep, shearing on the central horizontal plane. The vertical force is provided by weights on a hanger that straddles the box, while the shear force is developed by a small electric motor driving the bottom part of the sample box at a constant chosen speed, through a gear box. The top part of the sample box is almost prevented from moving by a stiff steel proving ring. The very small compression of this elastic ring is proportional to the compressive force, so that a dial gauge reading across its diameter can be converted by a calibration factor into the shear force F. The bottom part of the sample box is pushed along by the motor, so that a dial gauge pushing on the outer box effectively offers a reading of x, the relative shear displacement of the two halves of the sample box that forces the soil contained within it to rupture along plane PP. The 60 mm square sample can be sheared up to 9 mm: larger samples require proportionally larger displacements. Once the hanger weight N is fixed, readings of F and x can be taken as the motor shears the sample.

Most soils are required to act under water and it is therefore normal to use saturated soil and to fill the outer box with water before testing, so that the soil cannot dry out. The speed at which the test is conducted is chosen with regard to the permeability of soil: a sand can be tested at 1 mm/min, a silt at perhaps 0.01 mm/min and a clay at only 0.001 mm/min. If these speeds are not exceeded then water will probably be able to escape around the moving soil particles and ultimately reach the boundaries of the sample without causing changes in water pressure. Such a test is called 'drained'. The water pressure will then remain at zero, the head of water above the rupture plane being negligible.

Although the ideal shear force/deflection relationship would be curve 1 in figure 4.5, a typical drained shear-box test on a medium dense soil gives a

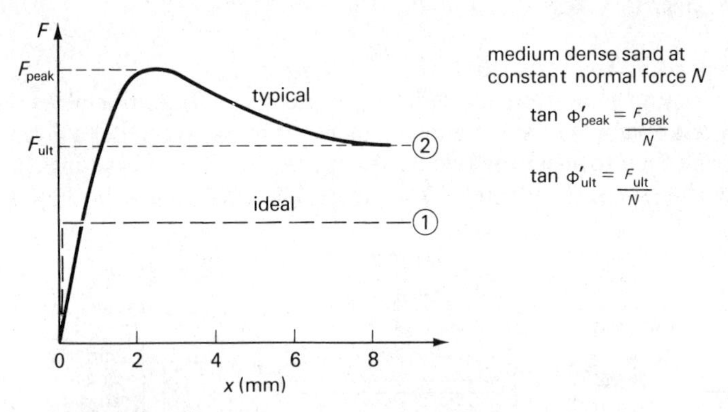

Figure 4.5 *Drained shear-box test*

relationship which more resembles curve 2. We are not concerned with the initial small deformations at this stage, but rather with the limiting values of shear force which cause large deformations. When the shear force reaches F_{peak} the soil has mobilised all the friction it can: the corresponding angle of friction ϕ'_{peak} is the greatest possible inclination of force across the rupture plane in that particular soil, at that density, and that magnitude of stress. Once the peak is passed the resistance of the soil begins to fall. If the shear force were provided by weights on a hanger, like the normal force, the peak would be a point of catastrophe after which the top part of the shear box would accelerate away as a result of the external force F_{peak} being increasingly greater than the internal resistance F provided by the soil. Since the lateral displacement of the shear box is positively controlled by the motor, however, the portion of the curve between peak and ultimate can be determined quite soberly. The ultimate angle of friction ϕ_{ult} simply represents the lowest likely resistance of the soil: it can represent a safe and conservative lower bound to the strength of the soil in service. It is a useful parameter since it is usually almost independent of the initial density of the soil and of the magnitude of the applied stresses. The displacement of a conventional shear box may be too small to achieve the constant plateau of ultimate strength.

For accurate work it is necessary to use much larger samples and to test them more slowly in accordance with the need for perfect drainage. In order to acquire a comprehensive picture of the friction behaviour of a particular soil, the engineer will normally carry out a suite of tests, with various initial densities and normal forces. A typical group of results for a well-graded sand is portrayed in figure 4.6, where the soil density is characterised by its void ratio e; loose

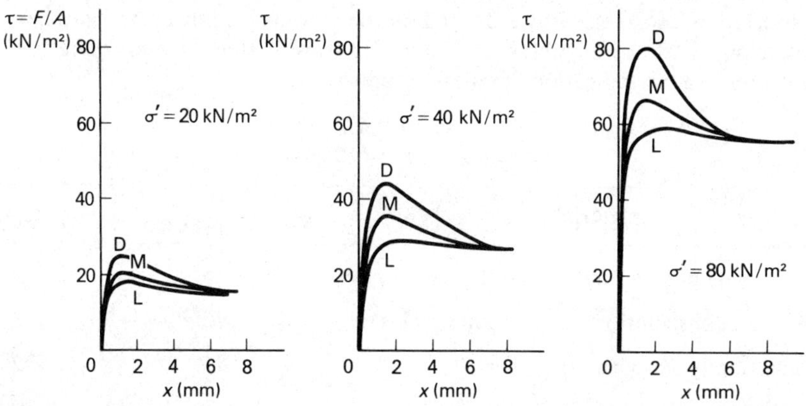

Figure 4.6 *Typical averaged drained shear test results for a sand: D dense, M medium, L loose*

$e = 0.65$, medium $e = 0.55$, dense $e = 0.45$. In order to make the results of more direct use the engineer will normally convert forces to stresses: $\tau = F/A$ and $\sigma' = \sigma = N/A$ where the area of the rupture surface in the conventional shear box is $A = 36\ \text{cm}^2$. It is also normal to abstract the peak and ultimate values from the shear force/deflection curves and replot them all on a shear stress/normal stress

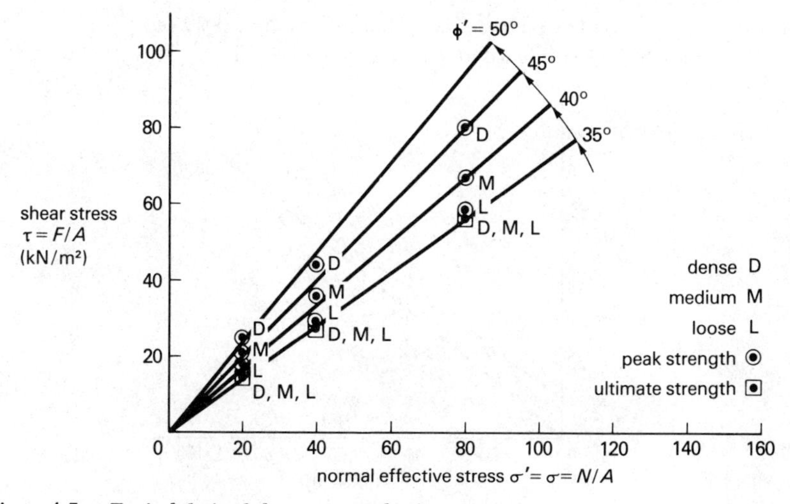

Figure 4.7 *Typical drained shear test results for a sand*

graph, as in figure 4.7. In this typical example we can see that

(1) the ultimate friction angle is almost independent of stress and density, at 35°.

(2) the peak friction angle depends on both stress and density, and lies in the range 35° to 50°. Peak friction rises as the void ratio falls, and also as the normal effective stress decreases. It is very important, therefore, to obtain or reproduce the stress and density in a laboratory sample which will be applicable in a field situation. It is never advisable to rely on a guess for the angle of friction of a soil since it will depend on the particle grading curve, particle mineralogy, particle roughness, void ratio and stress. Table 4.1 must therefore be viewed only as an indication of common values.

TABLE 4.1
Typical Angles of Friction

Material	ϕ'_{peak}	ϕ'_{ult}
Dense well-graded sand or gravel, angular grains	55°	35°
Medium dense uniform sand, round grains	40°	32°
Dense sandy silt with some clay	47°	32°
Sandy silty clay (glacial)	35°	30°
Clay-shale, on partings	35°	25°
Clay (London)	25°	15°

Some authors use the approximation

$$\tau_{peak} = c'_{peak} + \sigma' \tan \phi'_{peak}$$

to fit the data of rupture of a soil at a certain initial void ratio, and a certain range of stress; see figure 4.8, where the line is $\tau = 10 + 0.7\sigma'$. This can be most

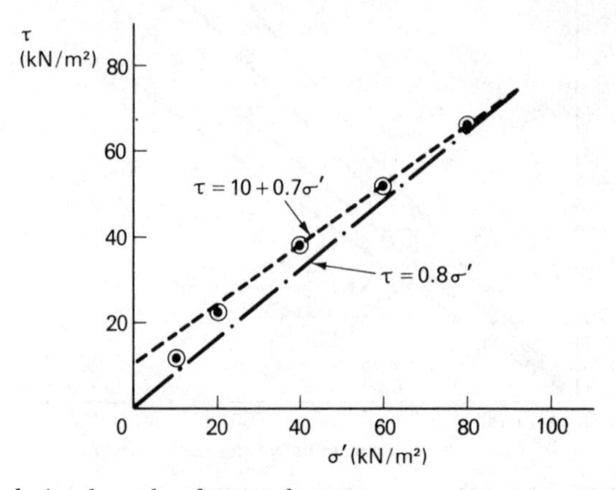

Figure 4.8 *Analysing data of peak strength*

misleading, as it may be assumed by the unwary that the material must have some strength c'_{peak}, other than its frictional strength, which can still be developed when the normal effective stress is zero. It is much more satisfactory to retain the simple 200-year-old friction law

$$\tau_{\text{peak}} = \sigma' \tan \phi'_{\text{peak}}$$

and choose a value for ϕ'_{peak} which looks reasonable; the example shows $\tau = 0.8\sigma'$, which safely underestimates all the data. The engineer must always be on his guard against facile mathematical concepts such as the fitting of 'the best straight line' to data, which will always leave some data on the unsafe side of the line, and which may result in the unwelcome birth of a bogus parameter such as c'. Particles of long-standing sands or clays do sometimes become bonded together by recrystallisation, or cemented, by the oxides of iron, manganese or calcium, or by silica, salt, or other agents. In such a case there may be an intercept c' on the strength axis at zero effective stress, which will represent the strength of the cement. Some soils engineers call this intercept the 'true cohesion' of the soil, but 'cementation' might be a more appropriate label. As soil becomes cemented it is necessary to adopt rock-mechanics versions of our simple models, in which the joints and fissures between cemented blocks take on a great significance. Such materials may collapse in the field solely by sliding on discontinuities with a small angle of friction ϕ'_{ult}: this may come as a surprise to an engineer who has only measured the large cemented strength of intact nuggets.

4.5 Dilation, Contraction and Critical States

The variation in ϕ' with stress and density which is typified by figure 4.7 is of great significance to the engineer. A number of theoretical models have been developed to explain the shear box observations, which are generally as follows

(1) A particular soil has a roughly unique ultimate angle of friction no matter what its initial density.

(2) Initially dense soils have a peak strength that is higher than their ultimate strengths. Beyond the peak, thin rupture zones can be detected in the previously uniform soil within which the density falls towards some 'critical' value below which it will not drop. The soil sample dilates (becomes larger) due to the reductions in density.

(3) Soils that are initially very loose take a good deal of straining in order to mobilise their ultimate strength: no peak is observed. The soil sample contracts (becomes smaller) as it is sheared, and the density rises towards the same order of 'critical' density as that observed in the rupture zones of dense samples.

(4) The magnitude of the 'peak' for dense soils is related to the rate at which the soil dilates.

The new theories have much in common, not least that they attempt to take account of the particulate nature of soil in comparison with the bricks and blocks used by Coulomb, and essentially the ability of particles to change their packing and in particular their density.

The notion of a 'critical state' is central to a better understanding of soil friction. A simple critical state model can be described thus

> When it is sheared a soil eventually comes into a critical state with a unique ratio of stress $\tau/\sigma' = \tan\phi'_{\text{critical}}$ on planes of shearing and a critical void ratio e_{critical} in the zones of shear which is a logarithmic function of the stresses.

The subscript c as in ϕ'_c will be used in this book to denote 'critical' conditions: the 'ultimate' state of soil at the end of a shear-box test will be approaching a 'critical' state. It is useful to depict the lines of critical state values of τ, σ' and e on a pair of connected graphs of τ/σ' and e/σ', which are the projections of a unique line in the three dimensions, as shown in figure 4.9. The third graph of e

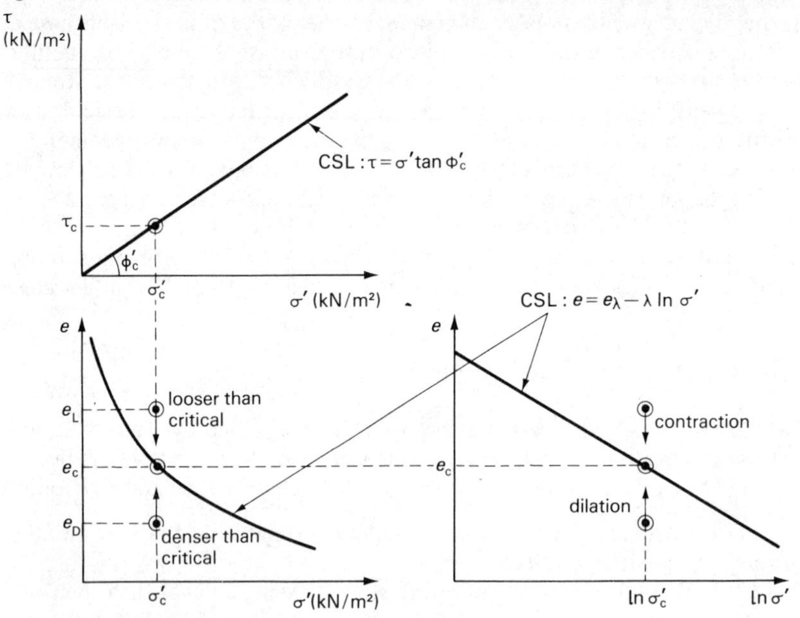

Figure 4.9 *Critical state line (CSL) in (τ, σ', e) space*

against ln σ' is used simply because it almost invariably linearises the e/σ' critical state line over the required range of stresses. Any logarithmic base could have been used. I have chosen the exponential or 'natural' base so that mathematical differentiation is made easy, and because the overwhelming majority of engineers now use pocket calculators rather than slide rules or tables of logarithms to base 10.

The triplet (τ_c, σ'_c, e_c) which is marked on the graphs is only one of the many critical states into which the soil could come after much distortion. In a particular drained shearbox test with a constant vertical effective stress of σ'_c, the 'ultimate' or 'critical' strength τ_c will be independent of the *original* void ratio, which could be e_L (loose) or e_D (dense). As the soil is sheared, the loose soil must *contract* in order to reach the 'critical' void ratio, while the dense soil must *dilate* so that e_D can increase to e_c, which is the preordained void ratio at a critical state, in which the normal effective stress is fixed at σ'_c.

The critical state model does much to tie together the data of density changes and the uniqueness of the 'ultimate' conditions: indeed it is hardly other than a creative restatement of the observations! It does not, as presently stated, explain the phenomenon of the 'peak' strength of dense soils. A 'stress/dilatancy' model is additionally called for. Reconsider the implications of the dilatancy demanded before dense soils can reach their critical states. The shear box lid and hanger load for such a soil are being driven upwards as they travel sideways, as depicted in figure 4.10. At any particular point S, the lid will be rising at an

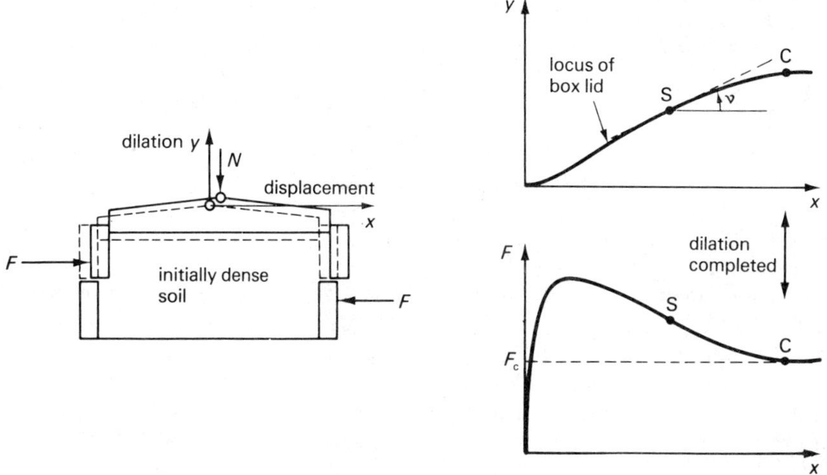

Figure 4.10 *Drained shear test on dilatant soil*

angle ν to the horizontal as it is sheared. If Coulomb had been watching the test very closely he would have been able to observe the upward motion (ν can be as great as 20°) and might have said: 'I hope you are not going to calculate the ratio F/N on a horizontal plane, because I see that the true plane of sliding is at an angle ν to the horizontal'. If you had shown him the brass shear box in its two halves and insisted on the value to you of a ratio F/N on the predetermined horizontal plane of gross sliding then he might have rephrased his objection thus: 'I now see that the horizontal plane is dictated by the test as the macroscopic failure plane. However, I must insist that the hanger load N is being driven up into the air which will affect the magnitude of F. Imagine the toothed saw blades depicted in figure 4.11: the top blade rides over the lower blade when it is pushed along. This causes a very simple alteration to the observed angle of friction $\phi_S = \phi'_c + \nu$ where ϕ'_c is the inherent angle of friction when smooth parallel blades slide'. In this imaginary interview I have had Coulomb choose the symbol ϕ'_c as the angle of friction between the saw blades so that it matches the critical state angle of friction. This is quite logical. At a critical state the soil density is constant so that $\nu = 0$ and the observed angle of friction ϕ'_S must indeed be ϕ'_c.

This dilatancy model predicts that the maximum or peak angle of friction is mobilised when the angle of dilation ν is greatest, and that

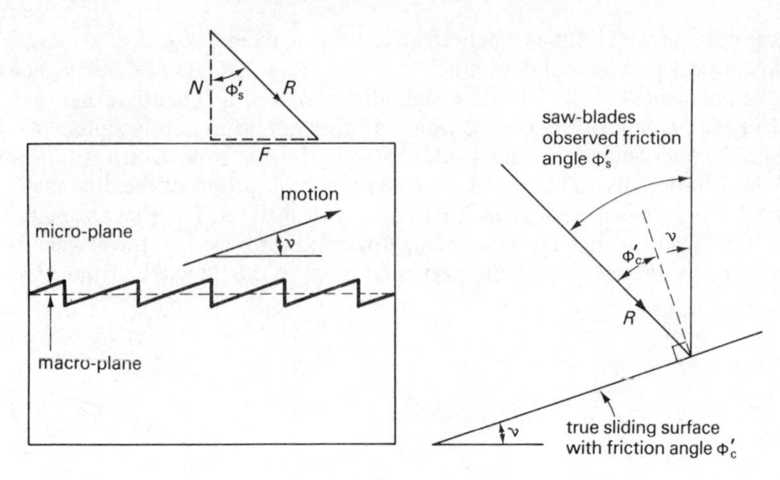

Figure 4.11 *Saw-blade dilatancy model*

$$\phi'_{max} = \phi'_c + \nu_{max} \tag{4.2}$$

Careful laboratory work substantiates the assertion that the peak is controlled by the angle of dilation, but also shows that equation 4.2 somewhat overestimates its contribution. Nevertheless, the critical state model in conjunction with this simple dilatancy model offers the engineer a powerful group of easily remembered concepts, which may be very helpful in sorting out the physics of complicated problems. In this task, figure 4.12 may help to fix the

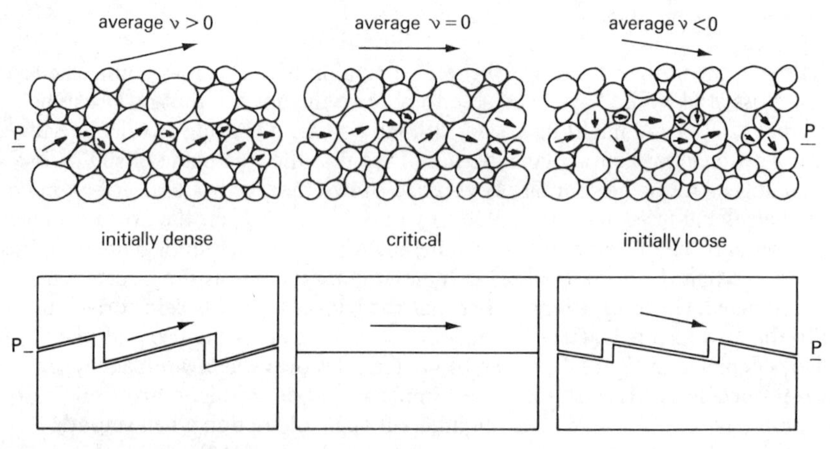

Figure 4.12 *Crude visualisation of dilatancy*

ideas. When soil is initially denser than the critical state which it must achieve, then as the particles slide past each other owing to the imposed shear strain they will, on average, separate. The particle movements will be spread about some mean angle of dilation ν. When soil is initially looser than the final critical state, then particles will tend to get closer together as the soil is disturbed, and the average angle of dilation will be negative, indicating a contraction. If the density of the soil does not have to change in order to reach a critical state then there is zero dilatancy as the soil shears at constant volume.

The variation in critical density with the applied stress is most significant. It is important to realise that a critical state is only reached when the particles have had full opportunity to juggle around and come into new configurations. If the confining pressure is increased while the particles are being moved around then they will tend to finish up in a more compact state. In order to know which critical void ratio in figure 4.9 is the appropriate ultimate condition of the soil in a particular situation, it is necessary to know the final stress σ'. This is fairly easy in the laboratory, where the total stress σ can be applied directly, and where the pore-water pressure can be kept at zero by submerging the sample in water and testing it slowly enough to ensure that its tendency to change in volume can be satisfied by the requisite flow of water (in or out) at a very small hydraulic gradient. If the hydraulic gradient were allowed to be significant, then (unknown) water pressure differences would exist which would alter the effective stress σ'. Not only would this alter the final strength τ, it would also bring the soil towards a critical state at a different density. The response to stress of an element of soil deep within a possible landslide is rather more difficult to predict: consider how you might discover

(1) its present void ratio and state of stress
(2) its critical state line
(3) the future additional stresses
(4) its present water pressure
(5) its surrounding drainage system
(6) its future water pressure

The prediction and control of landslides usually requires a very expensive site investigation without which any sophisticated theoretical models are stillborn.

If our hypotheses are qualitatively correct, then we can divide the diagrams of figure 4.9 into various dilation zones, as shown in figure 4.13. If the initial state of soil were above the critical state line on the e/σ' graph then it would be looser than critical, and would tend to contract ($\nu < 0$) when it was sheared. If the soil were denser than critical then it would tend to dilate ($\nu > 0$) when disturbed. Angles of dilation greater than 20° have exceptionally been recorded in very dense well-graded sands with strong angular grains: such soils are also usually impossible to set up in a condition looser than critical! Some sands with uniform smooth grains, most silts, and all clays, can be found in conditions looser than critical if they have been undisturbed: such soils are troublesome due to their collapsible structure and slow mobilisation of frictional resistance.

It is vital to understand why the dilatancy/contraction contrast is the most important discrimination between soils after the grain size contrast of sand/clay. This will become clear after we have worked carefully through figure 4.14, which traces the stress–strain curves of a soil (sand or clay) subjected to a drained shear test at two relative densities but at the same initial stress σ'. The 'loose' (looser than critical) state follows line LQC, whereas the 'dense' state follows line DPC. The critical state hypothesis leads us to propose, firstly, that because the vertical stress σ' is identical throughout each test the ultimate critical state

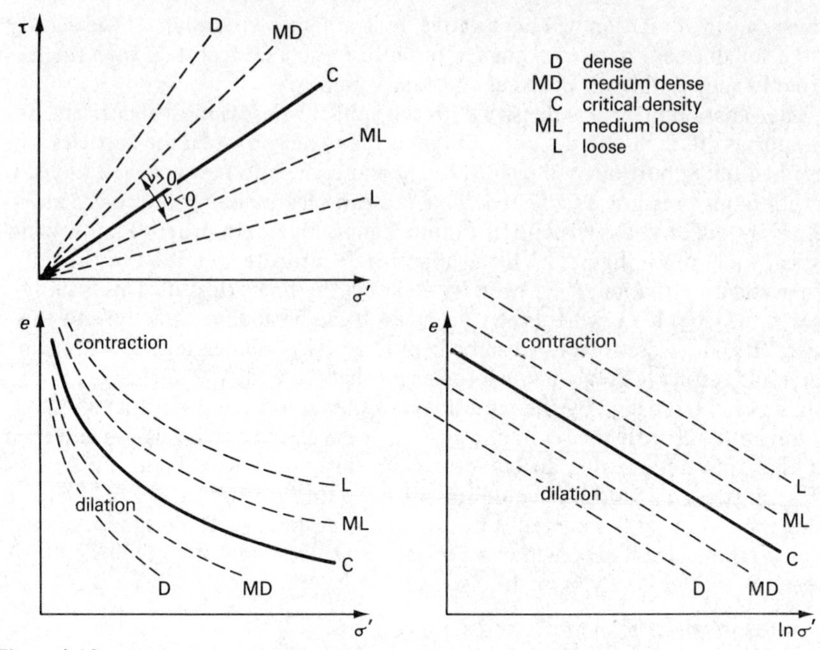

Figure 4.13 *Zones of dilation ($\nu > 0$) and contraction ($\nu < 0$)*

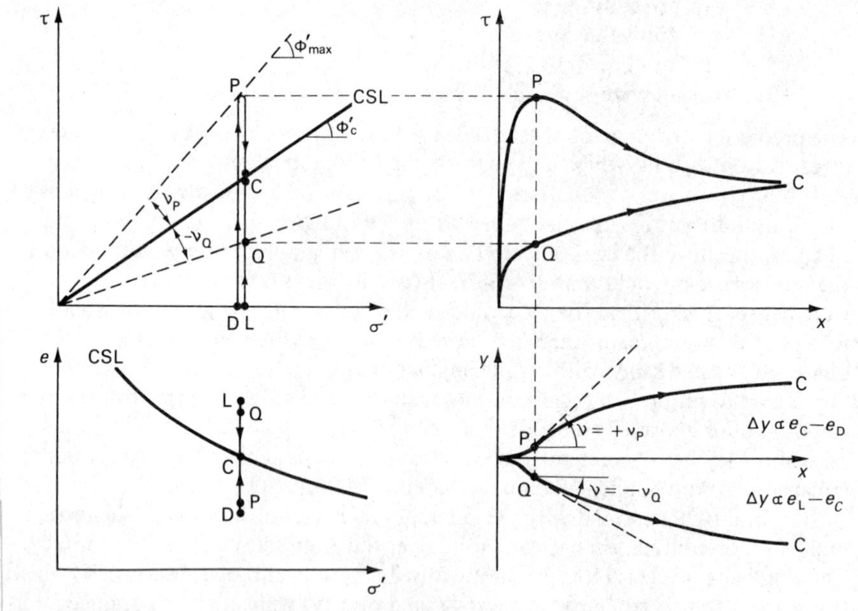

Figure 4.14 *Stress–deformation curves for loose and dense soil samples slowly sheared at the same constant normal stress*

must be identical, namely point C. The e/σ' diagram makes clear that the loose sample contracts ($\nu < 0$) while the dense sample dilates ($\nu > 0$). Because it dilates, the dense sample will be able to mobilise an angle of internal friction

$$\phi' = \phi'_c + \nu \geqslant \phi'_c$$

The magnitude of ν_{max} will depend on the potential total expansion $(e_C - e_D)$ and on the grain characteristics of the soil. The outcome is that a peak of strength P is reached at that point on the y/x diagram when the gradient is steepest: this is marked as ν_P on the y/x diagram, and is transferred to the τ/σ' diagram. Beyond point P the angle of dilation drops, so ϕ' drops, until, when $\nu = 0$ at C, $\phi' \rightarrow \phi'_c$.

The loose soil shows no peak. When the dense soil is at P the loose soil is at Q, contracting at a rate $-\nu_Q$, which when transferred to the τ/σ' diagram merely indicates that the soil is taking a long time to reach its critical state.

The consequences of the peak/no-peak contrast for the behaviour of soil are enormous. When a dense soil is being distorted (sheared) it eventually passes point P on its stress/strain curve. As it does so certain zones within the soil body will, perhaps accidentally, be distorting more than others: these zones will pass P first. As soon as they have done so, they will be weaker ($\tau < \tau_P$). They will therefore tend to continue distorting. The distortion will tend to be localised, with shear taking place in thin rupture zones between more-or-less rigid intact blocks within which the peak P was never reached. In other words the soil continuum segregates into a rubble of dense sliding blocks separated by an increasingly weak 'mortar' of the same soil, which is rapidly dilating towards its critical state. The sites of ancient landslides are particularly troublesome to highway engineers. The disturbed and very weak slip zone might be only a few millimetres thick, compared with the many metres of good dense soil on top. Only the expert geological interpretation of aerial photographs can warn of the possibility and location of such landslips: looking for the disturbed zone in boreholes could be less rewarding than searching the proverbial haystack for the needle. The use for design purposes of the 'shattered' or 'remoulded' or 'disturbed' or 'ultimate' strength of the soil – what we have come to call its critical state – is usually the correct and safe approach if the peak strength may not apply throughout the soil body in question. Exceptions, qualifications and complications are dealt with in chapter 8.

If, on the other hand, a loose soil is being sheared then each small zone hardens and strengthens as it distorts. This encourages the distortion to spread to neighbouring zones which have not yet caught up. The distortion is likely to spread uniformly through the whole mass of soil, which contracts uniformly throughout its volume.

Just as the strain-softening engineering materials (cast iron, glass, concrete) require a greater attention to detailed design than the strain-hardening materials (mild steel, aluminium), so 'dense' soils require more care than 'loose' soils. Fortunately they repay the extra care by settling less, causing less lateral pressure, and standing more steeply than do their 'loose' counterparts. Furthermore the fully-softened critical state strength of a dense soil is still a dominant fraction of the peak strength, whereas cast iron explodes to fragments and even concrete suffers a relatively dramatic fall in its load-carrying capacity after the peak. Nevertheless, the distinction between the rupturing of 'dense' soils and the ductile deformation of 'loose' soils remains one of the most significant contrasts that the soils engineer can observe.

In the work that follows I shall continue to refer to 'loose' and 'dense' meaning 'above' and 'below' the critical state line on an e/σ' diagram, and I shall refer to the angle of shearing resistance ϕ' which you now know can be visualised in two parts, the friction component ϕ'_c and the dilatancy component ν.

4.6 A Landslide Model: The Interaction of Friction and Seepage

4.6.1 The Model

The friction model of Terzaghi proposes that the strength of soil depends on water pressures rather than on water content, while the seepage model describes how water pressures exist in relation to flow regimes. A combination of the two models provides the engineer with a hypothesis for predicting the possible movement of soil particles due to the flow of water. The simple interaction mechanism that will now be described can be used to explain and predict the following observations

(1) A trench is dug in waterlogged ground consisting of sandy peat and silt layers. The trench sides repeatedly fall in.

(2) A dam is constructed out of compacted silty clay. No special drains are laid. The downstream slope erodes away, starting at the extreme toe.

(3) A landslide occurs in a steep valley after a month of exceptionally wet weather.

The mechanism is sketched in figure 4.15, where an infinite slope is considered to fail on one of a family of parallel slip planes so that a rigid flake of surface

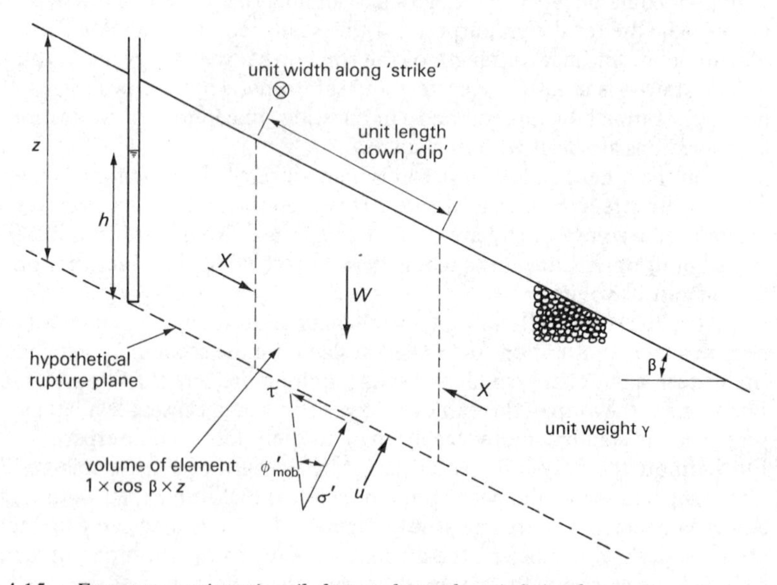

Figure 4.15 *Forces on a prismatic soil element beneath an infinite slope*

soil slides down the slope. The water pressure u is allowed to vary with depth, but is considered to be constant along any one of the possible slip planes. A section of unit width (along the 'strike' of the slope) is considered, and a block of soil of unit length (down the 'dip' of the slope) is taken to be representative. The forces on the block are shown: they must be in equilibrium. The stresses on the slip plane are considered to be σ' and u normal, and τ tangential. Since a unit area of plane is being considered these stresses also represent the forces on the block. The weight of the block is $W = \gamma z \cos \beta$, while the forces on the vertical sides of the block are said to be X. Since any point on the slope is indistinguishable from any other the forces X must be equal, and must therefore cancel each other out when the equilibrium of the block is considered. There are no forces on the surface of the slope.

Resolving forces on the block along the slip plane

$$\tau = W \sin \beta \tag{4.3}$$

Resolving forces normal to the slip plane

$$\sigma' + u = W \cos \beta \tag{4.4}$$

It follows that

$$\frac{\tau}{\sigma'} = \frac{W \sin \beta}{W \cos \beta - u} = \frac{\tan \beta}{\left(1 - \dfrac{u}{W \cos \beta}\right)}$$

So let us define

$$\tan \phi'_{\text{mob}} = \frac{\tau}{\sigma'} = \frac{\tan \beta}{\left(1 - \dfrac{u}{\gamma z \cos^2 \beta}\right)} \tag{4.5}$$

where ϕ'_{mob} is a mobilised angle of shearing resistance implied by the angle of the slope and its hidden pore-water pressures. Of course it is necessary to ensure that ϕ'_{mob} can indeed be safely mobilised by the soil concerned: by what margin is $\phi'_{\text{mob}} < \phi'$ at the relevant density and stress? We therefore arbitrarily define a factor of safety $F = \tan \phi' / \tan \phi'_{\text{mob}}$ so that when $F > 1$ the slope is safe, when $F = 1$ the landslide occurs, and $F < 1$ implies an impossibly steep slope. It follows from equation 4.5 that

$$F = \frac{\tan \phi'}{\tan \beta}\left(1 - \frac{u}{\gamma z \cos^2 \beta}\right) \tag{4.6}$$

Very often it will be convenient to deduce the water pressure u from the height h to which water would rise in a standpipe with its open end on the supposed slip surface at depth z below ground level, that is, $u = h\gamma_w$. Equation 4.6 can then be rewritten

$$F = \frac{\tan \phi'}{\tan \beta}\left(1 - \frac{h\gamma_w}{z\gamma \cos^2 \beta}\right) \tag{4.7}$$

In assessing the stability of a slope it is necessary to calculate the factor of safety of every possible slip plane, since the angle of friction ϕ' and the ratio h/z are likely to vary with depth. Soil samples must be recovered from various depths and subjected to drained friction tests: the regime of groundwater pressures must likewise be determined. Only then can equation 4.7 be used to find the potential slip surface with the smallest factor of safety.

4.6.2 Particular Cases

The dry sandy slope: $u = 0$

From equation 4.6

$$F = \frac{\tan \phi'}{\tan \beta}$$

The failure surface is simply the one with the smallest angle of friction ϕ', and the slope will fail ($F = 1$) when its slope $\beta = \phi'$.

The saturated slope sited over a drainage blanket

Figure 4.16 demonstrates that a simple rectangular flownet satisfies the case of a

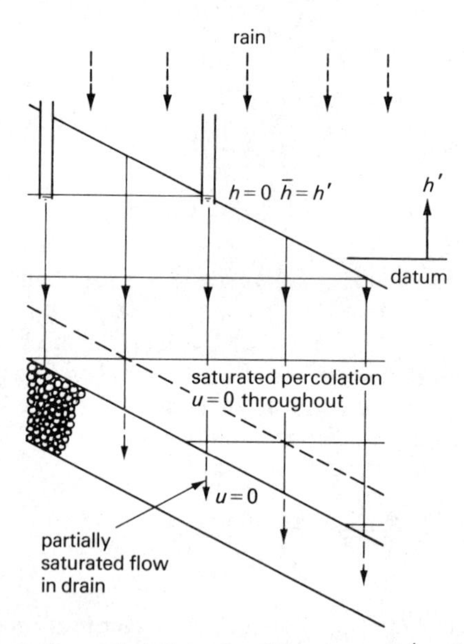

Figure 4.16 *Percolation into relatively permeable stratum: $h/z = 0$*

plane rain-sodden slope which is perfectly drained at its base: in particular the top flowline condition $\bar{h} = h'$ is observed, which was introduced in section 3.5.3. We have already seen in figure 3.6 that such a flownet generates zero pore-water pressures throughout, so that although the slope is saturated with water it is mechanically identical to the previous case of a dry sandy slope, and it accordingly shares the slope angle at failure $\beta = \phi'$.

The waterlogged slope: flow parallel to slope as in figure 4.17

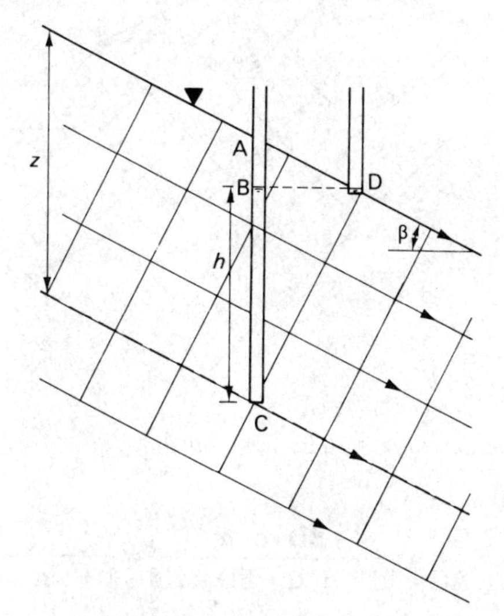

Figure 4.17 *Flow parallel to waterlogged slope: $h/z = \cos^2 \beta$*

Since the flowlines are parallel to the slope, the equipotentials are normal to it. The definition of an equipotential is that water in standpipes placed with their open ends at any point along it will rise to the same level in each one. If the equipotential passing through the point of interest C is followed to the top flowline at D, this level becomes apparent: the head of water required is seen from figure 4.17 to be $h = z \cos^2 \beta$. Making this substitution in equation 4.7

$$F = \frac{\tan \phi'}{\tan \beta}\left(1 - \frac{\gamma_w}{\gamma}\right) \tag{4.8}$$

Once again the safety factor is not explicitly dependent on the depth of the slip surface: F only varies if ϕ' varies. If typical values of $\gamma_w = 10$ kN/m^3 and $\gamma = 20$ kN/m^3 are used, we obtain

$$F = \frac{1}{2} \frac{\tan \phi'}{\tan \beta}$$

so that the critical slope angle ($F = 1$) occurs when $\tan \beta = (\tan \phi')/2$. These slopes can only stand at roughly one-half of their angle of friction.

The waterlogged slope: flow in a general direction

Figure 4.18 shows a long slope of inclination β through which water is flowing at inclination α. Once again the intersection of the equipotential with the top flowline betrays the relevant equipotential level. Trigonometry then demands that

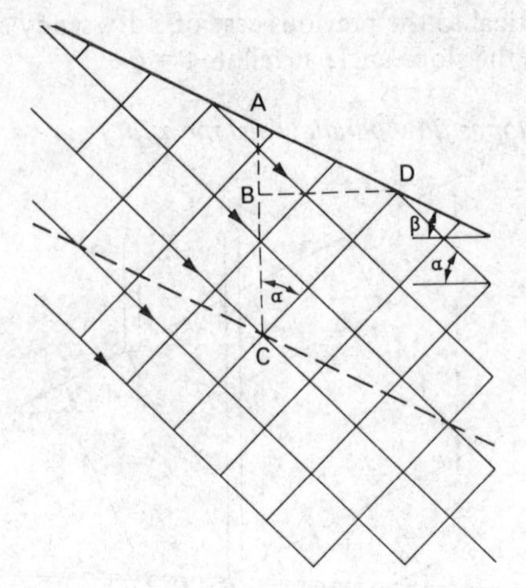

Figure 4.18 *Uniform flow: h/z = 1/(1 + tan α tan β)*

$$\frac{h}{z} = \frac{\mathrm{BC}}{\mathrm{AC}} = \frac{\mathrm{BD} \cot \alpha}{\mathrm{BD} \cot \alpha + \mathrm{BD} \tan \beta} = \frac{1}{1 + \tan \alpha \tan \beta}$$

So from equation 4.7

$$F = \frac{\tan \phi'}{\tan \beta}\left(1 - \frac{\gamma_w}{\gamma \cos^2 \beta (1 + \tan \alpha \tan \beta)}\right)$$

which can be written

$$F = \frac{\tan \phi'}{\tan \beta}\left(1 - \frac{\gamma_w (1 + \tan^2 \beta)}{\gamma (1 + \tan \alpha \tan \beta)}\right) \tag{4.9}$$

This formula is, perhaps, best utilised by returning to equation 4.5 and using the mobilised stress inclination ϕ'_{mob} to obtain

$$\tan \phi'_{mob} = \frac{\tan \beta}{\left(1 - \dfrac{\gamma_w (1 + \tan^2 \beta)}{\gamma (1 + \tan \alpha \tan \beta)}\right)} \tag{4.10}$$

Figure 4.19 shows how ϕ'_{mob} varies with α and β for the particular case of $\gamma = 2\gamma_w$. It will allow you to enter the slope and water details α and β and obtain the implied mobilised stress inclination ϕ'_{mob}, which ought to be less than the angle of friction ϕ' if the slope is to be in equilibrium. Alternatively it allows the choice of a safe slope angle in the light of a safe ϕ'_{mob} and a likely α. It is interesting to see that a given soil slope gets increasingly unsafe as α decreases from 90° (vertically downward flow) when $\phi'_{mob} = \beta$, through $\alpha = 0°$ (horizontal outward flow) when $\phi'_{mob} = 2\beta$, and that it then must inevitably come to failure as α becomes increasingly negative (upward and outward flow). Other γ_w/γ ratios give different charts.

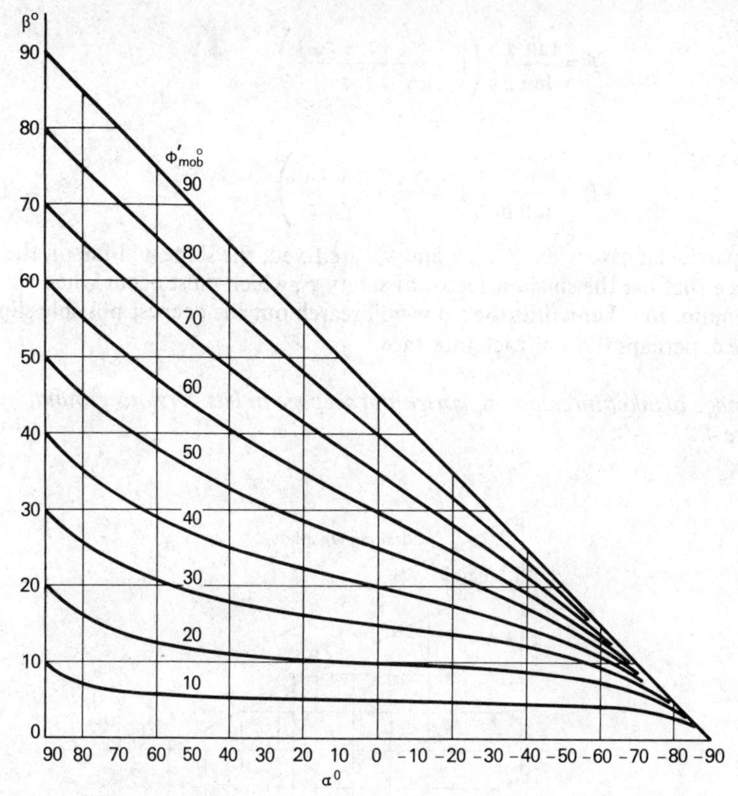

Figure 4.19 *Contours of ϕ'_{mob} with an infinite slope, for $\gamma = 2\gamma_w$*

The depressed water table: flow parallel to slope as in figure 4.20

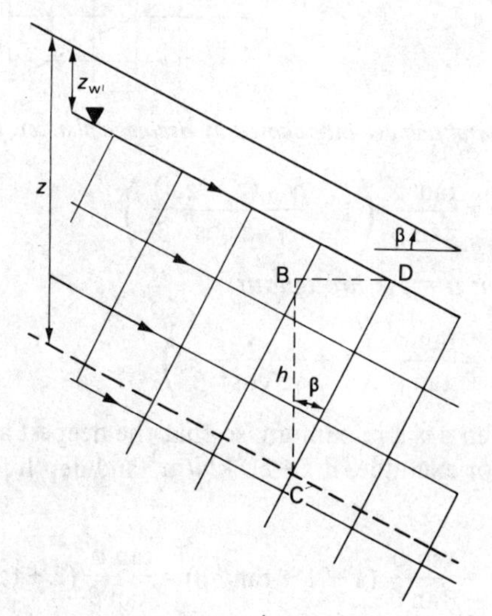

Figure 4.20 *Depressed parallel water table: $h/z = \cos^2 \beta(z - z_W)/z$*

$$F = \frac{\tan \phi'}{\tan \beta}\left(1 - \frac{\gamma_w}{\gamma}\frac{(z - z_w)}{z}\right)$$

or

$$F = \frac{\tan \phi'}{\tan \beta}\left(1 - \frac{\gamma_w}{\gamma} + \frac{\gamma_w}{\gamma}\frac{z_w}{z}\right) \tag{4.11}$$

In a particular case when β, ϕ', γ and z_w are fixed, the slope will fail on the slip surface that has the smallest factor of safety F, which must occur when z is a maximum. In other words the slope will search out the deepest possible slip surface, perhaps the soil/rock interface.

Drainage blanket intercepting horizontal seepage in less pervious ground, figure 4.21

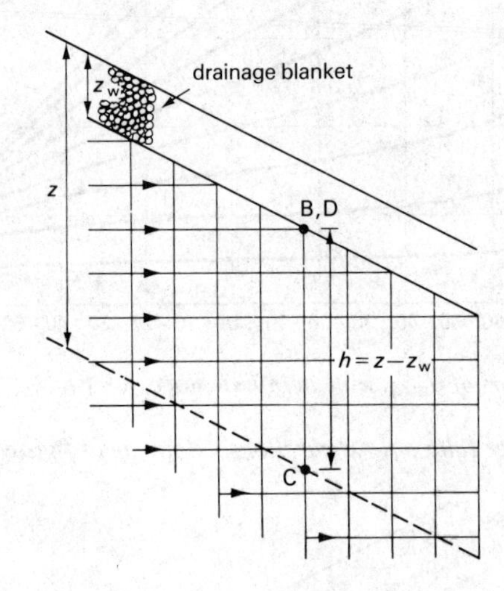

Figure 4.21 *Horizontal seepage intercepted by drainage blanket: $h/z = (z - z_w)/z$*

$$F = \frac{\tan \phi'}{\tan \beta}\left(1 - \frac{\gamma_w}{\gamma}\frac{(z - z_w)}{z \cos^2 \beta}\right) \tag{4.12}$$

Soil slope in suction: $u = -s$ throughout

$$F = \frac{\tan \phi'}{\tan \beta}\left(1 + \frac{s}{\gamma z \cos^2 \beta}\right) \tag{4.13}$$

F is a minimum when z is a maximum, so that the deepest available slip surface will be activated. For example, if $s = 50$ kN/m^2 and depth to bedrock $z = 2.5$ m then

$$F = \frac{\tan \phi'}{\tan \beta}(1 + 1 + \tan^2 \beta) = \frac{\tan \phi'}{\tan \beta}(2 + \tan^2 \beta)$$

so

$$\phi'_{\text{mob}} \approx \frac{\beta}{2}$$

and the slope would fail at roughly double its angle of friction.

4.6.3 *A Case-study: Stabilising a Slope*

The owner of a small factory wished to have his car park extended. He employed an architect who instructed a local plant hire firm to trim back the slope behind the factory, sketched in figure 4.22. The architect felt he had little choice,

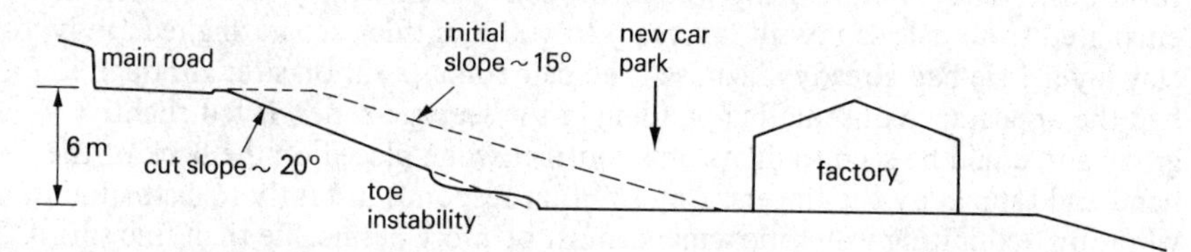

Figure 4.22 *Case study*

because the factory was established on a fairly narrow bench half way up a major hillside. On a higher bench ran a main road; the architect's new cutting ran just inside the boundary fence to the footpath. Immediately after the cutting and grading of the slope was completed, springs of water appeared near the toe of the slope. These triggered small soil slippages, and the disturbed soil soon mixed with the spring water and turned into a slurry. Before long the movement of soil at the toe threatened the whole hillside, and especially the main road. The architect decided to consult a specialist soils engineer.

The engineer visited the site and carried out the following steps

(1) fixed the site on his Ordnance Survey map so that he could later refer to geological maps

(2) talked to the driver who had trimmed the slope: he had also excavated various pits in the area and knew something of the soil and rock conditions

(3) inspected the whole site and sketched the relevant details, working partly from the architect's plans

(4) took samples of soil from various shallow pits which he asked to be dug, above and below the 'spring-line' which was about one-third the way up the slope from the toe

(5) inspected a deep pit which he asked the digger-driver to excavate near the toe of the slope

(6) drove slowly around the area, finding a disused limestone quarry about 300 m up the hill, but seeing no signs of other landslips

(7) made a note to ring the local highway authority so that a check could be made on the integrity of the sewers and drains under the main road: leaky pipes often cause landslides due to the raising of groundwater levels

When he began to piece together his information he made the following discoveries

(1) The local structure of the bedrock was rather complicated. The geological map showed the site to be the location of the feathering out of the limestone on top of an underlying sequence of sandstones and mudstones. This explained the many limestone boulders that had been encountered in the glacial silts and sandy clays that formed the slope in question. The deep pit at the toe of the slope encountered a weathered mudstone at a depth of 2 m. The digger driver reported other pits he had seen in the general area which only encountered rock at 4 m or more.

(2) The highway drains were reported to be intact.

(3) The soils on the slope were in irregular layers which dipped gently toward the valley. Before being wetted the soil was extremely stiff. The 'springs' emanated from pale sandy silt layers up to 100 mm thick separating red sandy clay layers. He had already diagnosed the pale soil as a silt on site: although it had the appearance of semolina pudding in the seepage zones, it felt slightly gritty and could be seen to dilate and contract when placed on the back of the hand and tapped by the fingers. Any soil that responds instantly to distortion or vibration, exhibiting water movement, must be more permeable than fine silt. A group of particle-size analyses in the laboratory had confirmed the observation.

(4) The effective angles of shearing resistance of the various soil samples that he had taken lay in the region 34° to 38°. He knew that these values were somewhat pessimistic, because he had simply compacted the soils lightly into the shear boxes, subjected them to a fairly high vertical stress of 50 kN/m^2, submerged them, and slowly (24 h tests) sheared them. They exhibited little or no peak strength, confirming that the measured angle of friction approached its critical state value. He found that the unit weight of the soil samples was in the region of 20 kN/m^3.

The engineer knew that the slope itself provided an indication of the average angle of shearing resistance ϕ' of the soil near its toe. Since it had actually failed, although rather slowly, the factor of safety of the toe must be less than unity, though marginally. The mobilised stress inclination ϕ'_{mob} must be a rough measure of the average angle of shearing resistance ϕ'. The engineer knew that the toe of the slope was waterlogged and that seepage was taking place at the ground surface. He guessed that the seepage was taking place parallel to the bedding of the silts, which he estimated to be horizontal. Using $\beta = 20°$, $\alpha = 0°$ in figure 4.19 he saw that the mobilised inclination of stress $\phi'_{\text{mob}} = 40°$. He knew that the soil had found a slightly more favourable failure mechanism than sliding in a parallel flake: this meant that his estimate of $\phi' \approx 40°$ might be a fraction smaller than the correct value. The correspondence with the shear box tests on the loosened soil was good enough.

The engineer proceeded to find what slope angle might be safe if seepage remained horizontal and the toe of the slope remained waterlogged. He decided rather arbitrarily that a mobilised inclination of stress of 31° would be safe with

$$F = \frac{\tan 40°}{\tan 31°} = 1.40$$

Returning to figure 4.19 he saw that with $\alpha = 0°$, $\phi'_{mob} = 31°$, the safe slope angle β was 15½°, just the angle it had initially been before the extension. He felt that any encroachment on the newly created car park was unlikely to satisfy the client, so he turned his thoughts to drainage.

If he could draw down the groundwater table so that it never appeared on the slope, and so that the situation would be similar to that sketched in figure 4.21, he might achieve stability. Using equation 4.12 with $F = 1.40$, $\phi' = 40°$, $\beta = 20°$, $\gamma_w = 10$ kN/m^3 and $\gamma = 20$ kN/m^3 he obtained $z_w/z = 0.31$. He must draw down the water table roughly one-third of the way to bedrock. The engineer was prepared to assume that rock that was sufficiently well interlocked to resist a sliding failure was no deeper than 3 m in the zone of instability. He also saw that slip surfaces deeper than 3 m would be at a great disadvantage on his particular slope, which was only 6 m high; the non-infinite extent of his slope would effectively 'lock' deep slip surfaces. (In this intuition, the engineer had the advantage of being well read and practised in slope stability studies using slip circles.) He concluded that groundwater should be drawn down roughly 1 m below ground surface.

It would, of course, be quite uneconomic and wasteful to strip off 1 m of soil from the whole slope and replace it with 1 m of high-quality drainage material (clean crushed rock or gravel). The art of economic land drainage is to provide intermittent deep and narrow drains which draw down the water table between them to a sufficient extent. 'Buttress' drains, which are trenches full of gravel, perhaps 0.75 m wide and running straight up the slope at 10 m intervals, would be called for in this case. The depth of the trenches would have to exceed 1 m due to the tendency for the water table between the drains to be drawn down less than that under the drains: a typical scheme is sketched in figure 4.23. The dimensions of the drains were chosen partly by trial and error, although 0.75 m was the width of a convenient dragline bucket. The decision-making tool was the flownet for an assumed thin horizontal sheet of silt confined between clayey layers: the final solution for drains 1.75 m deep is shown in figure 4.24 (plan). With a slope angle of 20° the 1.75 m deep drains encroach 4.80 m into the horizontal plane. These drains, and the front face of the slope, were treated as sinks at zero water pressure head as far as the flow of water in the thin lamina was concerned. There is an initial uniform flow zone in the interior of the slope, corresponding to a plan view of a horizontal plane through figure 4.21, which would persist to the atmospheric boundary in the absence of the drains. The drains attract the flow to themselves symmetrically. The flownet for this spacing of drains shows that roughly 30 per cent of the original flow still finds its way on to the front face of the slope, although the wider spacing indicates that the hydraulic gradient is reduced by a factor of 2.5 or so. The worst situation occurs at the deepest possible slip surface on the central plane of symmetry between two drains: here the water pressures are highest.

Neglecting any support from its shoulders, the stability of the central plane between drains can be estimated using equation 4.7. The point representing a depth $z = 3$ m, which was the deepest slip surface the engineer could imagine, is marked on the flownets in figure 4.24. It is 8.2 m along the horizontal plane, and has a head 5.3 equipotential drops above datum in the with-drains net and 8.2 drops above datum in the without-drains net. Datum in each case is the

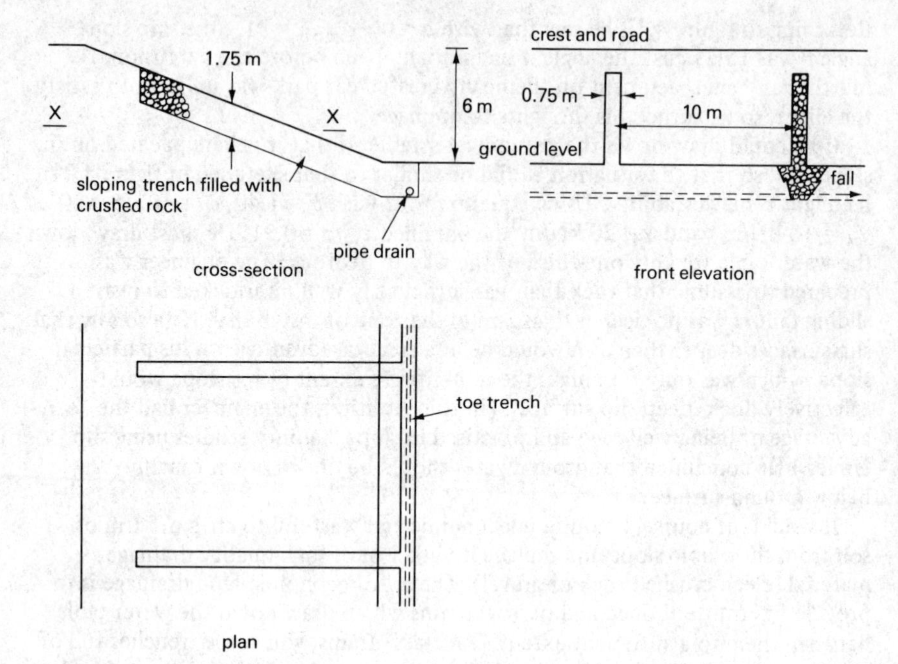

Figure 4.23 *Drainage design*

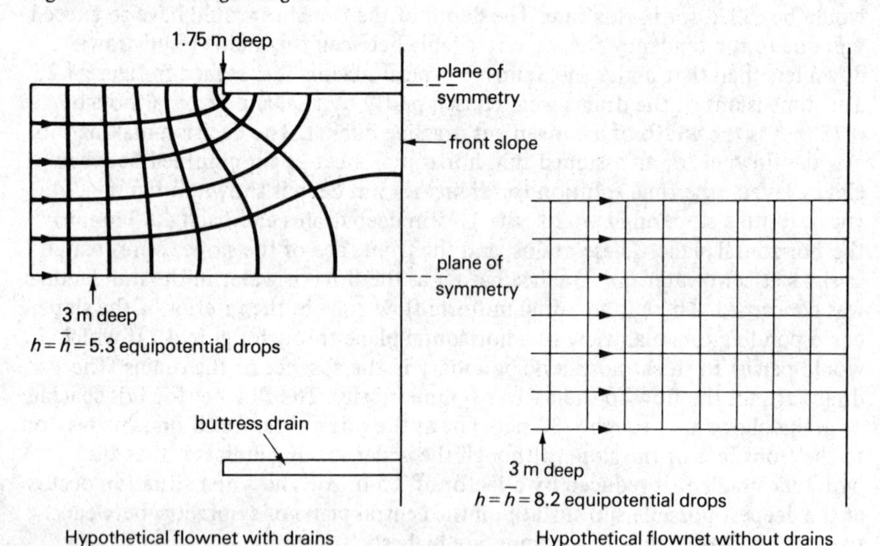

Figure 4.24 *Plan view of flow in typical sheet lamina* XX *of silt*

elevation of the plane itself; a standpipe placed on the slope surface exhibits exactly zero head. Figure 4.25 makes clear that the original hypothetical flownet, viewed in elevation, has a 3 m water head at 3 m depth: this generated the instability. The proposed buttress drains encourage flow out of the plane of figure 4.25 and it was evident from the plan view in figure 4.24 that they

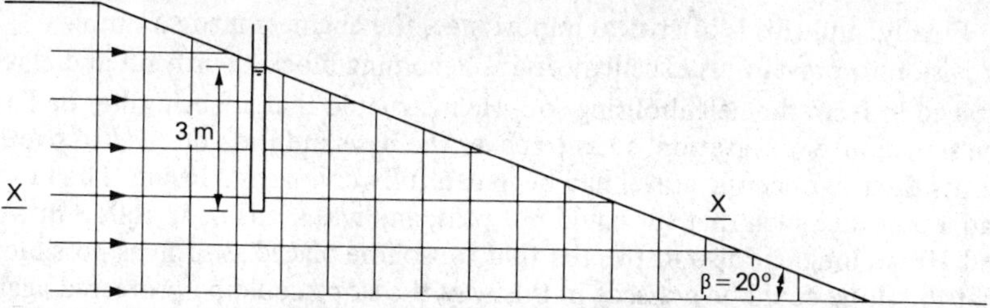

Figure 4.25 *Cross section of hypothetical flownet without drains*

reduced the head at 3 m by a factor of 5.3/8.2 or 0.65: this leaves a head of 1.95 m at 3 m depth. Using equation 4.7, the stability of a 3 m deep slice is

$$F = \frac{\tan \phi'}{\tan \beta}\left(1 - \frac{h\gamma_w}{z\gamma \cos^2 \beta}\right)$$

$$= \frac{0.809}{0.365}\left(1 - \frac{1.95}{3}\frac{10}{20}\frac{1}{0.883}\right)$$

$$F = 1.41$$

which was the desired value.

It would now repay you greatly if you attempted to repeat the trial-and-error design process for drains at 6 m intervals. Try three sensible trench depths and plot a rough curve of the factor of safety of the central slide against the depth of the drain, thereby choosing an appropriate value.

Before he quitted the problem the engineer turned his attention to the drains themselves. They should be (and remain) sufficiently permeable to conduct their water away without generating any important pressure heads. The critical condition is at the base of the buttress drain, where the flow of water may be D_F deep, calculable if the permeabilities can be estimated. The engineer specified a crushed rock or gravel for the drainage material, and included an injunction against material that would pass a 1 mm aperture sieve: he felt that such a material would not have a permeability less than 10^{-2} m/s by Allen Hazen's correlation of equation 3.8. He guessed that the soil, in the slope would not, at any section, have an average permeability exceeding 10^{-5} m/s. The greatest flow which he could foresee a single drain collecting over its entire height, was that emanating from a uniform horizontal flow 10 m wide and 6 m high with a hydraulic gradient of $\tan \beta$ or 0.365: that is, $10 \times 6 \times 0.365 \times 10^{-5} = 2.19 \times 10^{-4}$ m^3/s. The hydraulic gradient in the drain would also be $\tan \beta$ or 0.365, being the gradient of the bed of the drain and therefore (roughly) of the water surface in the drain. The depth D_F of flow is then given by

$$0.75 \times D_F \times 10^{-2} \times 0.365 = 2.19 \times 10^{-4}$$

$$D_F = 0.08 \text{ m}$$

This is sufficiently small to be ignored: the engineer would have had to be an order of magnitude too optimistic before D_F became very important. The engineer nevertheless deepened the trenches to 1.85 m in order to cope with the expected maximum flow (leaving the water level in the deeper drains roughly where the previous trench bottom was located).

Finally, and this is of critical importance, the engineer made a simple provision to prevent his excellent drains becoming blocked with silt and clay washed in from the neighbouring soil. He instructed that a fabric filter be laid in the trench after excavation, so as to cover the base and the sides and to overlap across the top once the gravel had been carefully compacted inside. The fabric had a pore size such that silt could not pass, and was available in rolls 8 m wide and 100 m long. He also instructed that topsoil be placed as soon as possible, so that the slope could be grassed: in this way the slope could be protected against infiltration and erosion from above.

4.7 Problems

(1) A silty sand stratum uniformly 5 m deep overlies sandstone, and slopes at 25° on an extensive hillside. What is the factor of safety of the slope when the groundwater table can be encountered in pits 1 m deep? Use the following data: saturated $\gamma = 20$ kN/m^3, potential capillary rise 2 m; ϕ' at appropriate density = 45°.

Answer 1.29.

(2) Turn back to chapter 3, figure 3.17, and by determining the stability of various 'flakes' of soil on the downstream face, find the minimum safe angle of friction for the dam if no part of that face is to have a safety factor of less than 1.3 against being 'washed away'. Assume that the bulk weight of the soil is 20 kN/m^3, so that you may use figure 4.19.

Answer 66°, hence the need for seepage control!

(3) Find the critical slope angle for completely waterlogged silty peat ($\gamma = 12$ kN/m^2, $\phi' = 40°$). In what circumstances would this represent the steepest side slopes in an excavation to bury a pipeline?

Answer 8°.

(4) Modify figure 4.15 and equation 4.6 to apply to the case of a slope fully submerged under water with hydrostatic conditions applying throughout, and show that it would just stand at its angle of friction ϕ', like a dry slope. Hint: *either* use Archimedes' principle in this hydrostatics problem *or* continue the slice boundaries upwards to the water surface, and allow for the superimposed weight of water, and the new imbalance in lateral forces due to the water pressure increasing down the slope.

(5) Why would the general tipping of 1 m of rubble on the clay-silt slope referred to in the case study have been dangerous?

(6) Confirm that sandy buttress drains which bottommed out in an exceedingly permeable zone of shattered rock might well prove entirely useless in drawing down the groundwater which could emanate from it.

(7) Would the buttress drains in the case study need redesigning if the contractor discovered that a medium grained sandstone was exposed at a depth of roughly 2 m beneath the slope?

5

Cohesion

5.1 Undrained Strength

While nineteenth century engineers made good use of Coulomb's friction model for permeable soils with low groundwater levels, they found two groups of problems quite intractable because they did not have the advantage of knowing the crucial role that is played by pore-water pressure, a role only clarified in the 1930s. The first was 'running sand', which was naturally encountered when excavating sand and water beneath the water table: in section 4.6 we saw that the angle of repose of a sand slope streaming with water is likely to be roughly half its angle of friction, whereas damp sand above the water table might even be prepared to stand vertically if the suction due to surface tension were strong enough. The contrast between damp and waterlogged sand, although sharp, is now understood. The second problem was 'cohesive soil', which seemed to disobey the friction law. Clods of clay stood in cliffs rather than at an angle of repose. Furthermore it seemed unlikely that the friction law could account for the evident difference in strength of a patch of clayey ground, soft and muddy in wet weather and hard after some sunshine. We now know that this strange behaviour can be described in terms of the changes in water pressure within the soil. Large negative water pressures which can exist in fine-grained soils cause the effective stress in the soil to be higher than expected, which means that the frictional strength is higher than expected. Large positive water pressures have the opposite effect. Without this understanding it was necessary to invoke some extra source of strength or 'cohesion' in dense or dry clays, while denying the basic friction of soft clays or muds.

Such was the mystique that surrounded 'cohesive' soils that Terzaghi's disciples felt it necessary to attempt an exorcism. Having discovered that the strength of most clayey soils in most circumstances could be well explained using the effective friction model, they decided to rename the temporary constancy of strength in a clayey soil its 'apparent cohesion'. The addition of the word 'apparent' certainly had the required effect: we are such complex creatures that we implicitly assume that when someone says that something is apparent, he means that it is not in fact there! It was but a short step to the discovery that for some clays the friction line on the effective stress diagram seemed not quite to pass through the origin (see figure 4.8, for example) and for the observed intercept c' of strength in the absence of effective normal stress to be pronounced 'true cohesion'. Thus does one religion supplant another. Unfortunately, 'true cohesion' is not usually present to any important degree in clayey soils and is almost always absent from soft clays and recent muds. And yet any infant is well aware that mud pies behave rather differently from sand pies, irrespective of their sharing effective friction graphs of identical form each with zero 'true cohesion'. The point about 'apparent cohesion' is that it *is* apparent. The paradox about 'true cohesion' is that it is usually *not* apparent, causing otherwise sober engineers to withdraw pocket magnifying glasses so that they may do it the honour of measuring it. It now seems to many that the high priests of the new religion of 'true cohesion' have been responsible for as many misunderstandings as were those of the old. Perhaps you, too, are beginning to lose your grasp?

Let us recap. Terzaghi's effective friction model, described in the last chapter, has been proved to be as applicable to natural clays as it is to natural sands. But although the strength of fine-grained soils can be explained by effective stress analysis when the crucial pore-water pressures are *measured*, they prove rather difficult to *predict*. In particular, the transient pore-water pressures generated in a mud pie when it is thrown at a wall are infinitely complex and yet the outcome is seemingly so reliable and simple. Is it not possible to include clay in that list of materials such as plasticene, butter, concrete and steel which have a unique strength if certain conditions are fulfilled? Many soils engineers affirm that this *is* possible and express the conditions thus.

If a homogeneous block of soil such as that depicted in figure 5.1 is forced to rupture along a slip surface while its void ratio is constrained to remain constant,

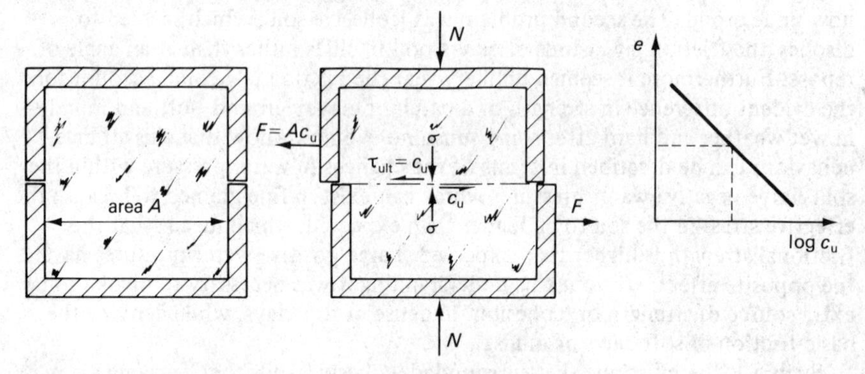

Figure 5.1 *Shearing soil at constant void ratio*

the ultimate shear stress τ_{ult} developed on the slip surface should be a function solely of that void ratio: it may be called the undrained strength c_u of the soil at that void ratio. The graph of e/log c_u is very often a straight line for soil samples that share a similar history but which possess different void ratios: this indicates that the undrained strength of soil usually increases exponentially as the void ratio decreases.

The undrained strength defined in these terms retains the strong predictive spirit of the old notion of cohesion and does nothing to contradict the friction model, although it is by no means as universal. Since the void ratio is supposed to remain constant during the rupturing process, this model should not be applied to partially saturated soils, in which air bubbles must compress according to Boyle's law if the total stress σ is allowed to increase. Neither can the void ratio remain constant for very long when a sand or silt is sheared in the field, even when it is saturated: water movement will take place fairly quickly and allow the soil skeleton to change its density of packing. With a fine-grained saturated soil of low permeability, however, the model will apply for an appreciable period to a block of soil under stress. If the soil skeleton is saturated, and the permeability is negligibly small, the void ratio of a soil element must be sensibly constant. The cohesion model then proclaims that until water is able to drain in or out the strength must remain constant.

This lends respectability to a very common design procedure. An engineer wishes to construct a spread foundation on a stratum of homogeneous clay. He is aware of an effective stress calculation based on angle of friction and water pressures but although he knows the long-term water pressures, he is unable to say how the pore-water pressures will react in the short term to the very foundation that he is planning. He therefore asks that samples of the clay be recovered from a number of boreholes, and that each sample be sealed immediately after being taken. He then instructs a laboratory to conduct an undrained shear test on each sample, probably by containing each block of soil inside a rubber sheath during the test. The engineer then invokes the cohesion model by assuming that the samples had remained at constant void ratio and that the clay in the field would also remain at the same constant void ratio, until his foundation was completed: the laboratory undrained shear strength should then apply to the clay in the ground immediately after construction.

If he were able to take good saturated samples of a sand or silt, and if he could test them at constant volume, the engineer could equally well report on their undrained strength. This would have less significance, however, because these permeable soils allow water movement to take place so rapidly that the changes in volume (which always tend to accompany changes in stress) can be completed in the field in hours or days. The soil would repack and the building settle as it was being built, so that the strength of the soil at its original void ratio would not be applicable at any future time. This problem is made more severe by the fact that the $e/\log c_u$ line for sand is very flat; a small change in void ratio offers a large change in strength. Only in the dynamics (earthquakes, explosions, traffic loads) of sands and silts might it be wise to consider the use of their undrained strength.

Before I demonstrate how to use the undrained strength of a soil in design, I must re-emphasise that this cohesion model is controversial. There are senior soil mechanicians who discount it completely. Some say that almost all 'clays' in the field are made so permeable by sand and silt partings that drainage always takes place fairly quickly. Others attack the whole philosophy of constant strength at constant void ratio by invoking the concept of grain 'structure' or 'fabric', which means that a soil could be constructed at the same void ratio but with a different geometry of packing to give it a different strength. A third group of critics believe that the undrained strength of a soil increases as the rate of shearing increases due to viscosity in the soil water system, so that the results of fast (1 hour or 1 day) undrained laboratory tests are unsafe when applied to rather slower (1 week or 1 month) field situations, which are nevertheless fast enough relative to the soil permeability to deserve the label 'undrained'. All these critics claim strong field and laboratory evidence gathered on particular soils to support them. Nevertheless, the majority of civil engineers believe that clay at constant volume can be talked of as having a shear strength, in exactly the same fashion as concrete or steel, although requiring a greater margin of safety. Since they believe it, act on it, and succeed in satisfying many clients and making a profit, it will be a brave student who ignores the cohesion model entirely! It will likewise be a brave man who relies unquestioningly upon it.

And quite apart from these informed criticisms there is the awful confusion generated by the terminology. I am presently tracing the powerful notion of 'undrained strength', which is often called 'apparent cohesion', but which ought perhaps to be called simply 'cohesion', and which gives the adjective 'cohesive' to those soils which can temporarily demonstrate it. I am saying nothing whatever about the weak and confusing notion of so-called 'true cohesion', defined as the intercept c' on an effective stress graph. I remarked in section 4.4 that the c' intercept might be a purely mathematical effect caused by the fitting of a tangent to a curve: in such cases I have advised that an acknowledgement that ϕ' may vary would be of greater value than the coining of a new variable. I also remarked that the cemented soil for which the word 'rock' has long been in use, can display an intercept c', and that the parallel discipline of rock mechanics should be of some value. I am afraid that you will have to try to remember that anyone referring to the presence of a great deal of 'true cohesion' is probably talking about rock rather than soil.

5

5.2 Application: A Foundation on Firm Clay

An engineer wishes to found a long wall on a deep stratum of firm clay. The typical finished cross section is depicted in figure 5.2. His desire to place the

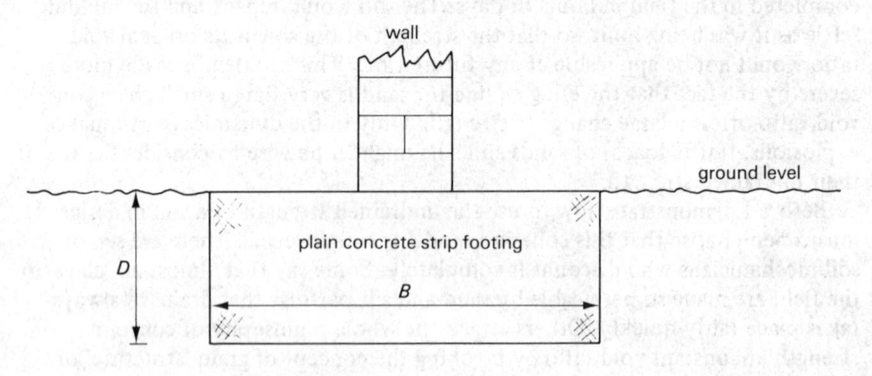

Figure 5.2 *Founding a wall on firm clay*

foundation below the zone of seasonal moisture variations will lead the designer to choose a depth D of the order of 1 m. The width B of the footing will depend on the load to be carried and the nature of the soil. Suppose that the wall will deliver a force Q per unit length to the foundation block and that the soil has been found to be homogeneous, with an undrained shear strength of c_u.

If he wishes to employ the cohesion model described here, then the designer must appreciate

(1) that a rupture mechanism has to be postulated

(2) that the soil may discover a failure mechanism which the designer failed to discover!

(3) that a margin of safety against failure in service must be provided, so that a design strength of c_u/F must be used where a value of 3 might be chosen for the safety factor F.

A simple slip-circle failure mechanism is depicted in figure 5.3. The foundation

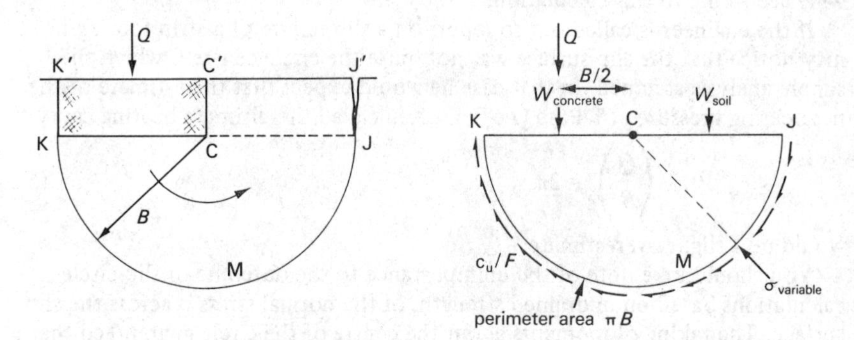

Figure 5.3 *Section of unit length through strip footing*

block CKK′C′ is shown as rotating about C, causing a semicylinder of soil CJMK to pivot about its axis C. The foundation sinks into the ground while a neighbouring patch of soil heaves. The strength of the soil is discounted in the seasonal variation zone above foundation level (that is, in block CC′J′J) so that the only forces acting on the semicylinder of soil for design purposes are Q the structural load, the weights W_{concrete} and W_{soil} of the two zones CKK′C′ and JCC′J′ respectively, and the summed stresses on the slip surface KMJ.

The engineer may now assume

(1) that as the soil slips, the shear stress on the rupture surface could in practice achieve the value c_u determined by previous undrained shear tests

(2) that the building will be erected so quickly relative to the speed of drainage of the clay that the undrained strength c_u is applicable to the post-construction stability of the strip footing

(3) that the eventual strength of the soil under the footing will be greater than the present strength c_u because water will tend to be expelled from the clay under the footing, which will cause a reduction in void ratio and an increase in strength (we shall see later that this is not necessarily the case)

(4) that the factor F will guard against small errors and uncertainties.

He may then take moments about the pivot C for a unit length of the semicylinder of soil CJMK. If he neglects the small difference between the unit weights of concrete and soil, then the moments of the blocks CKK′C′ and JCC′J′ will cancel, W_{concrete} equalising W_{soil}. He will then obtain

$$Q \times \frac{B}{2} = \frac{c_u}{F} \times \pi B \times B \tag{5.1}$$

$$B = \frac{FQ}{2\pi c_u} \tag{5.2}$$

An alternative formulation of this result is that the safe net bearing pressure Q/B is roughly equal to $2c_u$ if a safety factor of $F = 3$ is chosen. For a firm clay ($c_u = 50$ kN/m^2) the safe bearing pressure might be roughly 100 kN/m^2, so that a simple masonry wall 0.3 m thick and 10 m high, which would deliver a force $Q = 75$ kN/m along the footing, would require a foundation strip roughly 0.75 m wide according to this calculation.

If the engineer is called out to report on a slip failure of a strip footing he may notice that the slip surface was not quite the circle centre C which this simple analysis assumed. In that case he would expect that the estimate of the net bearing pressure at failure ($F = 1$), often called the ultimate bearing capacity

$$\left(\frac{Q}{B}\right)_f = 2\pi c_u \tag{5.3}$$

would be a slight overestimate.

You should take note of the unimportance to the outcome of slip circle calculations based on undrained strength, of the normal stress σ across the slip surface. The taking of moments about the centre of the circle guaranteed that σ had no effect since it had no leverage, so that the subdivision of σ into its components σ' and u becomes irrelevant.

5.3 Cohesion and Friction

The 'cohesion' of 'cohesive soils' which may be referred to by general civil engineers acquires the name 'undrained strength' in the hands of soils engineers. The concept of undrained strength rests on the (presumed) definite shear strength c_u of soils at constant volume. Small hand samples of clayey soil will require many minutes before they become dry (decrease in volume) or wet (increase in volume). They will therefore display their cohesion when remoulded in the hand, being described as soft (c_u 20–40 kN/m^2), firm (c_u 40–80 kN/m^2) or stiff (c_u 80–160 kN/m^2). As you will discover in chapter 7, larger blocks of soil will take much longer to drain, according to the square of their scale. Calculations based on the undrained strength of clayey soils may well be applicable many months after the strength was measured.

The evident 'friction' of 'granular soils' is known to depend on two concepts, effective soil stresses τ and σ' and friction angle ϕ'. The effective normal stress σ' is the total normal stress σ minus the water pressure u acting on the potential slip plane. Dry slopes of sandy soil will stand at roughly their effective friction angle ϕ'.

You must appreciate that the discrimination between 'cohesive' soils to which c_u calculations are often applied and 'granular' soils to which ϕ' calculations are often applied is one of grain size controlling the permeability and speed of drainage. Friction calculations can usefully be applied to clay if the water pressures are known: undrained strength calculations might be applied to silt or sand if the void ratio were known to be perfectly constant. In order to make clear the relationship between the undrained strength model and the friction model I shall set out below some results that would be achieved if *both* undrained and friction analyses were applied to *both* undrained and drained shear box tests on an idealised clayey soil which obeyed both models.

Consider a firm saturated clay with an ultimate undrained strength $(c_u)_{ult}$ of 50 kN/m^2 and with an ultimate effective friction angle ϕ'_{ult} of 27°, each measured by a series of shear box tests in which any early peaks of strength were ignored (see figures 4.6 and 4.7). We should assume that, having reached the ultimate plateau in all these shear box tests, the clay at the end of a test is in a constant condition and that no further changes in void ratio (in a drained test) or pore-water pressure (in an undrained test) were taking place when the 'ultimate' strengths were recorded.

If both models are valid then

$$\tau_{ult} = (c_u)_{ult} \tag{5.4}$$

and

$$\tau_{ult} = \sigma'_{ult} \tan \phi'_{ult} \tag{5.5}$$

so that

$$\sigma'_{ult} = \frac{(c_u)_{ult}}{\tan \phi'_{ult}} \tag{5.6}$$

or

$$(c_u)_{ult} = \sigma'_{ult} \tan \phi'_{ult} \tag{5.7}$$

Equation 5.6 indicates that in an undrained test in which $(c_u)_{ult}$ is fixed, the effective normal stress σ' on the rupture plane must also achieve a fixed value which depends only on the properties of the clay and which is completely independent of the applied normal stress σ. Equation 5.7 indicates that in a drained test in which σ'_{ult} is fixed and equal to the applied normal stress σ, the shear strength of the clay *when it has finished changing in volume* must also be fixed at a value dependent on the normal stress and completely independent of the initial undrained strength at the initial void ratio. These deceptively complicated statements may become clearer if numerical values are chosen.

Consider three undrained shear box tests on the firm clay, carried out with different loads on the hanger.

Test A No hanger, so the total normal stress σ must be zero. The ultimate strength $(c_u)_{ult}$ must be 50 kN/m^2, since the volume remains constant. So using equation 5.6 the ultimate effective normal stress σ'_{ult} must be 50/tan 27° or roughly 100 kN/m^2. The ultimate pore-water pressure on the slip plane

$$u = \sigma - \sigma'$$

must therefore be −100 kN/m^2. It is water pressures such as these that make friction calculations on clays difficult to apply to undrained problems.

Test B Weights provide

$$\sigma = 100 \text{ kN/m}^2$$

Once again

$$\sigma'_{ult} = 100 \text{ kN/m}^2$$

So

$$u_{ult} = 0 \text{ kN/m}^2$$

Test C Weights provide

$$\sigma = 200 \text{ kN/m}^2$$

Once again

$$\sigma'_{ult} = 100 \text{ kN/m}^2$$

So

$$u_{ult} = +100 \text{ kN/m}^2$$

It should now be clear that the soil can choose its own water pressure so that it can obey both the cohesion model and the friction model. Extra normal stress is simply resisted by equal extra water pressure in the saturated undrained soil.

Now imagine the results if the same tests had been carried out with full drainage. The water pressure in the soil would have been forced to remain at zero, on the assumption that the stressed block of soil was submerged in water which was itself at atmospheric pressure.

Test A' No hanger, so

$$\sigma = 0$$

But

$$u_{ult} = 0$$

So

$$\sigma'_{ult} = 0$$

So using the friction law $\tau_{ult} = 0$

Test B'

$$\sigma = 100 \text{ kN/m}^2$$

But

$$u_{ult} = 0$$

So

$$\sigma'_{ult} = 100 \text{ kN/m}^2$$

But

$$\tau_{ult} = \sigma'_{ult} \tan \phi'_{ult}$$
$$\approx 50 \text{ kN/m}^2$$

Test C′

$$\sigma = 200 \text{ kN/m}^2$$
$$u_{ult} = 0$$
$$\sigma' = 200 \text{ kN/m}^2$$

So

$$\tau_{ult} \approx 100 \text{ kN/m}^2$$

The clay in test A′ started with a strength of 50 kN/m^2 and finished with no strength at all: a great deal of water must have been drawn into the sample. The clay in test B′ did not change in strength, it must therefore not have changed in volume even though this was permitted. The clay in test C′ has grown stronger: water must have bled out of it, allowing the void ratio to fall and the strength to rise. If it had been possible to 'switch on' some drainage after the ultimate undrained conditions in test A had been obtained, then the tester would have seen the suction of −100 kN/m^2 slowly dissipate as water entered the sample: eventually the clay on the slip plane would have turned to mud. Likewise if drainage were allowed at the end of test C, the positive pore-water pressure would have been seen to drive water out of the clay into the surrounding reservoir, thereby dissipating itself as the clay doubled in strength. The clay in test B was remarkable in that its combination of undrained strength, angle of friction and applied normal stress led to it being volumetrically stable when it was sheared: perhaps you recognise this condition as a 'critical state'.

I should make clear that the tests described above have not actually been conducted. The conventional direct shear box test is not quite able to provide the data. Water pressures on the slip plane are not easily measured, and are not necessarily equal to those in the body of the sliding blocks above and below the slip plane. The plane itself is an abstraction: in fact a finite but irregular zone of shear disturbance is usually observed inside which the water movements due to shear take place. Most importantly drainage cannot be controlled in a conventional shear box: an 'undrained' test is simply carried out rather quickly while a 'drained' test is carried out slowly. The lack of precise control of drainage makes validation of the principles expounded above very difficult. Nevertheless, confirmation of the value in certain circumstances of the combination of cohesion and friction in the manner described has been obtained by many research workers using special shear testing apparatus.

My main objective in this example has been to demonstrate that it is possible to visualise undrained strength without making any reference to special interatomic forces. I have found it to be impossible to overstate the importance of grasping that what engineers of the older school describe as the cohesion of clay can be generated solely by friction between the particles. Truly cemented or bonded clays can be difficult materials to work with: a special class of very dangerous cemented clays is discussed in section 8.7, where I disclose that neither the simple friction model of chapter 4 nor the present undrained strength model has much relevance.

5.4 Cohesion and Critical States

The simple critical state model described in section 4.5 and summarised in figure 4.9 contains a version of 'cohesion' as a corollary: indeed the critical state model can be viewed as the superposition of elementary 'friction' and 'cohesion' models. Figure 5.4 illustrates the numerical example contained in section 5.3.

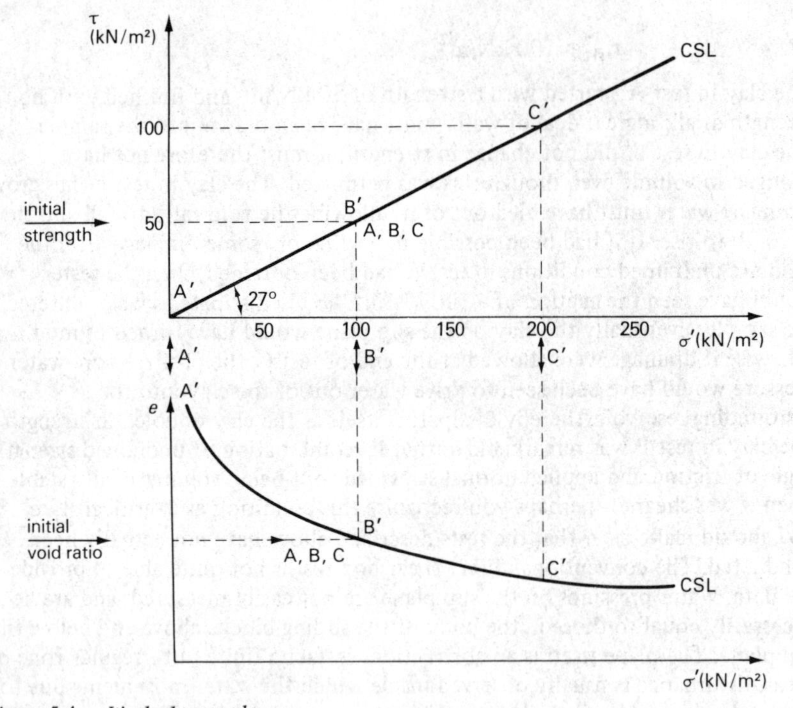

Figure 5.4 *Worked example*

Please work through the tests A, B, C, A′, B′, C′ once more, seeing how the critical state diagrams can help.

Tests A, B and C being undrained remain at constant void ratio. A horizontal line ruled on the e/σ' diagram at that void ratio shows that only one critical state and one strength, $c_u = 50$ kN/m^2 of course, is available. The friction line shows that the *effective* vertical stress σ' at failure is therefore 100 kN/m^2, using $\tan \phi'_c \approx 0.5$ since ϕ'_c is known to be 27° or so. The water pressure when the soil has reached an undrained critical state is simply the total vertical stress applied via the loading hanger, minus the effective vertical stress displayed on the diagram.

Tests A′, B′ and C′ being drained remain at constant (zero) pore-water pressure, so that the effective vertical stress equals the applied total vertical stress. Test A′ with $\sigma' = 0$ invites the soil to suck in so much water that the critical state can be at the origin of the friction line. The other drained tests provide strengths which can also be read off the τ/σ' diagram directly. If the equation of the e/σ' critical state line had been given, the volume of water draining into the soil in test A′ and out of the soil in test C′ would have been capable of prediction.

At the initial void ratio, an effective vertical stress of zero implies that the soil is in a state denser than critical (below the line). It must either attempt to dilate or to generate pore suction in order to reach a critical state (which it must if it is to shear a great deal). At the same void ratio, an effective vertical stress of 200 kN/m^2 implies that the soil is in a state looser than critical (above the line). It must either attempt to contract or to generate pore pressure in order to reach a critical state. These concise statements of simplified soil behaviour are summarised for the general case in figure 5.5. According to this simplified model

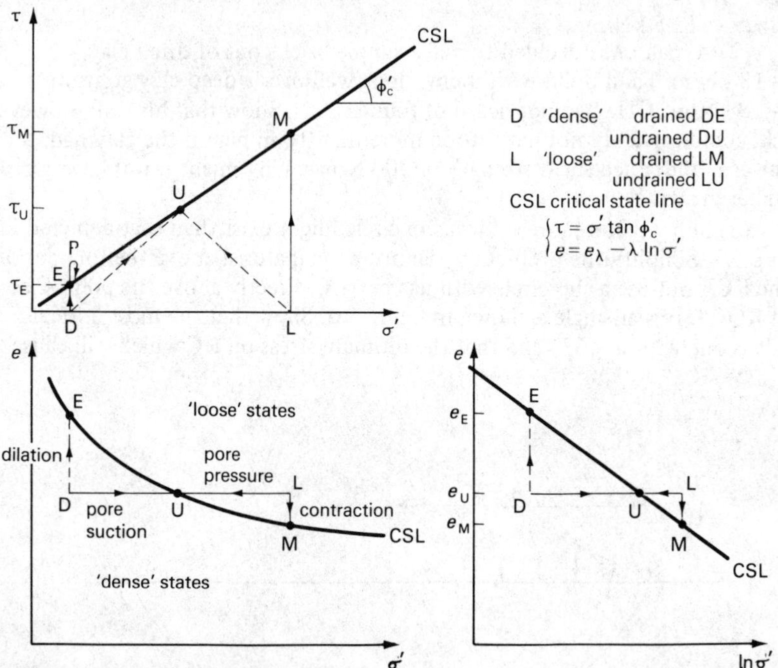

Figure 5.5 *Simple interpretation of soil strength*

the tests all eventually reach a critical state on the critical state line. For the undrained tests ($\nu = 0$) and the drained test on 'loose' soil ($\nu < 0$) the critical state represents the greatest shear stress that was achieved throughout the test.

Only the drained test on 'dense' soil ($\nu > 0$) would give a peak strength above the ultimate critical state, due to the dilatancy effect discussed in section 4.5.

It takes a good deal of effort to imagine that the provision of a very small effective confining stress is just as useful a way of producing dilatant soil as the provision of compaction to make it dense. It is, perhaps strangely, easier to understand that the provision of a very large effective confining stress is just as useful a way of producing 'contracting' soil as the provision of a very loose and open initial packing. You will find the struggle to understand figure 5.5 worth while, although you will discover in chapter 8 that even simple soils are more complicated than figure 5.5 implies.

5.5 Problems

(1) A man weighing 500 N, the sole of whose boot has an area of 0.02 m^2, goes hiking. He is attempting to climb up a 30° slope in soft clay and finds that when he lifts his back foot in order to move it forwards his front foot just slips backwards. Assuming that the concept is applicable here, what is the undrained strength of the clay on the surface of the slope? If the clay were uniformly strong throughout its depth, show that it cannot be deeper than 1.5 m if it weighs 19 kN/m^2.

Answer 12.5 kN/m^2.

(2) A man once decided to make simple bricks out of dried clay (γ = 18 kN/m^3) and build with them a long wall over a deep clay stratum (γ = 18 kN/m^3). He had not heard of foundations. Show that his wall, however thick, could certainly not have stood more than 10 m high if the clay had a uniform undrained shear strength of 30 kN/m^2. Why might it not have stood even so high?

(3) Show that a more critical slip circle might exist than that depicted in figure 5.3. Simplify the problem by ignoring the material above the foundation plane KCJ, but use a slip circle with its centre Q directly above the point C so that KQC forms an angle α shown in figure 5.6. Show that the most critical circle is one with α = 67° and that the ultimate stress on KC which will cause

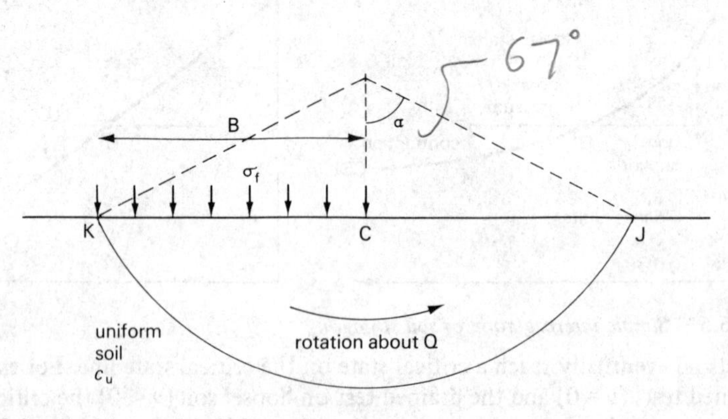

Figure 5.6 *Search for critical slip-circle*

this particular circle to rotate is only $5.7c_u$ rather than the $6.3\ c_u$ previously obtained.

(4) A summer sleigh was designed by an Eskimo for use on the recently unfrozen mud expanses. It possessed two runners each 3 m long and 0.01 m wide, and was to be pulled by six huskies, each of which could exert a thrust of 100 N. The Eskimo intended that the total mass of the sleigh and its load should be 300 kg. Complete adhesion can be expected between the runners and the mud. Confirm the Eskimo's experience that if the mud was soft the sleigh sank into it and that when it was firm enough not to sink the dogs couldn't pull it! Show that his brother's advice 'Make the runners broader' was completely unsound. Advise him on a remedial measure that would overcome the problem.

(5) In order to gain some understanding of the behaviour of a clayey sludge ($G_s = 2.70$) from a mining operation, an engineer conducted three slow shear box tests each over a period of 1 day and three fast shear box tests each over a period of 2 min, in a standard 60 mm square direct shear apparatus. In each of the two groups of tests the saturated and submerged samples were initially allowed to reach a drained equilibrium under vertical loads N of 10, 20 and 30 kg respectively, which were kept constant during the subsequent shear tests. The ultimate shear forces F_{ult} were recorded, and immediately afterwards a moisture content sample was taken from the centre of the rupture zone. The results were

	Slow			Fast		
$N(kg_f)$	10	20	30	10	20	30
$F_{ult}(kg_f)$	5.3	10.5	15.6	4.2	8.0	12.0
m	0.351	0.313	0.295	0.360	0.326	0.306

(a) Interpret these results in terms of a simple critical state model, declaring the critical state parameters ϕ'_c, λ and e_λ. Find the ratio c_u/σ'_0 for the undrained strength c_u of this soil when it has been subjected to one-dimensional virgin compression up to an effective stress σ'_0.

(b) Predict the likely undrained strength of the soil when at a moisture content of 0.25, and the likely bounds on the prediction.

(c) The sludge is to be allowed to settle in 10 m deep lagoons. Devise and use a numerical technique to determine the eventual profile of undrained strength with depth in the soil under each of two conditions of drainage:

(i) pore water is hydrostatic beneath shallow ponds on the surface (ii) the lagoon has shallow surface ponds but its base is drained to atmospheric pressure by a blanket of sand over gravel. (You will become aware that water merges indistinctly into soil, and that a working definition of the 'surface' is required. Why not use the criterion $e < e_\lambda$ to distinguish 'soil' from what is effectively dirty water: in which case imagine a dirty pond resting over 'soil' which starts with a void ratio $e = e_\lambda$.)

Answers (a) $\phi'_c = 28°$, $\lambda = 0.14$, $e_\lambda = 1.41$, $c_u/\sigma'_0 = 0.39$ (b) $c_u = 97$ kN/m^2 between 90 and 110 kN/m^2 (c) as follows

z(m)	1	2	4	6	8	10
(i) c_u(kN/m^2)	3	6	13	20	27	34
(ii) c_u(kN/m^2)	7	14	29	44	59	75

6

Small Strains

6.1 Modes of Deformation

We have reviewed models in chapters 4 and 5 which aim to describe the limiting statical equilibrium of soil bodies. When the inclination of the resultant effective stress (τ, σ') on a particular plane exceeds the appropriate friction angle ϕ', the soil concerned will divide, according to the friction model, into two rigid blocks sliding relative to each other on the rupture plane. The cohesion model predicts a similar catastrophe when the shear stress on a plane exceeds the undrained shear strength of the soil. If soil obeyed these models to the letter then every soil movement, whether landslide or foundation failure, would occur without warning. Perhaps fortunately, soil deforms a good deal before its limiting strength is reached: the shear tests depicted in figure 4.6 have already indicated this. Although the deformations leading to collapse are a useful warning of danger, quite small ground movements can be costly. While few clients will be concerned if an earth slope creeps at a rate of 20 mm/year, the same client would soon be hammering on the engineer's door if his foundations were settling at the same rate. Bricks, mortar, concrete and glass are brittle materials which crack when their supporting building settles by only a few centimetres. Soil deformation must, therefore, be predicted.

Solids exhibit only two fundamentally distinct modes of deformation: volumetric strain which is change of volume with constant shape and shear strain which is change of shape with constant volume. They are depicted for a unit cube in figure 6.1. Solid matter is hard to represent on two-dimensional pages, so

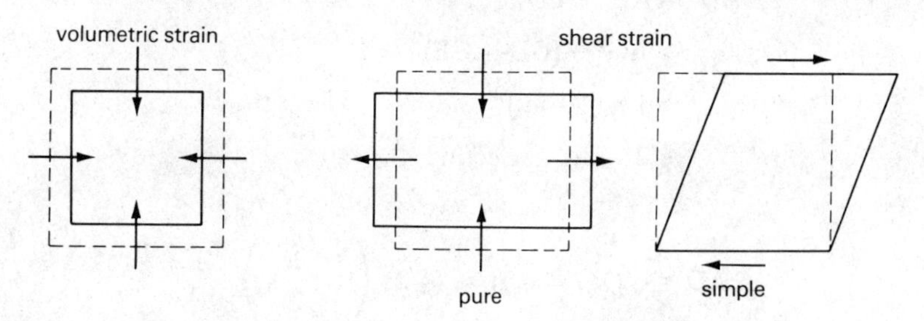

Figure 6.1 *Modes of solid deformation: views of one face of a unit cube*

figure 6.1 represents the view of only the front face of the cube. You must imagine that the third dimension of the cube also contracts in volumetric compression, and that all three pairs of faces of the cube can be subjected to shear strains pulling them out of square in the same fashion as the single face shown. You will see that the shear deformation has been characterised in two ways. The pure shear and simple shear modes are not independent but are simply views from different angles of different features on the same deformed face.

If you obtain a square rubber sheet and draw a square within it which is orientated at 45° to the boundaries of the sheet, as shown in figure 6.2, the

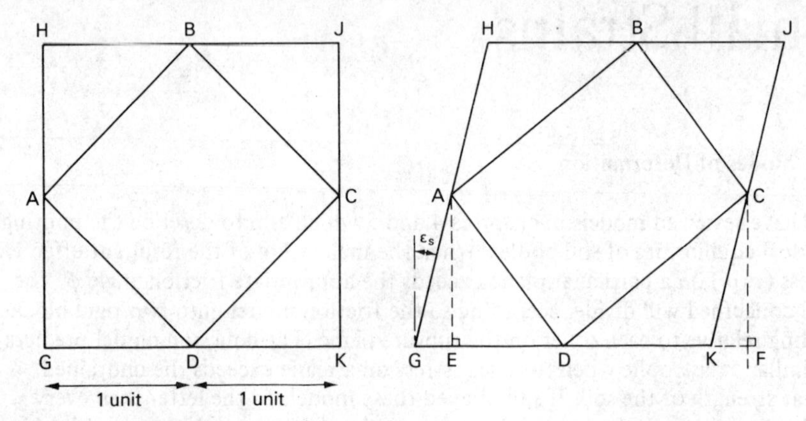

Figure 6.2 *Simple shear generating pure shear*

drawn square will be seen to exhibit pure shear if you deform the sheet in simple shear (and vice versa). If a simple shear angle ϵ_s occurs as shown, then AD shortens while AB lengthens, the angle DAB remaining 90°. The only pairs of orthogonal lines that can be drawn on the sheet so that they remain perpendicular when the sheet is subjected to simple shear are the lines running parallel to AB and AD. An alternative specification for the deformation ϵ_s in simple shear is, therefore, the linear strain (change in length/original length) of these so-called principal axes of strain AB and AD. If the length GD was originally 1 unit, then both AB and AD were $\sqrt{2}$. After the simple shear ϵ_s the lengths GE and FK must be 1 sin ϵ_s, which is identical to ϵ_s for small angles. Now

$$\begin{aligned} AD^2 &= AE^2 + ED^2 \\ &\approx AG^2 + (GD - GE)^2 \\ &\approx 1 + (1 - \epsilon_s)^2 \\ &\approx 2 - 2\epsilon_s \text{ neglecting } \epsilon_s^2 \end{aligned}$$

So

$$AD \approx \sqrt{[2(1 - \epsilon_s)]} \approx (\sqrt{2})\left(1 - \frac{1}{2}\epsilon_s\right)$$

Similarly

$$AB \approx (\sqrt{2})\left(1 + \frac{1}{2}\epsilon_s\right)$$

The principal compressive strain

$$\frac{\Delta AD}{AD} = \frac{\epsilon_s}{2}$$

and the principal tensile strain

$$\frac{\Delta AB}{AB} = \frac{\epsilon_s}{2}$$

Taking the sign convention compression positive (as we always do in soil mechanics, and already have done with stresses), the principal strains are $\epsilon_s/2$ and $-\epsilon_s/2$. In pure plane shear the principal strains must always have the same magnitude and opposite signs, otherwise some change of volume would be indicated.

The conditions for pure volumetric strain are indicated in figure 6.3, which

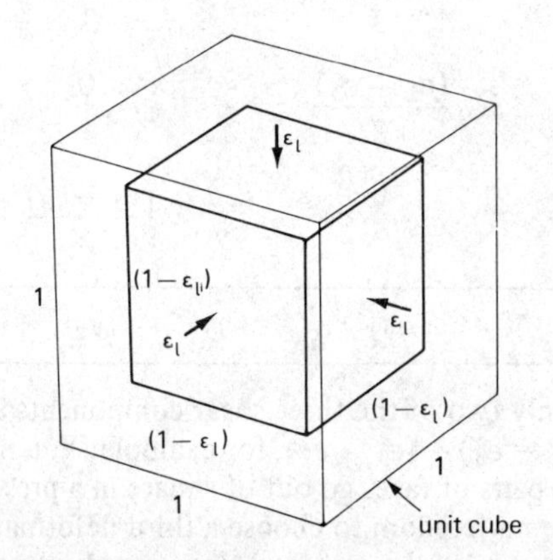

Figure 6.3 *Unit cube subjected to volumetric strain*

depicts a unit cube subjected to isotropic compression. If the linear strain (reduction in length/original length) is ϵ_l, it must be of this magnitude and sign in every direction.

$$\text{Initial volume} = 1 \times 1 \times 1$$

$$\text{Final volume} = (1 - \epsilon_l)(1 - \epsilon_l)(1 - \epsilon_l)$$

$$\text{Volumetric strain} = \frac{\text{reduction in volume}}{\text{original volume}}$$

So

$$\epsilon_v \approx 3\epsilon_l \text{ for small strains}$$

Any uniform deformation can be considered to be made up from independent components of volumetric strain and shear.

It is much easier to keep a three-dimensional strain calculation in good order if the elementary cube displaying the overall deformation of the body is initially orientated so that each of its faces will deform in pure shear; that is, if its edges are parallel to the principal axes of strain and therefore remain mutually perpendicular. By convention we refer to ϵ_1, ϵ_2 and ϵ_3 as the principal compressive strains in descending order of magnitude, so that ϵ_1 is the major compression of the 1-axis, ϵ_2 is the intermediate compression of the 2-axis, and ϵ_3 is the minor compression (often negative, indicating tensile strain) of the 3-axis. The linear strains of the sides of the carefully oriented cube can then be written $(\epsilon_1, \epsilon_2, \epsilon_3)$. Their fundamental components are shown in table 6.1. You

TABLE 6.1

Strain components	1-axis	2-axis	3-axis
Volumetric	$\frac{(\epsilon_1 + \epsilon_2 + \epsilon_3)}{3}$	$\frac{(\epsilon_1 + \epsilon_2 + \epsilon_3)}{3}$	$\frac{(\epsilon_1 + \epsilon_2 + \epsilon_3)}{3}$
Pure shear 1–2 plane	$\frac{(\epsilon_1 - \epsilon_2)}{3}$	$-\frac{(\epsilon_1 - \epsilon_2)}{3}$	0
Pure shear 1–3 plane	$\frac{(\epsilon_1 - \epsilon_3)}{3}$	0	$-\frac{(\epsilon_1 - \epsilon_3)}{3}$
Pure shear 2–3 plane	0	$\frac{(\epsilon_2 - \epsilon_3)}{3}$	$-\frac{(\epsilon_2 - \epsilon_3)}{3}$
Principal	ϵ_1	ϵ_2	ϵ_3

should notice that only two of the three shear components are independent, since $(\epsilon_2 - \epsilon_3) = (\epsilon_1 - \epsilon_3) - (\epsilon_1 - \epsilon_2)$, for example. You may wish to imagine a cube in which two pairs of faces go out of square in a prescribed fashion, leaving the third face no freedom to choose a third deformation. It is convenient to refer to the simple sum of the principal strains as the volumetric strain component

$$\epsilon_v = \epsilon_1 + \epsilon_2 + \epsilon_3$$

without using the divisor 3 which appears in the table. The reason is that if the sides of a unit cube are strained by small amounts ϵ_1, ϵ_2 and ϵ_3 respectively, the change in volume is indeed equal to their sum, repeating the logic which followed figure 6.3. Likewise, it is convenient to refer to the principal strain differences as components

$$\epsilon_s = \epsilon_1 - \epsilon_3$$

and so on, again omitting the divisor 3. Once again, this definition happens to hold a physical significance in that $\epsilon_1 - \epsilon_3$ is the greatest angular strain in the

1–3 plane, reversing the logic of figure 6.2. The value of table 6.1 is chiefly that it may enable you to see the concept of strain in a slightly different light.

The following are useful examples of the shorthand

(1) (0.01, 0, 0) Pure uniaxial compression in the 1-direction of 1 per cent, with no strains in the 2 and 3-directions. The volumetric strain is the sum of the linear strains, and is therefore 0.01. The whole deformation can be thought of in three components

volumetric	(0.0033, 0.0033, 0.0033)
pure shear in the 1–2 plane	(0.0033, –0.0033, 0)
pure shear in the 1–3 plane	(0.0033, 0, –0.0033)

(2) (0.01, 0, –0.01) Pure shear in the 1–3 plane with no movement in the 2-direction. The simple shear angle of the 45° lines in the 1–3 plane is 0.02 or 1.15°.

(3) (0.02, 0.01, 0) Pure shear in the 1–3 plane, that is, (2) above, with an additional 3 per cent of pure volumetric compression strain

volumetric	(0.01, 0.01, 0.01)
(2) above	(0.01, 0, –0.01)
total	(0.02, 0.01, 0)

(4) (0.02, 0, 0) Pure shear in the 1–3 plane, that is, (2) above, with a plane uniform contraction of the 1–3 plane

1–3 plane contraction	(0.01, 0, 0.01)
(2) above	(0.01, 0, –0.01)
total	(0.02, 0, 0)

Clearly (1) and (4) are identical strain patterns, although of different magnitude, so that their respective components are entirely equivalent. If you were engaged in a plane strain calculation, in which it was impossible for the material to move in the 2-direction, you might employ the components described in (4), omitting any mention of the intermediate strain ϵ_2. If, on the other hand, you had to handle a fully three-dimensional problem with a completely general theory, you might prefer to visualise the components of (1).

The special plane strain condition $\epsilon_2 = 0$ and the even more specific case of one-dimensional strain $\epsilon_2 = \epsilon_3 = 0$ are very significant to the engineer. If a newly constructed embankment or strip footing trench is so long compared with its cross-sectional area that it might be considered endless, then the associated soil cannot move lengthways along the 2-axis due to the requirement for symmetry. Why move to the right, if right and left are indistinguishable? A similar argument applies to a piece of soil shallowly buried under a very extensive and uniformly loaded foundation: if the north, east, south and west edges of the foundation are all effectively at infinity then only vertical movement is possible by symmetry, and the soil element must conform to one-dimensional compression. When dealing in plane strain it is tempting to refer to a plane contraction component

$(\epsilon_a, 0, \epsilon_a)$ as being 'volumetric' when it is nothing of the sort. Nevertheless, plane strains written as (ϵ_1, ϵ_3) with their two components $[(\epsilon_1 + \epsilon_3)/2, (\epsilon_1 + \epsilon_3)/2]$ and $[(\epsilon_1 - \epsilon_3)/2, -(\epsilon_1 - \epsilon_3)/2]$, and depicted in figure 6.4, will prove most useful, as you will shortly observe.

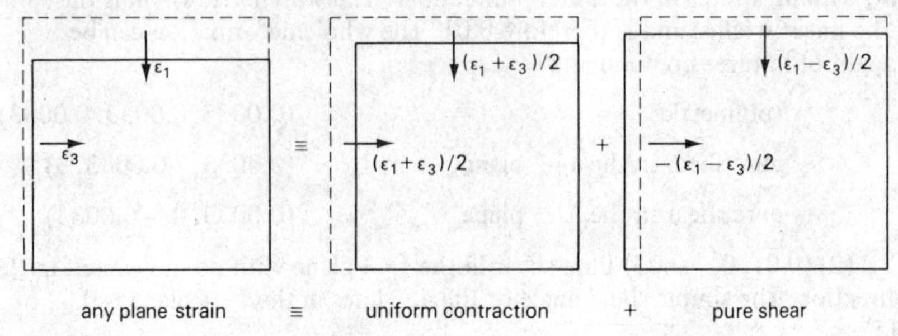

Figure 6.4 *Plane strain components*

6.2 Stresses

We have seen in earlier chapters that both shear stress τ and normal stress σ can act on any plane, and that in porous soil the normal stress σ can be split into two components u in the pore fluid and σ' carried through the soil grains. When we were considering the equilibrium of soil blocks with respect to their sliding along a rupture surface, we were able to treat each possible slip plane one at a time. If we are to associate the deformations of a body with the stresses acting on its boundaries, however, we must deal with a unit square in the case of plane deformation or a more general unit cube if the third dimension is relevant.

Consider the state of plane stress shown in figure 6.5*a*, supposing the element

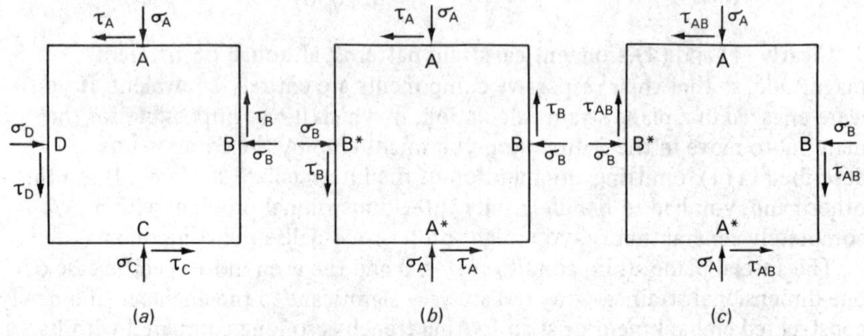

Figure 6.5 *A state of plane stress on a unit cube*

to be in static equilibrium. Our convention is

positive normal stress causes compression

positive shear stress causes anticlockwise rotation

Clearly the eight stresses cannot be independent. If the stress is uniform across the element, that is, along both the AC axis and the BD axis, then

$$\sigma_A = \sigma_C$$

$$\sigma_B = \sigma_D$$

$$\tau_A = \tau_C$$

$$\tau_B = \tau_D$$

leaving the components shown in figure 6.5*b*. The square can be seen to be in force equilibrium along the AA* and BB* directions, but a further condition on the shear stresses is found from taking moments in the AB plane. Each of the shear stresses τ_A act on unit areas, combining with a unit lever arm to form an anticlockwise couple of magnitude τ_A: shear stresses τ_B provide a further anticlockwise couple τ_B. If the square is in equilibrium

$$\tau_A + \tau_B = 0$$

or

$$\tau_B = -\tau_A$$

so that only three independent stress components are concerned in a plane stress problem. The reversal of sign is often expressed explicitly in the stress sketch as shown in figure 6.5*c*.

If the shear stress τ remains zero on the boundaries of a square element of an isotropic material, then the sides of the element must remain orthogonal. Similarly, if the normal stresses on the faces are always equal, then the element must remain square although it may change in area as the magnitude of the normal stresses changes. These considerations should lead us to propose separate stress components associated with volumetric and shear distortions. The analogues to the plane strains of figures 6.1 and 6.4 are depicted in figure 6.6.

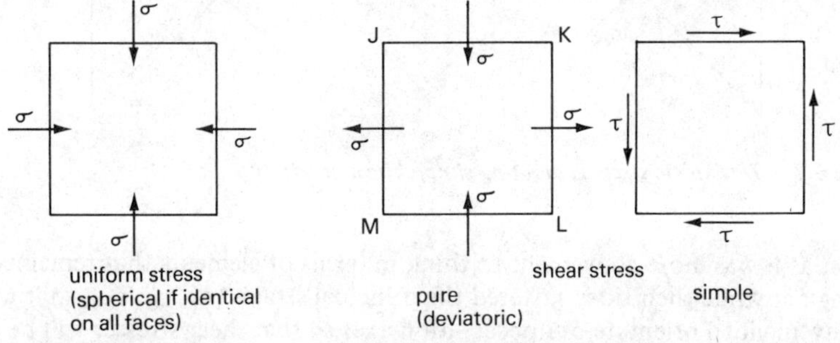

Figure 6.6 *Modes of stress: views of one face of a unit cube*

It will come as no surprise that the pure shear (more often called deviatoric) stress mode and the simple shear stress mode are equivalent ways of describing the same state of stress, but using elements rotated through 45° from each other. For example, consider the stress on diagonals MK and JL of the element JKLM subjected to pure plane shear stress in figure 6.6. They are redrawn for convenience in figure 6.7. Take triangle JKM and resolve the forces acting on it in the σ_{MK} direction

$$\sigma_{MK}\sqrt{2} = \sigma \times 1 \times \frac{1}{\sqrt{2}} - \sigma \times 1 \times \frac{1}{\sqrt{2}}$$

in the τ_{MK} direction

$$\tau_{MK}\sqrt{2} = \sigma \times 1 \times \frac{1}{\sqrt{2}} + \sigma \times 1 \times \frac{1}{\sqrt{2}}$$

so that $\sigma_{MK} = 0$, $\tau_{MK} = \sigma$.

Now take triangle JKL and resolve the forces acting on it in the σ_{JL} direction

$$\sigma_{JL}\sqrt{2} = \sigma \times 1 \times \frac{1}{\sqrt{2}} - \sigma \times 1 \times \frac{1}{\sqrt{2}}$$

in the τ_{JL} direction

$$\tau_{JL}\sqrt{2} = -\sigma \times 1 \times \frac{1}{\sqrt{2}} - \sigma \times 1 \times \frac{1}{\sqrt{2}}$$

so that $\sigma_{JL} = 0$, $\tau_{JL} = -\sigma$.

We have proved that the normal stresses $(\sigma, -\sigma)$ on the square element are equivalent to simple shear stresses $(\tau, -\tau)$ on the 45° planes, where $\tau = \sigma$.

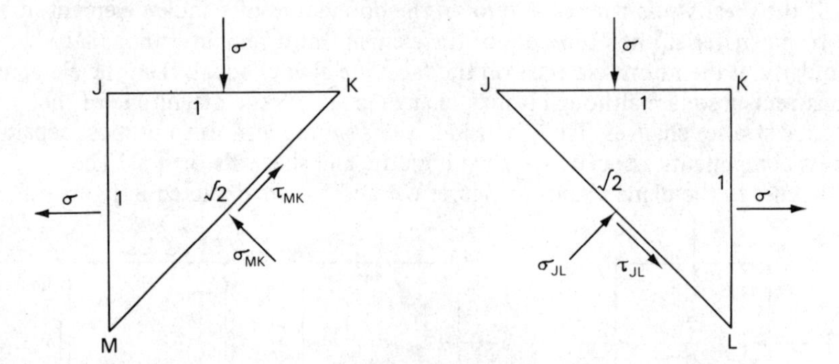

Figure 6.7 *Deviatoric stress generating simple shear stress*

Just as it was more convenient to think in terms of elements that remained rectangular while their sides suffered the principal strains $(\epsilon_1, \epsilon_2, \epsilon_3)$, so it will be convenient to orientate our observation axes so that shear stresses will be observed to be 'pure' rather than 'simple'. The orthogonal stresses are referred to as 'principal' stresses $(\sigma_1, \sigma_2, \sigma_3)$, and they must be split into the two components of principal effective stress $(\sigma_1', \sigma_2', \sigma_3')$ and pore-water pressure (u, u, u). The principal effective stresses are solely responsible for the deformations of the soil skeleton and can be thought of as being made up from the four components listed in table 6.2, analogous to the strain components of table 6.1. As with the strains, only two of the pure shear components are independent since any third can be seen to be the difference between the other two. Also, as before, special symbols are given to the more important groups

$$q' = \sigma_1' - \sigma_3'$$

as the analogue to

$$\epsilon_s = \epsilon_1 - \epsilon_3$$

and

$$p' = (\sigma_1' + \sigma_2' + \sigma_3')/3$$

as the analogue to

$$\epsilon_v = \epsilon_1 + \epsilon_2 + \epsilon_3$$

TABLE 6.2

Stress components	1-axis	2-axis	3-axis
Spherical	$\frac{(\sigma_1' + \sigma_2' + \sigma_3')}{3}$	$\frac{(\sigma_1' + \sigma_2' + \sigma_3')}{3}$	$\frac{(\sigma_1' + \sigma_2' + \sigma_3')}{3}$
Pure shear 1–2 plane	$\frac{(\sigma_1' - \sigma_2')}{3}$	$-\frac{(\sigma_1' - \sigma_2')}{3}$	0
Pure shear 1–3 plane	$\frac{(\sigma_1' - \sigma_3')}{3}$	0	$-\frac{(\sigma_1' - \sigma_3')}{3}$
Pure shear 2–3 plane	0	$\frac{(\sigma_2' - \sigma_3')}{3}$	$-\frac{(\sigma_2' - \sigma_3')}{3}$
Principal	σ_1'	σ_2'	σ_3'

The pore-water pressures are solely responsible for the flow of water, and being equal in every direction, can simply be referred to as u. Of course, the analyst usually arrives first at the total stress components

$$q = \sigma_1 - \sigma_3$$

$$p = (\sigma_1 + \sigma_2 + \sigma_3)/3$$

by virtue of resolving some forces over some areas, and has to remind himself to deduct pore pressures before continuing. You will see that

$$q' = q$$

$$p' = p - u$$

so that as with τ it is unnecessary to use the dash over q. It may be hard to remember that this rather formidable list of algebra is simply a string of definitions which engineers find useful. Note, for example, that the definition for p included the divisor 3 because it was convenient to speak of p as the average stress whereas the physical significance of volumetric strain ϵ_v allows its use as the analogue parameter without the divisor 3. The purpose of the chapter so far has been to reveal some of the underlying form of stresses and strains within a body: the main purpose of the remainder is to demonstrate how calculations are made.

Considering how difficult it has been to describe three-dimensional stresses and strains, it will not surprise you to learn that engineers usually limit their calculations to two dimensions, using plane stress and plane strain. This does not, of course, mean that square foundations oblige the designer by straining the ground other than in a complex three-dimensional fashion! Only the central portion of long, that is, almost endless, strip footings, embankments, slopes or trenches can really be said to deform in plane strain since lengthways movement would then suggest some sort of asymmetry. Even this does nothing to indicate that intermediate *stress* is irrelevant to the behaviour.

6.3 Mohr's Representation of Plane Stresses and Strains

To Otto Mohr we owe a simple visualisation of plane stress and strain, discovered as late as 1914, in which the equivalent states of stress and strain of elements of any orientation taken from a uniformly stressed and uniformly deformed plane are represented on a circle. It is normal to describe these diagrams in terms of the 1–3 plane upon which the greatest shear stresses act and of which the greatest

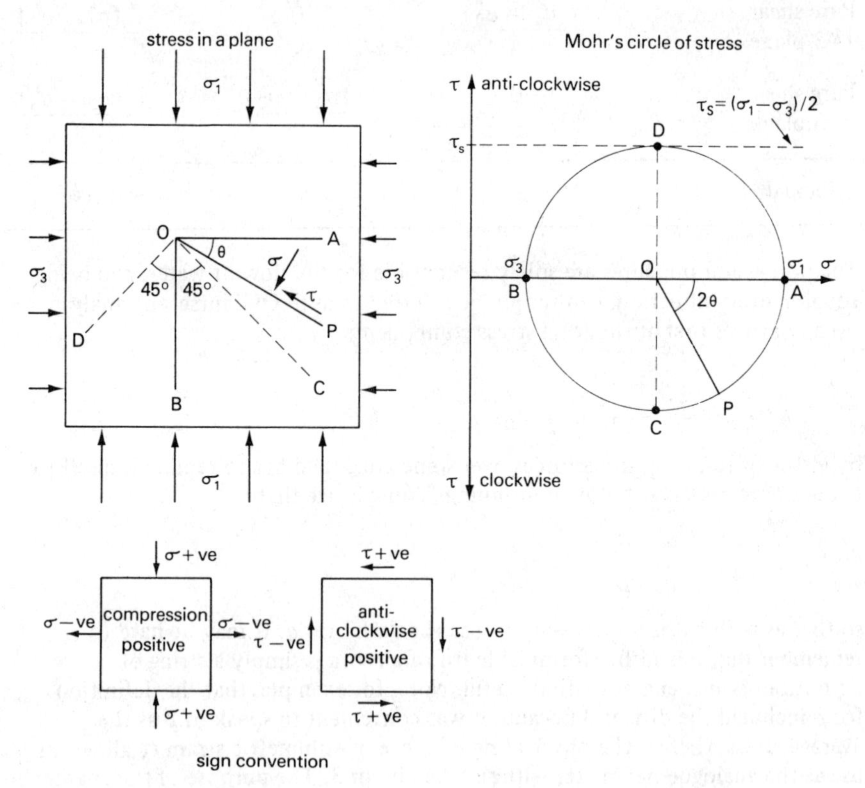

Figure 6.8 *Mohr's notation for stress: σ normal stress, τ shear stress*

distortion can be expected: it is important to remember that the extreme stresses and strains are allocated to the 1-axis (major) and 3-axis (minor). Mohr's stress diagram is shown in figure 6.8.

The perimeter of the circle ACBD contains all the stresses (τ, σ) which act across every hypothetical wound OP made in the 1–3 plane. Once the circle is drawn, it is simply necessary to have associated the stress on the circle (for example at A) with a particular line in the material (that is line OA): the stress on any other line OP which is θ clockwise from the line OA can be found by rotating an angle 2θ clockwise on the Mohr diagram. The minor principal stress is, of course, found by rotating 180° on the circle and 90° on the plane. We have already hinted that the axes of greatest shear stress τ_s are at 45° to the principal stress axes, and that $\tau = \pm\sigma$ on these 45° planes when $\sigma_1 = +\sigma$ and $\sigma_3 = -\sigma$: this is confirmed by the Mohr construction which, in that particular case, would be a circle of radius σ centred on the origin. The formal proof of the equation of the Mohr stress circle is given in most elementary textbooks on stress analysis; it is simply accomplished by repeating the analysis under figure 6.7, with general stresses acting on JK and JM, and with an arbitrary inclination θ replacing the 45° angle JKM.

Two important series of Mohr stress circles are drawn in figure 6.9,

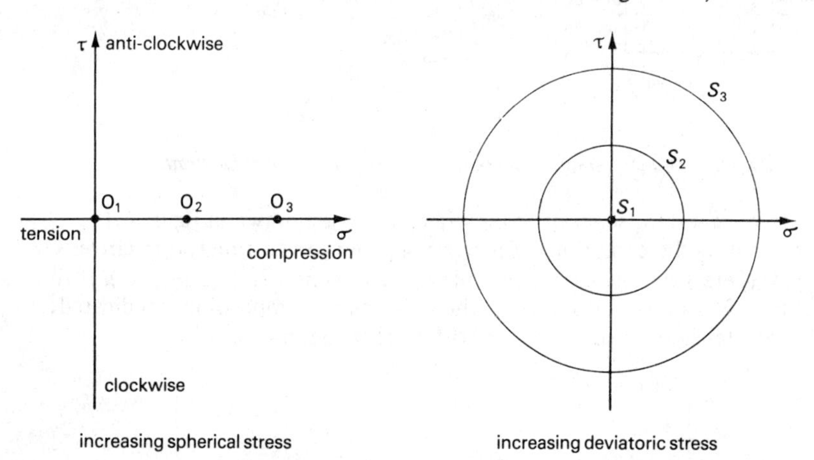

Figure 6.9 *Modes of increasing plane stress*

representing the conditions of increasing uniform plane stress (point circles O_1, O_2, O_3, etc.) and increasing pure plane shear stress (S_1, S_2, S_3, etc.). It is also clear that the movement of the centre of a Mohr stress circle indicates an overall rise in normal stress on every line in the plane, while the growth of its diameter indicates a rise in shear stress.

Mohr's strain diagram appears in figure 6.10, in which the circle joins points (ϵ, $\gamma/2$) which are the linear strains and angular strains of square figures drawn at various inclinations θ to the principal strain axis OA on the strained plane. We have already indicated the equivalence of strain pairs C, D and A, B in figure 6.2, where we showed that principal strains in pure shear were $\pm\epsilon_s/2$, corresponding to a Mohr circle of strain centred on the origin. ϵ_s is now seen to be the maximum value of the angular strain γ, at 45° to the principal axes.

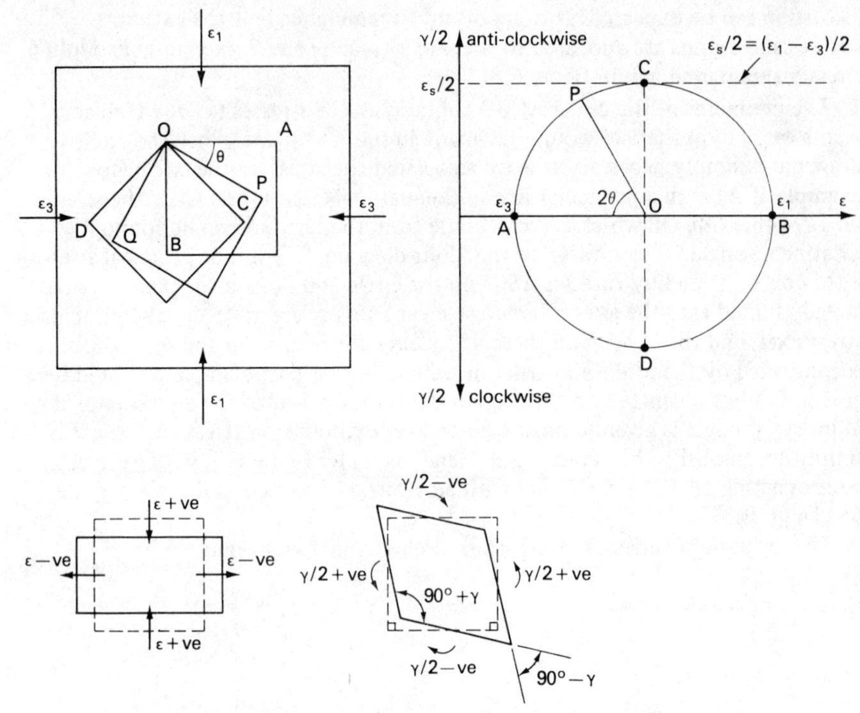

Figure 6.10 *Mohr's notation for strain: ϵ direct strain, γ angular strain*

Two important series of Mohr strain circles are drawn in figure 6.11, representing the conditions of increasing plane contraction (point circles O_1, O_2, O_3, etc.) and increasing pure plane shear strain (S_1, S_2, S_3, etc.). If the centre of a strain circle moves to the right, plane compression is indicated; if to the left, tension; shear distortion will cause it to grow in size.

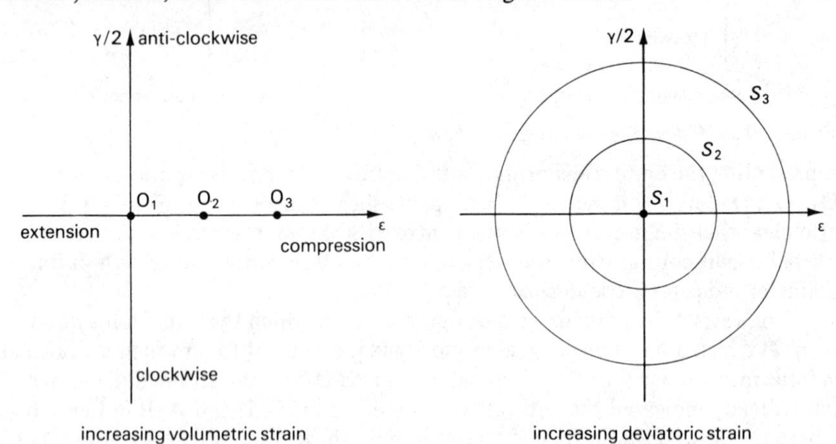

Figure 6.11 *Modes of increasing plane strain*

Mohr's representations of plane stress and strain can be used in a rather stilted fashion to describe on two-dimensional paper the nature of three-dimensional stresses and strains, by the expedient of drawing one circle for each of the 1–2, 2–3, 1–3 planes. Remembering that the magnitude of the principal stress or strain in the 2-direction must be intermediate between the other two, the outcome of drawing the circles always takes the general form shown in figure 6.12. It should be clear that the intermediate strain ϵ_2 is vital if the

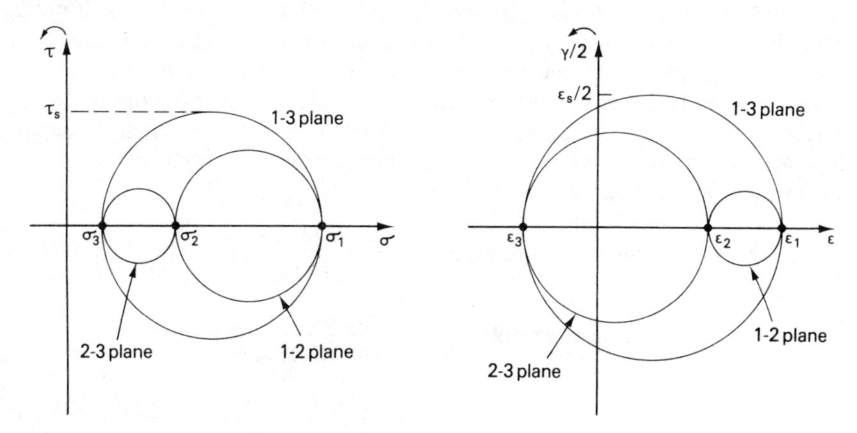

Figure 6.12 *Stress and strain circles for faces of principal cube*

change in soil density (and therefore volume) is important, since the volumetric strain $\epsilon_v = \epsilon_1 + \epsilon_2 + \epsilon_3$ contains ϵ_2. Likewise the intermediate stress σ_2, contributing as it does to the spherical stress component, must also be likely to have a bearing on the volumetric strain of a soil element.

The overwhelming majority of Mohr stress circles in soil mechanics will be drawn with respect to effective stresses τ and σ', and the effective principal stresses σ_1', σ_2', σ_3'. Only these effective stresses can cause a disturbance of the soil skeleton according to Terzaghi's principle.

6.4 Elasticity

Mohr's circle of strain contains the inevitable geometric relationship between lines drawn on a sheet of material which is uniformly deformed without tearing: such a relationship is called a condition of compatibility. Mohr's circle of stress represents the inevitable relationship between stresses on various planes cut through a sheet of material which is uniformly stressed and in static equilibrium: such a relationship is called a condition of equilibrium. The engineer wants to know the movement of ground when loads are applied. Loads can be converted to stresses, and movement is simply the integration of strains. What is yet required is the connection between stress and strain, the so-called constitutive condition of the deforming material being the third of the three conditions which must be fixed before problems of structural mechanics can be solved. The most well-known constitutive relations are based on the theory of elasticity.

The classical elasticity model allows the rate of volumetric strain per spherical stress to be different from the rate of shear strain per shear stress while providing for the principal axes of stress and strain to coincide. This sensible provision recognises the fact that the resistance a material can generate against spherical stress may have little to do with its resistance against shear deformation: consider water, which is almost as incompressible as it is deformable. A second condition on a proper elastic material is that it should not suffer a permanent deformation when the stresses acting on it are cycled and returned to their initial value. Finally, if the elasticity is linear, the rates of strain per stress remain constant up to any magnitude, so that the strains due to any group of stress increments can be superimposed. The most tangible features of the linear elastic model are its stiffness moduli E, which are the ratio of stress to ensuing strain in any particular mode of deformation. There are as many of them as there are modes but one pair is of fundamental significance

$$\text{bulk or volumetric modulus } E_v = \frac{\text{spherical stress } p}{\text{volumetric strain } \epsilon_v} \tag{6.1}$$

$$\text{simple shear modulus } E_s = \frac{\text{simple shear stress } \tau_s}{\text{simple shear angle } \epsilon_s} = \frac{\text{radius of Mohr stress circle}}{\text{diameter of Mohr strain circle}} \tag{6.2}$$

Two common elastic parameters which serve as an alternative to E_v and E_s are derived from the effects of uniaxial stress

$$\text{Young's modulus } E_y = \frac{\text{uniaxial stress}}{\text{corresponding axial strain}} \tag{6.3}$$

$$\text{Poisson's ratio } \mu = \frac{\text{negative lateral strain}}{\text{axial strain}} \tag{6.4}$$

A most useful property of these practical alternative parameters is that they can easily be used to generate the equations of three-dimensional principal strain

$$\left.\begin{aligned} \epsilon_1 &= \frac{1}{E_y}(\sigma_1 - \mu\sigma_2 - \mu\sigma_3) \\ \epsilon_2 &= \frac{1}{E_y}(-\mu\sigma_1 + \sigma_2 - \mu\sigma_3) \\ \epsilon_3 &= \frac{1}{E_y}(-\mu\sigma_1 - \mu\sigma_2 + \sigma_3) \end{aligned}\right\} \tag{6.5}$$

by superposing the effects of σ_1, σ_2 and then σ_3.

It will be of value later to be able to convert the practical parameters into the fundamental parameters. Consider volumetric compression, with $\sigma_1 = \sigma_2 = \sigma_3 = \sigma$ (say) and $\epsilon_1 = \epsilon_2 = \epsilon_3 = \epsilon$ (say). Using equations 6.5

$$\epsilon = \frac{\sigma}{E_y}(1 - 2\mu)$$

whereas equation 6.1 gives

$$E_v = \frac{\sigma}{3\epsilon}$$

so that

$$E_v = \frac{E_y}{3(1 - 2\mu)} \tag{6.6}$$

Now consider pure shear, where $\sigma_1 = \sigma$, $\sigma_2 = 0$ and $\sigma_3 = -\sigma$ while $\epsilon_1 = \epsilon$, $\epsilon_2 = 0$ and $\epsilon_3 = -\epsilon$. Equations 6.5 give

$$\epsilon_1 = \frac{\sigma}{E_y}(1 + \mu) = \epsilon$$

$$\epsilon_3 = \frac{\sigma}{E_y}(-\mu - 1) = -\epsilon \qquad \text{which checks}$$

$$\epsilon_2 = 0 \qquad \text{which checks}$$

whereas equation 6.2 gives

$$E_s = \frac{(\sigma_1 - \sigma_3)/2}{(\epsilon_1 - \epsilon_3)} = \frac{\sigma}{2\epsilon}$$

so that

$$E_s = \frac{E_y}{2(1 + \mu)} \tag{6.7}$$

Before escaping from the realm of algebra, one further mode of strain must be expressed in terms of the practical elastic parameters E_y and μ, namely one-dimensional compression. Consider the compression of an elastic material, rigidly confined at its sides so that $\epsilon_2 = \epsilon_3 = 0$. Define its axial stiffness in one-dimensional strain E_0 as σ_1/ϵ_1. Using equations 6.5

$$\epsilon_1 = \frac{1}{E_y}(\sigma_1 - \mu\sigma_2 - \mu\sigma_3) \tag{6.8}$$

$$0 = \frac{1}{E_y}(-\mu\sigma_1 + \sigma_2 - \mu\sigma_3) \tag{6.9}$$

$$0 = \frac{1}{E_y}(-\mu\sigma_1 - \mu\sigma_2 + \sigma_3) \tag{6.10}$$

From equations 6.9 and 6.10 we obtain

$$\sigma_2 = \sigma_3 = \frac{\mu}{(1 - \mu)}\sigma_1 \tag{6.11}$$

which, substituted in equation 6.8 gives

$$\epsilon_1 = \frac{\sigma_1}{E_y}\left[1 - \frac{2\mu^2}{(1-\mu)}\right]$$

which can be factorised thus

$$\epsilon_1 = \frac{\sigma_1}{E_y}\frac{(1+\mu)(1-2\mu)}{(1-\mu)}$$

so that

$$E_0 = E_y \frac{(1-\mu)}{(1+\mu)(1-2\mu)} \tag{6.12}$$

An interesting table of proportions has been drawn up in table 6.3. It may now be clear why only the range $\mu = 0$ to 0.5 was considered. If a material has

TABLE 6.3

Equation	μ	0	0.1	0.2	0.3	0.4	0.5
6.6	$\frac{E_v}{E_y}$	0.33	0.42	0.56	0.83	1.67	∞
6.7	$\frac{E_s}{E_y}$	0.50	0.45	0.42	0.38	0.36	0.33
6.12	$\frac{E_0}{E_y}$	1.00	1.02	1.11	1.35	2.14	∞
6.11	$\left(\frac{\sigma_3}{\sigma_1}\right)_0$	0	0.11	0.25	0.43	0.67	1.00

$\mu < 0$, then a wire made out of it would get wider as it was stretched while a compressed barrel would get thinner. If $\mu > 0.5$, an all-round increase in compression would result in an increase in volume. Well-behaved elastic materials must have intermediate Poisson's ratios, therefore. The majority of materials have μ in the range 0.1 to 0.3 (at which E_v/E_s ranges from 1 to 2) including steel, aluminium, concrete and many drained soils.

6.5 Elasticity Applied to Soil

The classical model of linear elasticity has been modified for use with soil in two important ways. Firstly, the presence has been acknowledged of pore fluid inside the elastic skeleton, and secondly the non-linearity of the stress–strain curves of a soil skeleton has been incorporated. Even with these major modifications, accurate predictions of small deformations are very difficult to make due chiefly to the lack of knowledge of the stiffness moduli as they change from point to point within a body of soil. Lack of isotropy (the stiffness varies with orientation) due to the bedding of the soil in horizontal layers is also blamed. If small deformations are poorly predicted, large deformations are quite unpredictable by these methods. Nevertheless, these quasi-elastic models are all that most engineers presently have to rely on in the prediction of the settlement of foundations.

Consider firstly the role of the pore fluid in an otherwise ideal linear elastic soil. If the full-scale soil construction is perfectly drained, so that the rate of loading of the soil body is so slow that pore-water pressures remain constant at their initial values, then the elastic analysis must proceed in terms of the effective stress parameters. All changes of stress due to loading ($\Delta\sigma$, $\Delta\tau$) will be absorbed by the soil as changes of effective stress ($\Delta\sigma' = \Delta\sigma$, $\Delta\tau' = \Delta\tau$) and the appropriate elastic parameters will be $E'_v = \Delta p'/\epsilon_v$ and $E'_s = \Delta\tau_s/\epsilon_s$ or E'_y and μ' measured in an appropriate suite of fully drained soil tests. If the soil were perfectly saturated and remained undrained at constant volume, then other parameters based simply on the change of total stress ($\Delta\sigma$, $\Delta\tau$) can be used. The volumetric stiffness E_v of a fully saturated soil with no drainage would be infinite were it not reduced by the finite volumetric stiffness of the pore water ($E_v = 2 \times 10^6$ kN/m^2) and soil solids ($E_v = 10^7$ to 10^8 kN/m^2), which nevertheless indicate very small strains when used in conjunction with typical engineering stress changes (0–1000 kN/m^2). The shear modulus with respect to changes of total stress E_s, on the other hand, must be identical to the shear modulus with respect to effective stress E'_s, since it is defined with respect to the difference in principal stresses (see equation 6.2) and $\sigma_1 - \sigma_3 = \sigma'_1 - \sigma'_3$. It can alternatively be pointed out that the laboratory test for the shear modulus might typically be a simple shear test at constant volume, the result of which could obviously be applied in an undrained field situation, extending the application to a drained situation only after reasoning that the shear stress in the test was an effective stress which could only be carried by the soil skeleton. If the drained parameters are E'_v and E'_s, therefore, the engineer's undrained parameters are $E^u_v = \infty$ and $E^u_s = E'_s$. Had the engineer been using the alternative parameters E'_y and μ', in the drained condition, he would have to use equations 6.6 and 6.7 in sequence to deduce that the appropriate undrained parameters would be

$$\left.\begin{aligned} \mu^u &= 0.5 \\ E^u_y &= E'_y \frac{1 \times 5}{(1 + \mu')} \end{aligned}\right\} \qquad (6.13)$$

The specially simple total stress parameters which can be used for elastic soil at constant volume should not be allowed to shake your belief in the power of effective stress analysis to determine soil strains in any and every situation. Once the effective elastic moduli E'_v and E'_s have been found, all elastic deformation problems can be solved if the changes of effective stress are known. If an increment of stress ($\Delta\sigma_1$, $\Delta\sigma_2$, $\Delta\sigma_3$) is applied to an elastic soil element so slowly that volume changes are effectively unresisted by the viscous drag of the moving water, then the process is 'drained', $\Delta u = 0$, and $\Delta\sigma'_1 = \Delta\sigma_1$, etc. The strains are then calculable using E_v' and E_s'. If, on the other hand, the increments of stress $\Delta\sigma$ are applied while the volume of the element is held constant, a pore-water pressure Δu will be generated. Fortunately Δu can be calculated from the zero-volume-change condition. Using the definition of the effective volumetric modulus E_v' it can be seen that for an 'undrained' elastic condition there must be no change of average effective stress

$$E_v' = \frac{\Delta p'}{\epsilon_v} = \frac{\Delta\sigma_1' + \Delta\sigma_2' + \Delta\sigma_3'}{3\epsilon_v}$$

and if

$$\epsilon_v = 0$$

then

$$\Delta p' = 0$$

but

$$\Delta p = \frac{\Delta\sigma_1 + \Delta\sigma_2 + \Delta\sigma_3}{3}$$

and

$$\Delta p = \Delta p - \Delta u$$

so

$$\Delta u = \Delta p = \frac{\Delta\sigma_1 + \Delta\sigma_2 + \Delta\sigma_3}{3} \tag{6.14}$$

The unsurprising conclusion is that in an 'undrained' soil the pore pressure will change exactly as the average total applied stress changes. The effective stress changes $\Delta\sigma_1'$, etc. are then known, but not needed because the volume was already known to be constant and the changes of shape are dictated by the shear stress equations, for example in the 1–3 plane

$$\epsilon_s = \frac{\tau_s}{E_s} = \frac{(\Delta\sigma_1 - \Delta\sigma_3)}{2E_s} \text{ or } \frac{(\Delta\sigma_1' - \Delta\sigma_3')}{2E_s'} \tag{6.15}$$

As I have already remarked, the shear distortion of a soil element is immediate and can be calculated as soon as the total stress changes $\Delta\sigma$ are known, no matter whether the soil tester had in mind to do a 'drained' test or an 'undrained' test.

Now consider the uncertainties caused by the evident non-linearity and hysteresis which can be displayed by small soil elements subjected to stress–strain tests. Turning to figure 6.13, you will see typical results for a one-dimensional compression test on a saturated sample of soil which was retained inside a steel ring and subjected to a uniform compressive strain by the application of vertical load, drainage being permitted through upper and lower porous stones as depicted in figure 6.14. This test is the most common stress–strain test used by foundation engineers in their task of estimating settlements. Commonly a sampling tube will be driven into the natural soil exposed in a trial

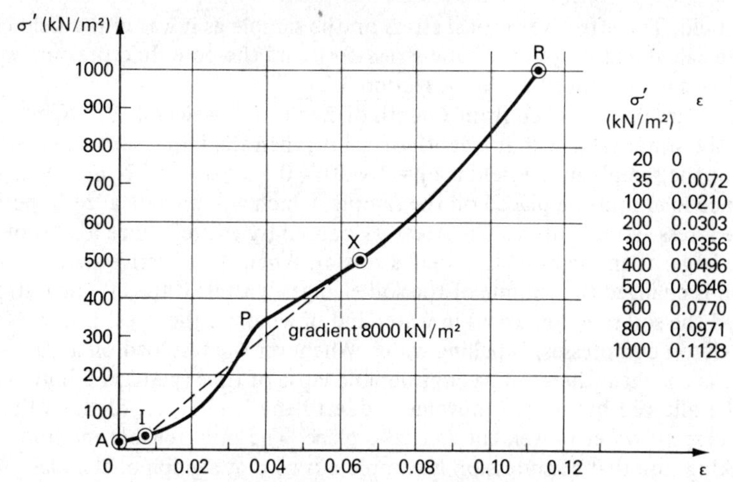

Figure 6.13 *One-dimensional stress–strain curve for a typical firm clay of low plasticity*

pit and lifted out with the soil in place. In order that the soil should remain in the tube it must be fairly clayey. The exposed ends of the soil sample can then be waxed to prevent drying, so that when it is subsequently jacked into the laboratory testing apparatus the sample ideally has the same void ratio as it had

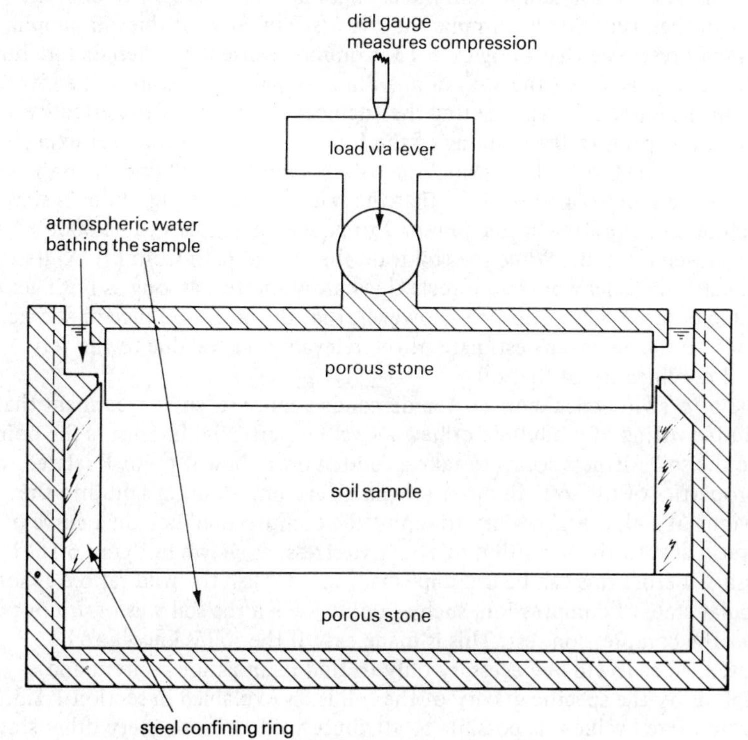

Figure 6.14 *Oedometer for one-dimensional compression*

in the field. The effective vertical stress on the sample as it was in the field can only be calculated if both the total stress on it and the pore fluid pressure within it are known: we looked at this in section 3.2.

If the sample was taken from a depth of 2 m in soil weighing 20 kN/m^3 and in which the water table was at a depth of 1.5 m, then the initial vertical effective stress of the sample in the field is $\sigma'_I = 2 \times 20 - 0.5 \times 10 = 35$ kN/m^2. A small reference load must be placed on the sample, which will provide a 'zero' point for the stress–strain curves. The stress σ'_A caused by the reference load should not be larger than the estimated field stress σ'_I. When the effective stress σ' on the soil is changed the volume of the soil changes with it: if the effective stress is reduced the soil skeleton draws in water and unwinds; if the stress is increased the skeleton compresses, expelling water. Whenever the test load on a clay sample is changed, therefore, a considerable lapse of time (perhaps 6 hours) must be allowed before the movement due to the load is recorded, so that the necessary water movement can take place. The only reliable method of making sure that compression is complete is to plot a graph of the dial gauge reading against time in order to determine when movement has ceased. Only then can the tester be sure that the pore water has attained zero gauge pressure, which is that of the surrounding water jacket: he can then say that the applied load divided by the surface area of the sample in the ring is equal to the effective vertical stress in the sample (that is, after drainage $\sigma' = \sigma = N/A$). Knowing the initial thickness of the sample and the changes in thickness due to load, the tester can then very slowly compile the stress/strain curve of the soil sample, which may resemble that in figure 6.13. Common sense must then dictate how to abstract a sensible stiffness E'_0 in one-dimensional compression. The pivotal point on the curve is I, representing the engineer's estimate of the effective vertical stress prior to the building of the foundation. If the engineer expects to increase the stress to σ'_X by 'asking' the soil to support a building, then the soil ought to strain up to a point X, so that the relevant stiffness modulus is simply the gradient of the straight line joining I to X, which happens to be 8000 kN/m^2 in the chosen example. While the soil took the curved path from I to X its mathematical model would go direct: this hardly matters as long as both get to the same point. You will observe, however, that the correct choice of stiffness depends on the engineer's estimate of the relevant stresses, due to the very marked nonlinearity of the soil.

The σ'/ϵ 'switchback' curve often demands a more sensitive treatment than the simple fitting of moduli described above. Of particular interest is the point P where the soil stiffness seems to take a sudden drop: how is point P related to the properties of the soil? In order to gain a very important insight into the behaviour of soil, it is necessary to replot the compression data on a curve of void ratio against the logarithm of effective stress, as shown in figure 6.15. It is necessary, before this can be accomplished, to establish the void ratio e at some particular state of compression, such as point I when the soil was *in situ* or point R after the compression test. This is made easy if the soil is known to be saturated, since it will be necessary only to find its moisture content and multiply it by the specific gravity of the solids, as explained in section 1.1.3. From that fixed value it is possible to attribute void ratios to every other state of compression. Assuming that the sample was L^* in thickness when the void

$$\frac{1+e}{L} = \frac{1+e^*}{L^*} \qquad (6.16)$$

ratio was e^*, then when it was L in thickness the void ratio must be e where by the definition of void ratio, in this case of the one-dimensional compression of a fixed amount of soil solids. You will always find reference to table 1.1 useful in such cases.

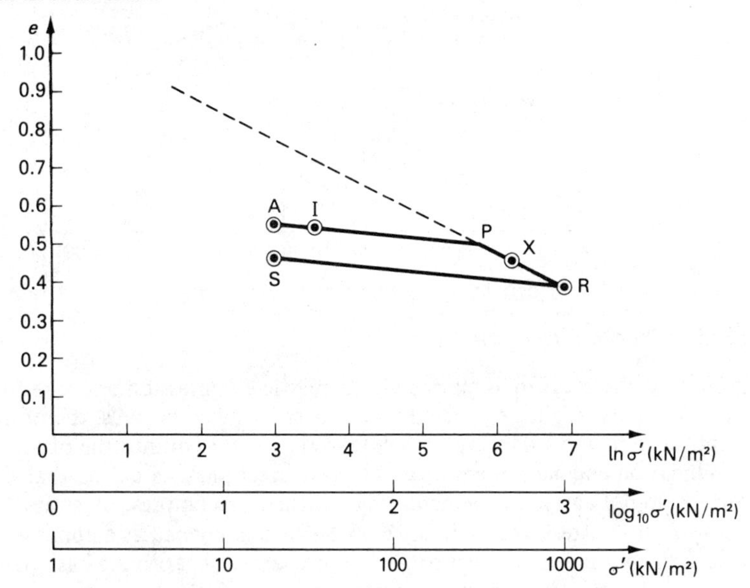

Figure 6.15 *The e/ln σ′ plot for one-dimensional compression*

It is remarkable that the $e/\ln \sigma'$ graph almost invariably linearises the compression curve of soils of whatever type. Section AIP of the curve in figure 6.15 is roughly elastic (although logarithmic rather than linear) in the sense that a cycle of stress from σ'_A to σ'_P and back to σ'_A would evoke very little permanent change of void ratio. At P the soil yields, and the section PXR is one of plastic compression of volume which cannot be recovered by a simple relaxation of stress, so that the point S would have been reached after point R if the vertical stress had been reduced to its original level. Figure 6.16 brings together the concepts and definitions that have been invoked to describe this elastic/plastic behaviour. The plastic compression line is usually called the 'virgin' or 'normal' consolidation line. It is the locus of state points (e, $\ln \sigma'$) representing the slow on-going compression of a soil starting off as mud (in the region $e \approx e_{\lambda_0}$, $\sigma' \approx 1$ kN/m^2) and ending as a very densely packed soil (in the region $e \approx 0.3$). Above this line is a large 'impossible' zone: if soil is artificially forced into a higher void ratio for a given stress than that indicated by the virgin consolidation line it merely collapses back on to the line. If, at any point such as P, the process of compression is reversed the soil expands along the line PA, which is a roughly elastic swelling and recompression line. Soil in this lower elastic region is said to be 'overconsolidated', and to have a 'precompression stress', σ'_P in this example. If the stress is kept below σ'_P the behaviour is crudely

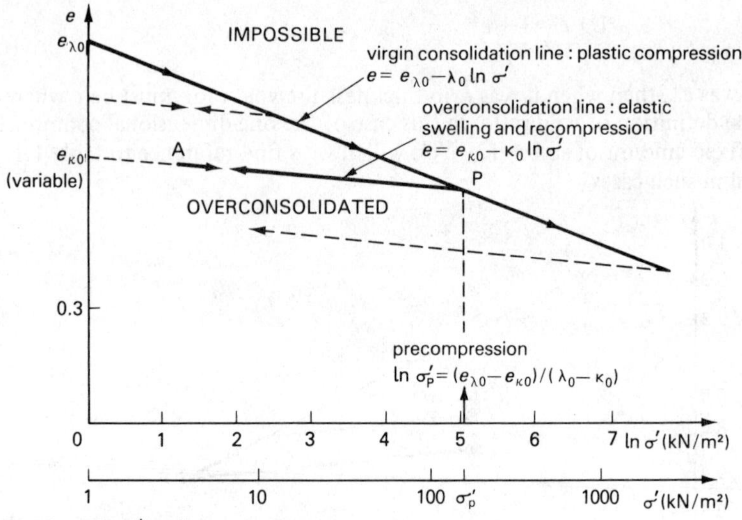

Figure 6.16 *The e/ln σ' plot: definitions*

logarithmically elastic: if σ'_P is exceeded, the plastic compression line is resumed. While elastic theory might be tailored to suit overconsolidated soils, therefore, it would appear to be most dangerous to extend this treatment into the plastic zone. Foundation engineers are prepared to take great pains to ensure that their foundations do not carry soil elements beyond their precompression stress.

The logarithmic stress axis of figure 6.15 unfortunately causes a number of confusions, since it has no perfect origin and no perfect logarithmic base. When $\sigma' = 0$, $\ln \sigma' = -\infty$, which seems to suggest that if all the effective stress could be removed from a soil element, then the void ratio of that element would be indefinitely large. This is somewhat of an overextrapolation. Nevertheless, sand on the sea-bed is unstressed and is certainly easily disturbed: the void ratio in the vicinity of the top grain is also very large! A further demonstration comes from what is known as the slaking test, when a soil sample is totally immersed in a bucket of water. All well-behaved soils, whether sands or clays, turn fairly quickly into a slurry. Those that retain a dense and strong configuration even at zero effective stress should be classified as rocks, since they clearly possess some sort of intergranular cement to hold them together. It is, as a matter of fact, very difficult to confine a soil under zero effective stress owing to the weight of overlying material: 10 mm of overlying soil is worth perhaps 0.1 kN/m^2, while a 3 mm steel lid is worth 0.25 kN/m^2. For this reason it is rarely meaningful to talk of effective stresses smaller than 1 kN/m^2.

The magnitudes of the flexibility parameters λ_0 and κ_0, which are the gradients of the plastic and elastic one-dimensional compression lines, are significant. Although the swelling lines on the $e/\ln \sigma'$ graph are steeper at higher void ratios, they are often idealised as being parallel and at a slope $\kappa_0 \approx \lambda_0/4$. The plastic flexibility index λ_0 may take the following values

$\lambda_0 > 0.30$ 'very highly plastic' or 'very highly compressible' clay

$0.30 > \lambda_0 > 0.15$ high plasticity clay

$0.15 > \lambda_0 > 0.075$ medium plasticity clay

$0.075 > \lambda_0$ low plasticity clay, silts and sands

If the one-dimensional elastic swelling lines were to conform exactly to

$$e = e_{\kappa_0} - \kappa_0 \ln \sigma' \tag{6.17}$$

then the one-dimensional stiffness modulus E'_0 could easily be deduced for an infinitesimal stress increase $d\sigma'$

$$E'_0 = \frac{d\sigma'}{d\epsilon} = 1 \bigg/ \left(\frac{d\epsilon}{d\sigma'}\right) = 1 \bigg/ \left(\frac{d\epsilon}{de}\frac{de}{d\sigma'}\right) \tag{6.18}$$

The rate of one-dimensional strain with change in void ratio is deduced from the definition of void ratio

$$d\epsilon = \frac{dL}{L} = -\frac{de}{1+e}$$

where de is an incremental increase of void ratio, so that equation 6.18 becomes

$$E'_0 = -(1+e) \bigg/ \left(\frac{de}{d\sigma'}\right) = \frac{(1+e)\sigma'}{\kappa_0} \tag{6.19}$$

after the differentiation of equation 6.17. The stiffness E'_0 of the soil is inversely proportional to its flexibility index κ_0. If the elastic compression line on the $e/\ln \sigma'$ graph had a gradient $\kappa_0 = 0.05$, for example, this would indicate $E'_0 = 20\,(1+e)\sigma'$, so that in the region $e = 0.5$ the stiffness would be roughly proportional to the effective stress, $E'_0 \approx 30\sigma'$. Since the vertical effective stress in ground increases almost linearly with depth because of gravity, the stiffnesses of most soils also increase linearly with depth. Sometimes, in a shallow zone, it is convenient to assume that the stiffness is constant. On the other hand it is sometimes necessary to acknowledge the hardening of soils with depth and to use piles or piers to search out deep, stiff strata to support heavy loads.

You will have become aware that the one-dimensional compression test as I have described it gives a measure of only one elastic parameter, E'_0. Since the steel confining ring is rarely instrumented to measure the lateral pressure increment accompanying the vertical increment, equation 6.11 cannot be used to determine μ'. This makes it impossible to deduce the parameters E'_y, μ', E'_v or E'_s. It is conventional, however, to assume that μ' for any soil is in the region 0.10 to 0.30. From the proportions given in table 6.3 it then follows that, once E'_0 has been determined for a particular range of effective stress

$$\left.\begin{aligned} E'_y &= 0.85\,E'_0 \pm 15 \text{ per cent} \\ E'_v &= 0.50\,E'_0 \pm 20 \text{ per cent} \\ E'_s &= 0.35\,E'_0 \pm 25 \text{ per cent} \end{aligned}\right\} \tag{6.20}$$

6.6 Uniform Compression: An Example

The simplest application of elasticity and the one-dimensional compression test is to the case of uniform compression in the field, such as that imposed by the dead weight of a layer of compacted fill on the underlying soil. Figure 6.17

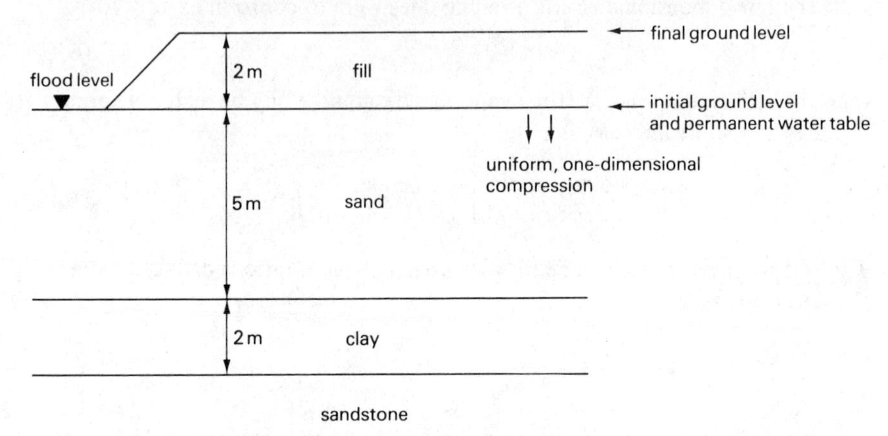

Figure 6.17 *Uniform compression: an example*

depicts a land reclamation scheme in which fill had to be placed over a 10 km^2 site that was presently subject to flooding. Five metres of medium-dense sand overlay 2 m of firm clay above a sound base of sandstone. A civil engineering consultant had been asked to supervise a scheme for placing fill to a finished level of 2 m above top sea level, which coincides with the original ground level. The consultant appreciated that if the natural soil were to settle 0.2 m, for example, due to the weight of fill, then a total of 2.2 m of fill would have to be placed rather than 2 m. The engineer decided to conduct a small site investigation with the purpose of estimating the compression of the ground.

Firstly, a 100 mm diameter borehole was shelled out inside steel casing which was driven ahead in order to prevent the hole collapsing. The chips of soil from the advancing shell were sufficient to assess the crude nature of the soil, but a standard penetration test (SPT) was additionally conducted for every metre advance through the sand. This consisted of counting the number of blows needed to hammer a special thick-walled tube into the sand below the extending borehole. The engineer in charge knew that this exercise would have been wasted if the sand in the region of the test had been disturbed, and so he took particular care to see that the borehole was fully charged with water at all times and that the shell of excavated soil was raised rather slowly immediately preceding an SPT. If the borehole had been kept with a lower water level than that in the ground, a steady seepage into the bottom of the steel casing would almost certainly have reduced the sand to a boiling slurry. Although the sand underneath the tube was relatively undisturbed, the sample recovered in the SPT tube certainly was greatly deformed and would have changed in density. Nevertheless, these disturbed samples were kept in bags and subsequently brought back to the laboratory. When the clay was encountered and recognised a

100 mm sampling tube took an 'undisturbed' sample which was waxed and sent back to the laboratory. Shortly afterwards the sandstone was encountered and its depth recorded. This general procedure was repeated a number of times at different locations. Figure 6.17 depicts the site of the borehole at which the clay was of the greatest thickness, while the sand was at its loosest with an SPT number of 10 to 15, indicating, according to Terzaghi and Peck who popularised the test, that it was medium dense.

The engineer chose this location for a crude settlement analysis. Firstly, he guessed that the unit weight of compacted fill would be 20 kN/m^3 so that the surcharge pressure would be 40 kN/m^2 for a 2 m layer. He recognised that the long-term water table would remain at the present level, so that the increase in vertical pressure of 40 kN/m^2 would eventually be seen overall as an increase in effective stress. Under the uniform blanket of fill no lateral strain would be possible by symmetry: the conditions would be the one-dimensional ideal if the inevitable spacial variation in soil properties was ignored. The engineer decided to base his estimate of settlement on one-dimensional compression tests.

The SPT number had indicated that the sand was medium dense on the scale of relative densities, which the engineer took to mean that its void ratio was roughly half way between the loosest possible packing he could devise artificially in the laboratory and the densest. He therefore chose a void ratio at which to conduct tests on the disturbed bag samples of sand. In order to make sure that the sand was overconsolidated, which he felt it must be, he initially cycled the pressure in the machine through 200 kN/m^2 and back to 10 kN/m^2. He then increased the stress up to 90 kN/m^2 in increments of 20 kN/m^2, recording the strain. He had expected to apply the results of the 10–50 kN/m^2 range to the upper zone of sand and the 50–90 kN/m^2 range to the lower zone. When he inspected the results, however, he found that all the moduli fitted within the bounds

$$E_0' = 40\,000 \text{ kN/m}^2 \pm 30 \text{ per cent}$$

which he felt was sufficiently tight to be applied to the whole depth of sand.

The clay sample left less scope for artistic interpretation! The average range of effective stress had to be 60 kN/m^2 to 100 kN/m^2, taking the unit weight of all soils as 20 kN/m^2 and making allowance for water pressures. When tested, a modulus

$$E_0' = 4000 \text{ kN/m}^2$$

was recorded.

Now the settlement of the surface was composed of the separate compressions of the sand and clay

$$\begin{aligned}
\rho &= \rho_{\text{sand}} + \rho_{\text{clay}} \\
&= \epsilon_{\text{sand}} \times 5 \text{ m} + \epsilon_{\text{clay}} \times 2 \text{ m} \\
&= \frac{\Delta\sigma'}{E_{0\,\text{sand}}'} \times 5 + \frac{\Delta\sigma'}{E_{0\,\text{clay}}'} \times 2 \\
&= \frac{40 \times 5}{4 \times 10^4} + \frac{40 \times 2}{4 \times 10^3} \\
&= 5 \text{ mm} + 20 \text{ mm}
\end{aligned}$$

So $\rho = 25$ mm, which is negligible compared with likely irregularities of the terrain and errors of surveying. Nevertheless, 25 mm of fill over 10 km^2 amounts to 250 000 m^3!

Only on completion of the calculation is it obvious that the clay provides the majority of the settlement, thereby justifying the engineer's relatively crude estimate of the modulus of the sand. The difficulty of recovering good undisturbed sand samples makes *in-situ* testing an attractive alternative. At best the engineer is then performing a model foundation experiment rather than attempting to discover a fundamental soil modulus. At worst, he gets hopelessly bogged down in a welter of empirical fiddle-factors. These issues are discussed later, in the context of foundation design.

6.7 Isolated Loads on the Surface of an Elastic Bed

In 1885 Boussinesq determined the stresses and strains within a homogeneous isotropic linear-elastic half-space due to the application of a point load on its surface. This feat of applied mathematics was achieved by simultaneously satisfying the dictates of equilibrium, compatibility, and the elastic constitution of the material, at all points within the infinite 'bed'. The formal derivation is quite outside the scope of this book but can be found, for example, in Timoshenko and Goodier (1951). Boussinesq's solution immediately provided, by integration, the stresses and strains due to every pattern of applied load. Furthermore, the method has been extended to include the provision of differing elastic layers in the bed, stiff loading pads, and the application of general loads and moments within the bed or on its surface. Problems to which algebraic solutions are not yet available have been solved numerically. Many of these solutions appear in detail in the very useful compilation of Poulos and Davis (1974), and a few are condensed below.

Many shallow pad footings can be crudely represented by a rough, rigid, circular pad of diameter B resting on the surface of an infinite homogeneous elastic bed, the behaviour of which has been elucidated by a number of applied mathematicians whose results are listed by Poulos and Davis. The vertical movement ρ_v due to vertical force Q_v, the horizontal movement ρ_h due to horizontal force Q_h, and the rotation θ due to moment M are given by

$$\rho_v = \frac{Q_v}{E_y B}(1 - \mu^2) \approx \frac{Q_v}{E_y B} \qquad (6.21)$$

$$\rho_h = \frac{Q_h}{E_y B}\,\frac{(7 - 8\mu)(1 + \mu)}{8(1 - \mu)} \approx \frac{Q_h}{E_y B} \qquad (6.22)$$

$$\theta = \frac{6M}{E_y B^3}(1 - \mu^2) \approx \frac{6M}{E_y B^3} \qquad (6.23)$$

the reduced approximations holding to within 25 per cent irrespective of Poisson's ratio. To within the same degree of approximation, the relationship for settlement is equally valid for a square footing of side B or a rectangular footing X by Y of equivalent diameter $B = (XY)^{1/2}$. Reference should be made to

Poulos and Davis (1974) for better numerical accuracy or for the effects of extreme aspect ratio X/Y, the burial of the footing, the limited depth of the soil, or the imposition of moments or other actions on bodies of various shapes. You may notice that the function of the theory in this case is only to supply the numerical constant. It was inevitable in a linear elastic material that the settlement ρ_V, for example, should be proportional to the load Q_V and inversely proportional to the modulus E_Y. In order to achieve the appropriate dimension of length for the settlement, it was then inevitable that the ratio Q_V/E_Y should be divided by the diameter B of the footing: no other length dimension was available for the division! The actual constants of proportionality contained Poisson's ratio, which is itself dimensionless of course. The magnitude of the movement of the pad may come as a surprise. If the pad had simply been placed on a column of identical soil of length L and diameter B, the settlement would have been

$$\rho_V = \frac{Q_V L}{(\pi B^2/4)E_Y} \approx 1.3 \frac{Q_V L}{E_Y B^2}$$

so that the depth of pedestal equivalent to the *infinite* bed is only $L = 0.8B$. Clearly the bed must spread the load over a much greater cross-sectional area of material than that which lies directly beneath it.

This outward diffusion of load can also be inferred from the extra vertical stresses

$$\Delta\sigma_V = \Delta\sigma \left\{1 - [1 + (B/2z)^2]^{-3/2}\right\} \tag{6.24}$$

induced under the centre line of a circular patch of surcharge pressure $\Delta\sigma$, corresponding to a flexible raft. This stress distribution is depicted in figure 6.18.

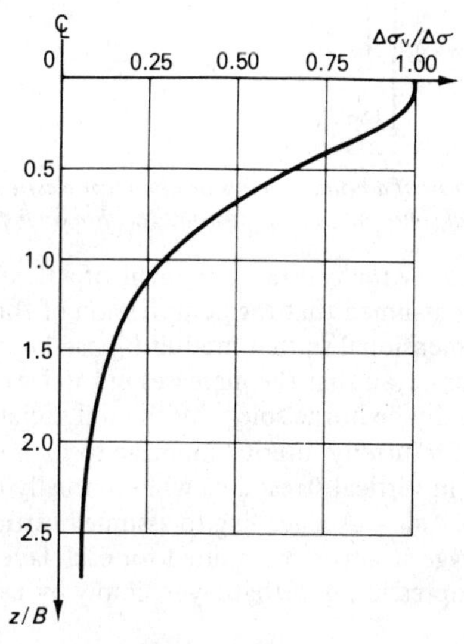

Figure 6.18 *Distribution of vertical stress $\Delta\sigma_V$ beneath axis of flexible circular patch of diameter B carrying surcharge $\Delta\sigma$ over infinite ideally elastic bed*

It is quite clear that the surface load has little effect below a depth of two diameters, a fact which has implications for site investigators. An interesting feature of the typical distribution of vertical stress recorded in figure 6.18 is that the algebraic expression contains no elastic parameters. This led people to assume that the vertical stresses would not be greatly affected by a little non-homogeneity or anisotropy in the elastic bed, an intuition which has since been proved quite sound by numerical analysts. The layered sequence of most soil deposits, and in particular the presence of a horizon of very stiff bedrock, is therefore often assumed to have little effect on the *stress* distribution, although it would make the choice of an average modulus rather difficult and therefore militate against the simple use of whole solutions such as those of equation 6.21. The engineer may therefore feel justified in assuming a stress distribution, then applying whatever local moduli appear reasonable so as to deduce a strain distribution, which can easily be integrated to give settlements. Figure 6.19

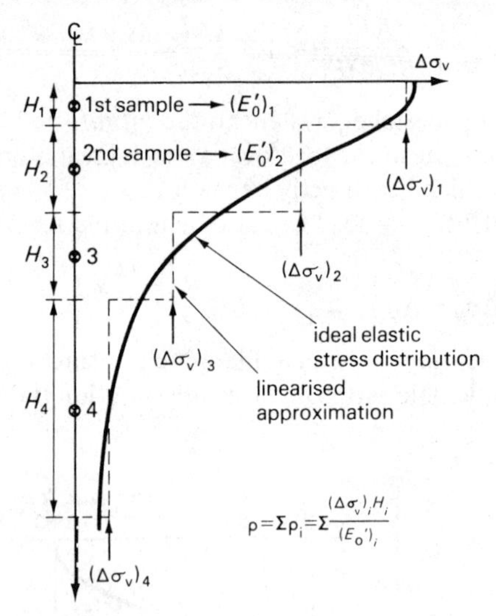

Figure 6.19 *Discretisation of a homogeneous or nonlinear elastic soil bed: stress distribution based on ideal solution, moduli measured within each supposed layer*

depicts such an analysis for the central settlement of a flexible circular patch in which the engineer has assumed that the compression of the axis of symmetry will be roughly one-dimensional so that moduli E_0' can be used. The analysis is carried out on the assumption that the increases in total stress caused by the foundation will eventually be borne solely by the soil skeleton, and that groundwater levels will return by internal drainage to their original positions. In that case the increases in vertical stress $\Delta\sigma_v$ will eventually cause increases in effective vertical stress $\Delta\sigma_v' = \Delta\sigma_v$ leading to assumed vertical strains $\epsilon_v = \Delta\sigma_v/E_0'$. The average strains ϵ_v computed for each layer of soil of thickness H then lead to the compression ρ of that layer simply by using the definition of strain $\epsilon_v = \rho/H$.

This technique, of using an elastic stress distribution to predict increases in vertical stress and one-dimensional compression tests to convert these stress increases into strains and settlements, is made much more versatile by Newmark's chart. The chart, figure 6.20, represents a plan view of the surface of an infinite

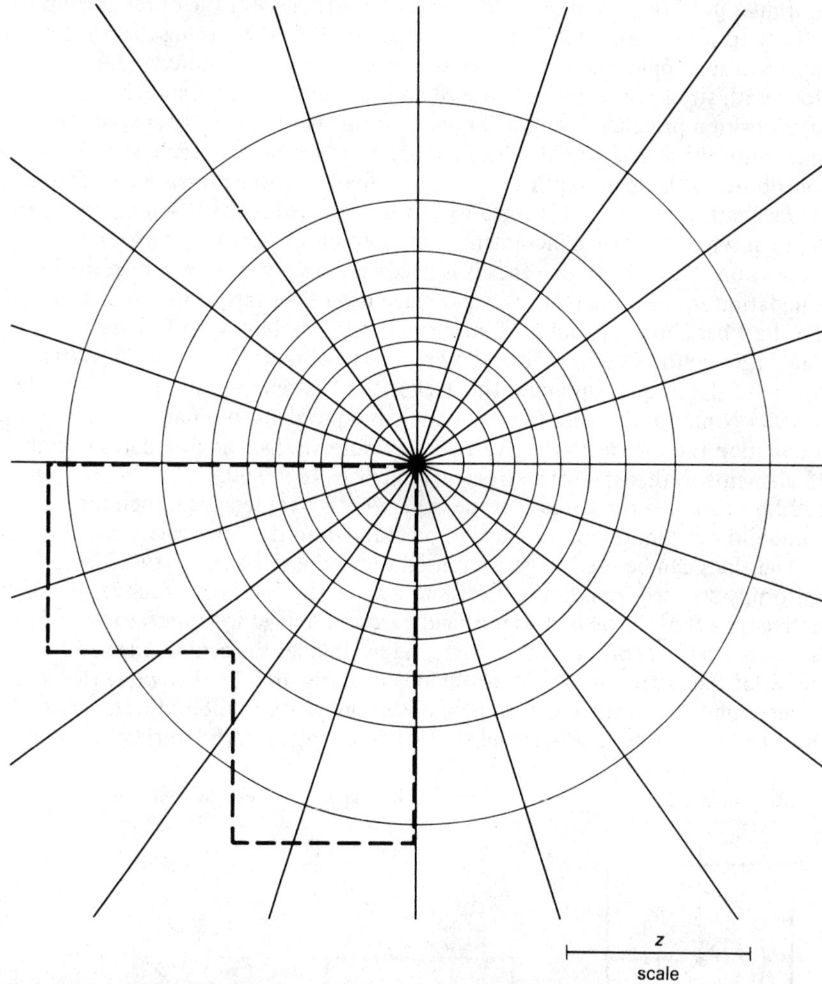

Figure 6.20 *Newmark chart for vertical stress with 200 elements*

homogeneous elastic bed which is marked off into rather small sectors, each of which has the same potential for increasing the stress at a point beneath the centre of the chart. Consider the circular core: its diameter B is such that when loaded with a pressure $\Delta\sigma$ the effect is to increase the vertical stress at depth z by 0.1 $\Delta\sigma$. Using equation 6.24

$$0.1\Delta\sigma = \Delta\sigma\left\{1 - \left[\frac{1}{1 + (B/2z)^2}\right]^{3/2}\right\}$$

So

$$\frac{B}{z} = 0.54$$

The whole of the second circular area is similarly proportioned so that when loaded with $\Delta\sigma$ the net effect at depth z is an increase in stress of $0.2\Delta\sigma$. This continues until the last circle, which causes $0.9\Delta\sigma$, leaving the outer area up to infinity to provide the final increment of stress. It follows, using the principle of superposition applicable to linear elastic systems, that each annular area loaded alone with stress $\Delta\sigma$ would give rise to an increment $0.1\Delta\sigma$. Using the superposition principle again, it should be clear that the equally spaced radial lines must divide each annulus into twenty identical sectors, each of which would contribute $0.005\Delta\sigma$ at depth z when it was loaded with a pressure $\Delta\sigma$. The scale of the chart in figure 6.20 is fixed by the line marked z which will represent the depth at which the stress increment is to be found. The radii are marked off in proportion to this length, so that it is necessary simply to draw a plan of the foundation to the same scale as z and place it on the chart so that the centre of the chart lies above the point of interest. The figure depicts an L-shaped foundation with sides 20 m long drawn so as to obtain the increase in vertical stress at a depth of 10 m under the heel of the L when the foundation is loaded to 100 kN/m^2. In this case the scale line z is equivalent to 10 m, and the foundation is drawn accordingly. You may observe that the foundation occupies 45 elements so that the total stress increment is $45 \times 0.005 \times 100$ kN/m^2, that is, 22.5 kN/m^2. If the stress at some other depth were required, then the foundation would, of course, have to be redrawn to the new scale.

The chart can be used to gain an insight into the settlement caused by a uniformly stressed foundation. Consider a large square flexible foundation resting on a thin elastic bed over a rigid base and loaded with uniform intensity $\Delta\sigma$: in particular, consider the stresses induced under the centre, edge and corner of the square. It should be obvious from figure 6.21 that while the stress increase under the centre is $\Delta\sigma$, the increase under the mid-point of an edge is $\Delta\sigma/2$ and at a corner $\Delta\sigma/4$. If the elastic bed is isotropic, the settlements under

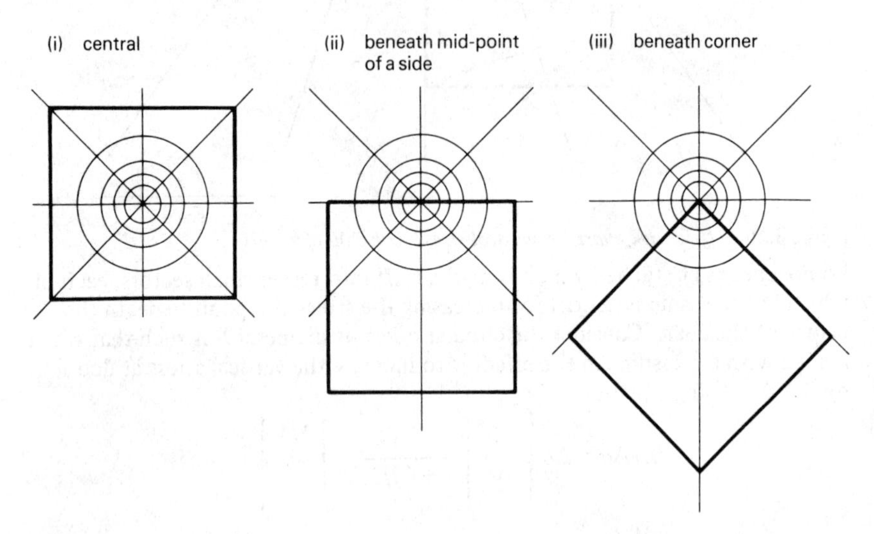

Figure 6.21 *Square foundation against a 40-element chart: stresses under various points in a shallow layer*

the centre, edge mid-point and corner would also be in the proportion $1:\frac{1}{2}:\frac{1}{4}$. The average settlement might be 0.8 times the central settlement. A conventional concrete raft foundation could well be rather stiffer than the soil: in such a case the whole raft would tend to settle monolithically (to roughly 0.8 times the calculated central settlement assuming the raft was flexible), but the stresses would be concentrated around the edges.

Consider as a further example the floor of a warehouse, 20 m x 45 m, which is to carry a uniformly distributed load of up to 50 kN/m^2. The underlying soil appears, from the evidence of three trial pits and two boreholes, to be

0 to 0.5 m topsoil

0.5 m to 7 m glacial clay, firm to stiff

7 m to 20 m weathered Keuper mudstone

20 m to 22 m and inferred great depth, Keuper mudstone

The results of three compression tests were averaged at each of the following horizons, and the tangent moduli at the estimated initial effective stresses are recorded below.

$$1 \text{ m} : E'_0 = 15\,000 \text{ kN/m}^2$$

$$2 \text{ m} : E'_0 = 20\,000 \text{ kN/m}^2$$

$$4 \text{ m} : E'_0 = 40\,000 \text{ kN/m}^2$$

$$6 \text{ m} : E'_0 = 40\,000 \text{ kN/m}^2$$

$$10 \text{ m} : E'_0 = 80\,000 \text{ kN/m}^2$$

$$15 \text{ m} : E'_0 = 100\,000 \text{ kN/m}^2$$

$$20 \text{ m} : E'_0 = 200\,000 \text{ kN/m}^2$$

Looking back to equation 6.21, it is possible to estimate the settlement of the floor on the assumption that it is rigid, by reference to the settlement of an equivalent circular raft of diameter $B = \sqrt{(20 \times 45)} = 30$ m. Most of the settlement arises from the compression of the soil up to a depth of one diameter or 30 m, so a modulus must be chosen between 15 000 kN/m^2 and 200 000 kN/m^2. A conservative average for E'_0 might appear to be 40 000 kN/m^2. Guessing that $\mu' = 0.2$, then from table 6.3 we can deduce that E'_y might be 40 000/1.11 or roughly 36 000 kN/m^2. Then using equation 6.21

$$\rho_{\text{rigid}} = \frac{20 \times 45 \times 50}{36\,000 \times 30}$$

$$= 0.042 \text{ m}$$

It will be worth while to make the prediction again with more formality. Figure 6.22 offers a subdivision of the ground upon which a more accurate analysis can be based. The zones are chosen with regard to the sampling frequency, and the knowledge that upper layers are more significant than deeper layers. The average increase in stress in each layer is characterised by the increase at a key depth chosen to be at, or slightly above, the middle height of the layer. The central settlement can be found by using Newmark's chart: you should now confirm yourself that the stresses I list are correct. The calculated settlement is $\sum \frac{\Delta\sigma H}{E_0'}$, that is

$$\frac{50 \times 2.5}{20\,000} + \frac{49 \times 4}{40\,000} + \frac{42 \times 8}{80\,000} + \frac{15 \times 35}{200\,000} \text{ m}$$

$$= 0.0063 + 0.0049 + 0.0042 + 0.0026 \text{ m}$$

$$= 0.018 \text{ m}$$

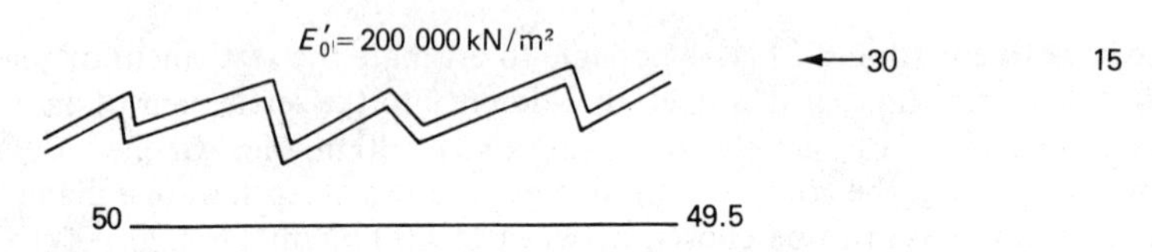

Figure 6.22 *Subdivision of soils under warehouse floor*

If you are concerned about the material below the 50 m mark you should now establish the settlement due to the next 30 m and check that it is rather small even when the pronounced rate of increase of stiffness with depth is ignored. The difference between 0.042 m and 0.018 m mirrors the difficulty in guessing a meaningful average modulus in circumstances such as these.

Now while the central settlement has been predicted as 0.18 m, the settlement at the mid-point of the short edges can similarly be found to be

$$\frac{25 \times 2.5}{20\,000} + \frac{24 \times 4}{40\,000} + \frac{20 \times 8}{80\,000} + \frac{9 \times 35}{200\,000} = 0.009 \text{ m}$$

which is roughly half the central settlement, as expected. The settlement of the mid-points of the long edges would be similar. The average settlement will be roughly 0.8 x 0.018 m, or 0.014 m.

This leaves the designer aware that the distortion of the raft could not exceed 0.009 m over a 20 m base length, which will comfort him greatly. It does not inform him whether the raft he will design will be so stiff as to eliminate even this distortion at the expense of increasing the stresses at its edges and therefore the central bending moments. Even less does the calculation offer him the answer to the case of partial loading: the client would not want a half-loaded warehouse to develop a 'step' in it close to the edge of the load! All the calculation demonstrates is that the engineer is not wasting his time in pursuing these further problems.

It is important to appreciate that the use of the one-dimensional modulus E'_0 is an approximate device for calculating settlements which has been shown not to depart too seriously from the 'correct' analytical solution in circumstances where such a solution exists. Unfortunately its use blurs the distinction between drained and undrained settlement: the E'_0 calculation offers a sort of compromise between the two. Reconsider the settlement of a pad footing as recorded in equation 6.21. If the immediate undrained settlement ρ^u were required, the engineer would substitute E^u and μ^u; if the final drained settlement ρ' were required he would substitute E' and μ'. It is interesting to calculate the ratio of the two and to use equations 6.13 for the ratios of the parameters.

$$\frac{\rho^u}{\rho'} = \frac{(1-\mu^{u2})}{E^u_y}\frac{E'_y}{(1-\mu'^2)} = \frac{(1-\mu^{u2})}{(1-\mu'^2)}\frac{(1+\mu')}{(1+\mu^u)}$$

$$= \frac{(1-\mu^u)}{(1-\mu')}$$

so that

$$\frac{\rho^u}{\rho'} = \frac{0.5}{(1-\mu')} \tag{6.25}$$

This result is roughly applicable to all foundations on linear elastic beds of soil, and has a narrow range of 0.5 to 1.0, depending on the value of μ'. In other words, the major proportion of the settlement of a foundation is due to shear displacements and is immediate.

6.8 An Infinite Strip Load on an Infinite Bed

Many textbooks on the strength of materials derive from first principles the plane strain solution to the problem of an infinite line load resting on the edge of an infinite homogeneous isotropic linear elastic half-space: see, for example, Timoshenko and Goodier (1951). When such line loads are made infinitesimal and placed side by side they become an infinite and perfectly flexible strip load carrying a pressure $\Delta\sigma$, and the principal stresses induced in the bed by the load can be written

$$\left.\begin{aligned} \Delta\sigma_1 &= \frac{\Delta\sigma}{\pi}(\alpha + \sin\alpha) && \text{bisecting angle } \alpha \\ \Delta\sigma_3 &= \frac{\Delta\sigma}{\pi}(\alpha - \sin\alpha) && \text{perpendicular to } \sigma_1 \\ \Delta\sigma_2 &= \frac{\Delta\sigma}{\pi}2\alpha\mu && \text{parallel to strip axis} \end{aligned}\right\} \quad (6.26)$$

where α is the angle subtended by the strip at the point in question, as shown in figure 6.23. The designation of these induced stresses is so comprehensive and so simple that they may be used to gain some insight into the general pattern of stresses and strains beneath foundations.

It is made clear in the figure that a circle passing through either edge of the strip would be the locus of points with a shared value of α and therefore a contour of any principal stress. For example, figure 6.24 plots the average increase in stress

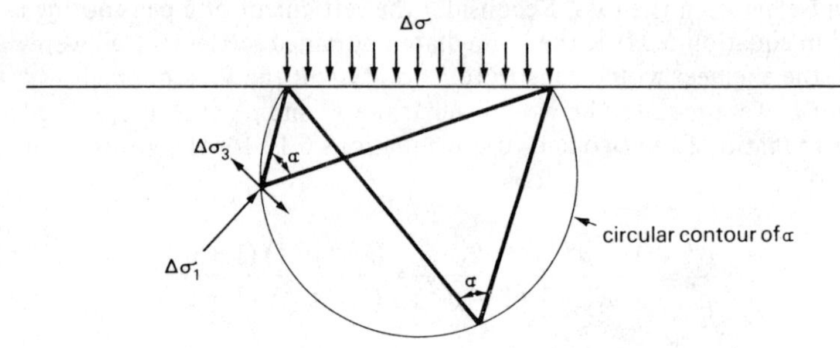

Figure 6.23 *Cross section through long flexible strip on infinite elastic bed carrying surcharge* $\Delta\sigma$

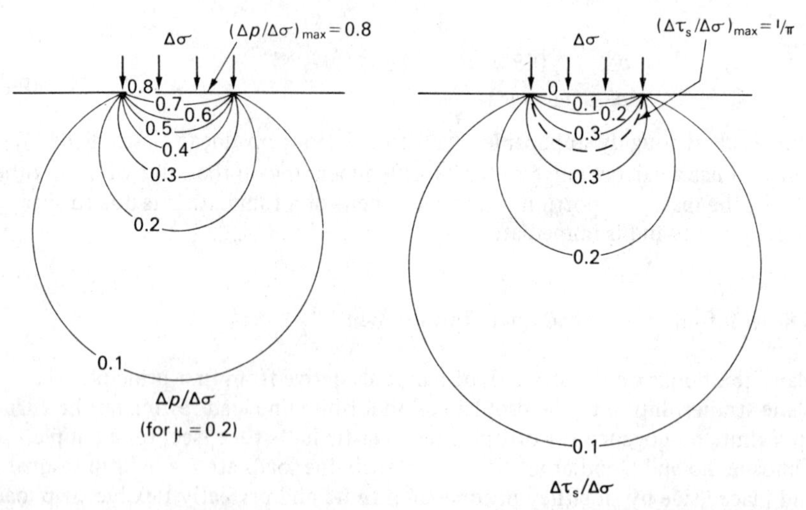

Figure 6.24 *Contours of stress components* Δp *and* $\Delta\tau_s$ *under long strip carrying surcharge* $\Delta\sigma$

$$\Delta p = \frac{\Delta\sigma_1 + \Delta\sigma_2 + \Delta\sigma_3}{3} = \frac{2\alpha}{3\pi}\,\Delta\sigma(1 + \mu) \tag{6.27}$$

and the greatest mobilised shear stress

$$\Delta\tau_s = \frac{(\Delta\sigma_1 - \Delta\sigma_3)}{2} = \frac{\Delta\sigma}{\pi}\sin\alpha \tag{6.28}$$

as contours equally spaced at intervals of 0.1 $\Delta\sigma$. Whereas the vertical increment of stress under the centroid of a footing of diameter B dropped to one-tenth of its surface value at a depth of $2B$, it is clear that the stresses under a strip load of width B require a rather greater depth to effect the same reduction: $\Delta\sigma_1$ under the centre line requires a depth of $6B$ to drop tenfold. This is due to the reduced capacity for lateral diffusion of the load: only one lateral dimension is available to the long strip, whereas the stresses under the circle could spread out symmetrically over the horizontal planes. Furthermore, the contours of the increment in spherical pressure Δp give an indication of the location of the volumetric compression of the bed which will take place only as the pore fluid is expelled. Likewise, the contours of the induced shear stress indicate the seat of the immediate shear strains which will cause the greater part of the settlement of the strip.

The induced stress distributions can also be used to determine whether the real soil can realistically be expected to remain elastic. For example, an engineer might wish to suppose that a saturated clay would be roughly linear elastic until some shear stress rose to equal c_u, its undrained shear strength. If he were fearful of the consequences of any zone yielding in this way he would clearly have to stipulate that

$$\tau_{s\,max} = \frac{\Delta\sigma}{\pi} \not> c_u$$

or

$$\Delta\sigma \not> \pi c_u$$

You may be interested to recall equation 5.3 in which we calculated that the net bearing pressure at the point of collapse of a long strip footing in clay was of the order of $2\pi c_u$, or double the pressure at which the first elements of clay first yielded. The shape of the idealised stress–strain curve of a uniformly stressed element is compared in figure 6.25 with the shape of the bearing stress–settlement curve derived from it. Designers who wished not to risk the consequences of extra settlement due to the onset of yielding would have to supplement their basic elastic settlement analysis with proof of a load factor of at least 2 against plastic collapse.

Similar considerations would arise if the analyst hoped to prove that the newly induced stress distribution would not disobey the friction model. Figure 6.24 demonstrates that large shear stresses are generated close to the edge of the strip load and at very shallow depth. The negligible overburden shown in the figure is insufficient to generate sufficient friction to make such shear stresses viable, and such a footing on the surface of a bed of soil would indeed cause the spread of plastic zones underneath its edges. For this reason, if for no other,

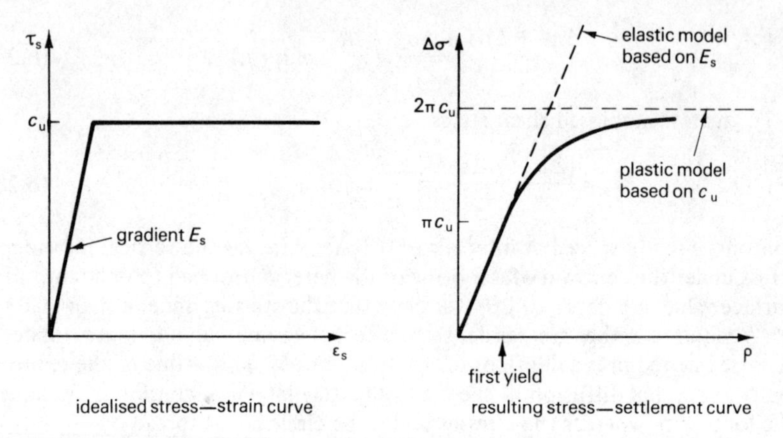

Figure 6.25 *Consequences of uneven stress distribution under a flexible strip carrying surcharge* $\Delta\sigma$

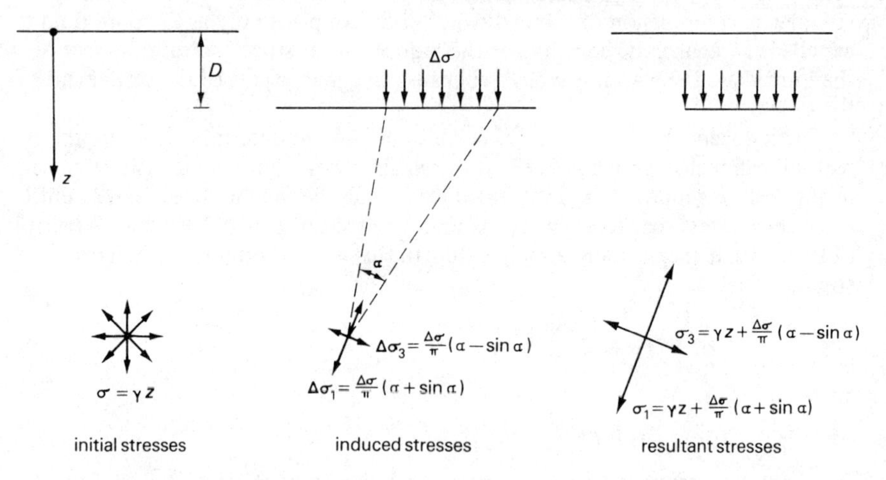

Figure 6.26 *Superposition of elastic stress increments on original gravity stresses (assumed spherical)*

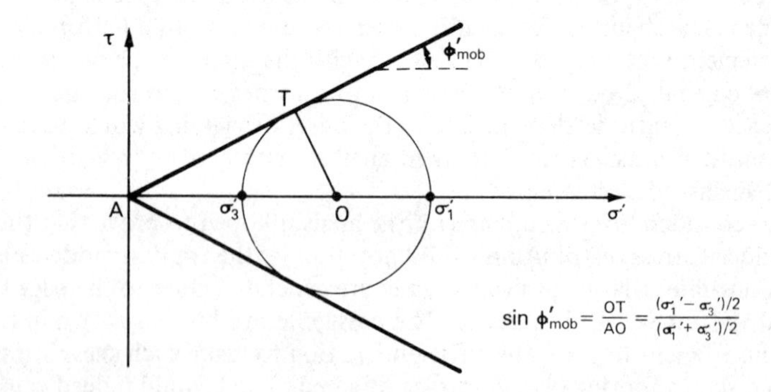

Figure 6.27 *Mobilised angle of shearing deduced from principal effective stresses*

foundations should be buried. Figure 6.26 depicts a bed of dry supposedly elastic sand in which the initial stress distribution has been assumed to be spherical and to increase linearly with depth. At some depth D a flexible strip pressure has been considered to generate extra stresses, on the pessimistic assumption that the soil above that level has no stiffness and therefore no effect Superposition has allowed the total principal stresses shown in the figure to be deduced: since the soil is dry its pore pressure can be taken to be zero so that the effective principal stresses are identical to the total stresses. Figure 6.27 demonstrates that the greatest mobilised angle of shearing can easily be deduced from the effective principal stresses mobilised at any point, and that

$$\sin \phi'_{\text{mob}} = \frac{(\sigma'_1 - \sigma'_3)/2}{(\sigma'_1 + \sigma'_3)/2} = \frac{\dfrac{\Delta\sigma}{\pi} \sin \alpha}{\gamma z + \dfrac{\Delta\sigma\alpha}{\pi}} \tag{6.29}$$

It is necessary to discover the largest mobilised shearing angle in order to check that the limiting angle has not been exceeded. Inspection of equation 6.29 reveals that the calculated mobilised shearing angle increases as the depth z decreases. The lowest meaningful value for z is D, so that the most dangerous condition can be seen to occur very close to the edge of the strip load. In that vicinity α can take any value between 0 and π, and by differentiation using $z = D$

$$\frac{\mathrm{d}(\sin \phi'_{\text{mob}})}{\mathrm{d}\alpha} = \frac{(\gamma D + \Delta\sigma\alpha/\pi)\dfrac{\Delta\sigma \cos \alpha}{\pi} - \dfrac{\Delta\sigma \sin \alpha}{\pi}\dfrac{\Delta\sigma}{\pi}}{\left(\gamma D + \dfrac{\Delta\sigma\alpha}{\pi}\right)^2}$$

This is zero when

$$\alpha = \tan \alpha - \frac{\pi\gamma D}{\Delta\sigma} \tag{6.30}$$

and it is easy to show by further differentiation that this particular condition for α implies a maximum value for $\sin \phi'_{\text{mob}}$. Now by substituting this condition into the denominator of equation 6.29, and using $z = D$, we obtain

$$\sin \phi'_{\text{mob}} = \frac{\dfrac{\Delta\sigma}{\pi} \sin \alpha}{\gamma D + \dfrac{\Delta\sigma}{\pi} \tan \alpha - \gamma D} = \cos \alpha$$

so that the critical value of α at which $\sin \phi'_{\text{mob}}$ is a maximum is simply

$$\alpha = \pi/2 - \phi'_{\text{mob}}$$

Equation 6.30 can then be written

$$\frac{\pi}{2} - \phi'_{\text{mob}} = \cot \phi'_{\text{mob}} - \frac{\pi\gamma D}{\Delta\sigma}$$

or

$$\frac{\Delta\sigma}{\gamma D} = \frac{\pi}{\phi'_{\text{mob}} + \cot\phi'_{\text{mob}} - \dfrac{\pi}{2}} \tag{6.31}$$

where ϕ'_{mob} then refers to the greatest mobilised friction angle on any plane anywhere in the hypothetical soil. If the engineer used $\phi'_{\text{mob}} = 30°$ because he felt that such an angle would be mobilisable without undue plastic strain, he would obtain $\Delta\sigma/\gamma D \approx 5$. The greater the pressure $\Delta\sigma$ he wished to apply, so the deeper would he be forced to bury the strip: at 1 m depth in normal soil he would be limited to a bearing pressure $\Delta\sigma \approx 100$ kN/m^2. We shall show in later chapters that very much larger pressures will be needed to cause collapse of such a footing. My main objective in having introduced this particularly tricky topic at this stage is to demonstrate that it is possible to define limits to the strict applicability of these linear elastic predictions. If the engineer knows that certain shear stresses or friction angles can definitely not be mobilised by the soil under his scrutiny, he can if he wishes translate these limits into limitations on the applied loads.

As a matter of fact, the onset of plasticity in the soil can be of some benefit to a designer. The stress distributions we have just analysed were based on a completely flexible strip load, which would have become 'dished' as it settled. All analyses based on Boussinesq's original solution for a point load, or on Newmark's chart, suffer the same fate: the bearing pressure is taken to be constant and the foundation becomes dished, settling roughly twice as much at the centre as at an edge. Most foundations will have appreciable bending stiffness, and many concrete strip footings will be so stiff as to be effectively rigid. The monolithic settlement of the rigid footing on elastic soil is nearly equal to the average settlement of the equivalent flexible footing, so no problems are immediately apparent. However, the structural engineer faces a new problem in that the uniform settlement is only achieved at the cost of increasing the contact stress at the edges while reducing it in the centre: this generates new bending moments in the foundation. Plastic zones under the edges of such a foundation may help to redress this outward migration of stress and recreate a more uniform contact pressure distribution.

6.9 Approximate Elastic Interaction of Pad Footings and Structural Frames

Equations 6.21, 22 and 23, with their numerical factors modified if necessary by reference to Poulos and Davis (1974) in order to suit the particular geometry of the foundation, can be used to extend the conventional elastic analysis of a framed structure. It might even be argued that the conventional analysis in terms of rigid foundations is valueless since the footings may in practice be the source of the most significant deformations in the frame. The effect of including the footing deformations will be to throw a greater fraction of the inevitable moments and shear forces on to columns which have footings of greater relative stiffness, and to increase the overall deformation of the structure. The redistribution of theoretical elastic stresses is of minor importance in a mild steel frame since any tendency towards a local overstress will simply be dissipated by harmless plastic flow, which will redress the balance of forces. Overstraining of a reinforced concrete frame is of greater concern owing to the possibility of cracking, followed by corrosion of the steel reinforcement. Likewise the serviceability of any framed structure in terms of static or dynamic deflections should be tested, taking full account of the capacity of footings to deflect and rotate: the natural frequency of vibration of such structures would be particularly sensitive to the degree of fixity of the column bases.

Unfortunately the extra degrees of freedom, and especially the vertical and horizontal sways, allowed by the independent movement of footings greatly increases the difficulty of reaching an *ad hoc* solution: even the smallest frames require solution by digital computer. This makes it important for the engineer to be able to assess whether or not he has a foundation interaction problem. This assessment can be made crudely but quite simply by comparing the relative magnitude of the deformation of a typical column due to its own bending with that due to the rotation and translation of its footing. Figure 6.28 shows a single

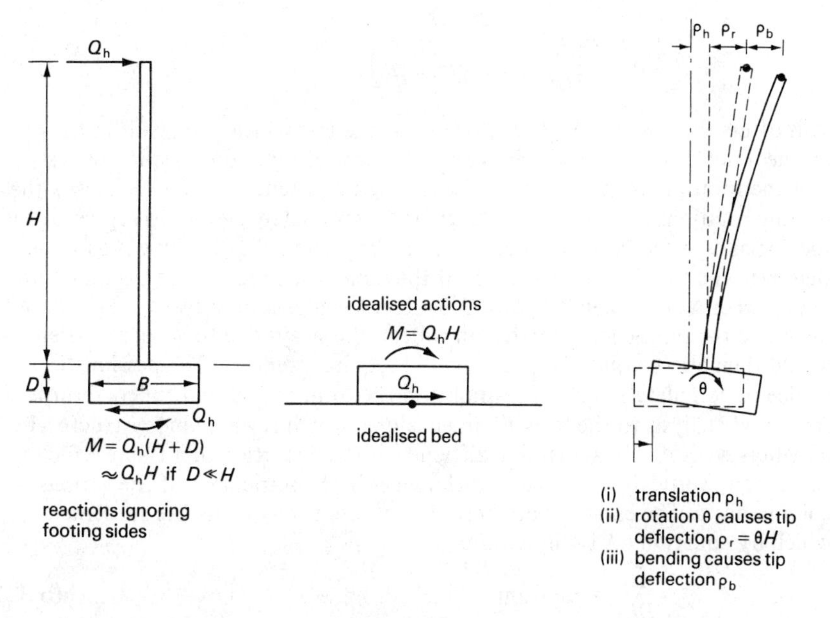

Figure 6.28 *Idealisation of square footing supporting cantilever column*

cantilever column with an effective bending stiffness (EI) supported by a shallow square footing with an effective stiffness $M/\theta = E_yB^3/3$ in rotation and $Q_h/\rho_h = E_yB$ in translation, and carrying a horizontal force Q_h at its tip. The elastic deflection due to bending is well known to be $\rho_b = Q_hH^3/3EI$, but the figure demonstrates that the total deflection

$$\rho = \rho_b + \rho_r + \rho_h$$

or

$$\rho = \frac{Q_h H^3}{3EI} + \frac{3Q_h H^2}{E_y B^3} + \frac{Q_h}{E_y B} \tag{6.32}$$

Now it is probable that

$$\frac{3H^2}{B^2} \gg 1$$

so that

$$\rho_r \gg \rho_h$$

and

$$\rho \approx \frac{Q_h H^3}{3EI} + \frac{3Q_h H^2}{E_y B^3}$$

or

$$\rho \approx \frac{Q_h H^3}{3EI}\left(1 + \frac{9EI}{E_y B^3 H}\right)$$

This offers the criterion $9EI/E_y B^3 H \ll 1$ as the test which will result in the engineer deciding to employ the simple concept of rigid footings. If this test fails the rotation of the footings is becoming important. If $9EI/E_y B^3 H \approx 1$ the footing rotations are dominant; the engineer may safely choose to represent the footing as a perfect pin which can carry no moment. If the footing is so weak that even the translation term ρ_h is of the same magnitude as the bending term ρ_b, in other words when $3EI/E_y BH^3 \approx 1$, the engineer may also safely choose to designate the pin as being carried on rollers: the sways due to vertical settlement would then also be quite large and would require attention. The pessimistic choice of completely relaxed restraints in the form of pins and rollers returns structural analysis to the back of an envelope, which is where most structural designers want it. The remaining difficulty is the distortion of a frame which will accompany vertical sways due to differences in the settlement of the various column bases. The escape route here is to attempt to equalise the settlements, which by equation 6.21 will require

$$\frac{Q_v}{BE_y} = \text{constant} \tag{6.33}$$

The designer of individual footings on a deep homogeneous bed should proportion their width (and *not* their area) to the load they carry. The effect of this will be to proportionately reduce the stress under larger footings. The designer will then have allowed for the fact that larger footings affect a proportionately deeper zone of soil.

6.10 Approximate Elastic Interaction of Soil against a Retaining Wall

The concepts of elasticity are useful whenever the deformation of a soil construction is so small that limiting stresses $\tau = c_u$ or $\tau = \sigma' \tan \phi'$ are unable to develop. Soil which lies adjacent to a structural component made of steel or concrete often exhibits this feature. The stress in the soil is of vital interest to the designer since it loads the structure and yet he will often not be prepared to allow the structure to deform sufficiently for the soil to mobilise its full strength. In such circumstances the designer may seek some whole elastic interaction analysis in which the soil and structure deform together, generating compatible elastic stresses and strains at their boundaries. Approximate solutions may be quite helpful if only to clarify behaviour and detect the more intransigent classes of problem.

Consider, for example, the placing of soil against a retaining wall. I shall analyse the problem of a saturated clay fill placed and buried at constant density behind a long, frictionless cantilever wall AB shown in figure 6.29, which is

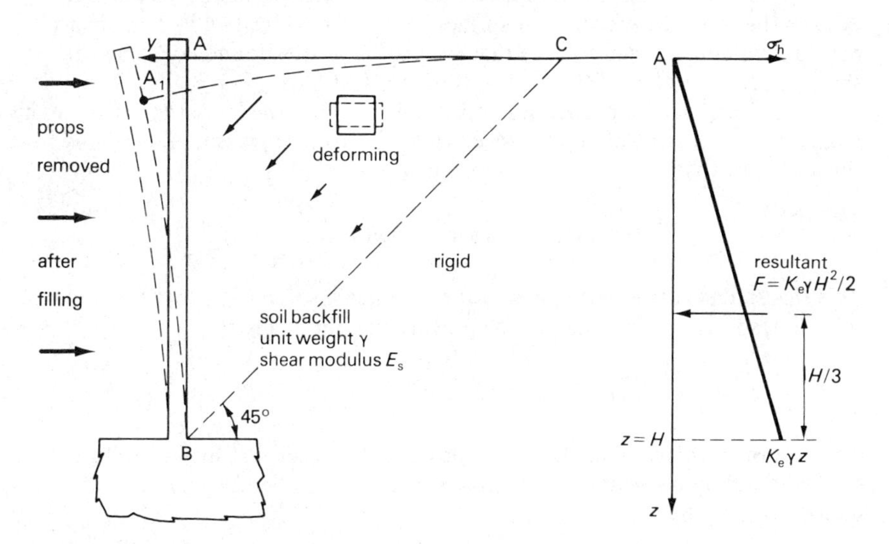

Figure 6.29 *Soil–wall interaction model*

perfectly founded so that there is no deflection or rotation at point B. Each of these conditions is judged to be pessimistic. For simplicity I shall also assume that the wall is a uniform cantilever, with a flexural stiffness EI, which is rigidly propped while the soil is placed behind it. What are the lateral pressures exerted by the soil on the wall immediately after the props are removed?

The initial burial of the clay behind a smooth propped wall conforms to the one-dimensional compression process which led to equation 6.11 as the ratio of lateral to vertical stress. Substituting the undrained value of $\mu^u = 0.5$ generates a spherical state of stress

$$\sigma_h = \frac{\mu}{1-\mu}\sigma_v = \sigma_v \tag{6.34}$$

at any depth z below the surface of the completed clay fill. If the cantilever wall were as rigid as the props that were supporting it, this lateral stress would remain when the props were removed. If the wall were perfectly flexible, however, it would deform so far that soil would tend to strain as if a Young's modulus test was being performed. In other words, the lateral stress in unsupported soil can drop to zero *if the soil remains elastic*, while the ratio of lateral expansion to vertical settlement tends to equal Poisson's ratio. Suppose that the actual behaviour of the wall is intermediate to these extremes and that the final lateral earth pressure is

$$\sigma_h = K_e \gamma z \tag{6.35}$$

where K_e is some constant elastic earth pressure coefficient which must now be estimated. Since there will be no shear stresses on the smooth wall or the surface of the fill, we will further assume that these are principal directions of stress and strain. It is then consistent to suppose that as the wall AB bends outwards to A_1B so the soil deforms on shear surfaces at 45° to the wall while remaining at constant volume. It then follows that the profile of the ground A_1C must be similar in form to that of the wall A_1B for small strains.

If the lateral earth pressure at depth z is $K_e\gamma z$, then the lateral thrust per unit length of wall exerted at depth z is equal to the average pressure $K_e\gamma z/2$ over the area z, so that

$$F = \frac{1}{2} K_e \gamma z^2 \quad \text{per unit length}$$

This thrust acts at the centre of pressure, at a height of $z/3$ above the point in question, so that the moment of the pressures at depth z is

$$M = \frac{1}{6} K_e \gamma z^3 \quad \text{per unit length}$$

The deformed shape of an elastic cantilever with a triangular load distribution can be found by elementary structural mechanics. The bending moment at depth z is given by

$$M = (EI)\frac{d^2 y}{dz^2} = \frac{1}{6} K_e \gamma z^3 \tag{6.36}$$

so by integration the slope at any depth is

$$\frac{dy}{dz} = \frac{K_e \gamma z^4}{24(EI)} + A$$

The constant of integration A is fixed using the condition $(dy/dz) = 0$ at $z = H$, so that

$$\frac{dy}{dz} = \frac{K_e \gamma}{24(EI)}(H^4 - z^4) \tag{6.37}$$

Figure 6.30 shows that the rotation dy/dz at any point is, following the assumed pattern of soil deformation, very closely related to the local shear strains. At some locality KJ the wall translates and rotates. The translation y

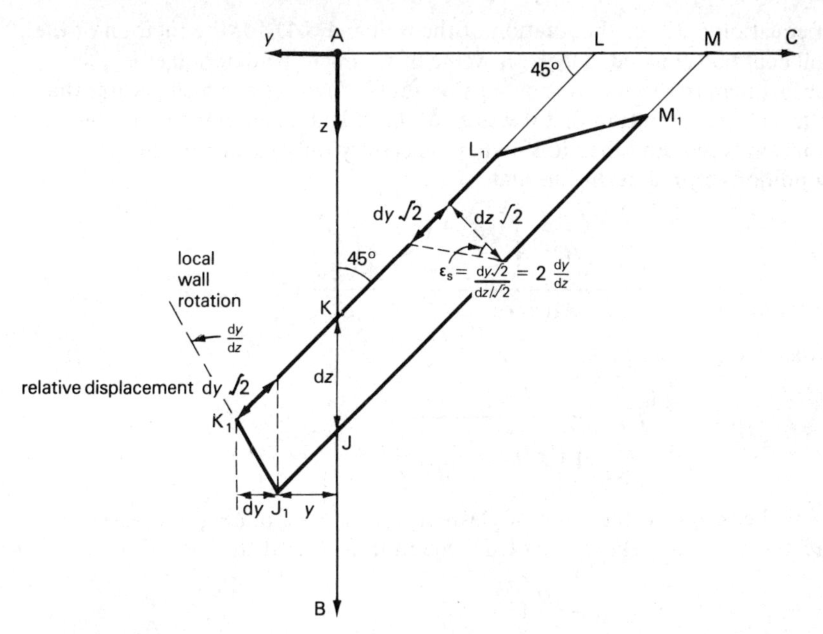

Figure 6.30 *Wall rotation and soil shear strain: a crude relationship assuming uniform small strains*

must be due to the deformation of the soil beneath JM. The rotation dy/dz, however, causes deformation within the band of soil JKLM. If KJ is dz, then the soil at K moves horizontally dy further than that at J. But we have assumed that the soil moves downwards parallel to KL, so the total magnitude of that relative movement must be $dy\sqrt{2}$. The thickness of the elementary band of shear JKLM is $dz/\sqrt{2}$ so the simple shear strain within it is

$$\epsilon_s = \frac{dy/\sqrt{2}}{dz/\sqrt{2}} = 2\frac{dy}{dz} \tag{6.38}$$

But the shear stress across these 45° slip lines must, in an elastic material, be

$$\tau_s = E_s\epsilon_s = 2E_s\frac{dy}{dz} \tag{6.39}$$

Furthermore these shear stresses correspond to the radius of a Mohr circle of stress which has already been fixed by equation 6.35

$$\tau_s = \frac{1}{2}(\sigma_v - \sigma_h) = \frac{1}{2}(1 - K_e)\gamma z \tag{6.40}$$

The two values in equations 6.39 and 6.40 for the shear stress on the assumed slip planes can be equated to generate a further condition for the rotation of the wall

$$\frac{dy}{dz} = \frac{1}{4}\frac{\gamma z}{E_s}(1 - K_e) \tag{6.41}$$

If equation 6.37 for the rotation of the wall and 6.41 for the rotation of the soil could be equated, a 'correct' value of K_e might be determined. Unfortunately the resulting expression for K_e involves z, which negates the originating assumption that it was a constant. If it be accepted that only a crude solution is required then it is perhaps necessary only to equate the two conditions at mid-height, so that

$$\frac{K_e\gamma\left(H^4 - \left(\frac{H}{2}\right)^4\right)}{24(EI)} = \frac{1}{4}\frac{\gamma\left(\frac{H}{2}\right)}{E_s}(1 - K_e)$$

whicn yields

$$K_e = \frac{1}{\left(1 + 0.31\dfrac{E_s H^3}{EI}\right)} \tag{6.42}$$

A check against the onset of plasticity would be required, of course. The greatest mobilised shear stress from equation 6.40 is at the base of the wall where

$$\tau_{max} = \frac{\gamma H}{2}(1 - K_e)$$

If the stress–strain relationship of the saturated clay were as simple as the bilinear model used previously in figure 6.25 it would follow that

$$\frac{\gamma H}{2}(1 - K_e) < c_u \tag{6.43}$$

was a necessary condition on the validity of the approximate elastic coefficient. The most stringent condition on the height of the wall then occurs when $K_e \to 0$, and is simply

$$H < \frac{2c_u}{\gamma}$$

Walls higher than this would be outside our present scope, but might be pessimistically treated with a coefficient $K = K_e$ down to a depth of $2c_u/\gamma$ and then $K = 1$ down to its base.

Even though equation 6.42 seems to offer the prospect of very small earth pressure coefficients, if the stiffness of soil relative to the wall $E_s H^3/EI$ is high enough, the retaining wall designer is unlikely to attempt to avail himself of coefficients much smaller than 0.25. He would be concerned lest the soil behind some very flexible wall was able to swell and achieve a fully drained condition of limiting friction at some future time. Figure 6.27 demonstrates that major and minor effective stresses corresponding to a Mohr circle which just touches the limiting friction line must satisfy

$$\sin\phi' = \frac{\sigma_1' - \sigma_3'}{\sigma_1' + \sigma_3'}$$

If this is applied to determining the horizontal stress σ_3 in a retained fill which has zero pore-water pressure, then $\sigma_h = \sigma_3 = \sigma_3'$ and $\sigma_v = \sigma_1 = \sigma_1'$ so that

$$\frac{\sigma_h}{\gamma z} = \frac{\sigma_3'}{\sigma_1'} = \frac{1 - \sin \phi'}{1 + \sin \phi'} \tag{6.44}$$

This ratio, usually known as the active earth pressure coefficient K_a', takes a value of 0.25 when $\phi' = 37°$. Few clay soils can sustain an effective friction angle higher than this, and therefore few walls retaining clay should be incapable of carrying an earth pressure coefficient of 0.25.

Figure 6.31 now displays the relationship between the elastic earth pressure

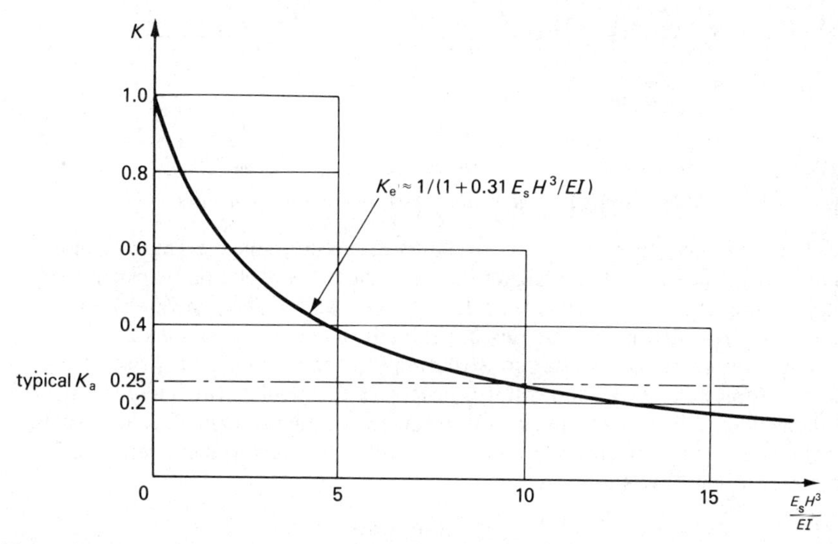

Figure 6.31 *Approximate elastic earth pressure coefficient* K_e

coefficient K_e and the relative stiffness ratio $E_s H^3/EI$. The active coefficient $K_a' = 0.25$ is superimposed on the diagram as a reminder that the designer may have criteria other than the immediate undrained earth pressures. The figure demonstrates that any value for $(E_s H^3/EI)$ greater than 15 is likely to remove the elastic interaction problem from the designer's considerations. The wall would then be flexible enough to generate long-term plasticity in the soil, and what would then remain is a plastic design problem based on future drainage conditions, a topic which will be discussed in later chapters. On the other hand, a relative stiffness less than 3 would demonstrate that the soil is incapable of supporting itself at the small level of deformation allowed by the wall: the prudent designer would consider treating it as a heavy fluid with $K = 1$.

Most forms of cantilever wall in steel or reinforced concrete which retain high-quality crushed rock, sand or sandy clay fills will prove sufficiently flexible to eliminate the elastic interaction problem up to heights of 8 m or so. Consider, for example, the problem of a reinforced concrete wall of height H with an idealised cross-section similar to that in figure 6.32, which is to be designed to

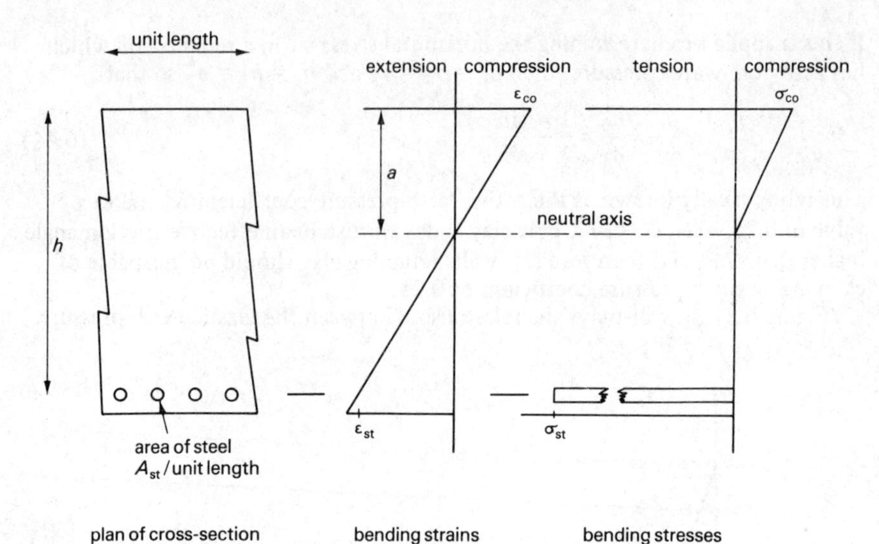

Figure 6.32 *Idealised section through reinforced concrete retaining wall*

retain a sandy clay fill with a shear modulus E_s, an undrained strength c_u and a friction angle $\phi' = 37°$. The designer intends that the wall should be capable of withstanding long-term stresses $\sigma_h = K_a'\gamma z$: will a wall which is strong enough to cope with these stresses also be flexible enough to allow any useful soil participation immediately after construction? In order to make a check on the elastic interaction of wall and soil it is necessary to calculate the likely stiffness of the wall, and it is therefore helpful to undertake an elastic method of design for the wall. Using suffixes co for concrete and st for steel, typical permissible stresses in the wall would be

$$\sigma_{co} = 7000 \text{ kN/m}^2 \text{ (compression)}$$

$$\sigma_{st} = 140\,000 \text{ kN/m}^2 \text{ (tension)}$$

while a typical modular ratio would be

$$m = \frac{E_{st}}{E_{co}} = \frac{200 \times 10^6 \text{ kN/m}^2}{13 \times 10^6 \text{ kN/m}^2} = 15$$

Consider a design in which the extreme zone of steel, a distance $(h - a)$ below the neutral axis, reaches its permissible stress σ_{st} in tension at the same instant that the extreme zone of concrete, a distance a above the neutral axis, reaches its permissible stress σ_{co} in compression. Assuming that the steel and concrete are well bonded, and that plane sections remain plane, it must follow that the strain gradient across the composite section is unique, so that

$$\frac{\epsilon_{co}}{a} = \frac{\epsilon_{st}}{h - a}$$

and

$$\frac{7m}{a} = \frac{140}{h - a}$$

This fixes the proportional depth of the neutral axis

$$\frac{a}{h} = \frac{m}{20 + m} = 0.43$$

It is conventional to assume that the concrete cracks in tension, so that the remaining stresses are within the block of compressive concrete and the steel bars in tension. For longitudinal equilibrium with no longitudinal force

$$A_{st} 140 = \frac{1}{2} 7a$$

$$A_{st} = 0.025a$$

It is then easy to show that the second moment of area of a section of the wall with the concrete transformed to steel is

$$I_{st} = \frac{1}{m} \frac{a^3}{3} + A_{st}(a - h)^2$$

$$= 1.77 \times 10^{-3} h^3 + 3.49 \times 10^{-3} h^3$$

$$= 5.3 \times 10^{-3} h^3$$

The bending stiffness then becomes

$$EI = E_{st} I_{st} = 1.05 \times 10^6 h^3 \text{ kN m}^2/\text{m} \tag{6.45}$$

Now the relationship between stresses σ at distance y from the neutral axis and bending moment M is

$$\frac{\sigma}{y} = \frac{M}{I}$$

or

$$M = \frac{\epsilon(EI)}{y}$$

The permissible strain gradient is

$$\frac{\epsilon}{y} = \frac{\epsilon_{co}}{a} = \frac{\epsilon_{st}}{h - a} = \frac{1.23 \times 10^{-3}}{h}$$

so the permissible bending moment must be

$$M = \frac{1.23 \times 10^{-3}}{h} 1.05 \times 10^6 h^3 \text{ kN m/m}$$

or

$$M = 1.29 \times 10^3 h^2 \text{ kN m/m} \tag{6.46}$$

Now the designer wished the wall to support an earth pressure of $0.25\gamma z$ which would generate a moment

$$M = \frac{1}{6} 0.25\gamma H^3$$

at the base of the wall. This gives a condition for the effective thickness h of the wall at its base, from equation 6.46

$$1.29 \times 10^3 h^2 = \frac{1}{6} 0.25 \gamma H^3$$

$$h = 5.7 \times 10^{-3} (\gamma H^3)^{1/2}$$

which in turn fixes the bending stiffness (EI) at the base, from equation 6.45

$$EI = 0.19 (\gamma H^3)^{3/2} \text{ kN m}^2/\text{m} \tag{6.47}$$

If, following figure 6.36 for the crude elastic interaction analysis for wall and soil, the designer believes he should satisfy the condition

$$\frac{E_s H^3}{EI} > 10$$

so that immediate earth pressures should not exceed their long-term values, then it follows from equation 6.47 that he must satisfy

$$E_s > 10 \times 0.19 (\gamma H^3)^{3/2} H^{-3} \text{ kN/m}^2$$

or

$$E_s > 1.9 (\gamma H)^{3/2} \text{ kN/m}^2 \tag{6.48}$$

or if

$$\gamma = 20 \text{ kN/m}^3$$

then

$$E_s > 170 H^{3/2} \text{ kN/m}^2 \tag{6.49}$$

If the wall were 4 m high the shear modulus of the soil should exceed 1350 kN/m^2 if the assumed long-term soil stresses are not to be exceeded in the short term: this is such a low value that only soft clays would disobey. If the wall were 8 m high the shear modulus of the soil should exceed 3850 kN/m^2, which only quite firm or stiff sandy clays would be able to obey. If the wall were 12 m high the modulus should exceed 7050 kN/m^2, which only exceptional very stiff undisturbed clays would be able to achieve. Finally, the designer would have to check that $H \leqslant H_p$ where

$$H_p = \frac{2 c_u}{\gamma 0.75} = 2.7 \frac{c_u}{\gamma} \tag{6.50}$$

in accordance with the rule in equation 6.43 to prevent the soil going plastic at the base. The standard descriptions for clayey soil are then easily translated into permissible heights of wall

soft: $c_u < 40$ kN/m^2 : ($H < 5$ m)

firm: $40 < c_u < 75$ kN/m^2 : (5 m $< H <$ 10 m)

stiff: $75 < c_u$: (10 m $< H$)

It is therefore clear that the separate requirements for stiffness and strength of undrained clays behind retaining walls coincidentally lead to rather similar restrictions on the heights of such walls, if they are to be designed only to resist small lateral pressures $\sigma_h = 0.25\gamma z$. But whereas failure to comply with the flexibility requirement would logically demand that the earth pressure coefficient *throughout* be replaced by a higher value, probably unity, failure to comply with the strength requirement would only demand such treatment below the plastic depth H_p.

Free-draining soils are rather less of a problem. At the outset of the clay–wall analysis we invoked equation 6.34 to infer that the lateral pressure coefficient K_0 behind a propped wall would be unity. The proper mode of analysis for free draining soils is effective stress analysis: in this case such an analysis would simply be based on zero pore pressures so that $\sigma' = \sigma$. If the largest effective Poisson's ratio likely to be encountered in a sand fill was thought to be $\mu' = 0.3$, the corresponding one-dimensional lateral pressure coefficient would be $K_0' = 0.5$. A more likely value of $\mu' = 0.2$ would lead to $K_0' = 0.25$. The one-dimensional compression envisaged by the user of equation 6.34 is successful in generating quite large supportive shear stresses in the drained soil even in the absence of lateral expansion, due to the shear strains mobilised as the skeleton compressed. Similar extra shear strains in the soil would likewise come to the aid of the wall when the props were removed, so that equations 6.42 and 6.49 would be grossly pessimistic if applied to granular soils. Nevertheless a fairly dense sand, gravel or crushed rock fill would certainly be expected to have an operational shear modulus exceeding 10 000 kN/m^2, so that even equation 6.49 would offer heights in excess of 15 m. Most designers restrict their calculations for the retention of free-draining soils to those concerning limiting strength based on an active earth pressure coefficient K_a' and their worst expectation of future pore-water pressures.

6.11 Problems

(1) Use the data of figure 6.13 in order to
(a) determine appropriate quasi-elastic moduli E_0' to cover the ranges of stress

$$35\text{–}100 \text{ kN/m}^2;\ 35\text{–}200 \text{ kN/m}^2;\ 200\text{–}300 \text{ kN/m}^2$$

(b) estimate the precompression of the soil sample which was tested
(c) estimate the *in-situ* tangent modulus of a similar element of soil at a depth of 15 m, on the assumption that it shared the same geological cycle of deposition and erosion which caused the test sample from 2 m to behave as it did.

Answers (a) 4710; 7140; 18 870 kN/m^2 (b) 330 kN/m^2 (c) 11 100 kN/m^2.

(2) The warehouse floor considered after figure 6.22 actually settled only 0.007 m at its centre when it had been fully loaded for 6 months: offer some possible explanations.

(3) Use the Newmark chart of figure 6.20 to contour a crude vertical stress distribution over each of two thin horizontal clayey strata which were found to exist at depths of 5 m and 20 m below the L-shaped foundation raft which has 20 m 'long' and 10 m 'short' sides.

(4) A 6.5 m deep sequence of soil strata above rock was sampled at 1 m intervals from the ground surface. The tangent moduli E_0' at the *in-situ* effective stresses were 4000, 2000, 1500, 2000, 2500, 10 000 kN/m^2. Calculate the harmonic mean tangent modulus E_0' for the stratum which might be applicable to a raft placed at 0.5 m depth. Find the settlement of such a raft due to a lowering by 1 m of the groundwater table.

Answers 2483 kN/m^2, 24 mm.

(5) A rigid 2 m square footing carries a load of 600 kN down to a 10 m deep stratum of firm clay on sound rock. If the clay has an average one-dimensional modulus $E_0' = 15\ 000\ kN/m^2$ and a Poisson's ratio $\mu' = 0.2$, predict the settlement, both immediate and long term, of the footing. Make the calculations again for the case where the modulus increases linearly from 5000 kN/m^2 to 25 000 kN/m^2.

Answers within 25 per cent, $\rho_u = 13$ mm, $\rho' = 21$ mm; $\rho_u = 25$ mm, $\rho' = 40$ mm.

(6) In an undulating area which is to be the site of a number of new buildings, the groundwater table is level at 50 m AOD, and a consistent stratum of clay is encountered between 47 m and 49 m AOD. Above and below the clay is an extensive deposit of stiff glacial sand. The surface of the sand, which corresponds to the level at which shallow footings will be founded, wavers between 50 m and 55 m AOD from place to place. The clay has been precompressed and its state can be considered to lie on the curve $e = 0.7 - 0.04 \ln(\sigma'\ kN/m^2)$. Plot the relationship between the average one-dimensional compression modulus of the clay and its depth of burial by the sand. Plot a similar relationship for the average maximum stress increase in the clay due to a 1 m wide strip footing carrying 100 kN/m^2. Plot a graph of the likely settlement of the footings as a function of the depth to the clay layer beneath them, ignoring the compression of the sand.

Answers footing 50 m AOD, settlement 47 mm; 51 m, 21 mm; 52 m, 12 mm; 53 m, 8 mm; 54 m, 5 mm; 55 m, 4 mm.

(7) Demonstrate that where a building frame is to be supported on isolated circular footings on homogeneous elastic soil, their diameter ought to be in proportion to the load they carry in order to minimise distortion of the frame. Are there any circumstances in which footings should be designed to carry equal working stresses?

(8) Demonstrate that the effect of using both high-strength steel and high-strength concrete in order to double their permissible stresses compared with the values used in the derivation of the stiffness criterion in equation 6.48, is to reduce the minimum permissible soil stiffness by a factor of 2.8, or to increase the permissible wall height by a factor of 2.

7

Transient Flow

7.1 Concept

The most significant discriminating factor in the last three chapters has been drainage. The friction model demands that water pressures be known, which often means that the drainage of water must be steady. The cohesion model requires a constancy of volume between the measurement of a soil strength and its future application, and this means that the rate of drainage of water in or out of the soil must be very slow in comparison with the speed of construction. The elasticity model can either be used to predict undrained stresses and settlements at constant volume, or drained conditions at some time when water pressures are known to have returned to stable values after an episode of change due to loading. This chapter is devoted to those transient flows of pore fluid which are caused by changes of stress, and in particular the time it takes before equilibrium is re-established. Only when he can calculate the likely duration of a transient flow, can the engineer make a decision regarding the relevance of an assumption of 'undrained' or 'drained' conditions in a particular field problem.

The simplest transient flow problem to understand is the one-dimensional compression of a saturated elastic soil with a modulus E_0' and an isotropic permeability k. Consider the sequence shown in figure 7.1. In (i) a small sample of the soil is shown retained in the conventional one-dimensional compression apparatus sometimes called an oedometer. It is submerged in water and is in equilibrium with a vertical stress $\sigma_v' = \sigma_v$. In (ii) a load increment $\Delta\sigma$ has just been placed. We considered in section 6.5 that undrained saturated soil was effectively incompressible. It follows that there can be no immediate compression of the sample. The vertical effective stress must therefore remain at its previous value, for if it were to have changed then some strain would have been inevitable. This means that the stress increment must instantaneously be totally supported by an increase in pore-water pressure $\Delta u = \Delta\sigma$. This excess water pressure sets up strong hydraulic gradients at the ends of the sample, since the water pressure at the extreme top and bottom boundaries must remain at zero (that is, atmospheric). The hydraulic gradients cause water to flow out of the soil: this causes the void ratio of the soil to fall, which inevitably must be accompanied by the appropriate increase in effective vertical stress. As water bleeds out under pressure, therefore, the soil skeleton begins to take up the

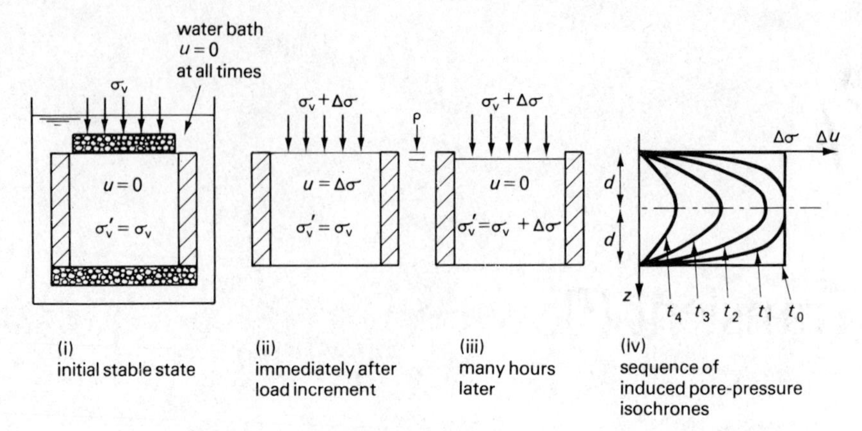

Figure 7.1 *One-dimensional consolidation*

stress increment as an effective stress, until eventually the water pressures have returned to zero, as in (iii), at which time the flow has stopped. The excess water-pressure contours at various times, called isochrones, are depicted in figure 7.1 (iv). A simple conceptual framework for transient flow can be demonstrated by a crude solution to the problem.

The total eventual settlement ρ of the layer can first be calculated, taking its thickness as constant at $2d$

$$\frac{\rho}{2d} = \frac{\Delta\sigma'_v}{E'_0} = \frac{\Delta\sigma}{E'_0}$$

So

$$\rho = \frac{2d\Delta\sigma}{E'_0} \tag{7.1}$$

The average rate of settlement might be very crudely estimated by considering the pore-water pressure distribution shown in figure 7.2. Each half of the sample

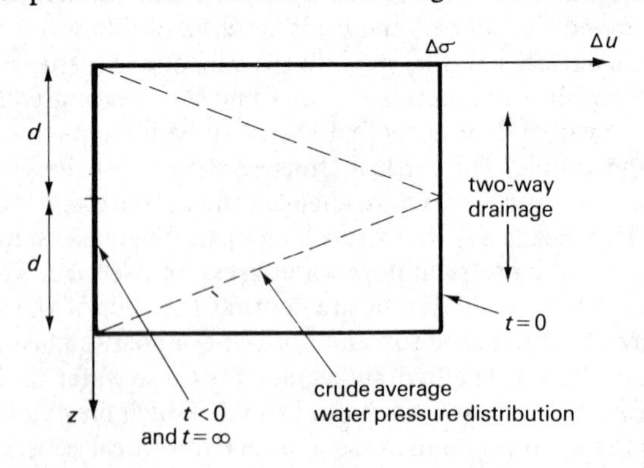

Figure 7.2 *Crude average isochrone*

is subjected to an outward hydraulic gradient of $(\Delta\sigma/\gamma_w)/d$. This indicates an average velocity of outflow of water of $(k\Delta\sigma)/d\gamma_w$ from both upper and lower surfaces. Now, the seepage velocity with which water flows out of a superficial area is equal to the velocity with which the soil skeleton is subsiding. This is made clear in figure 7.3. Considering both top and bottom surfaces, therefore,

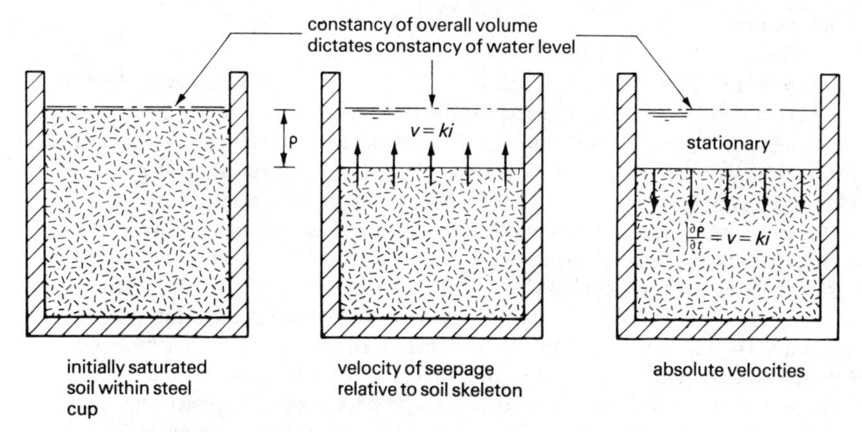

Figure 7.3 *Rates of seepage and settlement*

the total rate of settlement of the sample is, roughly

$$\left(\frac{d\rho}{dt}\right)_{\text{average}} = 2 \times \frac{k\Delta\sigma}{d\gamma_w} \tag{7.2}$$

The time t_c for completion of the transient drainage caused by the load increase is now given by equations 7.1 and 7.2

$$\frac{2d\Delta\sigma}{E_0' t_c} = \frac{2k\Delta\sigma}{d\gamma_w}$$

So

$$t_c = \frac{d^2\gamma_w}{E_0' k} \tag{7.3}$$

Terzaghi gave the group of parameters $(E_0' k)/\gamma_w$ the symbol C_v and called it the coefficient of consolidation, but for consistency I shall use the subscript 0 for one-dimensional, hence C_0. It has typical units of m^2/year rather than m^2/s. The stiffer the soil, and the more permeable, the smaller is the drainage time. Clays, which are neither stiff nor permeable, usually have consolidation times much longer than the construction time. For example, a typical stratum of uniform silty clay 10 m deep might have two-way drainage so that $d = 5$ m, $E_0' = 10\,000$ kN/m^2 and $k = 10^{-10}$ m/s, leading to

$$t_c = \frac{25 \times 10}{10^4 \times 10^{-10}} = 25 \times 10^7 \text{ s} = 8 \text{ years}$$

It is important to understand that the time for consolidation to finish is independent of the magnitude of the load increment: both the rate of consolidation and its eventual magnitude are proportional to the size of the stress increment, thereby cancelling out.

7.2 Causes

Any change in stress on or in a soil body is likely to cause a transient flow of pore fluid due to the excess pore-water pressures Δu which it has generated. If an element of saturated soil is in equilibrium before it is subjected to a change of total stress which involves an increase in the average normal stress $\Delta p = (\Delta\sigma_1 + \Delta\sigma_2 + \Delta\sigma_3)/3$, then the immediate response of the soil is to increase its pore-water pressure by Δp so as to leave the average effective stress unaltered. Any change $\Delta p'$ in the average effective stress would have meant an impossible instantaneous compression of the soil skeleton $\Delta\epsilon_v = \Delta p'/E'_v$ with instant drainage. If the water pressures at the surface of the soil in question remain constant, then hydraulic gradients will have been set up by the increment in spherical stress. These outward hydraulic gradients will cause, over a period of time, a reduction in the internal excess water pressures: eventually the water pressures throughout the soil will return to their original values, in equilibrium with the boundary conditions.

A reduction in spherical stress would cause a similar transient inflow of water from the boundaries, which will satisfy the corresponding negative water-pressure increment. The transient suction will eventually be relieved once the soil skeleton has been successful in swelling, so as to reduce the internal effective stresses to values which are in keeping with the new stable conditions.

If total stresses remain constant, an increase or decrease in water pressure at a boundary of a body of soil is slowly communicated to the rest of the soil in a similar fashion. If water is pumped from a cavity deep in a clay stratum, the soil in the immediate vicinity will suffer an increase in effective stress and will consequently compress. This will allow the wavefront of suction to be communicated to further more remote zones. Any increment in *total stress* is carried away from its source almost instantaneously, at the speed of sound in the material. Information about the relative balance between the two components of pore-water pressure and effective stress are propagated much, much, more slowly, following equations such as 7.3.

A stress increment Δq conforming to 'pure shear' which leaves the spherical stress component of the soil constant, will have no tendency to cause transient water pressures or changes of volume in an elastic porous medium. However, we saw in section 5.3 that a soil which obeys both the friction model and cohesion model is forced to adjust its internal pore-water pressures as it approaches failure at constant volume. If the soil is rather dense, large shear stresses will cause a transient suction in the failure zone, which will ultimately be satisfied by an inflow of water from the boundaries. If the soil is rather loose, then large shear stresses will cause a transient bleeding out of water.

It follows that two sorts of stress change cause transient flows within an element of elastic soil. Firstly, a change in water pressure at the boundary while the total stresses at the boundary remain constant, causes a slow migration of this information into the element. Secondly, a change in the total spherical stress component at the boundary $\Delta p = (\Delta\sigma_1 + \Delta\sigma_2 + \Delta\sigma_3)/3$ while the boundary water pressure remains constant, causes an almost immediate rise in the pore-water pressure throughout the whole element ($\Delta u = \Delta p$), which begins to dissipate at the boundaries. If the soil is not elastic but rather is approaching a frictional failure, then a third mode of stress change, pure shear, will cause water pressures Δu in the failure zone.

In order to make progress it is necessary to establish with greater precision the solution to the simplest transient flow problem, that of one-dimensional consolidation which was depicted in figure 7.1.

7.3 One-dimensional Consolidation of a Small Element: A Solution based on Parabolic Isochrones

The likely isochrones of excess water pressure in the one-dimensional consolidation problem were drawn out in figure 7.1 (iv). They are rather different in form from the crudely triangular distribution of figure 7.2, which was used to obtain an approximation to the average rate of settlement in equation 7.2. In order to obtain a better representation of the process it is necessary to choose a better shape for the isochrone, and to allow in the algebra for the pore pressures to dissipate.

It is possible to simplify the problem, firstly, by considering only half of the two-way drainage situation shown in figure 7.1. The central plane of the sample is a plane of symmetry, so that only one disc of thickness d, with one direction of drainage, need be considered. The new element is shown in figure 7.4,

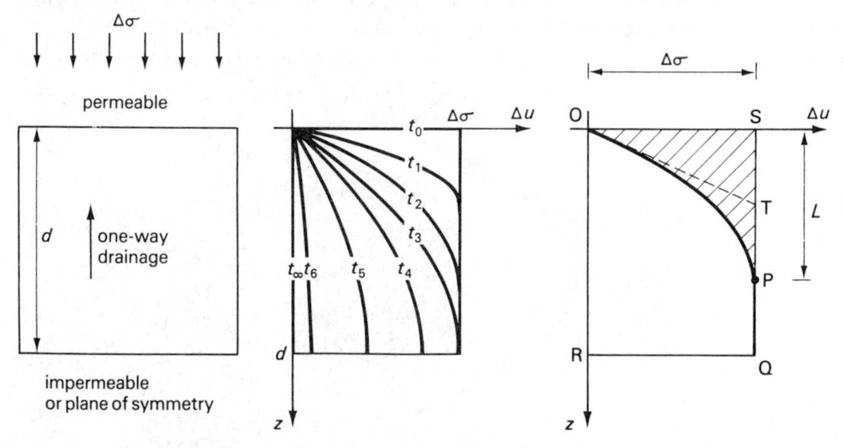

Figure 7.4 *Parabolic isochrones*

together with a family of parabolic isochrones depicting the dissipation of pore-water pressure. The initial rectangular excess pore-pressure distribution due to the load increment is marked t_0. As times t_1, t_2 and t_3 pass, the 'front' of reducing pore pressure advances to the base of the sample. As times t_4, t_5, t_6 and eventually t_∞ pass, the pore pressures throughout the soil fall eventually to zero.

Before advancing to a solution, it will be beneficial to study isochrones in general, and an assumed parabolic isochrone in particular. Two geometrical properties related to gradients and areas are important. Take the parabola OP in figure 7.4 as an example: time t_2 has elapsed and the pore-pressure distribution has changed from OSPQ to OPQ where the curve OP is a parabola with its apex at P. We will refer to length SP as the depth of penetration L after time t.

The first geometrical property is related to Darcy's law. Darcy's hydraulic gradient is

$$i = -\frac{\partial h}{\partial z} = -\frac{1}{\gamma_w}\frac{\partial u}{\partial z}$$

where $\partial u/\partial z$ is the gradient of the isochrone at any point, noting that the u/z graph is on its side. Strictly I should have written $\partial \bar{u}/\partial z$ or $\partial \Delta u/\partial z$ because I mean to analyse only those excess transient pore-water pressures that are caused by changes in total stress. The hydraulic gradients at the ends of the element are particularly important, because they determine the rate of outward seepage, which equals the rate of compression, as we saw in figure 7.3. The region PQ of the isochrone shows no change in water pressure, and therefore no seepage is indicated from the bottom of the element: this we had demanded by making the base impermeable. The hydraulic gradient at the top surface is related to the gradient OT of the curve OP at O. In fact

$$i_0 = \frac{1}{\gamma_w}\frac{\text{OS}}{\text{ST}} = \frac{1}{\gamma_w}\frac{\Delta\sigma}{\text{ST}} \tag{7.4}$$

A most useful property of the parabola apex P is that the gradient at any point O is half that of the chord OP. This is easy to prove with respect to the general parabola $y^2 = cx$ shown in figure 7.5 with its conventional orientation,

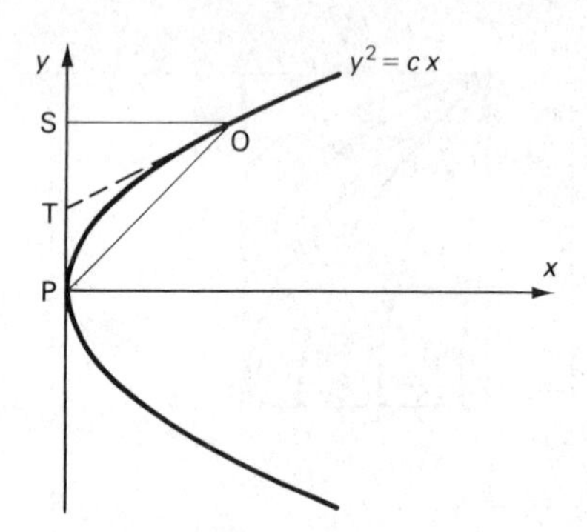

Figure 7.5 *A parabola apex P with conventional axes*

for which on differentiation

$$2y\frac{\mathrm{d}y}{\mathrm{d}x} = c$$

So

$$\frac{\mathrm{d}y}{\mathrm{d}x} = \frac{c}{2y} = \frac{y}{2x}$$

Using a corresponding geometry we have therefore proved that

$$\frac{\mathrm{ST}}{\mathrm{SO}} = \frac{\mathrm{SP}}{2\mathrm{SO}}$$

or that

$$\mathrm{ST} = \frac{1}{2}\,\mathrm{SP}$$

Substitution in equation 7.4 then yields

$$i_0 = \frac{1}{\gamma_\mathrm{w}} \frac{\Delta\sigma}{L/2}$$

so that the rate of compression of the element

$$\frac{\mathrm{d}\rho}{\mathrm{d}t} = v_0 = ki_0 = \frac{2k\Delta\sigma}{\gamma_\mathrm{w}L} \tag{7.5}$$

The second important property of isochrones refers to the area swept out by them. It is important to realise that the succession of isochrones in a conventional consolidation problem occur while the total stress remains constant. Only the balance of stress between the water and the soil skeleton is at issue once the load has been applied at time $t = 0$. Every pore-pressure reduction during consolidation is matched by an increase in effective soil stress. Each increment of effective stress contributes to the settlement ρ. Consider the two isochrones t_1 and t_2 in figure 7.6, and in particular the changes in stress in the

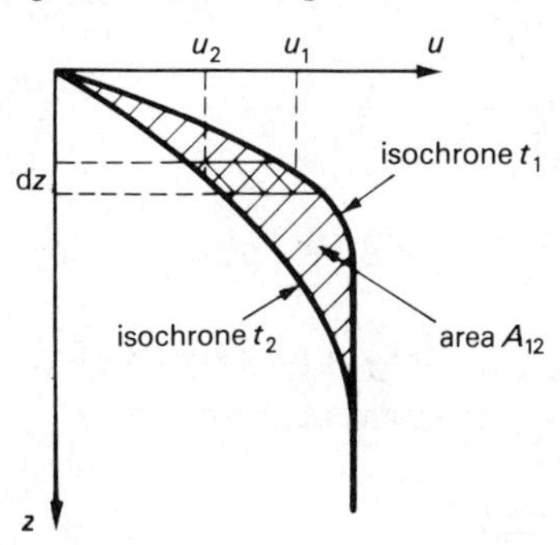

Figure 7.6

element dz deep during this time interval. The water pressure drops from u_1 to u_2: the effective soil stress must therefore rise an equal amount $(u_1 - u_2)$ in magnitude. This will cause a small compression dρ of the element dz so that by elasticity

$$\text{strain} = \frac{d\rho}{dz} = \frac{\Delta\sigma'}{E'_0} = \frac{(u_1 - u_2)}{E'_0} \tag{7.6}$$

The compression of the whole soil sample z deep during this time t_1 to t_2 can then be deduced by integration

$$\rho_{12} = \int \frac{d\rho}{dz} dz = \int \frac{(u_1 - u_2)}{E'_0} dz \tag{7.7}$$

or if E'_0 can be taken as roughly constant

$$\rho_{12} = \frac{1}{E'_0} \int (u_1 - u_2)\, dz = \frac{1}{E'_0} A_{12} \tag{7.8}$$

The settlement ρ up to any time has therefore been shown to be proportional to the area swept out by the advancing isochrone.

Once more, the parabola offers a simple piece of geometry. Area OSP in figure 7.4 swept out between time $t = 0$ and time $t = t_2$ in this case, is simply OS x SP/3 or $\Delta\sigma L/3$. This is easily proved in the case of the general parabola $y^2 = cx$ which can be written $y = \sqrt{(cx)}$. The area *inside* the half-parabola is

$$\int y\, dx = c^{1/2} \int x^{1/2}\, dx = c^{1/2} x^{3/2} \frac{2}{3} = \frac{2}{3} xy$$

Referring to figure 7.7, it is clear that in our example the area swept out is

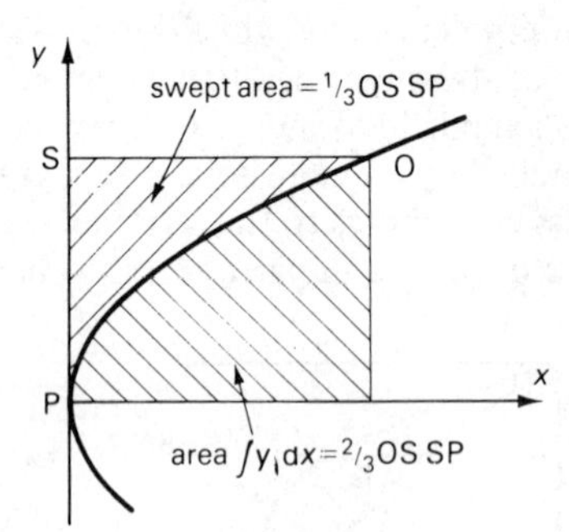

Figure 7.7

OS x OP/3. In the simple case of isochrone OP which has advanced a distance SP = L it follows that the settlement up to time t_2 is

$$\rho = \frac{1}{3} \frac{\Delta\sigma L}{E'_0} \tag{7.9}$$

The solution to the consolidation problem up to the time (t_3 in figure 7.4) when the advancing isochrone reaches the impermeable base of the sample can now be obtained from equations 7.5 and 7.9 in which ρ and L are the only unknowns.

Temporarily eliminating ρ by differentiating equation 7.9 and equating this to 7.5, we obtain

$$\frac{\Delta\sigma}{3E_0'}\frac{\partial L}{\partial t} = \frac{2k}{\gamma_w}\frac{\Delta\sigma}{L}$$

or

$$L\frac{\partial L}{\partial t} = \frac{6kE_0'}{\gamma_w} \tag{7.10}$$

This well-known class of differential equation has solutions of the form

$$L = at^b$$

so that

$$\frac{\partial L}{\partial t} = abt^{b-1}$$

$$L\frac{\partial L}{\partial t} = a^2bt^{2b-1} \tag{7.11}$$

Since t does not appear on the right-hand side of equation 7.10 we must eliminate it from equation 7.11 by fixing $2b - 1 = 0$ or $b = 1/2$

that is $$L\frac{\partial L}{\partial t} = \frac{1}{2}a^2$$

so that

$$a^2 = \frac{12kE_0'}{\gamma_w}$$

or using the consolidation coefficient

$$C_0 = \frac{E_0'k}{\gamma_w} \tag{7.12}$$

$$a^2 = 12C_0$$

so that

$$L = \sqrt{(12C_0t)} \tag{7.13}$$

while from equation 7.9

$$\rho = \frac{\Delta\sigma}{3E_0'}\sqrt{(12C_0t)}$$

$$= \frac{\Delta\sigma}{E_0'}\sqrt{\left(\frac{4}{3}C_0t\right)} \tag{7.14}$$

The final applicable parabola in this sequence is that for which $L = d$, so that

$$d = \sqrt{(12C_0t)}$$

$$t = \frac{d^2}{12C_0} \tag{7.15}$$

and at which

$$\rho = \frac{\Delta\sigma}{E_0'} \sqrt{\left(\frac{4}{3} C_0 \frac{d^2}{12C_0}\right)}$$

that is

$$\rho = \frac{\Delta\sigma d}{3E_0'} \tag{7.16}$$

We now have a clear picture of the sequence of pore-water pressures and settlements up to the time when the pressure front reaches the bottom of the sample. From that point onwards the isochrones will be as shown in figure 7.8.

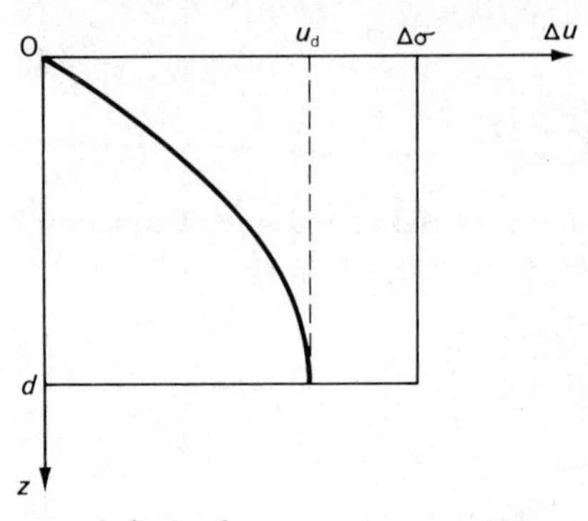

Figure 7.8 *Diminishing parabolic isochrone*

The analogue to equation 7.5 governing the rate of seepage becomes

$$\frac{d\rho}{dt} = \frac{2k}{\gamma_w} \frac{u_d}{d} \tag{7.17}$$

while the settlement calculation based on the swept area, and equation 7.8 is

$$\rho = \frac{1}{E_0'} \left(d\Delta\sigma - \frac{2}{3} du_d \right) \tag{7.18}$$

The ensuing differential equation is

$$\rho = \frac{1}{E_0'} \left(d\Delta\sigma - \frac{2}{3} \frac{d^2}{k} \frac{\gamma_w}{2} \frac{\partial\rho}{\partial t} \right)$$

which can be written

$$\frac{\partial\rho}{\partial t} + \frac{3E_0' k\rho}{\gamma_w d^2} = \frac{3\Delta\sigma k}{\gamma_w d} \tag{7.19}$$

The complementary function

$$\frac{\partial\rho}{\partial t} + \frac{3E_0' k}{\gamma_w d^2} \rho = 0 \tag{7.20}$$

provides the first part of the desired solution, which is of the form

$$\rho = A\mathrm{e}^{Bt}$$

so that

$$\frac{\partial \rho}{\partial t} = AB\mathrm{e}^{Bt}$$

and equation 7.20 becomes

$$AB\mathrm{e}^{Bt} + \frac{3E_0'k}{\gamma_\mathrm{w}d^2}A\mathrm{e}^{Bt} = 0$$

which yields

$$B = -\frac{3E_0'k}{\gamma_\mathrm{w}d^2}$$

The whole solution must therefore be of the form

$$\rho = A\exp\left(-\frac{3E_0'k}{\gamma_\mathrm{w}d^2}t\right) + C \tag{7.21}$$

so that in equation 7.19 we obtain

$$-\frac{3E_0'k}{\gamma_\mathrm{w}d^2}A\exp\left(-\frac{3E_0'k}{\gamma_\mathrm{w}d^2}t\right) + \frac{3E_0'k}{\gamma_\mathrm{w}d^2}\left[A\exp\left(-\frac{3E_0'k}{\gamma_\mathrm{w}d^2}t\right) + C\right] = \frac{3\Delta\sigma k}{\gamma_\mathrm{w}d} \tag{7.22}$$

If equation 7.22 must be true during a range of values of time t, then those parts of equation 7.22 which do not vary with time must be eliminated, that is

$$\frac{3E_0'k}{\gamma_\mathrm{w}d^2}C = \frac{3\Delta\sigma k}{\gamma_\mathrm{w}d}$$

$$C = \frac{\Delta\sigma d}{E_0'}$$

which allows equation 7.21 to be rewritten as

$$\rho = A\exp\left(-\frac{3E_0'k}{\gamma_\mathrm{w}d^2}t\right) + \frac{\Delta\sigma d}{E_0'}$$

or using

$$C_0 = \frac{E_0'k}{\gamma_\mathrm{w}}$$

$$\rho = A\exp\left(-\frac{3C_0t}{d^2}\right) + \frac{\Delta\sigma d}{E_0'} \tag{7.23}$$

It remains only to fix the initial isochrone in the new family, which must coincide with the last of the previous group represented by equations 7.15 and 7.16: making the substitution

$$\frac{\Delta\sigma d}{3E_0'} = A \exp\left(-\frac{3C_0}{d^2}\frac{d^2}{12C_0}\right) + \frac{\Delta\sigma d}{E_0'}$$

$$A = -\frac{2}{3}\frac{\Delta\sigma d}{E_0'}\exp(1/4)$$

so that

$$\rho = \frac{\Delta\sigma d}{E_0'}\left[1 - \frac{2}{3}\exp\left(\frac{1}{4} - \frac{3C_0 t}{d^2}\right)\right] \quad (7.24)$$

The most useful way of expressing the two settlement–time relationships of equations 7.14 and 7.24 is in terms of the proportional settlement $R = \rho/\rho_{ult}$, where $\rho_{ult} = \Delta\sigma d/E_0'$, and the non-dimensional group $T = C_0 t/d^2$. The relationships then become

$$R = \sqrt{\left(\frac{4}{3}T\right)} \quad \text{for} \quad T \leqslant \frac{1}{12}, R \leqslant \frac{1}{3} \quad (7.25)$$

and

$$R = 1 - \frac{2}{3}\exp\left(\frac{1}{4} - 3T\right) \quad \text{for} \quad \frac{1}{12} < T \leqslant \infty, \frac{1}{3} < R \leqslant 1 \quad (7.26)$$

The values of R and T are listed for convenience in table 7.1. If the parameters C_0 (which is $E_0'k/\gamma_w$) and d (which is the longest drainage path) are known, then T can be obtained for any period t. The proportional settlement R can then be predicted. The actual settlement ρ can be calculated if the ultimate settlement (which is $\Delta\sigma d/E_0'$) is known, which means that the load increment $\Delta\sigma$ must be known.

Terzaghi was able to solve the one-dimensional consolidation problem without assuming the shape of the isochrones: his 'exact' solution, which appears in Terzaghi and Peck, for example, would differ numerically from that derived above, but never by more than 0.04 in the R column of table 7.1. This difference is so small in comparison with the assumption which both methods make, namely

k constant, even though void ratio is changing

E_0' constant, even though soils are logarithmically hardening

that the 'exact' solution will be dispensed with. Indeed, by referring back to equation 7.3 you will see that the very crude 'average triangular isochrone' approach led to a prediction that drainage would be complete when

$$t = \frac{d^2\gamma_w}{E_0'k} = \frac{d^2}{C_0}$$

or

$$T = 1$$

which is itself quite satisfactory, if that were all that were required.

TABLE 7.1

Time factor $T = \frac{C_0 t}{d^2}$	Proportional settlement $R = \rho/\rho_{ult}$ Solution using parabolic isochrone		Terzaghi's 'exact' solution
0	0		0
0.01	0.12	advancing	0.12
0.02	0.16	parabola	0.17
0.04	0.23		0.23
0.06	0.28		0.28
0.08	0.33		0.32
0.10	0.37		0.36
0.15	0.45		0.45
0.20	0.53		0.51
0.25	0.60		0.58
0.30	0.65	reducing	0.62
0.40	0.74	parabola	0.70
0.50	0.81		0.77
0.60	0.86		0.82
0.70	0.90		0.86
0.80	0.92		0.89
0.90	0.94		0.92
1.00	0.96		0.94
∞	1.00		1.00

7.4 Applications of One-dimensional Consolidation Theory

7.4.1 Measurement of C_0

The one-dimensional compression test can be used to obtain a measurement of C_0 (and a test of the consolidation theory) if the settlement–time relationship is obtained after a load increment has been applied. The easiest method is to use a graph of settlement against $(\text{time})^{1/2}$, as shown in figure 7.9. The initial part of the curve ought, by equation 7.25, to be a straight line: extrapolated up to the line of completed settlement it will intersect at some time t_x. C_0 is then calculated as

$$C_0 = \frac{3d^2}{4t_x} \tag{7.27}$$

A typical small sample will initially be 18 mm thick, so with double drainage $d = 9$ mm. If the time t_x was 20 min, then C_0 would be almost exactly

3 mm²/min, which would be a typical laboratory value for a glacial sandy clay.

This direct approach to the prediction of consolidation rates is often preferred to the use of the fundamental relationship $C_0 = E'_0 k/\gamma_w$ since the permeability k of a fine-grained soil is so very small that it is very difficult to determine. The theoretical variation of C_0 is easily demonstrated in this way, however. A soil containing only clay minerals might have a permeability in the range 10^{-9} to 10^{-12} m/s. The engineer is unlikely to encounter the soil so soft that its stiffness E'_0 is less than 1000 kN/m² or so hard that its stiffness exceeds 100 000 kN/m². Bearing in mind that low permeability is associated with small void ratio and therefore high stiffness, it is unlikely that pure clay soils will have a consolidation coefficient C_0 outside the range 10^{-8} m²/s to 10^{-5} m²/s (0.6 to 600 mm²/min: 0.3 to 300 m²/year). A typical glacial 'clay' containing sand and silt and subjected to precompression during the ice ages, will probably have a modulus $E'_0 \approx 20\,000$ kN/m² and a permeability $k \approx 10^{-10}$ m/s so that C_0 will be roughly 2×10^{-7} m²/s (12 mm²/min, 6 m²/year). A silty sand might have $E'_0 \approx 50\,000$ kN/m² and a permeability 10^{-4} m/s so that C_0 will be roughly 0.5 m²/s (3×10^7 mm²/min, 1.5×10^7 m²/year). Following these guidelines, the engineer employing 18 mm thick samples in the standard cell is likely to observe times t_x (defined in figure 7.9) between 100 min for a soft fine clay and 0.12 ms for a silty sand which is, of course, unmeasurable.

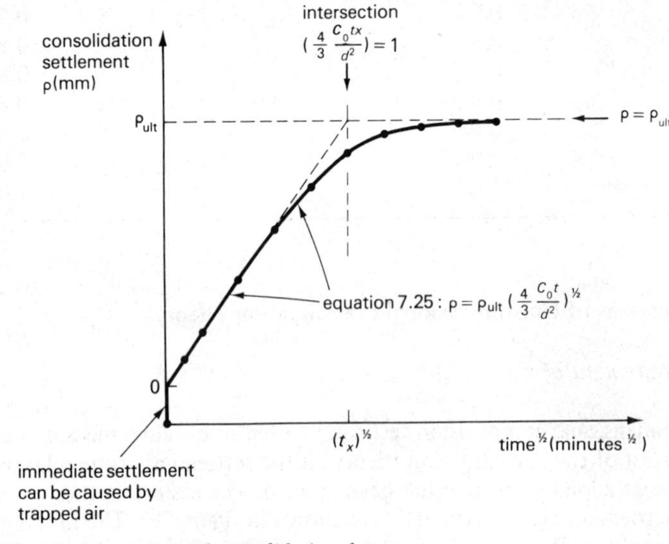

Figure 7.9 *Interpretation of consolidation data*

7.4.2 Rates of Strain for Shear Box Tests

The standard small shear box sample is 60 mm x 60 mm x 20 mm, and is assembled so that two-way drainage is permitted, and d = 10 mm. If a clay of known consolidation coefficient C_0 = 1 mm²/min, for example, were being tested then the times for 10 per cent and 90 per cent consolidation would be easily found from table 7.1: they are 40 s and 70 min in this case. If more than 40 s elapses under a new stress application then the clay sample as a whole will be 10 per cent of the way towards its new equilibrium void ratio, so that the assumption of 'undrained' behaviour may be suspect. If less than 70 min elapses before a so-called drained shear test is completed, then substantial average excess pore-water pressures may still be left, which the conductor of the test will later be forced to ignore. This situation is slightly modified when the overriding importance of the central plane in a shear-box test is recognised. Perhaps no great damage will have been done to a so-called undrained sample if the isochrone from the upper and lower boundaries has not yet reached the failure plane. This will occur when $T = 1/12$, which corresponds to 8 min in our example. On the other hand, the greater than average residual water pressures on the central plane may cause substantial errors to a 'drained' analysis even after 70 min have elapsed: if the central pore-water pressure is to drop to 10 per cent of its excess value then 90 min must elapse. A much greater alteration to elapsed times is made when the experimenter decides that no excess water pressure should arise during the course of the whole experiment. He may imagine the whole 10 mm shear displacement split into 1 mm steps and demand that each small step should be fully drained, so that the whole test occupies about 15 hours. Only if this were done could he successfully apply effective stress analysis at *any* stage of the test.

7.4.3 Prediction of the Rate of One-dimensional Field Consolidation

Let us reconsider the problem described in section 6.6, in which a sand and a clay stratum are to be subjected to one-dimensional compression by a uniform surcharge of 40 kN/m^2. It ought to be clear that the compression of the sand stratum (which was estimated as 5 mm) will occur as soon as the load is applied, so fast will be its consolidation rate. The clay, however, will take much longer to compress its 20 mm.

The field isochrone pattern for the clay is shown in figure 7.10. The baseline for water pressures in the field is the hydrostatic distribution, which is where the isochrones start and finish. Since the water pressures used in the foregoing model of consolidation were those transient pressures caused by stress changes which gave rise to hydraulic gradients, it follows that any static base line must be ignored. Only the water pressure changes which take place between $t = 0$ and $t = \infty$ after a change of boundary conditions are subject to the consolidation model: these excess pore pressures are displayed in figure 7.11. Suppose that a sample 40 mm thick was obtained from the clay, and that it was installed in a one-dimensional compression machine and brought into initial equilibrium under an effective vertical stress of 60 kN/m^2. When the increment of 40 kN/m^2 was added consider, with two-way drainage allowed, that a sequence of settlements over a period of 5 hours culminated in an ultimate settlement of 0.40 mm, and that the time t_x derived from a curve such as figure 7.9 was 80 min. From equation 7.27

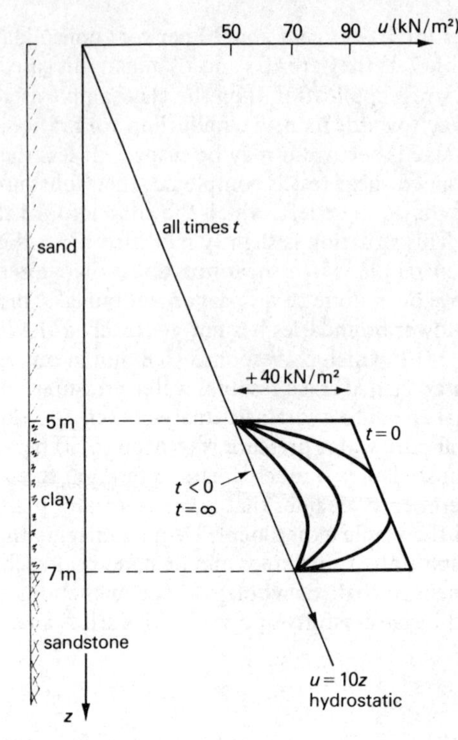

Figure 7.10 *Total pore pressures*

$$C_0 = \frac{3 \times 20^2}{4 \times 80} = 3.75 \text{ mm}^2/\text{min}$$

$$= 1.95 \text{ m}^2/\text{year}$$

Now assume that the permeability of sand above and the sandstone below the clay stratum in the field is relatively large, so that free two-way drainage will take place. The field drainage path being 1 m, the nondimensional time factor in the field

$$T = \frac{C_0 t}{d^2} = 1.95t \text{ (when } t \text{ is in years)}$$

If the surcharge is all placed very quickly at time $t = 0$, then after 1 week $t = 1/52$ so $T = 0.0375$ and the proportional settlement $R = 0.22$, while after 1 month $t = 1/12$, $T = 0.16$ and $R = 0.47$, the settlement being effectively complete after 6 months. If, as is much more likely, it takes a number of months to spread the required 2 m of fill over the whole area, then it may be of some comfort to the engineer to know that the clay will settle almost in pace with the construction. In land reclamation work, it is most embarrassing to leave the filled site level at the required elevation only to find that it has sunk back into the mire by the following year due to the slow consolidation of any clay or organic strata. In this case, of course, the settlements were negligible in any event.

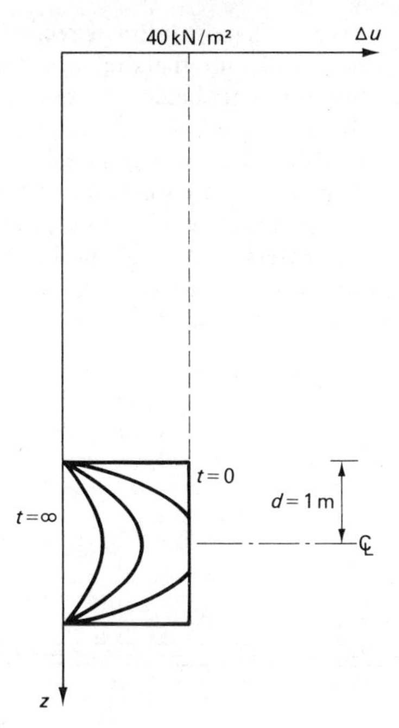

Figure 7.11 *Transient pore pressures*

7.5 Lateral Drainage

Consolidation rates are important in fine-grained soils because their response to changes in stress is so slow. Now although some of the ancient clay 'rocks' (London clay, Oxford clay, Gault clay, Weald clay) are consistent and uniform for many metres in depth, this is not true of the superficial glacial materials or the post-glacial drift which cover much of the country. Not only are laminae of silt or sand found at close intervals in glacial 'clay' strata which have been laid down in temporary lakes as the glaciers thawed, but often these laminated clays are themselves irregular. Gravel stream beds, sand banks, and bowls of peat are often found to interrupt the stratum, which may also be broken in places by fossil shrinkage cracks filled with sand or silt. Often, the hidden geometry will be so complicated that site investigation, the testing of appropriate soil samples, and the choice of 'drainage path d' for example, becomes a nightmare. A full scale site test of a prototype construction will often be justified in such a case. If a trial construction is not justified, then the engineer must have found an economic method of design and construction which can be used irrespective of consolidation rates in the superficial clay; for example, by supporting his structure on deep piles. Fresh graduates are often intrigued by their employers' evident lack of confidence in clever calculations, and amazed at the ingenuity of the experienced engineer who will so design his works that calculations will normally be redundant. I have gone out of my way here to emphasise that while the consolidation model is a useful mental guide by which to judge soil behaviour, it is associated with the soil's most intimate secrets concerning the nature, extent and interconnection of its internal drainage channels. It is very difficult indeed to translate laboratory data based on a few small samples into a good prediction of the rate of field drainage. Only in fairly uniform clayey strata will the most careful laboratory tests on large samples lead to a good order-of-magnitude estimate of the rate of consolidation in the field.

Having made this disclaimer I feel free to explain in greater depth the significance of internal horizontal sheets of more permeable material to the consolidation of a clayey stratum. Consider a section through the long foundation which is shown in figure 7.12, creating an initially rectangular pore-

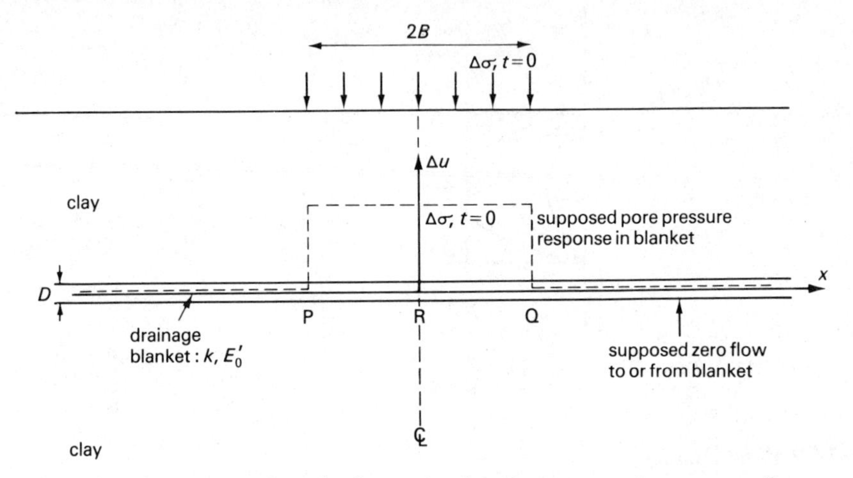

Figure 7.12 *Cross section through a long strip of surcharge $\Delta\sigma$ causing a supposedly rectangular excess pore-pressure distribution in an infinite drainage blanket*

pressure distribution within the saturated soils beneath, including the drainage blanket. We will ignore the fact that the initial effect of the strip load on the surface of the clay would not be a perfectly rectangular pressure distribution in the blanket. On the other hand, we will be able to explore how the water-pressure distribution in the drainage blanket changes with time.

Let us assume that the enveloping clay is very much less permeable than the drainage blanket. This will force us to the conclusion that the drainage blanket must remain at constant volume for the duration of our calculation. It ought to be clear that the large outward hydraulic gradients in the region of P and Q cause the isochrones to spread out. If $D \ll B$, the water pressure in the drain should be constant with depth, so that our example is a special case of two-dimensional consolidation in which the drainage is purely one dimensional. A reduction in water pressure ($-\delta u$) at any point in the blanket must then cause a settlement of the land above that point due to the accompanying compression of the blanket following the rise in average principal effective stress. If we ignore any tendency for horizontal displacement, then we can alternatively specify the vertical settlement $\delta\rho$ due to the change in vertical effective stress $\delta\sigma'_v = -\delta u$ (since there is no further change in the total stress supporting the overlying ground) using

$$\frac{\delta \rho}{D} = \frac{1}{E_0'} \delta \sigma_v' = -\frac{1}{E_0'} \delta u \tag{7.28}$$

This very close connection between a change in water pressure and the settlement at any point makes the 'constant volume' condition for the blanket easy to apply to the problem of drawing the isochrones. A change in volume would mean an average settlement (considering the whole drainage layer up to infinity): no change in volume means no average settlement and therefore no average change in water pressure. This means that the area of the isochrones must remain constant at $2B\Delta\sigma$ as they spread outwards. If

$$\int \delta u \, \delta x = 0$$

then by equation 7.28

$$\int \delta \rho \, \delta x = 0$$

so that the overall volume of the drain remains constant.

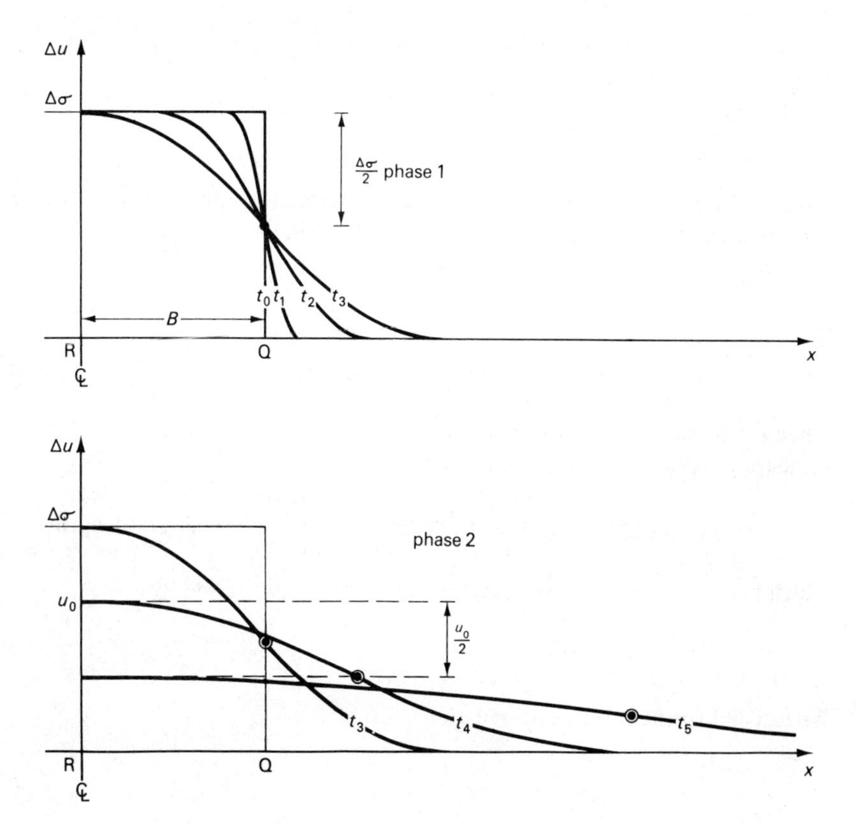

Figure 7.13 *Pairs of mirror-image parabolas with a constant enclosed area of* $B\Delta\sigma$

This equal-area condition suggests the use of a balanced pair of spreading parabolas to represent the isochrones: they are shown in figure 7.13. Consider phase 1, in which the fronts are travelling at identical speeds into the zone PQ as a pressure reduction, and out towards either infinity as a pressure increase. Figure 7.14 is a very close analogy to figure 7.4. Each shaded area is

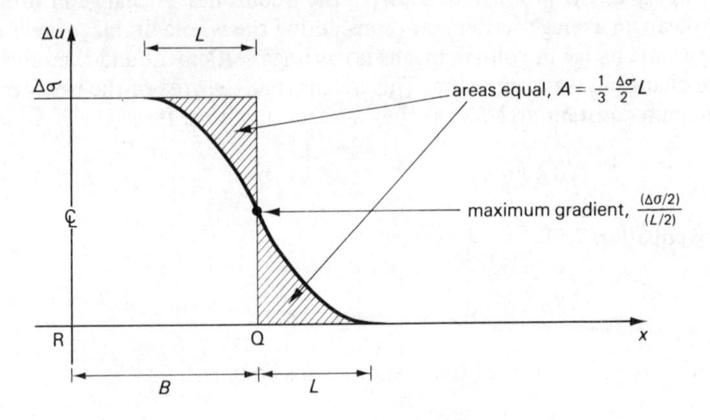

Figure 7.14 *Phase 1 isochrone: mirror symmetry about Q*

$$A = \frac{1}{3}\frac{\Delta\sigma}{2}L$$

which is a property of the parabola deduced in equation 7.9. This means that the average pressure ($\bar{u}$) in the loaded region RQ is given by

$$B\bar{u} = B\Delta\sigma - A$$

or

$$\bar{u} = \Delta\sigma - \frac{A}{B} = \Delta\sigma\left(1 - \frac{L}{6B}\right) \tag{7.29}$$

when the pressure front has advanced a distance L. Furthermore, the rate of drainage of water from underneath the loaded area RQ is, by Darcy's law

$$Q = Dki_Q = \frac{D}{\gamma_w}k\left(-\frac{\partial u}{\partial x}\right)_Q = \frac{Dk}{\gamma_w}\frac{(\Delta\sigma/2)}{L/2} \text{ per unit length}$$

which means that the rate of average settlement $\partial\bar{\rho}/\partial t$ is given by

$$B\frac{\partial\bar{\rho}}{\partial t} = \frac{Dk}{\gamma_w}\frac{\Delta\sigma}{L} \tag{7.30}$$

We can add further that using equation 7.28

$$\frac{\bar{\rho}}{D} = \frac{1}{E'_0}(\Delta\sigma - \bar{u}) \tag{7.31}$$

which leads us to say, using equations 7.31, 7.29 and 7.30 in sequence, that

$$\frac{B\partial\bar{p}}{\partial t} = -\frac{BD}{E_0'}\frac{\partial\bar{u}}{\partial t} = +\frac{BD}{E_0'}\frac{\Delta\sigma}{6B}\frac{\partial L}{\partial t} = \frac{Dk\Delta\sigma}{\gamma_w L}$$

or

$$L\frac{\partial L}{\partial t} = 6\frac{E_0'k}{\gamma_w} \tag{7.32}$$

the solution to which we already know is

$$L = \sqrt{(12\,C_0 t)} \tag{7.33}$$

when the substitution $C_0 = E_0'k/\gamma_w$ is made. The average water pressure is then, by equation 7.29

$$\bar{u} = \Delta\sigma\left[1 - \frac{1}{6}\sqrt{\left(\frac{12\,C_0 t}{B^2}\right)}\right] \tag{7.34}$$

until the front reaches the centre line R, when $L = B$, $t = B^2/12C_0$ and $\bar{u} = \frac{5}{6}\Delta\sigma$.

Now consider phase 2, which must be responsible for the remaining five-sixths of the consolidation and settlement under the loaded area. Figure 7.15 makes clear that the change-point T between the two parabolas which represent the

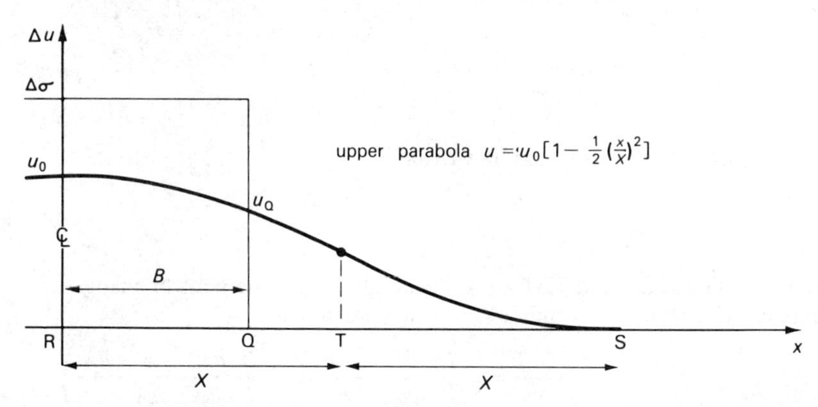

Figure 7.15 *Phase 2 isochrone: mirror symmetry about T*

isochrone must now move outwards and downwards, so as to remain at the mid-point of the peak water pressure u_0 at R, and the toe of the advancing pressure wave at S. The equations of the parabolas, based on a new origin at R, are given in the figure. Using the constant-area rule

$$u_0 X = \Delta\sigma B \tag{7.35}$$

while the average water pressure along RQ, $\bar{u}$, is given by

$$B\bar{u} = \int_0^B u\,\mathrm{d}x = u_0\left[x - \frac{x^3}{6X^2}\right]_0^B = u_0\left(B - \frac{B^3}{6X^2}\right) \tag{7.36}$$

So

$$\bar{u} = u_0 - u_0 \frac{B^2}{6X^2} \tag{7.37}$$

or, using equation 7.35

$$\bar{u} = \Delta\sigma \left(\frac{B}{X} - \frac{B^3}{6X^3} \right) \tag{7.38}$$

Now the hydraulic gradient at Q

$$i_Q = \frac{1}{\gamma_w} \left(-\frac{\partial u}{\partial x} \right)_Q = \frac{1}{\gamma_w} \frac{u_0}{2} \frac{2B}{X^2} = \frac{u_0 B}{\gamma_w X^2}$$

so the rate of outflow

$$Q = \frac{Dku_0 B}{\gamma_w X^2} \tag{7.39}$$

and the rate of average settlement $\bar{\rho}$ is given by

$$B \frac{\partial \bar{\rho}}{\partial t} = \frac{Dku_0 B}{\gamma_w X^2}$$

or

$$\frac{\partial \bar{\rho}}{\partial t} = \frac{Dku_0}{\gamma_w X^2}$$

which using equation 7.35 can be written

$$\frac{\partial \bar{\rho}}{\partial t} = \frac{Dk\Delta\sigma B}{\gamma_w X^3} \tag{7.40}$$

Now equations 7.28 and 7.31 are still valid, linking settlement to change of water pressure. Equations 7.38 and 7.40 can then be linked

$$\frac{\partial \bar{\rho}}{\partial t} = \frac{Dk\Delta\sigma B}{\gamma_w X^3} = -\frac{D}{E'_0} \frac{\partial \bar{u}}{\partial t} = +\frac{D}{E'_0} \Delta\sigma \left(\frac{B}{X^2} \frac{\partial X}{\partial t} - \frac{B^3}{2X^4} \frac{\partial X}{\partial t} \right)$$

It follows that

$$X \frac{\partial X}{\partial t} - \frac{B^2}{2X} \frac{\partial X}{\partial t} = \frac{E'_0 k}{\gamma_w} = C_0 \tag{7.41}$$

which can be integrated at once to give

$$\left[\frac{X^2}{2} - \frac{B^2}{2} \ln X \right]_{X=B}^{X} = C_0 (t - t_{(X=B)})$$

where $t_{(X=B)} = B^2/(12C_0)$, being the time at the end of the first phase. We therefore obtain

$$\frac{X^2}{2} - \frac{B^2}{2} - \frac{B^2}{2} \ln X + \frac{B^2}{2} \ln B = C_0 t - \frac{B^2}{12}$$

or

$$\frac{1}{2}\left(\frac{X^2}{B^2}-\frac{5}{6}-\ln\frac{X}{B}\right)=\frac{C_0 t}{B^2} \tag{7.42}$$

This is easily tabulated for values of X/B, which can also be used to determine the average remaining water pressure $\bar{u}$ using equation 7.38. After defining the proportion of consolidation in the usual manner, $R = 1 - \bar{u}/\Delta\sigma$, table 7.2 sets out the solutions to equations 7.34 and to 7.42 and 7.38 in numerical form.

The main lesson to be learnt from this exercise is that laminations of highly permeable material which spread continuously over an area much larger than the loaded area, can act as sinks for pore fluid as their water pressure drops due to lateral drainage. You will see from table 7.2 that the average water pressure in

TABLE 7.2

Water pressure ratio on centre line $u_0/\Delta u$	Spread ratio phase 1: $\frac{B+L}{B}$ phase 2: $\frac{2X}{B}$	Proportion of consolidation under the load R	Time factor $T = \frac{C_0 t}{B^2}$
1	1	0	0
1	1.2	0.03	0.0033
1	1.4	0.07	0.013
1	1.6	0.10	0.030
1	1.8	0.13	0.053
1	2.0	0.17	0.083
end of phase one and start of phase two			
0.91	2.2	0.22	0.14
0.83	2.4	0.26	0.21
0.77	2.6	0.31	0.30
0.71	2.8	0.35	0.39
0.63	3.2	0.42	0.62
0.56	3.6	0.47	0.91
0.50	4.0	0.52	1.24
0.40	5.0	0.61	2.2
0.33	6.0	0.67	3.5
0.25	8.0	0.75	6.9
0.20	10.0	0.80	11.3
0.10	20.0	0.90	48
0.05	40.0	0.95	198

the drainage layer under the load has dropped to 20 per cent of its original value (that is, $R = 0.80$) when the time factor $T = 11$. If the drainage layer had the following properties

$$E'_0 = 20\,000 \text{ kN/m}^2$$
$$k = 10^{-7} \text{ m/s}$$

that is

$$C_0 = 2 \times 10^{-4} \text{ m}^2/\text{s}$$

perhaps corresponding to a fine silt, *and no matter what its thickness might be*, it would be 80 per cent efficient as a potential internal drain under a strip load 20 m wide after a period

$$t = \frac{11B^2}{C_0} = \frac{1100}{2 \times 10^{-4}} = 5.5 \times 10^6 \text{ s} = 2 \text{ months}$$

if the surrounding clay were very much less permeable than the drain. The silt layer would also have to spread at least 100 m on each side of the loaded strip so that the spread ratio defined in table 7.2 could be the required factor of 10.

The qualification concerning the impermeability of the clay relative to its internal drainage layer is most significant. If much water begins to drain out of the clay and into the silty lamination while the latter is engaged in its own process of lateral drainage, then each isochrone profile in the lamination will have to remain in operation much longer so as to clear the excess water. The most severe departure from the assumption of isolation of the blanket would occur when sandy layers of thickness D occurred at such a regular close spacing H that the intervening clay was always at the same pore pressure as its neighbouring sand. If the properties of the sand layer were E'_S and k_S and those of the clay E'_C and k_C, it would be possible to view the *composite* as a horizontal drainage layer possessing a net permeability

$$k_N = (k_S D + k_C H)/(D + H)$$

and a net stiffness

$$E'_N = (E'_S D + E'_C H)/(D + H)$$

These net properties could be converted into a net consolidation coefficient

$$C_N = \frac{k_N E'_N}{\gamma_w}$$

prior to the use of table 7.2 to deduce consolidation times. Consider a typical example in which $D = 3$ mm, $H = 60$ mm, $k_S = 10^{-5}$ m/s, $k_C = 10^{-10}$ m/s, $E'_S = 20\,000$ kN/m^2, $E'_C = 10\,000$ kN/m^2: these figures refer to a silty clay with frequent silty sand layers. The net permeability k_N would be 5×10^{-7} m/s, representing effective flow through only that 5 per cent of the whole cross section which comprised sand. The net stiffness E'_N would be almost indistinguishable from that of the clay at 11 000 kN/m^2. Such an approach would neglect the leakage of water to the ground surface and would therefore probably be highly pessimistic in its prediction of the consolidation time, especially when the notional spreading of the pressure was large after long passages of time.

The only way of predicting consolidation rates accurately in a complex deposit is to test a great number of block samples recovered from various locations in the field, and to enter their properties into a sophisticated computer model which accounts accurately for two-dimensional or three-dimensional compression and drainage. If large shear stresses were being generated, then the resulting internal pore-water pressure changes should also be estimated and inserted in the computer model. Changes of stiffness and permeability due to reductions of void ratio would also be important. Rather than entertain such notions, the engineer will normally use crude estimates, such as those I have described, to decide whether the effect of the imposition of surface loads should be estimated by a trial construction in the field, or whether they should be avoided altogether, the omens being so bad that the expense of a trial construction was not warranted. These difficult decisions often afflict those engineers who are responsible for new roads, flood levees and embankment dams which, by their very nature, tend to be sited on undeveloped and difficult ground.

7.6 Transient Flow due to Construction

Almost any conceivable construction on or in clayey soil will cause a transient flow of pore water. This flow will not only cause ground movements due to changes of volume, but also changes in the soil's undrained strength or 'cohesion' c_u. Ground movements may continue to occur years after the construction company has finished its work, so slow is the movement of water in clayey soils. On the other hand, if this slowness of permeation is relied on in a temporary construction, the engineer may find that sand or silt partings in the clay reduce consolidation times so much that significant softening can occur before the job is finished. Every designer or constructor of earthworks should be able, at the least, to hazard a guess of the direction of transient flow – in or out: even if he cannot predict the speed he should know whether the internal water pressures are tending to rise or fall.

Figures 7.16 to 7.23 depict eight circumstances in which transient flows cause changes in strength and ground movements. The simple foundation construction on saturated clay is depicted in figure 7.16. The foundation stress $\Delta\sigma$ causes an increase in the average total stress of the soil and in its shear stresses. The immediate (speed of sound) effect of the increase in shear stresses is a shear deformation at constant volume which allows the foundation to settle. In

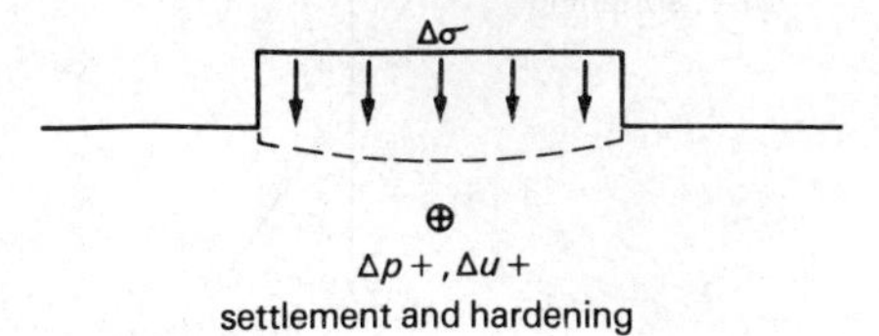

Figure 7.16 *Foundation on clay*

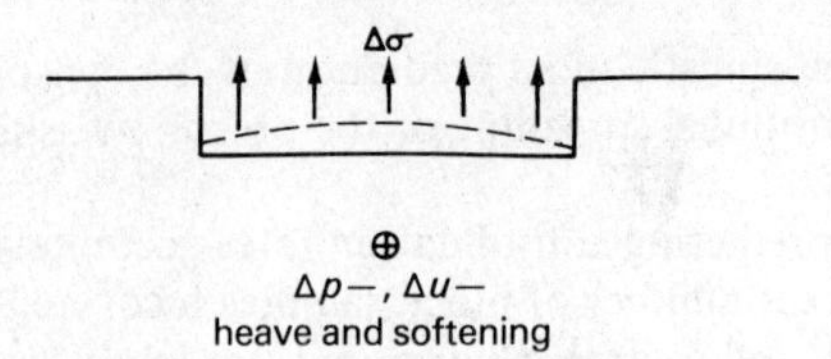

Figure 7.17 *Excavation in clay*

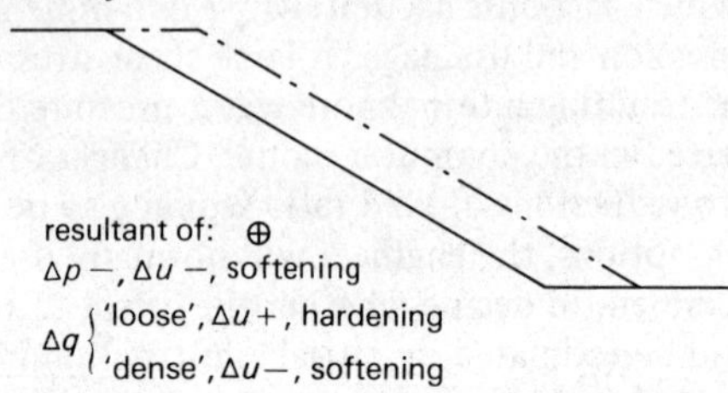

Figure 7.18 *Cutting in clay*

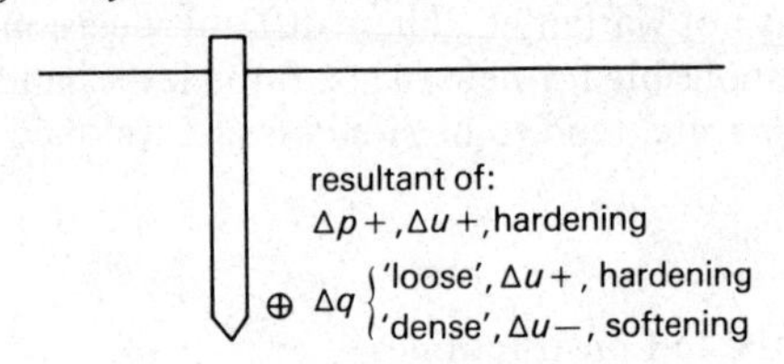

Figure 7.19 *Pile driving*

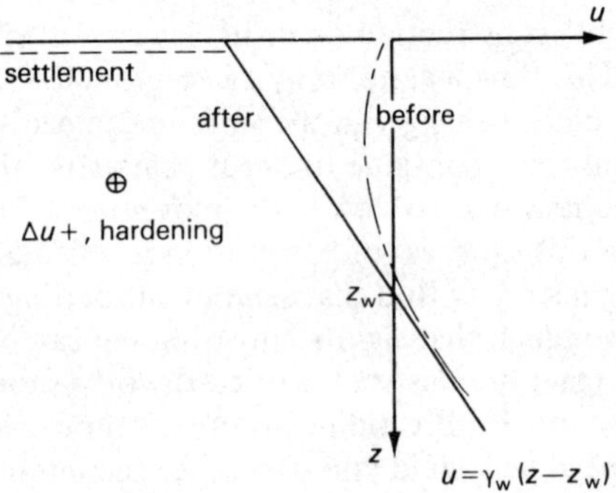

Figure 7.20 *Paving over clay in wet season*

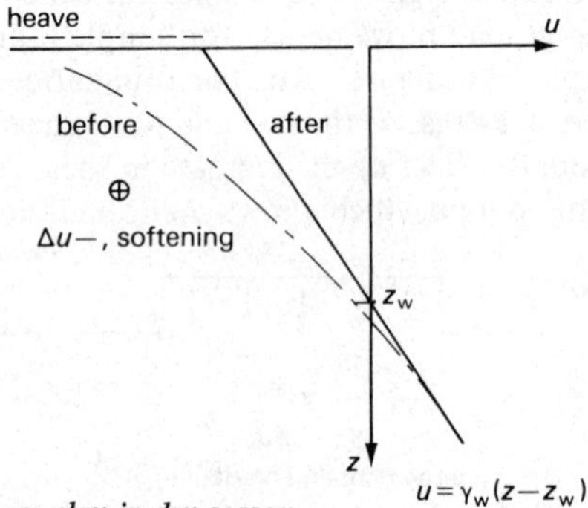

Figure 7.21 *Paving over clay in dry season*

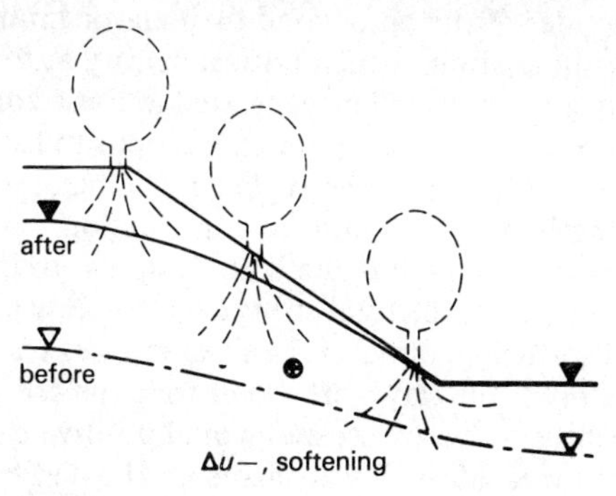

Figure 7.22 *Clearing vegetation*

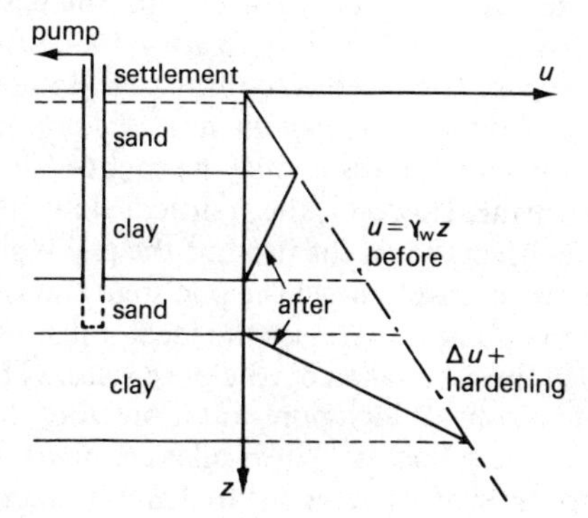

Figure 7.23 *Dewatering*

section 6.7 you saw that the immediate settlement on an infinite elastic bed could never be less than one-half the total settlement. The immediate effect of the rise in spherical stress is an increase in pore-water pressure. As time passes these excess water pressures will dissipate by transient flow as the soil compresses causing the ground to settle further. If the stress $\Delta\sigma$ is very large it may tend to cause such large shear stresses close to the edges of the foundation that the soil approaches its frictional limit: in that case 'loose' soil would tend to generate extra pore-water pressures in the shear zones while 'dense' soil would tend to generate suctions. It is quite likely that a small heavily loaded footing on stiff clay will slowly settle as the clay which supports it compresses, and while clay under the edges of the footing softens and expands.

The excavation depicted in figure 7.17 is analogous to the previous foundation. The majority to heave due to the reduction of stress would be generated by an immediate shear deformation at constant volume. The reduction in the average total stress would also cause an immediate drop in pore-water pressure, which would slowly be relieved by transient flow into the soil around the excavation. After the initial heave, therefore, a continuing heave will indicate that the ground is softening in order to come into equilibrium with the reduced stresses. If the soil is weak relative to the changes in stress caused by the excavation, then the immediate shear deformation may well carry the state of soil elements near the sides of the pit beyond their elastic limit. These elements are likely to be 'dense' in relation to their critical density at the newly reduced stress level, so that extra suctions will be generated in these zones, giving rise to extra softening with time. The tendency for excavations to cause softening of clay is a matter of serious concern to the engineer who has based his expectations on the undrained strength of the soil prior to construction.

The cutting back of a slope, as in figure 7.18, is similar to the problem of excavating a pit. Not only does the total stress decrease, causing an immediate suction and a delayed softening, but also the shear stresses tend to increase at the toe of the slope. Only if the clay were looser than the critical state appropriate to the eventual effective stresses would positive pore pressures be generated in the shear zones. Most clay elements will be denser than their critical state, and so a shear stress which exceeds the elastic limit will tend to give rise to suction. Cuttings, like excavations, tend to soften with the passage of time.

The process of pile-driving, symbolised by figure 7.19, causes quite complicated changes in stress. The pile has to displace soil in order to be driven downwards: the lateral ground pressure can be increased enormously. Indeed in dense sands and gravels the lateral pressure may be so great that skin friction is able to prevent further driving. However, the transient flow mechanism may be of great assistance to the driver. When the head of the pile is struck by the hammer, a compression wave travels down the pile stem towards the point. When the wave reaches the point it increases the local soil stresses enormously and this increased burden must be taken by the pore water. The pile finds it much easier to penetrate a zone of high pore-water pressure than it would to enter a zone of high effective soil stress. When piles are being driven through clay, the soil adjacent to the shaft is being sheared and horizontally compressed. If the clay is rather soft, these effects will combine to generate positive pore pressures: as these dissipate in the period after driving, the resistance of the pile will increase. If the clay is very stiff, the negative pore pressures due to shearing may predominate. In this case the resistance of the pile will decrease as the water pressures return to equilibrium.

Figures 7.20 to 7.23 depict processes which directly alter pore pressures. Paving a stratum of soil with an impermeable skin has the effect of inhibiting vertical drainage. The ultimate pore-pressure distribution in such a soil must be hydrostatic. If the soil concerned is fine-grained with a high capillary suction potential and a moderately shallow groundwater level, then the soil under the pavement is likely to remain saturated, with water pressures changing to become $u = \gamma_w(z - z_w)$, as shown in figures 7.20 and 7.21. Before paving, the soil is unlikely to have possessed a hydrostatic water pressure distribution. Figure 7.20 depicts an initial wet-season profile in which surface water is migrating into the ground. The pavement causes a reduction in water pressures by sealing the flow: the effective stresses therefore rise, inducing settlement. Figure 7.21 depicts a dry-season profile in which the initial pore pressures are lower than hydrostatic so as to generate an upward flow of water which would evaporate at the ground surface. When the drying zone is paved the water pressures rise, causing softening and heave which can threaten the integrity of the pavement.

Figure 7.22 is almost equivalent to 7.21. Vegetation sucks water upwards and transpires it. When this vegetation is cleared, the zone of suction can relax. If the water table had been in equilibrium with long-established trees then the local water table would rise when they were felled, causing a softening effect even at depth in the clay and a notable heave. The difference between 7.22 and 7.21 is purely one of time period. In 7.21 the suction was very local since it was generated by a hot season only a few months in duration. If the consolidation coefficient of the clay was 1 m^2/year then in 3 months a zone $L = \sqrt{(12C_0\ 3/12)} = 1.7$ m deep can be affected. The trees of 7.22 may have been pumping water upwards for years.

In figure 7.23 the effects of groundwater lowering are shown. Typical estuarial or deltaic deposits consist of alternating sands and clays. If an engineer wishes to make an excavation in one of the permeable sand strata he may choose to install pumps which reduce the local water pressures to zero (atmospheric). The water pressures in neighbouring sand strata may remain constant, due to their very great lateral extent. The final pore-pressure distributions are linear, representing uniform flow towards the pumped stratum. As the isochrones of pore-pressure reduction spread outwards from the pumped sand layer, they can be inferred at the surface by virtue of the wave of settlement caused by the increase in effective stress in the neighbouring clays.

The convention throughout figures 7.16 to 7.23 has been to label as $\Delta u+$ any process which implies that the pore pressures have been made greater than they will be eventually, either by the sudden superposition of surcharge as in figure 7.16 or by a change in boundary water conditions which implies that long-term pore pressures will be reduced as in figure 7.20. Likewise the label $\Delta u-$ has been used for any process which implies that pore pressures will slowly increase due to their present values being reduced in comparison with their future values, whether caused by a sudden reduction in total stress as in figure 7.17 or by the long-term increase in boundary water pressures of figure 7.22.

If, as is the general case in figures 7.16 to 7.23, the transient pore pressure Δu varies from point to point around the source of the perturbation, the foregoing relationships between R and T should be amended, based as they were on the simple one-dimensional relaxation of an initially rectangular isochrone. This can, of course, be done approximately by imposing new families of isochrones: Terzaghi and Peck (1967) also include a few initially triangular distributions which have been solved 'exactly'. No simple treatment of three-dimensional transient flow is yet available however, although the relationship of equation 7.13 may be a useful guide to the general propagation of pore pressure transients.

7.7 Problems

(1) A saturated sample of clay was recovered from the base of a 2 m deep pit in a 9 m deep stratum which overlay gravels with a groundwater level at a depth of 1 m below the ground surface. It was trimmed with as little disturbance as possible to a size of 76 mm diameter and 20 mm thick to suit a small oedometer. A lever loading mechanism was used to magnify the loads on a counterbalanced hanger by a factor of 11.5. The sample was initially allowed to come into equilibrium under a load of 1 kg. When the dial guage recording the

settlement had been set to zero an increment of load of 4 kg was applied to the hanger as a clock was started. The sequence of settlements was

t(min)	1/4	1	4	9	16	25	36	49	64	81	100	200
ρ(mm)	0.125	0.150	0.205	0.265	0.320	0.370	0.420	0.455	0.480	0.500	0.515	0.530

(a) determine a useful value for the one-dimensional consolidation coefficient
(b) find the stiffness E'_0 and deduce the permeability k
(c) estimate the time which must elapse before the whole stratum could have dissipated 80 per cent of the excess pore pressures caused by the addition of a relatively wide surface foundation
(d) estimate the proportion of settlement after the first year of this loading
(e) estimate the depth of the zone of clay which is likely to soften due to its suction being more than 50 per cent destroyed in a 3 month wet season
(f) are the estimates in (b) and (c) valid for the response of the stratum to a permanent drawdown of groundwater pressures in the underlying gravels?
(g) discuss the decisions you made in order to answer all these questions.

(2) A temporary 2 m deep trench cuts into an extensive horizontal stratum of organic silt which had previously come into equilibrium with a water table at the ground surface, and begins to draw the water table down to 2 m below ground level. The organic silt is itself 2 m deep and overlies a stiffer and very much less permeable inorganic clay. The one-dimensional consolidation coefficient of the organic silt from both vertical and horizontal cores was of the order of 20 mm^2/min while its stiffness was of the order of 2000 kN/m^2. Investigate the possible wave of settlement propagating from the trench by determining settlement profiles after 1 week, 1 month, and 4 months, on the assumption that no rainfall will percolate into the surface. Discuss all your assumptions.

(3) A 5 m deep stratum of post-glacial silty clay with a stiffness $E'_0 \approx 4000$ kN/m^2 and a consolidation coefficient $C_0 \approx 10$ mm^2/min overlies sandstone and possesses an initial groundwater table 1 m below the ground surface. A 10 m diameter steel oil tank is to rest on a 1 m thick mat of gravel placed over the clay and is to be quickly filled to a depth of 8 m with an oil of unit weight $\gamma = 8$ kN/m^3.

(a) Make simplifying assumptions which allow you to calculate the settlement of the tank after 1 day, 1 month, 1 year and ultimately.
(b) Repeat the calculation in the event of there being an extensive 0.1 m thick blanket of silty sand ($E'_0 \approx 40\,000$ kN/m^2, $k \approx 10^{-5}$ m/s) at the centre of the clay layer.

(4) The following problems are all based on the land reclamation problem discussed first in section 6.6 and considered with respect to its speed of drainage in section 7.4.3. The 10 km^2 site can be envisaged as a strip of land 2 km x 5 km.

(a) Repeat the analysis of section 7.4.3 but attempt to take better account of the rate of dumping, which is expected to last 50 weeks. Consider the imposition of the surcharge in five sudden equal increments with a gap of 10 weeks between each one and the next and take the end of construction

as equivalent to a time 5 weeks after the last. Find the proportional settlement at the end of the construction if it was considered to be the average of the proportional settlements under each of the five increments considered separately. Discuss the likely validity of the assumption.
(b) Repeat the analysis of section 7.4.3 but taking into account an extensive blanket of silty sand 0.1 m thick ($E_0' \approx 40\,000$ kN/m^2, $k \approx 10^{-5}$ m/s) which is now supposed to exist at the centre of the clay layer.

(5) Discuss the settlement–time curve which might be observed when a conventional 2 m square pad footing resting on a deep stratified deposit of stiff silty clay with sandy laminations is loaded over a period of 1 year. Include in your discussion the proportion of settlement which will be time dependent, the depth of the soil likely to be affected, the general pattern of isochrones, and the pattern of drainage likely to be observed.

Answers (1) (a) 1.2 mm^2/min (b) 4600 kN/m^2, 4×10^{-11} m/s (c) 18 years (d) 20 per cent (e) 0.4 m (f) no, $d = 9$ m here ignoring changes at the surface.

(2)

x(m)	1	2	4	6
t		ρ(mm)		
1 week	2.5	0	0	0
1 month	9	2.5	0	0
4 months	14	9	2.5	0

(3) (a) 6, 31, 95 mm (b) 11, 63, 105 mm (the lateral time factor T for the blanket is 100 when $t = 100 \times (10/2)^2/0.04$ s $= 1000$ min so the blanket can act as a drain). (4) (a) 0.85 (b) the lateral time factor T for the blanket is 100 when $t = 100 \times (2000/2)^2/0.04$ s $= 80$ years, so the drain is certainly not functioning well at 1 month. Better ignore the drain since when $T = 0.1$ the blanket itself will have just communicated the drainage to its centre and 1 month will already have elapsed.

8

The Deformation of a Soil Element

Previous chapters have been concerned with

small strains which can be considered elastic;
dilation and contraction due to large shear strains;
critical states in which infinite shear strains develop;
speed of drainage, which controls pore-water pressures;
friction and cohesion.

The principal objective of this chapter is to introduce the triaxial test apparatus in which each of these effects can be observed with varying degrees of precision, and the Cam-clay theoretical model which incorporates friction, cohesion, dilation, contraction, critical states, elasticity and drainage in a similarly fuzzy fashion. The triaxial test and its accompanying theoretical model do not replicate any one aspect of soil behaviour better than the individual models described so far. They simply fit all the aspects together in a logical framework: in this they are almost unique. Perspective is the aim here, not focus. Before you plunge into the applications of theory to practise in the later chapters, this present chapter gives you the opportunity of putting together most of the knowledge you have already gained. My hope is that you will then be able to use your new perspective to choose an appropriate model of behaviour to fit a variety of situations.

This development was foreshadowed in chapters 4 and 5 with the description of a critical state diagram in terms of τ, σ' and e. I pretended that the void ratio of soil on the rupture surface of a direct shear-box test was meaningful and measurable. This was a useful pretext to get across the meaning of dilatancy and cohesion. It ought to be clear, however, that the void ratio of a plane is a meaningless concept! I cannot even allow you to sustain the belief that many horizontal planes in the direct shear test share the same stress ratio and strain, and that we can speak of the void ratio of the intervening soil. The discrete step in the direct shear box after the test proclaims loudly that the strains somewhat above and below the central plane are highly nonuniform. This is the problem you must now face: chapters 3, 6 and 7 on seepage, elasticity and transient flow are based on continuum models in which soil properties vary uniformly from place to place, whereas chapters 4 and 5 on friction and cohesion are based on rupture planes which are mathematical singularities and at which 'strain' in particular is impossible to measure. Before the synthesis can start, therefore, a new perspective for the models of soil in a continuum is called for.

You must be most careful in the work that follows not to assume that the new models are 'right' and the old 'wrong'. It is simply a matter of convenience. The new models are more convenient to use with triaxial tests, which are more convenient than shear box tests in some circumstances and especially when a number of modes of behaviour are to be explored. The old models will remain the most convenient for interpreting individual aspects of behaviour in detail, using the shear box or the oedometer. 'Right' and 'wrong' mean 'useful' and 'not useful' to most engineers. An artillery officer in a battle would be better advised to employ Newtonian mechanics rather than attempt to follow Einstein in the prediction of the contraction of space–time around the various projectiles. Judgement is called for.

8.1 The Triaxial Test Equipment and its Uses

The triaxial test is based on cylindrical samples which are usually twice as long as their diameter. They can be trimmed to size on a small soil lathe, compacted into preshaped formers, or retrieved almost undisturbed in fine-grained soils by forcing a thin-walled sample tube of the correct internal diameter ahead of an advancing borehole. The tube can be sealed and stored until the laboratory is ready. This ease of sample extraction, storage and preparation explains the eagerness with which the construction industry has accepted the new (*circa* 1950) test. Standard diameters are 38 mm, 50 mm, 100 mm, 150 mm and 225 mm. The larger diameters offer much greater accuracy and are rarely performed commercially. The 38 mm diameter samples give relatively poor accuracy and are used almost exclusively by the industry due to their cheapness. This is not quite as unscientific as it seems: ten crude tests at £10 each on soil from different locations may offer more valuable information in some circumstances than one accurate test for £100 on the soil from one location.

The sample is placed on a pedestal within an enclosing cell so that it can be subjected independently to an all-round water pressure and to a vertical force acting through a piston. The various facilities are sketched diagrammatically in figure 8.1. Familiarity with any equipment only comes with use, so that I shall not attempt to do other than comment on the broad strategies. The sample is squashed either by cell pressure or by a motor which drives it upwards against a fixed beam. As it is being tested the pressure of the water in the cell can be measured either by a gauge or a transducer: it is usually kept constant. The pressure of the water in the pores of the soil can be measured by a transducer connected to the base of the sample via the pedestal. Drainage through the top cap can be prevented or, if allowed, measured by a constant-pressure burette or a volume change transducer. The force caused by the compression of the sample against the reaction beam can be measured by a load cell in the piston or, risking

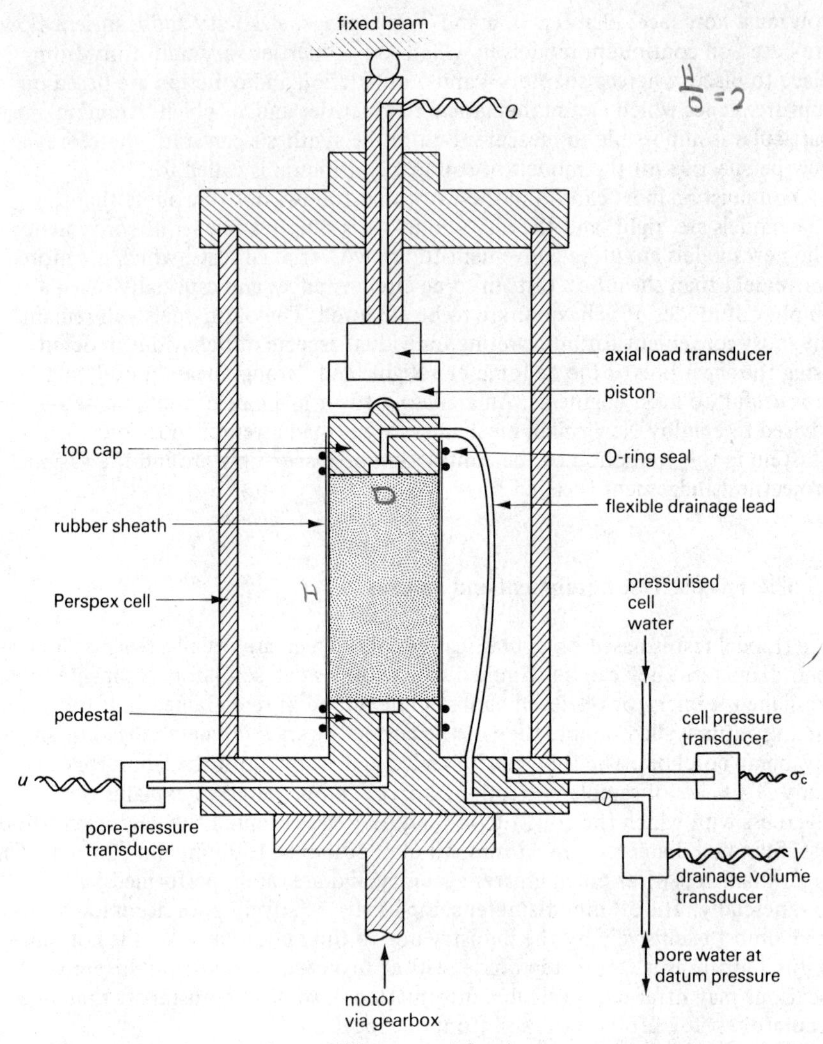

Figure 8.1 *Triaxial test equipment (diagrammatic)*

the danger of friction errors in the bush bearing, in a proving ring mounted directly under the beam. While it is being tested, the sample will be enclosed in an impermeable rubber membrane which is held tight against the top cap and the pedestal by undersized rubber O-rings and sealed against them with grease. In order to prevent restraint at its ends, the top cap and pedestal must be well lubricated at their junction with the sample. If the ends of the sample are not lubricated it will take up a barrel shape when it is squashed: it is just this sort of nonuniformity which the test was intended to prevent. The speed of the motor is vital. In an 'undrained' test with pore-pressure measurement the sample should not be tested so quickly that the pore pressures inside the sample do not have time to equalise from place to place: good practice might require the test to be spread over a working day. In a 'drained' test the motor should be set so slow that the transducer at the base always records zero excess water pressure above the level set at the top by the volume change indicator. A simple consolidation calculation can be performed to guide the choice of a suitable slow speed if the consolidation coefficient of the soil is known. A typical high-quality 'drained' test on a large sample would take a week rather than a day. The vertical strain at any time can either be deduced from the setting of the motor, or measured using a linear displacement transducer or dial gauge fixed to the reaction beam and sensing the upward movement of the cell lid. The lateral strain, if required, can be provided by a light clip with an electrical strain gauge bonded to it, placed across the central section of the sample.

Most errors are caused by clumsiness in setting the sample up. The most serious causes of error in undrained tests are

(1) the pore pressure lead has air bubbles in it which can be squashed, causing drainage

(2) air bubbles are trapped underneath the rubber sheath or against the ends of the sample

(3) the rubber sheath has a pinhole in it, causing drainage

(4) the pedestal or top cap has not been greased so that water can pass under the membrane and in or out of the soil sample at its ends

(5) the soil itself contains air, so that it cannot be tested at constant void ratio, and yet the test is reported as 'undrained' without comment.

If pore pressures during the undrained test are not required, the possibility of leakage can be reduced by placing a blank disc over the drainage pedestal: the top cap can also be blank rather than drilled if no drainage is required during the course of the test sequence.

Causes of errors in 'drained' tests are

(1) pore pressures are inadvertently caused through too-fast compression

(2) air bubbles in the system make the measurement of volume of drainage a gross underestimate

(3) leakage around the sample or through the rubber membrane causes the volume of drainage to be a gross overestimate.

Figure 8.2 shows the rationale behind the estimation of principal stresses and strains. If the ends of the sample are well lubricated, then no vertical or horizontal soil boundary suffers tangential stress. This means that vertical and horizontal stresses must be principal stresses. Similarly, if the sample remains cylindrical as it is compressed so that vertical faces remain vertical, then no shear strain has taken place with respect to horizontal and vertical axes, implying that the principal strains must also lie in these directions. In the axial compression test it should be clear that the principal compressive stresses and strains will be vertical, while the minor compressive stresses and strains will be horizontal. Normally the minor principal compressive strain will be so 'minor' as actually to be negative, that is, indicating an increase of diameter as the sample is squashed. The normal convention is to ascribe the subscript 1 to the major principal direction, 2 to the intermediate, and 3 to the minor direction. In a triaxial cell the 2 and 3 directions are equivalent, every horizontal axis being

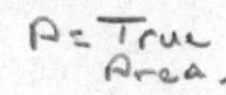

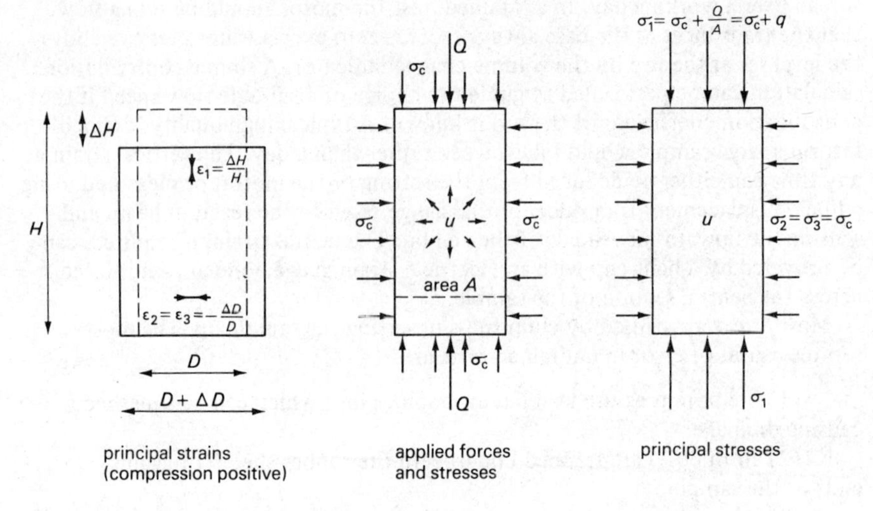

Figure 8.2 *Triaxial stress and strain*

equivalent due to symmetry. If symmetry evidently breaks down, for example due to a rupture surface inclined up-east/down-west then the relevance of any subsequent data may be challengeable due to the nonuniformity of strains.

The main forces and stresses which must be taken into account in the estimation of principal stresses are also shown in figure 8.2. The water pressure in the cell σ_c acts not only on the sides of the sample but also on the top surface. It therefore contributes equally to the major and minor principal stresses. The additional vertical compressive force Q must then be divided by the true (changing) area of the sample A in order to arrive at the deviatoric stress component q which is the stress difference between major and minor axes and therefore the diameter of the Mohr circle of stress. A subsidiary correction should be made to the minor stresses if very accurate results are required: the rubber membrane is forced into hoop tension as the sample is squashed, so that a 20 per cent increase in diameter might cause the equivalent of a 5 kN/m^2 cell pressure increase in the 2 and 3 directions, depending on the stiffness of the rubber.

The Mohr circles of stress and strain for the 1–3 plane are shown in figure 8.3. The circles can be drawn for any stage in any test, so long as the appropriate measurements have been taken. The most common practice is to plot the stress circles for a compression test when they reach their largest size, corresponding to the peak strength of the soil. Another practice is to plot the stress circles achieved when the vertical strain is 20 per cent: this may correspond roughly with a critical state if the measurements of strength, pore pressure or volume change indicate that a plateau has been reached.

In the following subsections I describe a number of standard types of triaxial test, referring to the tester's objectives, and making clear how our early models of soil behaviour are adapted so that useful parameters can be measured. Much greater detail can be found in the book by Bishop and Henkel (1962).

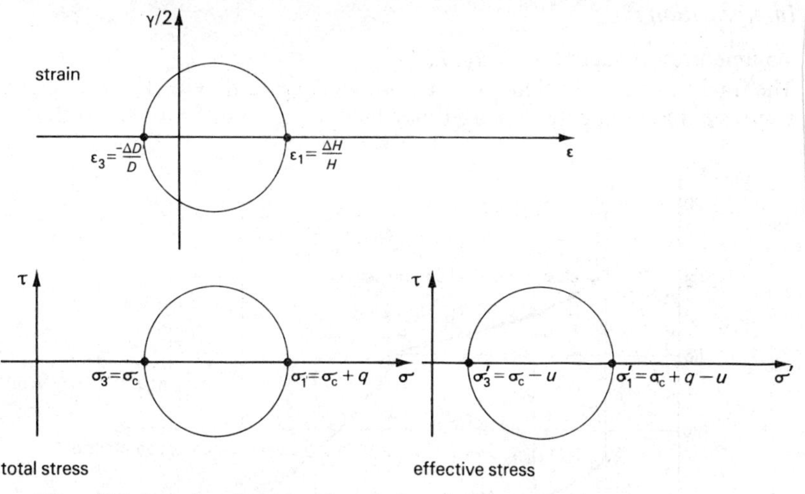

Figure 8.3 *Mohr's circles of stress and strain for triaxial compression*

8.1.1 Spherical Compression for Volumetric Stiffness E'_v, Volumetric Flexibility Indices κ_v, λ_v, and Volumetric Consolidation Coefficient C_v

Test sequence

(1) A saturated sample is obtained either artificially or from a borehole: its weight and measurements are taken.

(2) It is fitted as shown in figure 8.1, except that the piston is drawn upwards away from the sample for the duration of the test.

(3) An initial reference cell-pressure is applied, perhaps 50 kN/m^2, and the change of volume due to drainage is recorded when the drainage has ceased. In a clayey sample this may take more than a day. Water may either flow in or out of the sample, depending on whether its initial suction was greater or less than the applied cell-pressure. The pore pressure transducer reading should have returned to its datum level.

(4) An increment of cell-pressure is applied, perhaps 50 kN/m^2. The pore-pressure transducer should immediately show an equal increase. The volume change recorder should be read at regular intervals in order to chart the progress of consolidation against time: a $\Delta V/t^{1/2}$ graph will be drawn. When drainage has ceased and the pore-pressure transducer has returned to its datum reading, a further increment of cell-pressure can be applied, etc.

(5) When the pressure–volume curve has been extended in this fashion to a fairly high pressure, an exactly equivalent pressure–reduction sequence can be charted. Bearing in mind that each step will take at least 1 day even with a small sample, a whole sequence of isotropic loading and unloading in five steps would take two working weeks.

(6) At the completion of the test the sample will be stripped out, weighed, dried in an oven and reweighed so that its average moisture content and therefore its average void ratio can be determined for its final state.

Interpretation

Volume/pressure data for E'_v, κ_v, λ_v
The final equilibrium volume V under a number of different effective spherical stresses p' has been determined. It may take the form of figure 8.4. If the

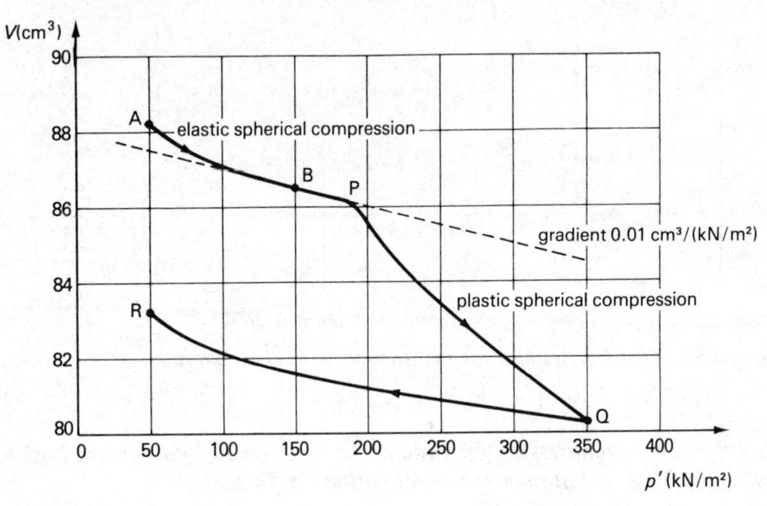

Figure 8.4 *Spherical compression data*

engineer wished to know the volumetric stiffness E'_v of this particular soil sample at 150 kN/m² pressure, point B, then he would find the volume of the sample at point B and divide by the gradient at that point

$$E'_v = \frac{dp'}{d\epsilon_v} = \frac{dp'}{dV/V} = \frac{V}{dV/dp'} = \frac{86.5 \text{ cm}^3}{(0.01 \text{ cm}^3)/(\text{kN/m}^2)}$$

So in the numerical example shown in the figure $E'_v = 8650$ kN/m² at 150 kN/m². Beyond point P the soil clearly becomes inelastic and will not unload along the same curve: for example, the swelling line from Q passes to R. Although the same numerical procedure as outlined above can be used for points on the plastic compression curve PQ in order to give a value which can be labelled E'_v, the practice is dangerous. Elasticity fails to hold along PQ and elastic methods and moduli do not apply. To take a simple example, at any particular point Q two rates of strain are possible: gradient of curve PQ at Q for loading, gradient of curve QR at Q for unloading. In practice even the so-called elastic swelling and recompression 'line' such as QR which we idealise as unique, often turns out to be a loop. If this effect is gross this also should be taken into account by the engineer. It is now obvious that the modulus of stiffness of soil depends not only on the effective stress but also on the stress-history of the soil element concerned.

Two neighbouring soil samples may give quite different V/p' curves owing simply to differences in their initial void ratio e. If they have the same mineralogy and loading history, however, their e/p' curves may be identical. It is always wise to reduce scatter where this can be done scientifically, so that graphs of e/p' are often preferred to graphs of V/p'. Suppose that the test described in figure 8.4 had been stopped at R, and that the moisture content of the saturated soil had been found to be 25.0 per cent. Assuming $G_s = 2.66$ for the soil grains, it follows from table 1.1 that the void ratio at R

$$e_R = 2.66 \times 0.25 = 0.665$$

so that the volume of soil grains (which must be constant throughout) can be calculated from

$$\left(\frac{V - V_s}{V_s}\right)_R = e_R$$

$$V_s = \left(\frac{V}{1 + e}\right)_R = \frac{83.2}{1.665} = 50.0 \text{ cm}^3$$

This means that the void ratio at any other volume can easily be calculated, for example

$$e_B = \frac{V_B - 50.0}{50.0} = \frac{36.5}{50.0} = 0.730$$

This allows an e/p' or an $e/\ln p'$ graph to be drawn as shown in figure 8.5, which

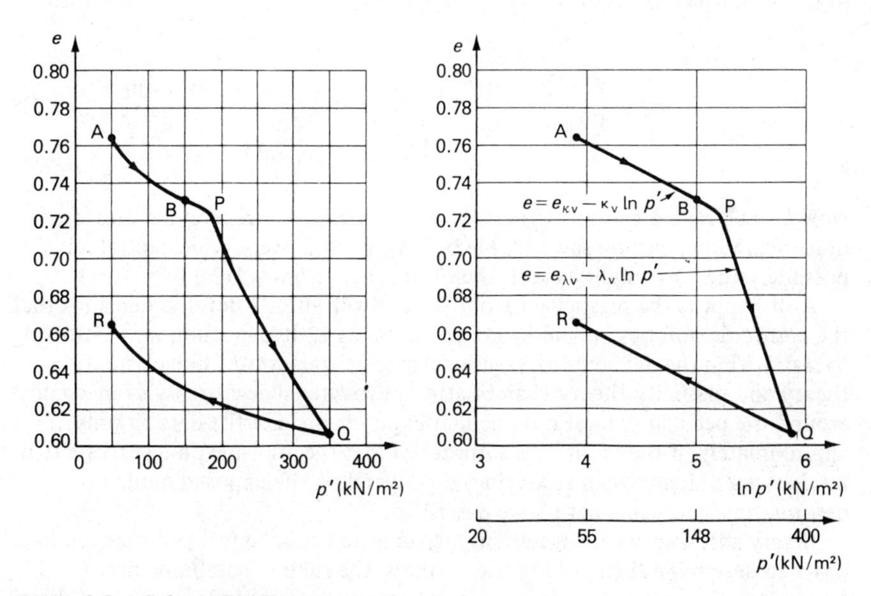

Figure 8.5 *Spherical compression graphs*

in turn allows the results of samples of different initial density to be superimposed. It should be clear that

$$d\epsilon_v = -\frac{de}{1 + e}$$

by considering the reduction in overall volume of a small reduction in void ratio. This means that

$$E'_{\mathrm{v}} = -\frac{\mathrm{d}p'}{\mathrm{d}e}(1+e) = -\frac{(1+e)}{\mathrm{d}e/\mathrm{d}p'}$$

which must yield the same answer as the previous method, of course!

There is clearly a close connection between figures 8.4 and 8.5 for spherical volumetric compression and 6.13 and 6.16 for one-dimensional compression. I have appealed to the same concepts throughout each description. You will notice however that although the form of the logarithmic graphs in figures 8.5 and 6.16 is identical, I have chosen symbols for the gradients of the elastic swelling lines κ_0 in one-dimensional compression but κ_{v} in spherical compression. I have done this to preserve the distinction between E'_0 and E'_{v}, the respective stiffness moduli, as you will see. In one-dimensional compression $e = e_\kappa - \kappa_0 \ln \sigma'$ so

$$E'_0 = \left(\frac{\mathrm{d}\sigma'}{\mathrm{d}\epsilon}\right)_0 = \frac{1}{\left(\dfrac{\mathrm{d}\epsilon}{\mathrm{d}\sigma'}\right)_0} = -\frac{(1+e)}{\left(\dfrac{\mathrm{d}e}{\mathrm{d}\sigma'}\right)_0} = \frac{(1+e)\sigma'}{\kappa_0} \tag{8.1}$$

In volumetric compression $e = e_\kappa - \kappa_{\mathrm{v}} \ln \sigma'$ so

$$E'_{\mathrm{v}} = \left(\frac{\mathrm{d}\sigma'}{\mathrm{d}\epsilon}\right)_{\mathrm{v}} = \frac{\mathrm{d}p'}{\mathrm{d}\epsilon_{\mathrm{v}}} = \frac{1}{\left(\dfrac{\mathrm{d}\epsilon_{\mathrm{v}}}{\mathrm{d}p'}\right)} = -\frac{(1+e)}{\left(\dfrac{\mathrm{d}e}{\mathrm{d}p'}\right)_{\mathrm{v}}} = \frac{(1+e)\sigma'}{\kappa_{\mathrm{v}}} \tag{8.2}$$

Now E'_0 and E'_{v} are elastic stiffness moduli which will in general have different numerical values; as you saw in table 6.3, $E'_0/E'_{\mathrm{v}} \approx 2$ over a wide central range of possible values. We might expect, therefore, that $\kappa_0/\kappa_{\mathrm{v}} \approx 1/2$.

As it happens the plasticity theory which I will shortly develop demands that the *plastic* flexibilities λ_0 and λ_{v} are equal. So be it. It is much more comforting to stay within the demands of existing elaborate frameworks like elasticity theory and plasticity theory than to attempt to wrap a new theory of mechanics around the peculiar data of our peculiar experiments. Let the data fit only approximately, if they must. As a matter of fact the available data suggests that λ-values *are* unique, whereas κ-values *do* depend on the imposed mode of deformation in the manner I have described.

If very stiff soils with large grains are subjected to spherical compression in order to determine their volumetric stiffness, the rubber membrane may be forced into the exposed voids in the soil, causing a relatively large error in the determination of the quantity of drainage.

Volume/time data for the consolidation coefficient C_{v}

The consolidation model of chapter 7 requires a slight adjustment before it can be made to fit the case of a triaxial sample draining to one end under the influence of an increment of all-round pressure. The parabolic isochrones of section 7.3 can be used without alteration. The rate of outflow of water remains equation 7.5, for the first family of parabolas

$$v_0 = \frac{2k\Delta\sigma}{\gamma_w L}$$

where $\Delta\sigma$ is the increment of cell pressure on the saturated sample. The rate of drainage therefore becomes

$$\frac{\partial \Delta V}{\partial t} = Av_0 = \frac{A2k\Delta\sigma}{\gamma_w L} \approx \frac{A_0 2k\Delta\sigma}{\gamma_w L} \tag{8.3}$$

where A_0 is the initial cross-section area. The change in volume at any time can be found by analogy with equation 7.9. The average increase in effective stress down to the shock wave front remains $\Delta\sigma/3$. This increase acts in all directions in this case, and causes a volumetric strain $\Delta\sigma/3E'_v$. The volume of soil so affected is approximately $A_0 L$. The change ΔV in volume since consolidation started is therefore given by

$$\Delta V = \frac{A_0 L \Delta\sigma}{3E'_v} \tag{8.4}$$

Differentiating 8.4 we obtain

$$\frac{\partial \Delta V}{\partial t} = \frac{A_0 \Delta\sigma}{3E'_v} \frac{\partial L}{\partial t}$$

which can be equated to equation 8.3 to give equation 7.10 once more, save that E'_v replaces E'_0. The solution $L = \surd(12C_v t)$ follows when we use

$$C_v = \frac{E'_v k}{\gamma_w} \tag{8.5}$$

as a newly defined volumetric consolidation coefficient. Substituting for L in equation 8.4 we obtain

$$\Delta V = \frac{A_0 \Delta\sigma}{3E'_v} \surd(12C_v t) \tag{8.6}$$

Now the ultimate change in volume due to additional cell pressure $\Delta\sigma$ is known to be

$$(\Delta V)_{\text{ult}} = A_0 H \frac{\Delta\sigma}{E'_v} \tag{8.7}$$

so

$$\Delta V/(\Delta V)_{\text{ult}} = \sqrt{\left(\frac{4}{3}\frac{C_v t}{H^2}\right)} \tag{8.8}$$

until the first group of parabolas reaches the base of the sample. This is identical in form to equation 7.25 in figure 7.9 which demonstrated how to find the consolidation coefficient. Clearly the same method applies, so that a curve of drainage ΔV against $(\text{time})^{1/2}$ will possess an initial tangent which cuts the ultimate drainage asymptote at time t_x such that

$$C_v = \frac{3}{4}\frac{H^2}{t_x} \tag{8.9}$$

The volumetric consolidation coefficient should be roughly half the one-dimensional consolidation coefficient owing to its containing E'_v instead of the stiffer E'_0 modulus. This derivation ignored the effects of the nonuniform reduction in cross-sectional area of the sample, and ignored the reduction in area entirely in comparison with the original area.

If the consolidation test is taken on to the plastic section PQ of the compression curve the stiffness modulus decreases markedly in conformity with the increase in the gradient of the e/p' curve. This may cause a reduction in the modulus, and therefore the consolidation coefficient, of a factor of about 4, being typically the factor between κ_v and λ_v. On swelling back, of course, the elastic modulus is resumed and the speed of consolidation returns to its much higher value.

The need to accelerate drainage often leads soils testers to promote radial rather than axial drainage by placing strips of soaked filter paper longitudinally under the rubber membrane. Gibson and Henkel (1954) showed that this would accelerate the consolidation of a conventional 2:1 cylindrical sample by a factor of 43 compared with drainage to one end.

8.1.2 Drained Compression Test for E'_y, μ', ϕ'_{max} and ϕ'_c

Test sequence

(1) A saturated sample of soil, or a perfectly dry sample of sand, is set up on the pedestal in the fashion of figure 8.1. A constant cell-pressure is applied, and the sample is given time to drain into equilibrium. The sample may alternatively have just completed a volumetric consolidation test.

(2) It is then allowed to drain while it is compressed so slowly that negligible excess pore-water pressures are generated at any stage of the test. The volume of drainage is measured. The pressure of the drain may be atmospheric or some higher constant datum pressure used to dissolve air bubbles if they are suspected of being present. The average void ratio of the sample can be traced throughout using the measured quantity of drainage. If rupture surfaces develop in the sample, the void ratio must then be expected to become nonuniform, with higher values tending to critical states within the zones of shear.

(3) The test will normally be terminated when the vertical strain has reached 20 per cent: this may well have taken 5 working days to achieve with a clayey soil. The final average moisture content of a saturated sample will fix its final average void ratio. The void ratio of dry sand samples would be deduced from their weight and volume. In either case the specific gravity of the soil particles must be measured (or assumed to lie in the range 2.65 to 2.75).

Interpretation

Stress–strain curve in compression for E'_y, μ', ϕ'_{max} and ϕ'_c

Before making any interpretations it is necessary to convert each reading of axial (deviatoric) force Q into deviatoric stress q by dividing it by the true area of the sample at that load. If a lateral strain device has been used this is a straightforward matter, because a running check can be made on the central diameter of the sample, working from a direct measurement of its initial value. If the diameter is not being measured, then changes in area ΔA in A must be inferred from the measurements of the changes in length (ΔH in H) and volume (ΔV in V) at any particular time.

$$\Delta V = AH - (A + \Delta A)(H - \Delta H)$$

$$(A + \mathrm{d}A) = \frac{AH - \Delta V}{H - \Delta H} = \frac{V - \Delta V}{H - \Delta H}$$

The true deviatoric stress then becomes

$$q = \frac{Q}{(A + \mathrm{d}A)} = \frac{Q}{(V - \Delta V)}(H - \Delta H) \qquad (8.10)$$

always assuming the sample remains cylindrical, as it should if its ends are well lubricated.

The data are conventionally presented in graphs of q against ϵ_1 and ϵ_v against ϵ_1; typical results for dilating 'dense' soil and contracting loose soil at the same cell pressure appear in figure 8.6. The general appearance of figure 8.6 is similar

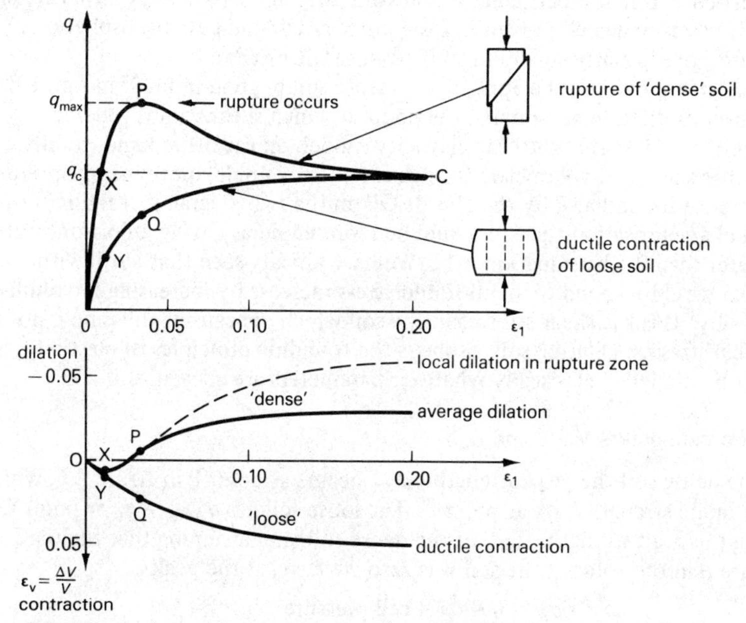

Figure 8.6 ***Drained triaxial compression on 'loose' and 'dense' samples of the same soil at the same cell pressure***

to that of figure 4.14 for shear-box results. The 'dense' soil has an initial response to compression, OX, to which the elasticity model can be applied. There follows an episode of increasing dilatancy XP up to a probable rupture at P: although the soil had been behaving as a continuum up to P, the dilation towards critical states along PC occurs only in those zones where the soil is continuing to shear, a rather small proportion of the whole volume of the sample. The 'loose' soil has an initial response OY to which the elasticity model can be fitted, followed by an episode of ductile contraction via Q towards a critical state (or a close approach to it) at C.

Elastic parameters E'_y and μ' with regard to effective stress

The initial gradient of a q/ϵ_1 graph corresponds to the Young's effective modulus of the soil E'_y, being a test in which the axial strain is measured as the axial stress (alone) is increased. The initial gradient of the ϵ_v/ϵ_1 graph can be used to find the Poisson's ratio μ' with regard to effective stresses. Since

$$\epsilon_3 = -\mu' \epsilon_1$$

and

$$\epsilon_v = \epsilon_1 + \epsilon_2 + \epsilon_3 = \epsilon_1 + 2\epsilon_3$$

it follows that

$$\epsilon_v = \epsilon_1(1 - 2\mu') \tag{8.11}$$

so that the initial gradient of contracting volume ϵ_v against compressive principal stress ϵ_1 has a magnitude of $(1 - 2\mu')$. These values of E'_y and μ' only refer to the particular soil at a particular initial void ratio and a particular initial spherical effective stress (the cell pressure, if the pores are draining to atmospheric pressure), over a fairly small range of strains (OX or OY).

Elasticity should not be applied to a dense soil beyond point X in figure 8.6. The onset of dilation at X marks the point at which shear strains generate volumetric strains: the isotropic elasticity model, on the other hand, totally uncouples shear and volumetric strains. It is, for example, quite inappropriate to characterise the soil at P by the chords OP on the two diagrams of figure 8.6. By such a characterisation the soil would be represented as having a Poisson's ratio μ' greater than 0.5 by equation 8.11. We have already seen that a soil with $\mu' > 0.5$ would respond to an all-round stress increase by increasing in volume: this is silly! It is the shear stresses in the soil which are causing dilatancy, not the spherical stresses. Dilating soil disobeys the founding principles of elasticity, and cannot be modelled elastically whatever parameters are chosen.

Friction parameters ϕ'_{max} and ϕ'_c

For the dense soil the peak strength q_{max} occurs at point P in figure 8.6, while the ultimate strength is q_c at point C. The loose soil has $q_{max} = q_c$ at point C. Taking the peak strength q_{max} of the dense soil, and assuming that the pore pressure u in the voids in the soil was zero we have, at the peak

$$\sigma'_3 = \sigma_3 - u = \sigma_3 = \text{cell pressure}$$

$$\sigma'_1 = q_{max} + \sigma'_3 = q_{max} + \sigma_3$$

At the approximate critical state C we have

$$\sigma_3' = \sigma_3 \text{ as before}$$

$$\sigma_1' = q_c + \sigma_3$$

The corresponding effective Mohr circles are drawn in figure 8.7, together with

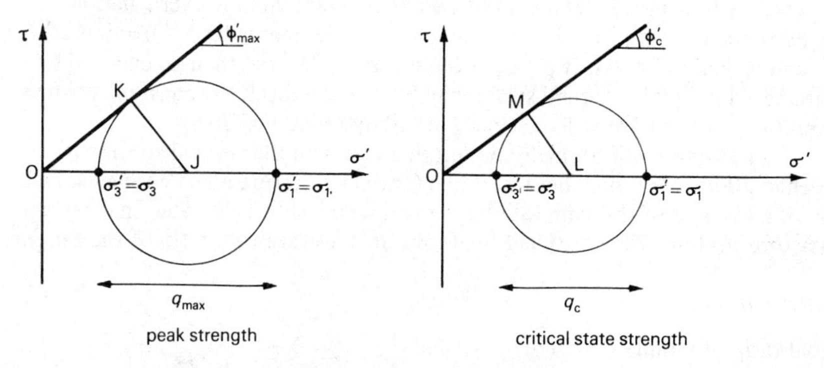

Figure 8.7 *Effective stress Mohr circles for drained strength*

a pair of tangents drawn through the origin. An interpretation of figure 8.7 using our friction model is that the maximum dilatant angle of shearing resistance ϕ'_{max} is a tangent to the peak strength circle, while the critical state angle ϕ'_c is a tangent to the ultimate strength circle. Therefore

$$\sin \phi'_{max} = \frac{JK}{OJ} = \frac{q_{max}/2}{\sigma_3' + q_{max}/2} = \frac{q_{max}}{2\sigma_3' + q_{max}} = \frac{q_{max}}{2\sigma_3 + q_{max}} \tag{8.12}$$

and

$$\sin \phi'_c = \frac{LM}{OL} = \frac{q_c/2}{\sigma_3' + q_c/2} = \frac{q_c}{2\sigma_3' + q_c} = \frac{q_c}{2\sigma_3 + q_c} \tag{8.13}$$

We have already seen in figures 4.13 and 4.7 that the peak angle of shearing resistance ϕ'_{max} depends strongly on the initial void ratio of the soil and its average effective stress. The critical state angle ϕ'_c is usually taken to be independent of stress and void ratio.

8.1.3 Undrained Compression Test for E_y^u, c_u and ϕ'_c

Test sequence

(1) A saturated sample of soil recovered from the field is trimmed, weighed and set up on the pedestal as shown in figure 8.1, except that the drainage facility through the top cap is dispensed with as unnecessary. Alternatively, the sample may have just completed a volumetric consolidation test, in which case the drainage lead will be in place and fully de-aired, although the tap leading from the drainage lead will remain closed throughout the following test. The cell

pressure is increased to some convenient value and then held constant. If the system was not fully saturated, the readings of the pore-pressure transducer connected to the pedestal would not have mirrored the raising of the cell pressure. If there is doubt as to the saturation of the system, the cell pressure should be increased in increments until the pore-pressure transducer responds by identical increments.

(2) The sample is then compressed at constant volume over a period of perhaps one working day. The minimum set of readings is Q, ΔH from which E_y^u and c_u can be found. If pore pressures u are measured then ϕ_c' can also be obtained. Readings of lateral strain may be used to check the constant volume condition, or to improve the estimation of cross-sectional area.

(3) The test will normally be terminated when the vertical strain has reached 20 per cent, or otherwise when (and if) a rupture plane has developed. The final weight of the sample should equal its initial weight. The final average moisture content will then yield the (constant) average void ratio of the sample.

Interpretation

Void ratio of sample

From table 1.1, for a saturated soil ($S = 1$)

$$\frac{\text{mass of sample}}{\text{volume of sample}} = \rho = \frac{e\rho_w + (1 \times G_s \times \rho_w)}{(1 + e)}$$

$$\text{moisture content} = m = \frac{e}{G_s}$$

Therefore

$$\rho(1 + e) = e\rho_w + \frac{e}{m}\rho_w$$

$$e = \frac{\rho}{\left(\rho_w + \dfrac{\rho_w}{m} - \rho\right)} \tag{8.14}$$

The void ratio of a saturated soil can therefore be obtained if its mass, volume and moisture content are determined. The presence of a plane of rupture may indicate that some internal drainage has allowed the soil to soften in the region of the plane at the expense of the remainder of the sample. Such results are difficult to interpret.

The constant-volume parameters E_y^u and c_u

Equation 8.10 at constant volume ($\Delta V = 0$) yields

$$q = \frac{Q}{V}(H - \Delta H) = \frac{Q}{A}\left(1 - \frac{\Delta H}{H}\right) = \frac{Q}{A}(1 - \epsilon_1) \tag{8.15}$$

from which the true deviatoric stresses can be obtained at various vertical strains. The appearance of the q/ϵ_1 diagram may be quite similar to that of figure 8.8 for a wide variety of soils, sand or clay, loose or dense. Certain

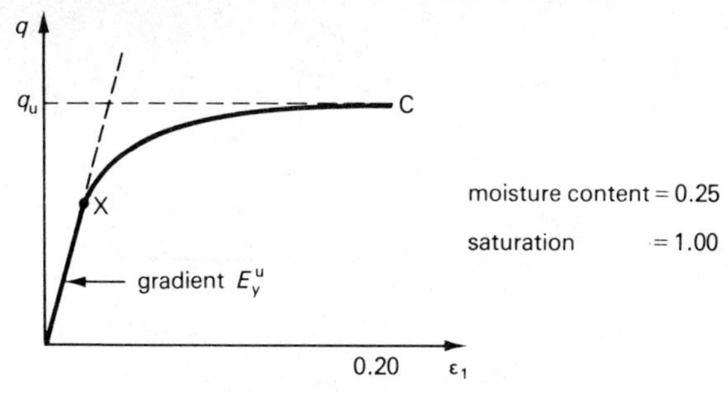

Figure 8.8 *Undrained triaxial compression*

cemented or sensitive soils may however offer a dramatic fall in strength from a high peak value as would rock; you will encounter this later in the chapter. The gradient of the initial part of this axial compression test OX may be referred to as the undrained Young's modulus E_y^u of the soil at that particular fixed void ratio. The ultimate strength q_u at C may be thought of as approaching a critical state. It certainly represents an ultimate shear strength $c_u = q_u/2$ by the Mohr's circle of figure 8.9 in terms of total stresses, being the shear stress developed on

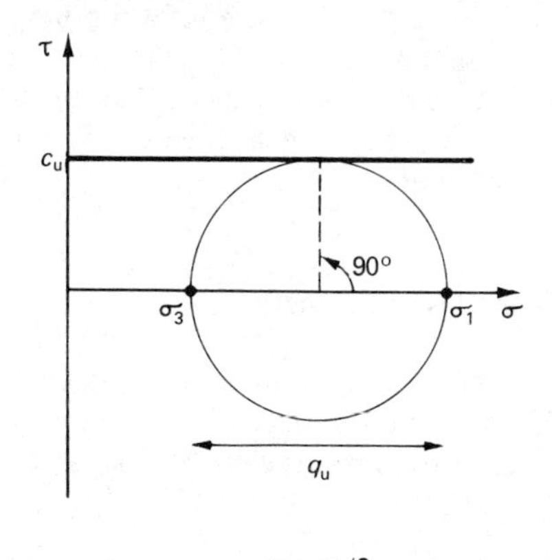

Figure 8.9 *Undrained strength: total stresses*

the 45° planes. This is a conclusion compatible with our previous cohesion model of chapter 5.

The effective angle of friction ϕ_c' at the presumed critical state C

If point C in figure 8.8 is assumed to be a critical state, and if the pore-water pressure eventually reached a constant value u_u at C, then the angle ϕ_c' can easily be obtained from the Mohr's circle with regard to effective stress (figure 8.10 drawn after the fashion of figure 8.3.) What has been calculated here is the maximum angle of obliquity of the resultant effective stress on any plane within the sample in its ultimate state. This interpretation is quite consistent with our previous friction model of chapter 4. We have simply used a Mohr's circle to display all the various stress obliquities on the various planes through the sample,

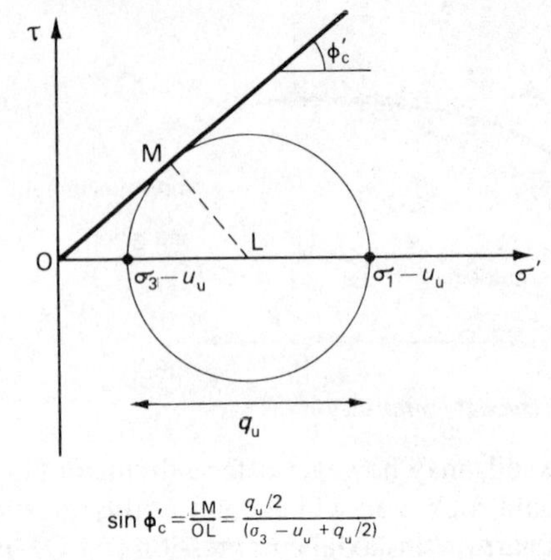

Figure 8.10 *Undrained strength: effective stresses*

and we have taken the fact that the ultimate strength has been reached as indicative that ϕ'_c has been reached on the plane with the steepest gradient of τ/σ', which is at point M in figure 8.10.

If the test had been interrupted at the onset of a rupture plane, then no plateau of pore pressure would be available. If such a test had been continued until the internal pore pressures became constant, then the internal void ratios would possibly be nonuniform – higher near the rupture plane. Such events may be difficult to interpret in detail.

8.2 Envelopes to Mohr's Circles of Strength

The practice has arisen of performing a number of triaxial tests on neighbouring samples, and plotting the Mohr circles together so that an envelope can be drawn which is roughly tangential to the various circles. This can be very interesting, and can teach you

(1) about the variability of soil in the ground, by using 'undisturbed' samples

(2) about the complexity of the mechanics of a particular soil, by continually reforming the same grains into new samples and then testing them under slightly different conditions.

It is vital not to confuse these two laudable aims; a very great deal of thought is necessary before reaching a conclusion about the choice of suitable parameters to cover a number of soil tests. I shall demonstrate this difficulty by making reference to the Mohr circles in figures 8.11 to 8.15.

Suppose that you were the soils testing engineer who had obtained the results of figure 8.11 from four separate samples obtained from a borehole. The samples were tested some weeks after they had arrived, being simply extruded from their sampling tubes, trimmed to size, and squashed at constant volume. In practice you may have twenty such results rather than four. What value do you quote for the strength of soil?

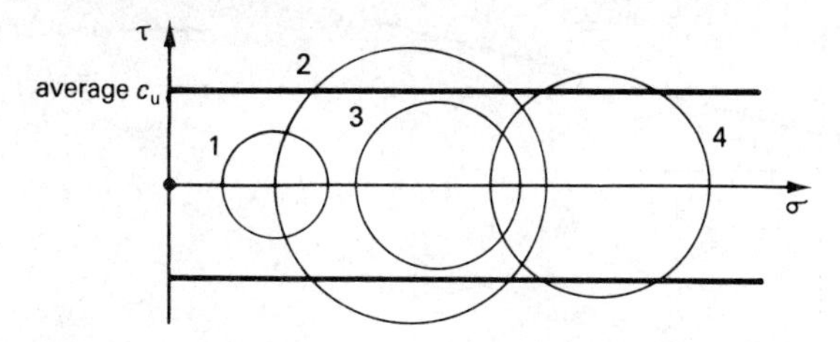

Figure 8.11 *Undrained tests on independent samples of undisturbed clayey soil*

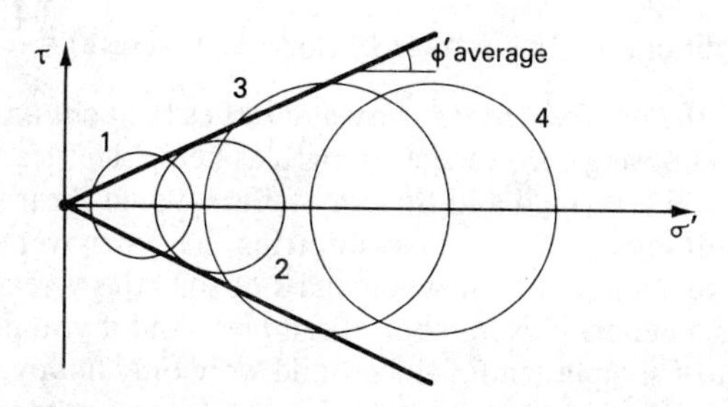

Figure 8.12 *Drained tests on independent samples of undisturbed clayey soil*

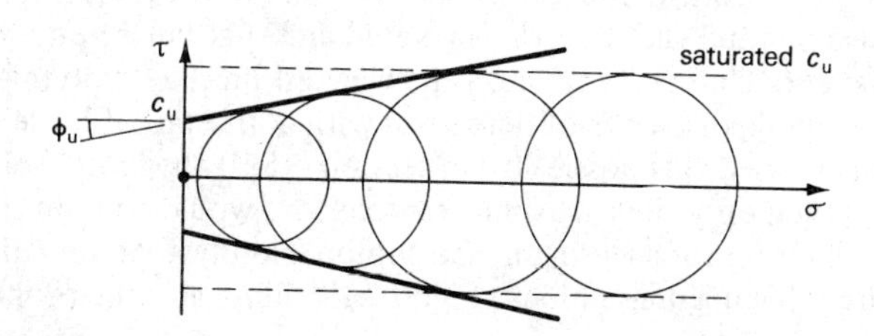

Figure 8.13 *Nominally undrained tests on samples of clayey soil compacted at a unique moisture content*

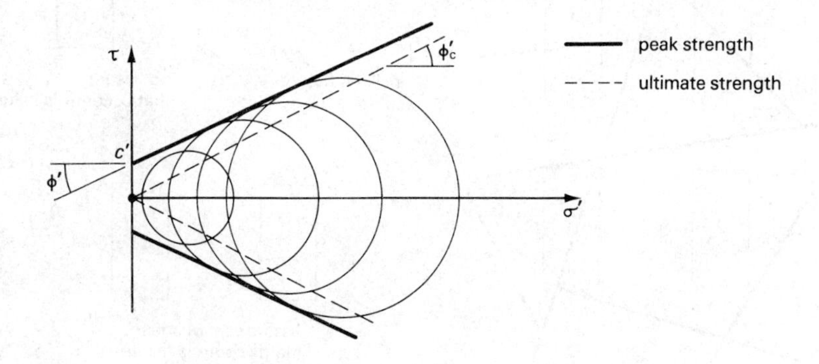

Figure 8.14 *Drained tests on similar samples of sand at a unique density*

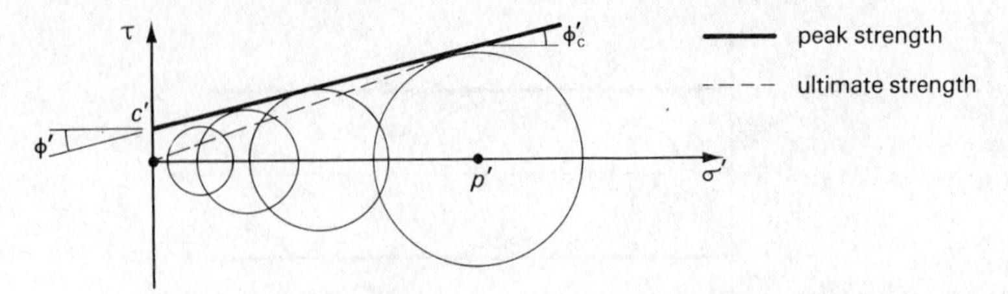

Figure 8.15 *Drained tests on similar samples of remoulded clay, precompressed to a unique spherical stress p'*

(1) You might choose the average strength c_u (average) $= \frac{1}{n} \sum_{i=1}^{n} c_{u(i)}$. You may prefer to 'fit' the parallel envelope by eye rather than obtaining it arithmetically. In choosing the average strength as being significant you may be implying that the soil is layered and that any failure would shear some hard layers and some soft layers. You are also inferring that your average strength is equal to the average strength of the whole mass of soil: this is relatively unlikely unless you possess an enormous number of samples. And if you possess an enormous number of samples and if the ground were only lumpy, rather than layered, then you are specifying that the failure surface must pass equally through the harder and softer lumps: but it would choose the softer zones and avoid shearing the harder lumps, would it not?

(2) You might choose the lowest strength on offer, c_{u1} in this example. You might have in mind that the soil is layered and that failure only need involve the weakest layer (a chain is as strong as its weakest link). Or you might conceive of the soil as a rubble of blocks of dense soil with a 'mortar' of loose soil between them: I raised this possibility in section 4.5. If the looser soil were only in thin fissures then only the occasional sample core would contain the fissure at its weakest inclination for shearing in the compression test. Some samples might be cored entirely from a dense block. Figure 8.16 illustrates the problem. You

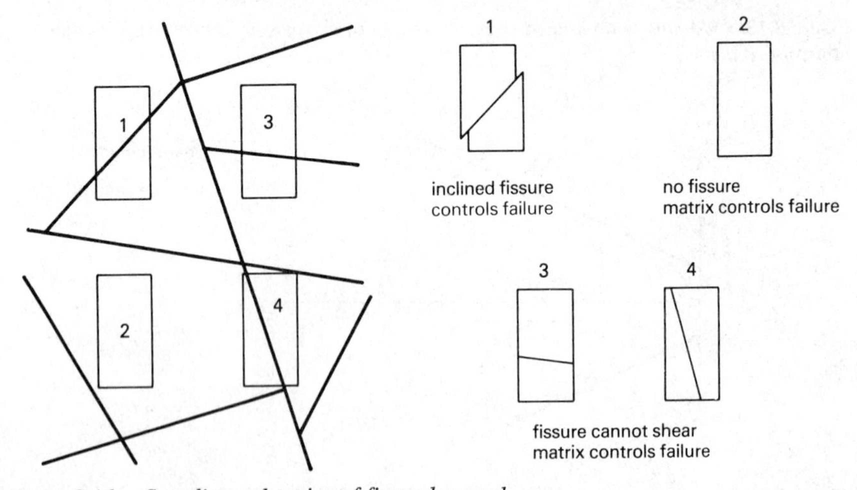

Figure 8.16 *Sampling and testing of fissured ground*

might be forced to assume that only the weakest sample contained the weak fissure at the worst possible inclination.

(3) You might eliminate certain results before adopting either of the previous strategies. Perhaps there was evidence that sample 2 in figure 8.11 had dried out before it was tested. Perhaps the rubber membrane in test 1 was suspected of leaking.

(4) You might use probability theory to estimate what the lowest strength would have been if you had been able to take more samples. Or you might go back to site and take better, larger and more samples and repeat the whole testing programme!

A very similar complex of options would face the interpreter of the effective friction results of figure 8.12, assuming that they were similarly recovered from site. Once again, the temptation may be to apply an average gradient as shown in the figure, whereas the more reasonable parameter for design purposes may be the lowest angle of friction on offer, from sample 4 in this case. Undisturbed samples of cemented soils and rocks can offer a sequence of Mohr's circles of effective stress whose tangent possesses the true cohesion or cementation intercept c'. Ancient materials such as the London clay are usually stiff, at least slightly cemented, and highly fissured due to tectonic activity over their geological life-span. They reproduce the fissure orientation problem of figure 8.16 but to a more severe degree. Bishop (1972) and Marsland (1972) describe how the use of 38 mm diameter samples artificially enhances the average laboratory strength, by making it more likely that the nuggets of lightly cemented clay can play a dominant role in the failure. The larger the samples, the lower is the measured strength: 100 mm diameter undisturbed samples of London clay usually contain sufficient natural fissures to ensure that failure can occur with respect to the uncemented fissures, as it would in the field. The Mohr envelope of effective stress for large samples of cemented but highly fissured soils will not depart very much from the friction ideal. Recent materials which are lightly cemented but which have remained unfissured offer even greater problems to the engineer: their distinctive and dangerous behaviour is introduced in section 8.7.1.

Figures 8.13 to 8.15 illustrate situations in which the soil tester has eliminated much of his normal variance by fabricating a number of 'identical' samples in the laboratory: when he tests them at various pressures he then regards stress as the key parameter. This sometimes leads to short-sightedness. In figure 8.13 you see some typical results from samples of a clayey soil which has been exhaustively mixed in the laboratory and compacted by a hammer to form samples which are almost identical. Their method of fabrication means that the samples will possess the same moisture content and the same proportion of air (perhaps 3 per cent of the sample volume). If the soil tester places the samples under various cell pressures before squashing them with no external drainage allowed, he will obtain this narrowing strength envelope rather than having each circle tangential to a parallel c_u-envelope. He may then be provoked into describing the 'strength' of his soil as

$$\tau_{max} = c_u + \sigma \tan \phi_u$$

which is a rough tangent to his low-pressure results. His use of the subscript u is the main cause of confusion: although no external drainage was taking place, internal drainage must certainly have been taking place to replace the air which was squashed when the cell pressure was applied. If the tester was far sighted enough to leave his samples 'soaking' under high pressure for some time, then he would dissolve the air and obtain the strength applicable to the *saturated* soil at that particular fixed moisture content. This 'saturated' c_u-value which I mark on fig. 8.13 is the only meaningful 'undrained' parameter which can be obtained from such results. Different (c_u, ϕ_u) envelopes would have been obtained for different speeds of test due to the different degrees of internal drainage: it is very difficult to apply (c_u, ϕ_u) envelopes sensibly to field problems. One answer to the problem of choosing parameters for partially saturated soil is to supplement the data of figure 8.13 at the specified moisture content by further tests on the soil when it has been saturated at a number of neighbouring void ratios. From this it will be possible to discover what the undrained strength of the partially saturated soil would be if the air were eventually replaced by water, the void ratio remaining the same. Void ratios can be estimated from measurements of bulk density and moisture content, if the specific gravity of the soil particles is known. Another useful step is to measure the effective friction angle of the soil: this should also be done when it is saturated. If the partially saturated soil is 'draining', then it would be useful to know the fully drained strength. In certain circumstances it could be safe to choose for design purposes the *lower* of

(1) $\tau = c_u$ (saturated at appropriate moisture content)

and

(2) $\tau = \sigma' \tan \phi'$; u at an equilibrium appropriate to location in the field.

Figure 8.14 refers to a series of drained tests, conducted at various cell pressures, on sand samples which had been carefully compacted to the same initial void ratio. The dotted envelope to the ultimate strengths of the samples might well pass through the origin and have a gradient of ϕ'_c. We have already remarked that the friction angle ϕ'_c in critical states, which are achieved only after considerable distortion, is usually independent of stress and initial density. The bold envelope to the peak strengths is unlikely to be straight, however, because the dilatancy component of strength is strongly influenced by the confining pressure. If the confining pressure is great enough all dilatancy would be suppressed, the soil would shear at constant volume, and the angle of shearing resistance would simply be ϕ'_c. Engineers are very frequently tempted to use the curve-fitting equation

$$\tau = c' + \sigma' \tan \phi'$$

to describe the shape of the envelope for peak strength over some middle range of stresses. This is often inappropriate. Although a pair of parameters c', ϕ' might satisfy the envelope of Mohr circles at a particular density, their values will alter if the density alters: much of this was explained in sections 4.4 and 4.5.

The engineer who possesses the data of figure 8.14 presumably has an enormous body of sand in the field to which he must attach a safe angle of effective friction. If the sand in the field is loose he must use ϕ'_c. If the sand in the field has loose layers or pockets, he should probably use ϕ'_c for the lowest possible angle of friction which might be encountered at any particular spot. If the sand is very deep, then ϕ'_c may well be appropriate due to the suppression of dilatancy with confining pressure. Only if the engineer knows that the sand never has a void ratio greater than e_{max}^{field} should he risk going to an envelope of strengths at initial void ratio e_{max}^{field}. He would then be well advised to determine the greatest conceivable effective stress σ'^{field}_{max} which could act on a rupture-surface through the sand, and move to the appropriate point on the envelope to determine ϕ'^{field}_{min}, the lowest conceivable angle of friction operative in the field. This procedure is sketched in figure 8.17. The double variation in density and stress

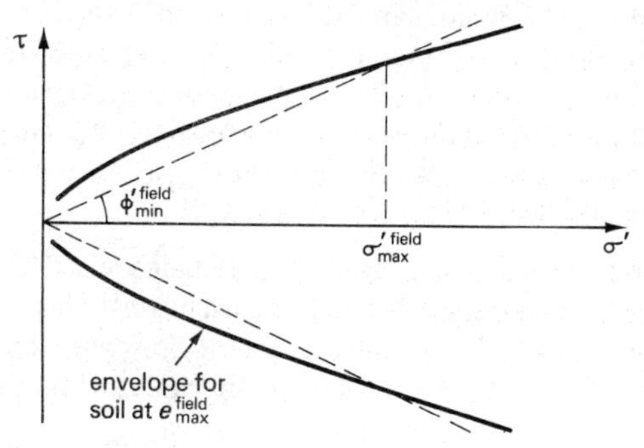

Figure 8.17 *Choosing a safe value for ϕ'*

in the field almost always makes curve-fitting with (c', ϕ') inappropriate.

Figure 8.15 presents much the same case, but for saturated samples of clay. Samples which are initially 'loose' with respect to their critical state line, or which are forced into a critical state by unrelenting compression, normally display a perfectly triangular envelope with the effective angle ϕ'_c. (Remember that the results are plotted in terms of effective stress: the tester has gone to some trouble to establish or measure pore-water pressures.) On the other hand samples which are densified by consolidation to some high pressure and which are then allowed to swell back, offer peak strengths before their strength falls back on to the 'critical' envelope. The peak envelope may well tempt engineers to fit

$$\tau = c' + \sigma' \tan \phi'$$

Exactly the same arguments must be used as in the preceding case: it will normally be sensible to choose a single safe angle of friction ϕ'^{field}_{min} applicable to the smallest conceivable density and the greatest possible effective stress which will be encountered in the field. If the engineer needs the strength of the soil at very small stresses, then he should apply a very small cell pressure and experimentally determine the value of q_{max} or q_{ult} as appropriate. This is precisely the value of the strength which he should use, not some 'true cohesion' intercept back-extrapolated from high pressure triaxial tests.

Although most engineers would be embarrassed to refer to the intercept c' for dense compacted sands in figure 8.14 as 'true cohesion' caused by 'cohesive bonds', very few would refrain from so describing the intercept c' for dense clays in figure 8.15. You may prefer to remark at the similarity in form of the drained strength envelopes of dense but remoulded sands and clays, and to remember the inevitable dilatancy which must enhance the strength of any 'dense' soil if it is to achieve a critical state.

8.3 Uniform Stresses and Strains: Cam-clay

Important elastic and then plastic strains can take place when soil is suffering no shear stress whatsoever: see curve AP and then PQ in figure 8.5. As shear stresses rise soil elements begin to distort, and quite gross changes of shape can take place before the soil has mobilised its entire strength (see figures 8.6 and 8.8). Triaxial samples are often compressed 20 per cent in order to achieve the strengths which were the subject of our discussions in the last section. Even the peak strength of dense soils may demand a 1 to 2 per cent reduction in length. If a 5 m deep stratum of soil under a foundation were to compress by 20 per cent (that is, 1 m) the building would probably collapse: if it compressed by 1 per cent (that is, 50 mm) ghastly cracks would probably develop in plaster or brick panels. It is obviously desirable, at least, to be able to assess

(1) the 'elastic' strains, assuming the soil remains 'elastic'
(2) the working soil strength, so as to guard against mass soil movements
(3) the range over which the 'elastic' assumptions are likely to be valid. In other words, to be able to fix points X and Y on figure 8.6 and point X on figure 8.8.

Codes of practice are a little help here; they usually require the designer not to mobilise more than a proportion of the soil 'strength' however it was assessed, so that

$$\tau_{\text{safe}} = \frac{1}{F} \times \tau_{\text{max}}$$

This safety factor F might be as low as 1.5 if strains were not very significant, 2.5 if strains were more important, and as high as 4.0 if strains were thought to be critical. If the stress q in such as figure 8.8 were only allowed to rise one quarter of the way up to the maximum value of q_u, it would be likely that the elastic range had not been exceeded.

Soil strains have another application. When soil rests against a structure, whether it be a wall, a pipe or a silo, the designer would often like to assume that the soil helps to support its own weight. It will often do this, but only if it is free to strain sufficiently: no shear strain, no shear stress, no self-help. This was the subject of section 6.10.

A most important development in the campaign to predict and control the stress–strain behaviour of soil was the development at Cambridge of a group of plasticity models, introduced in Schofield and Wroth's most philosophical and useful book *Critical State Soil Mechanics.* It will transpire that uniform 'dense' soils can remain quasi-elastic right up into the vicinity of the stress condition $\tau = \sigma' \tan \phi'_c$. The focus of the Cambridge models however, is the behaviour of 'loose' soils which yield and become inelastic (point Y, figure 8.6) at much smaller angles of inclination of stress. Perhaps even more significant, the plasticity models draw a comprehensive picture of 'compression' 'elasticity', 'yield', 'friction' and 'cohesion' which helps in the enormous task of gaining a perspective over the whole geography of soil mechanics.

Schofield and Wroth (1968) derived Cam-clay from work equations applied to a triaxial test sample suffering very slow drained increases in stress δq on q and

$\delta p'$ on p' where q is the deviatoric stress and p' the average effective stress

$$q = \sigma_1 - \sigma_3 = \sigma_1' - \sigma_3'$$

$$p' = \frac{\sigma_1' + \sigma_2' + \sigma_3'}{3} = \frac{\sigma_1' + 2\sigma_3'}{3}$$

Associated with the deviatoric stress component q is the shear strain of the 1–3 plane

$$\epsilon_s = \epsilon_1 - \epsilon_3$$

while the strain associated with the spherical stress component p' is the volumetric strain

$$\epsilon_v = \epsilon_1 + \epsilon_2 + \epsilon_3 = \epsilon_1 + 2\epsilon_3$$

We shall suppose that a small increase δq causes a vertical compression of the sample, so that $\delta \epsilon_s > 0$. Likewise a small increase $\delta p'$ causes a spherical compression so that $\delta \epsilon_v > 0$, the sign convention being 'compression positive' for all stresses and strains. These stress and strain components have been chosen so as to be compatible with the fundamental continuum mechanics of chapter 6. They will allow the progress of a triaxial test to be charted as a line on a q/p' diagram rather than as a progression of Mohr circles on a τ/σ' diagram.

The first step is to establish the work done by the stresses at the boundary of an element which can strain. Consider an elementary cube of soil of unit side which is aligned along the principal axes so that the only effective stresses acting on it are the principal stress σ_1', σ_2', σ_3'. Suppose that the 1-axis of the cube shortens by a small strain ϵ_1: the stress σ_1' acting on unit area moves a distance ϵ_1 x unit length. The work done per unit volume by the σ_1' stresses is therefore the force multiplied by the distance, $\sigma_1'\epsilon_1$. If the 2-axis shortens by ϵ_2 then the work done by the σ_2' stresses will be $\sigma_2'\epsilon_2$, while that in the 3-direction will similarly be $\sigma_3'\epsilon_3$. We have assumed here

(1) that the effective stresses are constant, and that no excess pore-water pressures are caused

(2) that the effect of the small strains on the areas and volume is negligible

The total work done per unit volume by the boundary stresses is

$$W = \sigma_1'\epsilon_1 + \sigma_2'\epsilon_2 + \sigma_3'\epsilon_3$$

We now require to translate the cubic system into our triaxial system with the 2 and 3-directions equivalent so that

$$W = \sigma_1'\epsilon_1 + 2\sigma_3'\epsilon_3$$

Following the principle of multiplying stresses by their corresponding strains, we shall attempt to translate W into the (q, p') system. Here we calculate

$$q\epsilon_s = (\sigma_1' - \sigma_3')(\epsilon_1 - \epsilon_3) = \sigma_1'\epsilon_1 - \sigma_1'\epsilon_3 - \sigma_3'\epsilon_1 + \sigma_3'\epsilon_3$$

and

$$p'\epsilon_v = \frac{(\sigma_1' + 2\sigma_3')}{3}(\epsilon_1 + 2\epsilon_3)$$

$$= \frac{\sigma_1'\epsilon_1}{3} + \frac{2\sigma_1'\epsilon_3}{3} + \frac{2\sigma_3'\epsilon_1}{3} + \frac{4\sigma_3'\epsilon_3}{3}$$

so that by inspection

$$W = \frac{2}{3}q\epsilon_s + p'\epsilon_v \tag{8.16}$$

In order to eliminate the inconvenient 2/3 factor which has appeared we shall adopt a new strain component associated with the deviatoric stress q in a triaxial test

$$\epsilon_q = \frac{2}{3}\epsilon_s = \frac{2}{3}(\epsilon_1 - \epsilon_3) \tag{8.17}$$

so that we can write simply

$$W = q\epsilon_q + p'\epsilon_v \tag{8.18}$$

which is the work done by the environment on a straining triaxial element of unit volume.

Now consider the effects of a small increase in stress δq in q and $\delta p'$ in p', supposing that they cause small changes in strain $\delta\epsilon_q$ in ϵ_q and $\delta\epsilon_v$ in ϵ_v. The average stresses during the increase were $(q + \frac{1}{2}\delta q)$ and $(p' + \frac{1}{2}\delta p')$ and the resulting strains were $\delta\epsilon_q$ and $\delta\epsilon_v$ so the increment of work done by the environment on the sample was

$$\Delta W = \left(q + \frac{1}{2}\delta q\right)\delta\epsilon_q + \left(p' + \frac{1}{2}\delta p'\right)\delta\epsilon_v$$

$$\Delta W = (q\delta\epsilon_q + p'\delta\epsilon_v) + \left(\frac{1}{2}\delta q\delta\epsilon_q + \frac{1}{2}\delta p'\delta\epsilon_v\right) \tag{8.19}$$

which can be written

$$\Delta W = \underset{\text{basic}}{\Delta W} + \underset{\text{probe}}{\Delta W} \tag{8.20}$$

and so

$$\Delta W \approx \underset{\text{basic}}{\Delta W} \quad \text{since } \delta q \ll q \text{ and } \delta p' \ll p' \tag{8.21}$$

where the basic stresses q and p' do the major work, while the main effect of the probe δq and $\delta p'$ is to cause strains for the basic stresses to work on.

Where has the work gone? The environment has fed in work ΔW to the element: the element must have stored some energy and dissipated some energy as heat:

$$\Delta W = \underset{\text{stored}}{\Delta W} + \underset{\text{dissipated}}{\Delta W} \tag{8.22}$$

Elastic systems store energy whereas plastic systems dissipate it. In order to make progress it is necessary to come to a fairly simple and crude decision regarding the storage and dissipation of energy in the sample. For Cam-clay the decision Schofield and Wroth made was

(1) All deviatoric energy is dissipated. This is equivalent to saying that no elastic shear energy is stored or that $E_s \to \infty$

(2) Volumetric energy is stored along κ-lines in such as figure 8.5, and dissipated when the soil state changes κ-lines. This is equivalent to our previous statement that κ-lines were 'elastic'.

Considering the $e/\ln p'$ diagram in figure 8.18, an arbitrary movement (A → Q)

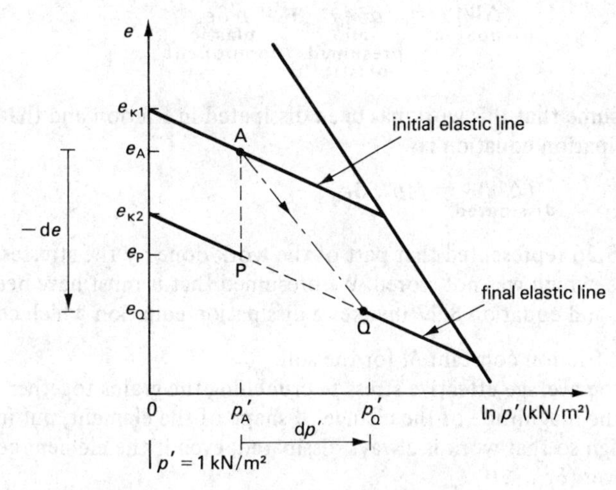

Figure 8.18 *Plastic and elastic volumetric strain components*

will be thought of as part plastic (A → P) and part elastic (P → Q). Suppose that A, P, Q are quite close, so that the void ratio e changes only infinitesimally. The total volumetric strain

$$\delta\epsilon_v = \frac{e_A - e_Q}{1 + e} = \frac{-\delta e}{(1 + e)} \tag{8.23}$$

Part of this is elastic and recoverable

$$(\delta\epsilon_v)_{\text{elastic}} = \frac{e_P - e_Q}{(1 + e)}$$

while the remainder is plastic and dissipative

$$(\delta\epsilon_v)_{\text{plastic}} = \frac{e_A - e_P}{(1 + e)}$$

Considering just the plastic component, you will see from figure 8.18 that

$$e_A - e_P = e_{\kappa 1} - e_{\kappa 2} = -\delta e_\kappa$$

so that the plastic volumetric strain

$$(\delta \epsilon_v)_{\text{plastic}} = -\frac{-\delta e_\kappa}{(1+e)} \tag{8.24}$$

which we might write as $\delta \epsilon_\kappa$, and the plastic volumetric work must therefore be

$$p'(\delta \epsilon_v)_{\text{plastic}} = \frac{-p' \delta e_\kappa}{(1+e)} = p' \delta \epsilon_\kappa \tag{8.25}$$

From equation 8.21 we can now deduce the magnitude of the plastic work which is fed into the element and presumably dissipated

$$\underset{\text{dissipated}}{(\Delta W)} = \underset{\substack{\text{all}\\\text{presumed}\\\text{plastic}}}{q \delta \epsilon_q} + \underset{\substack{\text{plastic}\\\text{component}}}{p' \delta \epsilon_\kappa} \tag{8.26}$$

We will assume that this work has been dissipated in friction and that the form of the dissipation equation is

$$\underset{\text{dissipated}}{(\Delta W)} = Mp' \mid \delta \epsilon_q \mid \tag{8.27}$$

Equation 8.26 represented that part of the work done by the stresses at the boundaries, which was not stored. We presumed that it must have been dissipated, and equation 8.27 invokes a dissipation equation which contains

(1) a friction constant M for the soil
(2) the average effective stress p' crunching the grains together
(3) the magnitude of the change of shape of the element, put inside a modulus sign so that work is always dissipated, even if the element gets longer and narrower ($\delta \epsilon_q < 0$)

By merging equation 8.26 with 8.27 we obtain the basic plastic work equation for Cam-clay

$$q \delta \epsilon_q + p' \delta \epsilon_\kappa = Mp' \mid \delta \epsilon_q \mid \tag{8.28}$$

Considering compression, $\delta \epsilon_q > 0$

so
$$q = p' \left(M - \frac{\delta \epsilon_\kappa}{\delta \epsilon_q} \right) \tag{8.29}$$

Equation 8.29 represents the following conditions

(1) Elastic; the state of the soil will reside on a single κ-line so that $\delta \epsilon_\kappa = 0$. Furthermore $\delta \epsilon_q = 0$ since deviatoric strain in Cam-clay is a purely plastic phenomenon. q is indeterminate.

(2) Plastic contraction; $\delta e < 0$ so $\delta \epsilon_\kappa > 0$ therefore

$$q < p'M$$

(3) Plastic dilation; $\delta e > 0$ so $\delta\epsilon_\kappa < 0$ therefore

$$q > p'M$$

(4) Plastic deformation at constant volume with unlimited shear strains $\delta e = 0$ and $\delta p' = 0$ so $\delta\epsilon_\kappa = 0$ and $\delta\epsilon_q \to \infty$ therefore

$$q = p'M$$

at a critical state.

The relationship between the Cam-clay plastic work equation 8.29 and the dilatancy–friction model of section 4.5

$$\tau = \sigma' \tan\left(\phi' + \frac{dy}{dx}\right) \tag{8.30}$$

is very close. The relationship $q = p'M$ demands that the diameter of Mohr's circles increases as the average soil stress rises. This is similar to the requirement for Mohr's circles to be tangential to a line $\tau = \sigma' \tan \phi'_c$. Furthermore, y was the measured rise in the lid of a shear box which is proportional to the plastic expansion of volume $-\delta\epsilon_\kappa$ while x was the measured shear displacement which will be proportional to the shear strains $\delta\epsilon_q$ within the soil. Although the Cam-clay model may be couched in unfamiliar terms, it is describing precisely the same quality of behaviour as that contained in section 4.5.

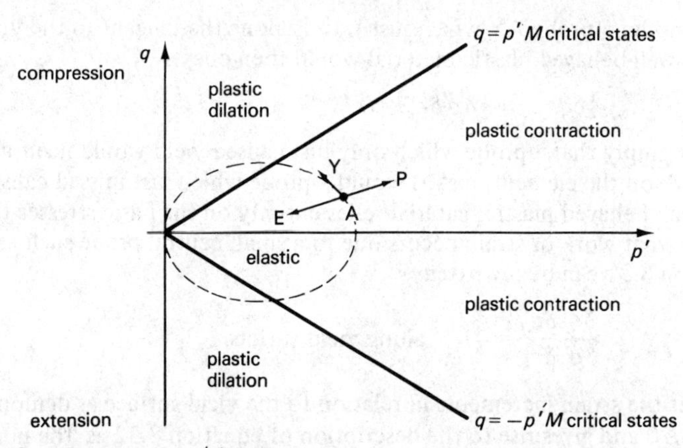

Figure 8.19 *Zones on the q/p' diagram*

Consider the q/p' diagram of figure 8.19, which replaces the previous τ/σ' diagrams of chapter 4. The figure includes compression ($q > 0$, $\delta\epsilon_q > 0$) and extension ($q < 0$, $\delta\epsilon_q < 0$), and displays a critical state line ($|q| = p'M$) with plastic contraction possible in the $|q| < p'M$ zone and plastic dilation in the $|q| > p'M$ zone. The figure also contains an elastic zone which has arbitrarily been drawn fairly close to the origin on the presumption that there are some small stresses which elicit elastic responses in a fairly dense soil. Can the boundary between the elastic and plastic zones be fixed?

The second major step in the derivation of Cam-clay is to invoke Drucker's concept of stable yielding on the elasto/plastic boundary which plasticians often call the yield surface. In equation 8.19 the last term $(\frac{1}{2}\delta q\delta\epsilon_q + \frac{1}{2}\delta p'\delta\epsilon_v)$ represented the work done by the probe $(\delta q, \delta p')$ itself. By analogy with equation 8.26 the plastic work done on the Cam-clay element by the probe will be $\frac{1}{2}\delta q\delta\epsilon_q + \frac{1}{2}\delta p'\delta\epsilon_\kappa$ where $\delta\epsilon_\kappa$ as before is defined as the reduction in void ratio caused by a shift in κ-line divided by the whole relative volume $1 + e$.

Consider a point A on the yield surface in figure 8.19. If a small probe $(\delta q, \delta p')$ causes the stress to move towards E in the elastic zone, the only response will be movement along a κ-line. We have specified that shear strains, proportional to $\delta\epsilon_q$, are purely plastic so that for AE $\delta\epsilon_q = 0$. We demand that the soil remains elastic, so e_κ = constant and $\delta\epsilon_\kappa = 0$. Therefore for any 'inward' probe from A, since $\delta\epsilon_q = \delta\epsilon_\kappa = 0$ it follows that there is no plastic work done by the probe

$$\tfrac{1}{2}\delta q\delta\epsilon_q + \tfrac{1}{2}\delta p'\delta\epsilon_\kappa = 0 \quad \text{for AE elastic}$$

If the small probe $(\delta q, \delta p')$ causes the stress to move towards P in the plastic zone then we must expect that $\delta\epsilon_q \neq 0$ and $\delta\epsilon_\kappa \neq 0$. If the probe actually does plastic work on the element then Drucker described the yield as 'stable'. The alternative, that plastic work was done by the element on the probe, is properly described as 'unstable' or even explosive. Imagine a heap of earth which kicks back harder than you kicked it!

$$\tfrac{1}{2}\delta q\delta\epsilon_q + \tfrac{1}{2}\delta p'\delta\epsilon_\kappa \geqslant 0 \quad \text{for stable yield AP}$$

Finally consider a probe AY which just travels along the tangent to the yield surface. A well-behaved plastic material would then obey

$$\tfrac{1}{2}\delta q\delta\epsilon_q + \tfrac{1}{2}\delta p'\delta\epsilon_\kappa = 0 \tag{8.31}$$

This would imply that a probe which only just caused yield would itself do no plastic work on the element, just as would a probe which just missed causing yield. A well-behaved plastic material relies entirely on the basic stresses to determine what work or strain occurs due to a small neutral probe such as AY.

Equation 8.31 can be rewritten

$$\frac{\delta q}{\delta p'}\frac{\delta\epsilon_q}{\delta\epsilon_\kappa} = -1 \quad \text{along yield surface} \tag{8.32}$$

which fixes the strain increments in relation to the yield surface as demonstrated in figure 8.20 and gives rise to the description of equation 8.32 as 'the normality

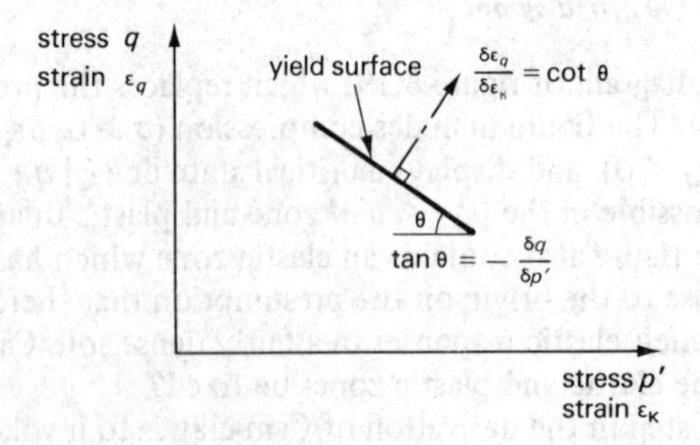

Figure 8.20 *The normality condition for associated strains*

condition'. If the strains and stresses are drawn against corresponding axes, the strain increment vector $(\delta\epsilon_q, \delta\epsilon_\kappa)$ caused by *any* small outward probe is at right-angles to the yield surface.

Taking the work equation in compression, equation 8.29, with the normality equation 8.32, the strain vector can be eliminated to give

$$\frac{q}{p'} = M + \frac{\delta q}{\delta p'} = M + \frac{\mathrm{d}q}{\mathrm{d}p'} \qquad (8.33)$$

to describe in calculus the yield surface in terms of its position and its gradient. Using the substitution $\eta = q/p'$

$$\eta = M + \eta + p' \frac{\mathrm{d}\eta}{\mathrm{d}p'}$$

$$\frac{\mathrm{d}p'}{p'} = -\frac{\mathrm{d}\eta}{M}$$

so

$$\ln p' = -\frac{\eta}{M} + \text{constant}$$

or

$$\frac{q}{Mp'} + \ln p' = \text{constant}$$

Now one point on the yield surface must be a critical state $q_c = Mp'_c$: this fixes the constant as $1 + \ln p'_c$. The equation of the yield surface reduces to

$$\frac{q}{Mp'} + \ln\left(\frac{p'}{p'_c}\right) = 1 \qquad (8.34)$$

which is depicted in figure 8.21. One of the most powerful aspects of the Cam-clay model is that it predicts that the extent of a κ-line is limited. There is a pressure p'_v which, if applied, causes plastic contraction even in the absence of deviatoric stress. We are already acquainted with the steeper λ-line which describes plastic compression after the previous preconsolidation pressure (p'_v) has been exceeded. Substituting $q = 0$ into the equation 8.34, we obtain

$$\ln\left(\frac{p'_v}{p'_c}\right) = 1 \qquad (8.35)$$

or

$$p'_v = 2.72\, p'_c$$

Cam-clay has predicted that the critical effective pressure is reached when the overconsolidation ratio (p'_v/p') is just less than 3. Ratios much less than 3 will indicate 'loose' states, ratios much greater than 3 will indicate 'dense' states.

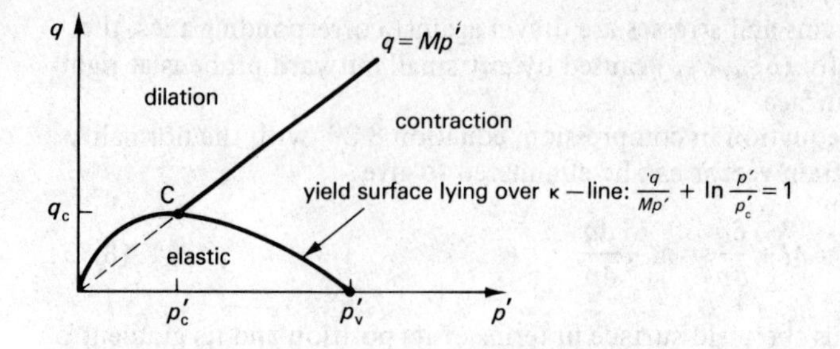

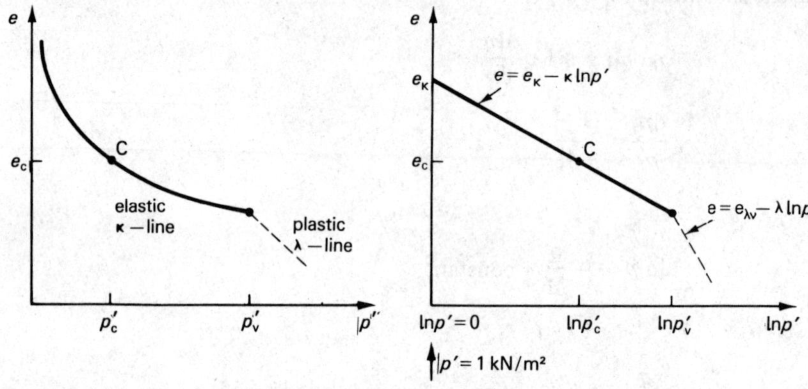

Figure 8.21 *Cam-clay yield surface strung over a κ-line*

Now, suppose that the critical states are known to lie on the line

$$e = e_{\lambda c} - \lambda \ln p' \tag{8.36}$$

For each critical state (e_c, p'_c) there is a point of plastic compression of volume (e_v, p'_v) a fixed distance out along a κ-line, which means

$$e_c - e_v = \kappa(\ln p'_v - \ln p'_c)$$

or

$$e_v = e_c - \kappa(\ln p'_v - \ln p'_c)$$

But we know from equation 8.35 that

$$\ln p'_v - \ln p'_c = 1$$

so

$$e_v = e_c - \kappa \tag{8.37}$$

If from equation 8.36

$$e_c = e_{\lambda c} - \lambda \ln p'_c$$

it follows that

$$e_v + \kappa = e_{\lambda c} - \lambda(\ln p'_v - 1)$$

or

$$e_v = (e_{\lambda c} + \lambda - \kappa) - \lambda \ln p'_v \tag{8.38}$$

This is the equation of a line of spherical plastic compression running parallel to the assumed logarithmic critical state line and $\lambda - \kappa$ higher in terms of void ratio, or $(\lambda - \kappa)/\lambda$ to the right in terms of the logarithm of stress.

The predicted (q, p', e) diagrams appear in figure 8.22 Every κ-line is limited

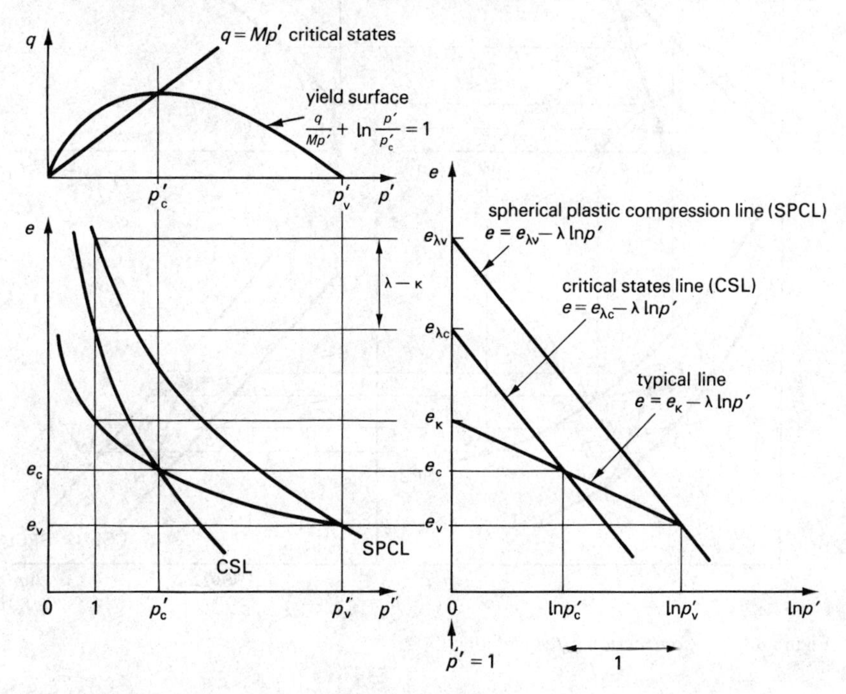

Figure 8.22 *Cam-clay's prediction of (q, p', e) space*

in extent by the plastic spherical compression line, and above each κ-line is strung one of the geometrically identical family of leaf-shaped yield curves

$$\frac{q}{Mp'} + \ln \frac{p'}{p'_c} = 1$$

which are scaled in proportion to p'_c, which is at the intercept between the κ-line and the critical state line.

Knowledge of the yield surface parameters detailed in figure 8.22 brings many varied aspects of soil behaviour under one roof. Consider some soil element with a void ratio e_A and which is under a pure spherical pressure p'_A shown at point A in figure 8.23. It is possible to enter this state A on the (e, p') and $(e, \ln p')$ diagrams. It is now possible to draw a κ-line VAC through point A on the $(e, \ln p')$ diagram; this line VAC can be transferred to the (e, p') diagram. An appropriate yield surface VC can then be drawn over the top of the curved line to appear in the (q, p') diagram. Three contrasting responses to a compression

($\epsilon_q > 0$) caused by a very slow drained change in deviatoric stress Δq and/or spherical effective stress $\Delta p'$ can then be charted.

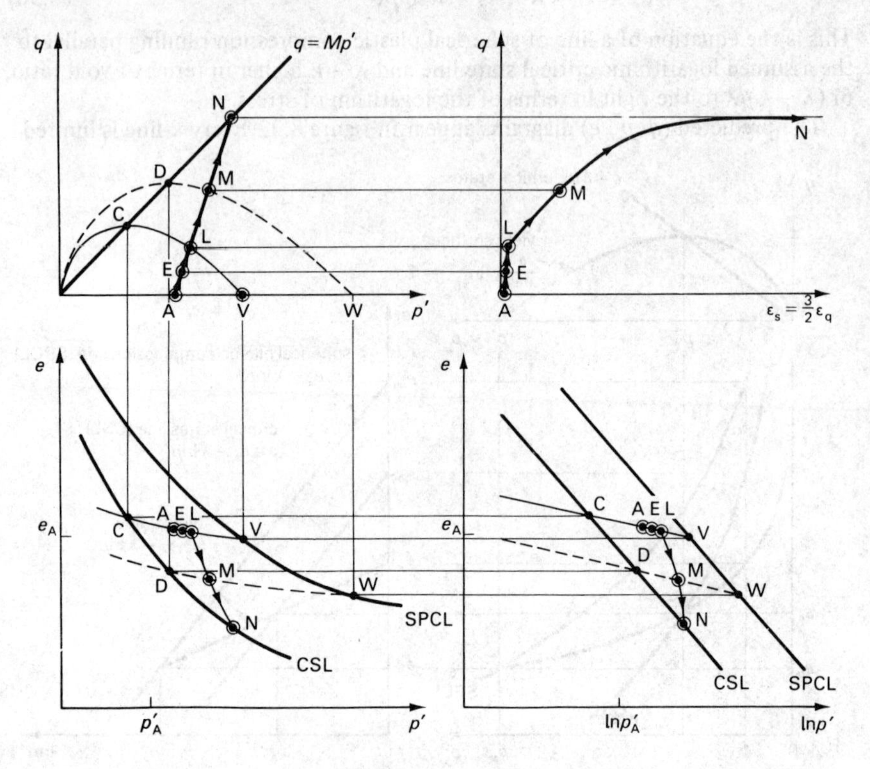

Figure 8.23 *Contraction after yielding*

(1) The soil will remain elastic if the vector (Δq, $\Delta p'$) drawn from the initial stress (0, p'_A) does not reach the initial yield surface, for example AE, figure 8.23.

(2) The soil will yield and contract if the stress vector shown as AL in figure 8.23 reaches the initial yield surface to the right of the critical state C. Contraction ($\delta\epsilon_\kappa > 0$) is mandatory either by equation 8.29 noting that $q < Mp'$, or by equation 8.32 noting that $dq/dp' < 0$. As the soil yields and contracts beyond L the κ-line sinks and becomes longer. This makes the yield surface grow to cover it. The loading vector (Δq, $\Delta p'$) can continue to increase and to cause successive hardening owing to the contraction of the soil skeleton. When the stress vector (Δq, $\Delta p'$) reaches any point such as M in figure 8.23, one of the family of geometrically similar yield curves can be drawn through M so as to fix points D and W. These points can be transferred down to the e/p' diagram and then out to the $e/\ln p'$ diagram where they fix a new κ-line. Since the effective stress p' at M is known, the point M can now be marked on the e/p' and $e/\ln p'$ diagrams. In this way, knowledge of the stress path (Δq, $\Delta p'$) and of the fact of yielding brings knowledge of the void ratio changes and the (e, p') path. Only if

the state of the soil eventually approaches the critical state line at N will the rate of contraction slow down; shear strains will then approach very large values.

(3) The soil will yield and dilate if the effective stress vector reaches the initial yield surface to the left of the critical state as shown by AQ in figure 8.24. Dilation ($\delta\epsilon_\kappa < 0$) is mandatory either by equation 8.29 noting that $q > Mp'$, or by equation 8.32 noting that $dq/dp' > 0$. As the soil dilates e_κ increases and the

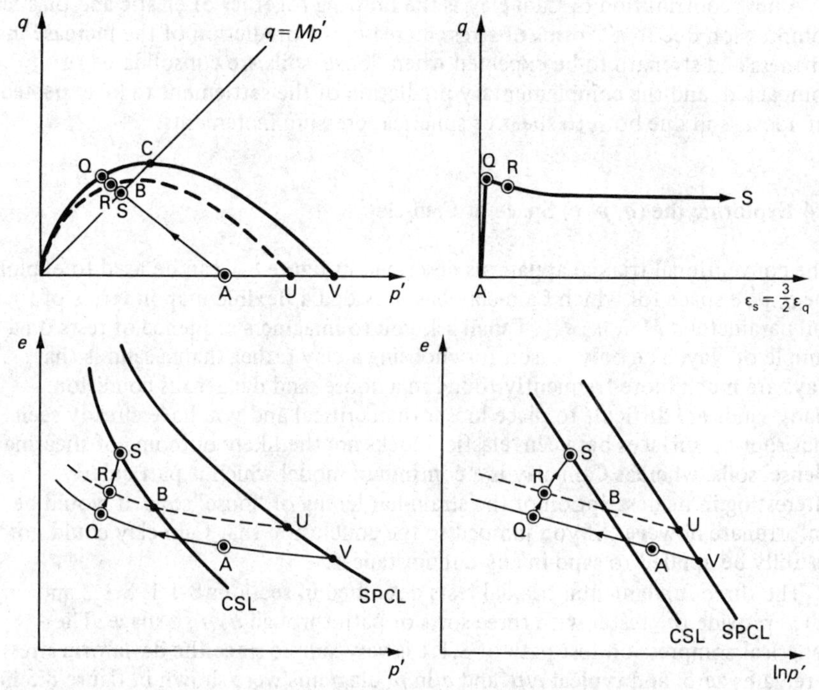

Figure 8.24 *Dilation after yielding*

κ-line moves upwards and becomes shorter. The yield surface shrinks, and unless the stress vector is rapidly retreated the sample will fail explosively. A point R on the q/p' diagram intermediate between the peak strength Q and the ultimate strength S can be used to plot the dilation on the e/p' diagram. A yield surface can be drawn through R since the specimen is known to have yielded already at Q: this new yield surface passes through the critical state line at B and the p' axis at U. Points B and U can be transferred down and across to the e/p' and $e/\ln p'$ diagrams to fix a new κ-line UBR which fixes the path QRS on the e/p' and $e/\ln p'$ diagrams. The vector must eventually shrink back until the soil has dilated towards a critical state $q = Mp'$ at S, near which very large shear strains will develop.

The natural intuitive reaction to the presence of the unstable strain softening zone of Cam-clay behaviour described in (3) is to doubt whether the continuum equations used in its derivation will be applicable. Surely the most likely mode of softening is the generation of a rupture surface, so that strains are severely localised in regions which happened to possess flaws or which are forced on to the soil by its boundary conditions. Furthermore, the normality and stability criteria appear to be in jeopardy. In this respect Cam-clay is able to warn its user only of a dilating region $q > Mp'$ which is unlikely to be well modelled by its own continuum theory. Fortunately the direct shear-box test and its interpretations are well suited to investigate the problem of rupture planes in 'dense' soil, as you saw in chapter 4.

A new contribution of Cam-clay is the drawing together of elastic and plastic compression due to all forms of stress increase, the prediction of the increase in stiffness and strength to be expected when 'loose' soils are consolidated or compacted, and the complementary prediction of the settlement to be expected on 'loose' soil due both to shear or spherical pressure increments.

8.4 Exploring the (q, p' e) Space of Cam-clay

The conventional triaxial apparatus described in figure 8.1 can be used to explore the q, p', e space for which Cam-clay has provided a flexible map in terms of four soil parameters: M, λ, κ, $e_{\lambda c}$. I shall ask you to imagine a sequence of tests on a sample of clay. The only reason for choosing a clay rather than a sand is that clays are much more frequently found in a 'loose' and dangerous condition. Many sands are difficult to place looser than critical and you have already seen that rupture surfaces between 'elastic' blocks are the likely outcome of shearing 'dense' soils, whereas Cam-clay is a continuum model which is particularly interesting in its description of the strain-hardening of 'loose' soils. It would be unfortunate however, if you jumped to the conclusion that Cam-clay could not usefully be applied to sand in any circumstances.

The three fundamental triaxial tests described in sections 8.1.1, 8.1.2 and 8.1.3 provide the tester with three sorts of path through q, p', e space. The spherical compression test path of 8.1.1 is very simple since the deviatoric stress q remains zero, and typical e/p' and $e/\ln p'$ diagrams were shown in figure 8.5 in which the gradients of κ and λ lines can easily be checked as constant and measured, together with $e_{\lambda v}$, the void ratio on the plastic compression line at $p' = 1$ kN/m^2. A spherical compression path through q, p', e space is shown in figure 8.25 to correspond with figure 8.5. The Cam-clay interpretation is that the soil starting at point A possessed a yield surface PC, so that an elastic compression was possible along a κ-line from A to P. At P the soil yielded and began to contract along the line PQ of spherical plastic compression (sometimes called 'virgin' or 'normal' compression). As the state moved from P to Q the soil was contracting through a succession of κ-lines. At Q when the effective pressure was reduced the yield surface had been pushed out to QD and so the soil responded elastically back to R while the yield surface remained at QD.

The drained axial compression test described in section 8.1.2 is a little more complicated to interpret. The specification for the test is to keep the cell pressure constant while the deviatoric stress is increased

$$\delta\sigma_3 = 0$$

$$\delta\sigma_1 = \delta q$$

Therefore

$$\delta p = \frac{\delta\sigma_1 + 2\delta\sigma_3}{3} = \frac{\delta q}{3}$$

This simply means that the average total stress p increases at one-third the rate of the deviatoric stress q. Since the pore-water pressure is drained to zero by running the test very slowly indeed it is possible to equate total to effective stresses

$$\delta p' = \delta p - \delta u = \delta p = \frac{\delta q}{3}$$

so that the observed effect of a drained axial compression test is a stress

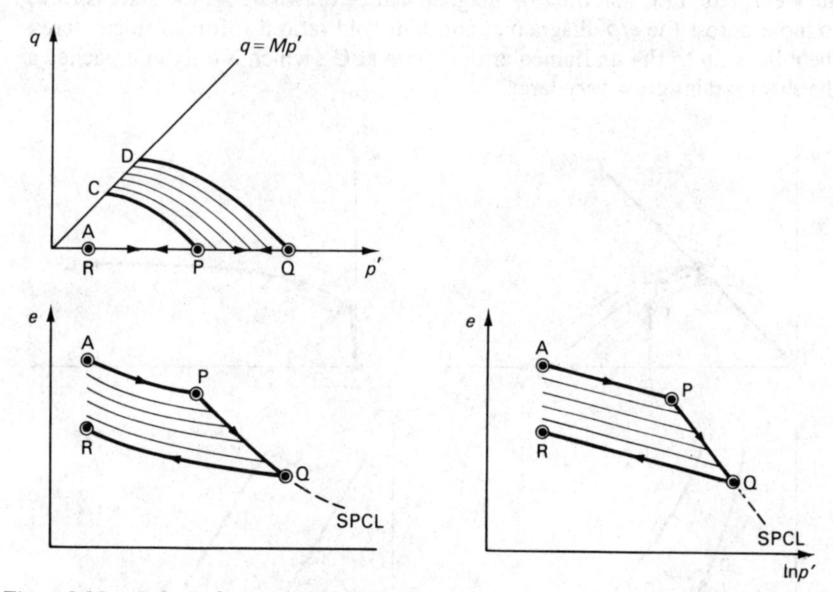

Figure 8.25 *Spherical compression path*

increment vector δq, $\delta p' = \delta q/3$ on the q/p' diagram and a measured volume of drainage fixing points on the e, p' and e, $\ln p'$ diagrams. I chose a slope of 3 for the vector ALMN in the q/p' diagram of figure 8.23 so that it would represent just such a test. Point A is fixed by the initial cell pressure and knowledge of the pore-water pressure datum together with the initial void ratio. The subsequent movement along ALMN is fixed experimentally by the measurement of deviatoric stress q and drainage (for e) and the knowledge that the slope of ALMN on the q/p' diagram must always be 3 while the drained axial compression test continues. It is clear that such a test offers evidence of the initial yield surface location, the pattern of yielding contraction LMN and the magnitude of the Cam-clay parameter M.

The undrained axial compression test described in section 8.1.3 can also be interpreted. Consider the path which would have been obtained if the drained test described above had been undrained, starting from the same initial state A in figure 8.23 transposed to figure 8.26. If the Cam-clay assumptions were to hold, the initial response AY of the clay must be elastic and therefore on the κ-line CAV; the void ratio in an undrained test must be constant, however, and so the effective spherical pressure p' must also remain constant up to Y. Cam-clay stipulates an infinite shear modulus, so the initial response on the q/ϵ_s curve is theoretically vertical until yield at Y (it would be quite a reasonable practical step to make it slope at E_s, nevertheless). At Y the stress path meets the initial yield surface VC. Yield must now continue at constant void ratio up to a critical state, so the horizontal path YZC″ on the e/p' diagram is inevitable. At any intermediate point Z the κ-line V′ZC′ can easily be fixed on the $e/\ln p'$ diagram and transferred to the e/p' diagram. The corresponding yield surface V′ZC′ can then be drawn on the q/p' diagram so as to fix Z on that diagram. In this way the yield path YZC″ on the q/p' diagram can be drawn in. As the state is forced to move across the e/p' diagram at constant void ratio it is forced to cut across the κ-lines up to the undrained critical state at C″, which is only approached as the shear strains grow very large.

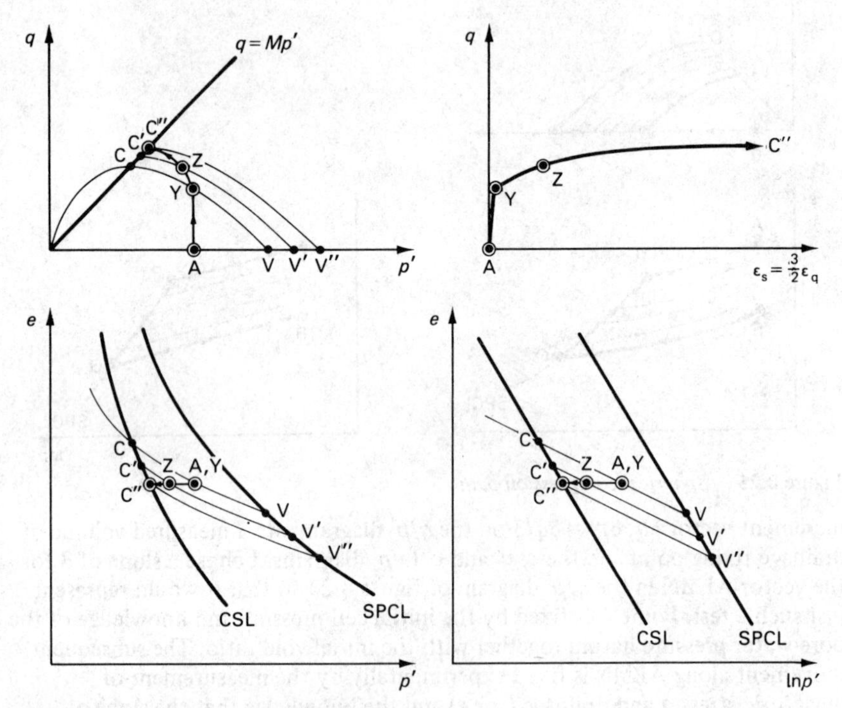

Figure 8.26 *Undrained axial compression path*

If the Cam-clay model were perfect (necessary and sufficient), then point C″ would fix parameter M while AYZC″ would fulfil the expectations we have laid down. The pore-water pressures u would have had to be measured, of course, and their prediction is contained in figure 8.27. Supposing that the pore-water pressure was zero at the start A of the undrained axial compression test, both p'

and p would be represented by point A on the spherical pressure axis. Cam-clay insists that the effective stress path is AYC″. But the triaxial test itself insists that $\delta p = \delta q/3$, as we saw in the previous account of the drained axial compression test. In this case, of course, excess pore-water pressures will exist, and they will be indicated by the horizontal difference $(p - p')$ between the total stress vector AT and the effective stress path AYC″. It is clear that the water pressure starts to increase dramatically at yield, reaching a constant value u_c towards the end of the test. Strictly, this is a transient pore pressure which should be given the symbol Δu_c: the closing of the drain tap in the undrained test means that the transient pore pressures cannot dissipate.

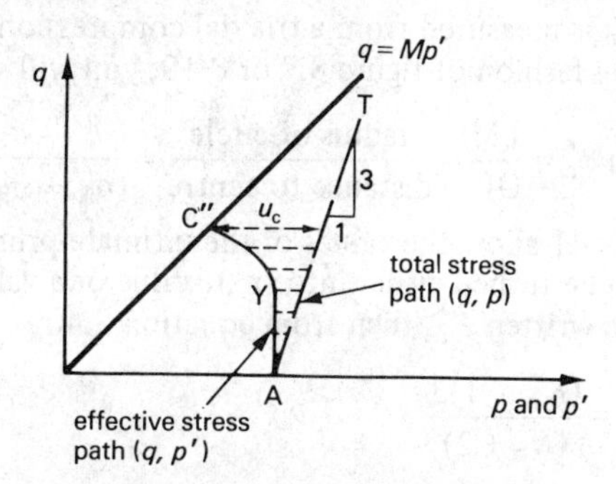

Figure 8.27 *Pore pressures in undrained axial compression*

Inevitably, the Cam-clay model is not perfect. Data of dilating yield fit it very badly. Data of contracting yield fit only approximately. Rupture frequently interrupts the smooth progress towards a critical state and destroys the uniformity of strains upon which Cam-clay is based. Nevertheless Schofield and Wroth demonstrate that with sensitive treatment the Cam-clay model for loose conditions in conjunction with a model of plane dilatant rupture for dense conditions can together form a formidable predictive arsenal. Drained and undrained triaxial tests of any type on sand and clay in loose or dense states at high or low pressure and with any required precompression all fall within the compass of predictions of strength, stiffness, pore pressure and volume change. The qualitative predictions are found to be remarkably sound: the actual numerical values often leave much to be desired. If the correct philosophy for soil in the field is to have a crude but robust conceptual framework against which to hold the variable results of crude but robust *in-situ* or laboratory tests, then it will be hard indeed to better Cam-clay.

8.5 Some Elementary Predictions using the Cam-clay Model

The two previous sections have introduced a mechanical model by which detailed stresses and strains can be predicted for a soil in a wide variety of conditions. Schofield and Wroth take the analysis of detailed behaviour somewhat further still, but they also hammer home some of the simpler facets of their innovation. I shall attempt to follow their example.

The first task is to establish a link between the angle of shearing resistance ϕ'_c at a critical state, and the friction parameter M, which is the ratio of deviatoric stress to average principal stress at a critical state

$$M = \frac{q_c}{p'_c} = \frac{(\sigma'_1 - \sigma'_3)_c}{(\sigma'_1 + \sigma'_2 + \sigma'_3)_c/3} \tag{8.39}$$

or in a triaxial test

$$M = \frac{(\sigma'_1 - \sigma'_3)_c}{(\sigma'_1 + 2\sigma'_3)_c/3} \tag{8.40}$$

Now suppose that ϕ'_c is measured from a triaxial compression test using a Mohr's τ/σ' diagram after the fashion of figure 8.7 or 8.10; you will see that

$$\sin \phi'_c = \frac{\text{LM}}{\text{OL}} = \frac{\text{radius of circle}}{\text{distance to centre}} = \frac{(\sigma'_1 - \sigma'_3)_c/2}{(\sigma'_1 + \sigma'_3)_c/2} \tag{8.41}$$

Equations 8.40 and 8.41 allow the values of the ultimate principal effective stresses σ'_1 and σ'_3 to be turned either into an M-value or a value for $\sin \phi'_c$. Let the ratio $(\sigma'_1/\sigma'_3)_c$ be written K'_c, then from equation 8.40

$$M = \frac{(K'_c - 1)3}{(K'_c + 2)} \tag{8.42}$$

while from equation 8.41

$$\sin \phi'_c = \frac{K'_c - 1}{K'_c + 1} \tag{8.43}$$

Equations 8.42 and 8.43 can then be inverted to give

$$K'_c = \frac{2M + 3}{3 - M} = \frac{1 + \sin \phi'_c}{1 - \sin \phi'_c} \tag{8.44}$$

which in turn gives either

$$M = \frac{6 \sin \phi'_c}{3 - \sin \phi'_c}$$

or

$$\sin \phi'_c = \frac{3M}{M + 6} \tag{8.45}$$

Although differences exist between the concept of limiting angle of inclination ϕ'_c and the concept of a dissipation constant M, equations 8.39 to 8.45 demonstrate how to translate between the two on an *ad hoc* numerical basis. When ϕ'_c measurements double from 20° to 40°, M-measurements increase from 0.77 to 1.64 by a similar factor, so an effect which increased ϕ'_c by 1 per cent would increase M by about 1 per cent also. The significant difference between the alternative ways in equation 8.44 of expressing the fact that a critical stress ratio K'_c has been reached, is that the M-expression contained the intermediate stress σ_2' from equation 8.39. The stress-inclination model described initially in chapter 4 and elaborated in section 8.1, however, was concerned only with the largest existing Mohr's circle, which could be drawn between the major stress σ_1' and the minor stress σ_3' irrespective of the size of the intermediate stress σ_2'.

It would not be inconsistent with what has gone before to use M with the implication that σ_2' matters if the soil is behaving as a continuum, and to use ϕ' for the stress inclination across a rupture plane should the soil be discontinuous in behaviour, with the implication that the applied stress σ_2' in the 'strike' direction is irrelevant. These alternative notions are symbolised in figure 8.28. This issue is important because σ_2' in the field will not in general be known and will not often be equal to the minor stress σ_3'. It is comforting, however, that the common triaxial apparatus always sets σ_2' at a conservative level: the intermediate stress can never be lower than the minor stress, by definition! With these provisos it is permissible to translate M-values into ϕ_c' values using equation 8.45.

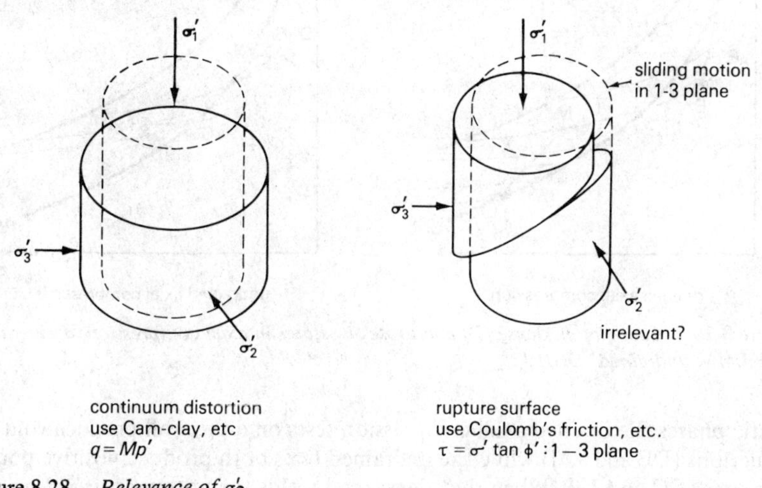

Figure 8.28 *Relevance of* σ_2'

8.5.1 The Contrast Between Soils that are 'Dense' and 'Loose' at Yield

Soils that are 'dense' with respect to critical states are often called heavily overconsolidated; they need a very large increase in spherical pressure if they are to undergo plastic compression. Without this extra spherical pressure they have a strong tendency to dilate and soften, by the agency of transient pore suction.

Soils that are 'loose' with respect to critical states are often called lightly overconsolidated or virgin; they need only a little extra pressure to undergo plastic compression. They not only compress readily under spherical pressures but also contract strongly when sheared, generating transient positive pore pressures. Once contracted they become stiffer and stronger.

The Cam-clay predictions for the path of drained and undrained triaxial compression tests on 'dense' and 'loose' soils appear in figure 8.29. In their

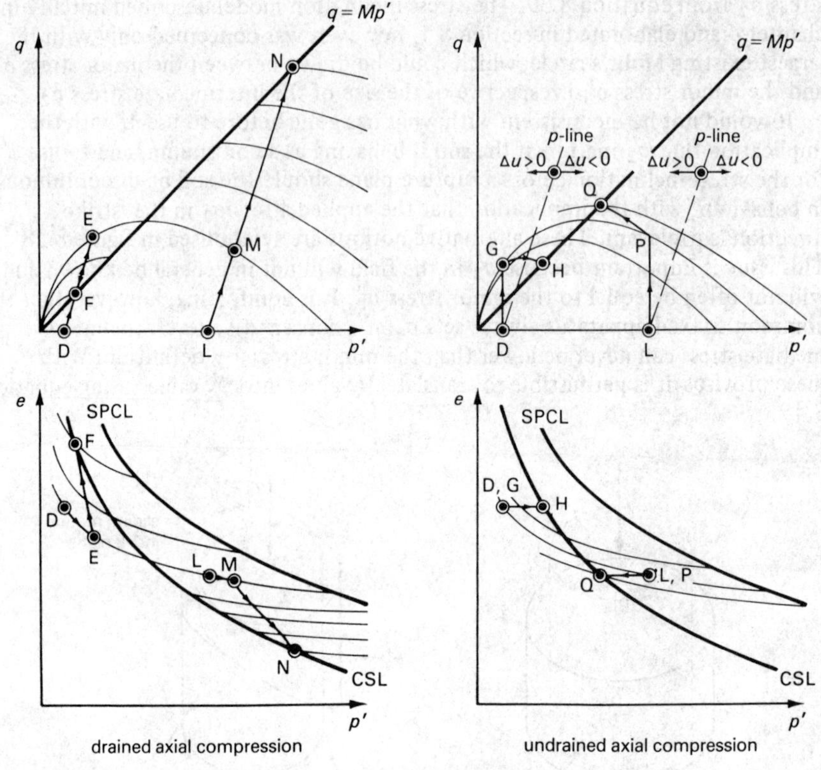

Figure 8.29 *Cam-clay in dense (D) and loose (L) states in axial compression; drained: DEF LMN; undrained: DGH LPQ*

elastic phases the drained axial compression tests on either soil produce void reductions (DE and LM) while the undrained tests both produce positive pore pressures (DG and LP). When the 'dense' soil yields the pore pressures drop in an undrained test (GH) and the voids dilate in a drained test (EF), while the strength drops. When the 'loose' soil yields the pore pressure rises dramatically in an undrained test (PQ) while in a drained test the voids contract (MN).

In 'dense' soils the elasticity model may well be valid up to $\tau = \sigma' \tan \phi'_c$, $q = Mp'$ on first loading; beyond this, dilation starts to generate the peak strength. 'Loose' soils begin to yield and contract at some shear stress $\tau < \sigma' \tan \phi'_c$, $q < Mp'$: if elasticity is to be applied to them the location of the yield surface must be established, and the loads analysed to prove that no stress increment will cause yield.

It may require an effort of will to remember that 'loose' soils can be very strong. Consider an element of ocean floor sediment which has suffered only a very gradual burial process since it was deposited. It must be in a state of plastic compression, which might be plotted somewhere between the critical state line and the spherical compression line, depending on the ratio q/p' which it is suffering, at a point not unlike P in figure 8.29. If it was buried under 100 m of similar sediment its vertical effective stress could be 1000 kN/m^2 while its horizontal effective stress might be 600 kN/m^2: point P would then be fixed by $q = 400$ kN/m^2, $p' = 730$ kN/m^2. To scale, therefore, the undrained strength of the sediment q_Q would be 530 kN/m^2, which would make it 'very stiff' according to convention. A conventional 38 mm diameter sample would require a force of 600 N, the weight of a man, to squash it! Nevertheless, the sediment at point P is in a rather dangerous condition. The weight of overlying material is on the point of causing further plastic compression. Any further loads which the engineer may want the clay to carry would cause large distortions followed later by large contractions in volume. Cam-clay is useful in sorting out the enigma of virgin soil. It correctly predicts that the undrained strength q_Q is proportional to the initial effective stress p'_P so that strength rises from zero near the surface of the soil to very large values at great depth. It also correctly warns that, irrespective of the magnitude of the strength, virgin soils can only be called on to take additional compression forces if large strains can be tolerated.

It happens, however, that the condition of one-dimensional plastic compression, which might well be thought to correspond to the situation of a level sea-floor sediment subsiding under its own weight, provides Cam-clay with a crucial test of numerical accuracy which it fails abysmally. It is widely accepted that the ratio σ'_3/σ'_1 in these circumstances, dubbed K'_0, is well predicted by the empirical formula $K'_0 \approx 1 - \sin\phi'$. Translating into Cam-clay language, the requirement is for $q/p' \approx M/2$ for the case $\epsilon_1 > 0$, $\epsilon_2 = \epsilon_3 = 0$. Schofield and Wroth easily demonstrate that when these strains are fed into the basic work equation, the ratio q/p' for most soils will be predicted to be 0; Cam-clay will not discriminate between spherical plastic compression and one-dimensional plastic compression. Roscoe and Burland (1968) have attempted to remedy this serious K'_0 defect by altering the dissipation function or the shape of the yield surface; any desired strain increment can then be achieved at any particular state of stress.

8.5.2 The Contrast Between Stress Paths for Loading and Unloading

We have already seen that if soil is loaded uniaxially while the horizontal pressure is kept constant, it compresses along a total stress path $\delta q/\delta p = 3$. If soil is unloaded horizontally by $\delta\sigma$ while the vertical pressure is kept constant, however, it compresses along a different stress path. For a triaxial sample

$$\delta q = 0 - (-\delta\sigma) = +\delta\sigma$$

$$\delta p = \frac{0 - \delta\sigma - \delta\sigma}{3} = \frac{-2\delta\sigma}{3}$$

Therefore

$$\frac{\delta q}{\delta p} = \frac{-3}{2} \text{ in direction } q+, p-$$

Likewise a triaxial test could be conducted in which the horizontal pressure remained constant while the axial pressure was reduced by $\delta\sigma$

$$\delta q = -\delta\sigma - 0 = -\delta\sigma$$

$$\delta p = \frac{-\delta\sigma + 0 + 0}{3} = \frac{-\delta\sigma}{3}$$

Therefore

$$\frac{\delta q}{\delta p} = 3 \text{ in direction } q-,\ p-$$

The final triaxial loading possibility would be to load horizontally by $\delta\sigma$, keeping the axial stress constant

$$\delta q = 0 - (\delta\sigma) = -\delta\sigma$$

$$\delta p = \frac{0 + \delta\sigma + \delta\sigma}{3} = \frac{2}{3}\delta\sigma$$

$$\frac{\delta q}{\delta p} = \frac{-3}{2} \text{ in direction } q-,\ p+$$

We will consider these various stress paths applied very slowly to a drained sample so that they can be treated as effective stress paths $\delta q/\delta p'$. Of course, two further effective stress paths are easily obtained, being those caused by increasing or reducing the pore-water pressure of the sample by allowing water at some fixed datum pressure to drain in or out of the sample as it requires.

These six stress paths are shown in figure 8.30, depicting a full Cam-clay yield

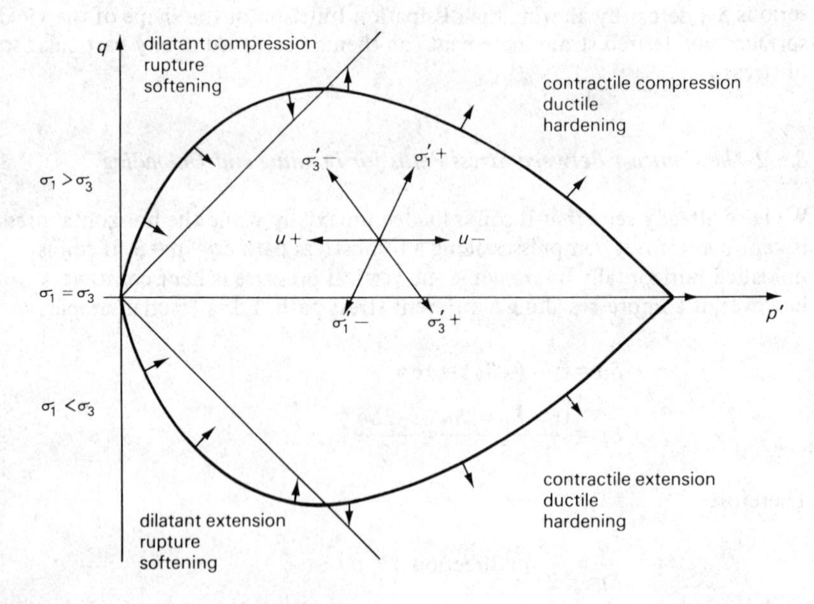

Figure 8.30 *Stress paths to yield Cam-clay*

surface around some initial stress point in an arbitrary position. Whereas the loading vectors *tend* to generate contractions the unloading vectors *tend* to generate dilatant ruptures, depending on the starting point. Likewise a reduction in pore pressure inevitably leads to contraction while an increase in pore pressure inevitably leads to dilatant rupture.

If soil in the field is exposed in a trench wall it may slowly follow the $\sigma_3'-$ vector: failure could be sudden and involve a wedge of trench-wall sliding down a rupture plane to crush the workers below. If a foundation is built to carry a heavy load, the loading vectors may be in the sector $\sigma_1'+$ to $\sigma_3'+$: settlements may well precede, and warn of, failure. If a slope is subject to inundation so that its water table rises, soil elements will be subject to the $u+$ vector, and the ensuing landslide might be sudden, involving the whole hillside in a movement on a thin rupture surface. If the base of a deep foundation excavation is not immediately blinded with a thin layer of concrete, the stress vector $\sigma_1'-$ will lead to dilation and softening if the soil can obtain the necessary volume of water from rain-pools; this can cause such deterioration in the intended foundation layer that the project must be totally redesigned. It is usually necessary to prevent groundwater from entering potentially dilatant zones for the same reason; water is then pumped from a pattern of specially bored wells around the critical zone, so as to reduce the local water pressures.

These various field problems only develop if the inevitable changes in total stress are allowed to be reflected as changes in effective stress by the mechanism of transient flow. Their prediction therefore rests as heavily on an intimate knowledge of the facility of drainage through the ground fabric as it does on knowledge of the mechanics of stress and strain: Rowe (1972).

8.5.3 The Contrast Between Drained and Undrained Stress Paths

We saw in figure 8.29 that an undrained test on 'dense' soil was offered a greater resistance than a drained test ($q_H > q_F$) due to the excess suctions developed at constant volume. Likewise we saw that a 'loose' soil was weaker when undrained than it was when drained ($q_Q < q_N$) due to the excess pore pressures that developed at constant volume.

Engineers who habitually think in terms of friction angles and simple groundwater levels are sometimes surprised to see soil standing in cliffs. This is nearly always due to pore suctions, which in turn are often caused by the engineer himself, who shears the ground suddenly by taking side support away from a 'cliff' which is thereby revealed. Trenching is a good example of this. If the soil is 'dense' and the tendency of the soil to shear or distort is strong, excess suction will be generated. The permeability of the ground will then slowly allow transient flow of any available water to dissipate the suction. Of course, if the engineer comes to expect dense soil to stand vertically in cliffs or trench walls, he may eventually leave such a cliff unsupported until the transient pore suction has been able to dissipate by drawing in groundwater. He will then see for himself the contrast between the 'undrained' and 'drained' behaviour of dense soils, and will complain to the court of inquiry that the soil mysteriously softened.

'Loose' soils present the engineer with a different problem: they may collapse and spread themselves violently around before they can be heaped at their angle of repose ϕ'_c. If a 'loose' sand is dry then if it wishes to contract it may do so very quickly and thereby offer the engineer its full drained strength: the air permeability of sand is very great, so the air offers little resistance. If a 'loose' sand or more especially a 'loose' silt or clay is saturated, then the process of contraction demands the outflow of pore water and may take minutes, hours or years. During the early part of this time the soil only offers to the engineer its 'undrained' strength, such as q_Q in figure 8.29. If the engineer had been expecting to obtain q_N, the drained strength after taking into account normal groundwater pressures, he may be very upset to see a catastrophic landslide. The pore pressure $u > 0$ setting point Q to the left of the drained loading vector is an excess transient water-pressure increment which must be added to any other likely water pressures acting in such a field situation if an effective friction calculation is to be made. This excess pore pressure is due entirely to the sudden shearing of the 'loose' soil.

Consider the response to an earthquake of a loosely tipped industrial spoil heap formed underneath a waste pipe which delivers a silty slurry from a washing process. If water is seeping through the growing heap of silt then the side slopes could not be much steeper than $\phi'_c/2$ by the analysis of section 4.6.2. The slopes would then be mobilising an angle of resistance roughly equal to ϕ'_c and their factor of safety would be approaching 1.0; the heap would be rather precarious. If the engineer deployed a complex of waste pipes he might be able to arrange the deposition of silt so that the slopes of the heap lay at an angle of

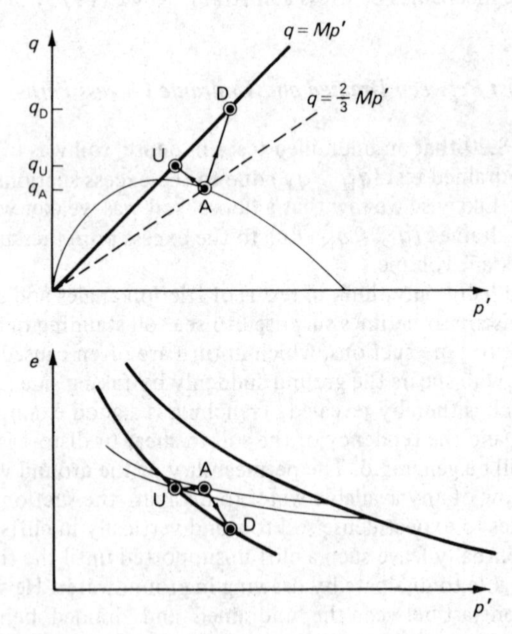

Figure 8.31 *Drained and undrained loading of an element contracting under a stress ratio $q/p' = 2M/3$*

$\phi'_c/3$. He might then feel that since he was mobilising only two-thirds of the available shearing resistance he would be justified in describing the slope as having a safety factor of roughly 1.5. Figure 8.31 shows a typical state point A for an element of soil within the regraded slope, assuming that Cam-clay applies. At point A the q/p' ratio is $\frac{2}{3}M$. The figure also shows that the undrained strength starting from A is q_u, which is only 20 per cent greater than q_A, whereas the drained strength q_D corresponding to pure uniaxial loading from A is 75 per cent greater than q_A. In no sense, therefore, does the slope have a unique safety factor of 1.50. If the regraded slope, standing at only $\phi'_c/3$, were to suffer an earthquake which temporarily and suddenly increased the shear stresses in the soil by more than 20 per cent the slope would begin to fail. If the quake were to last sufficiently long, or were to be much stronger than this, the slope could have developed such a large momentum that it would continue to slide away after the quake had finished. Bear in mind that $\phi'_c/3$ might be only 10° or a slope of 1:6, and that a fairly weak earthquake could easily 'tilt' the local net acceleration field 2° from the vertical for 10 s or more and have a good chance of causing a catastrophe.

8.6 The Onset of Rupture in a Dilatant Soil

Schofield and Wroth (1968) cite evidence in favour of the ability of Cam-clay to discriminate between dilating and contracting soil states, and to provide a crude but useful map of the detailed behaviour of contracting soil elements. They also admit that Cam-clay does not cope well with the detailed behaviour of dilating soil elements. This incapacity of the theory to mirror the behaviour of real soil would have been reflected in the previous section in the following ways.

(1) The stress vector in figure 8.24 would have reached a limiting failure condition before the Cam-clay yield surface was reached at Q. At a point such as R on the q/p' diagram the soil would have ruptured. This would have been preceded by plastic dilation, so that the path on the e/p' diagram would have 'cut the corner' and travelled from A to R without passing Q.

(2) In figure 8.29 the stress vector starting from a dense condition D would not have reached E in a drained test or G (nor yet H) in an undrained test. In either case a rupture would have interrupted the test at some point after the stress ratio q/p' had exceeded the critical value of M.

Considering the nature of the stability criterion represented by equation 8.32 it would have been surprising if the consequent yield surface for unstable dilation had proved reliable. There may be less reason for distrusting the basic work equation, 8.29.

The prediction of the dilation which precedes rupture and of the stresses which cause rupture in dense soils is still a matter for controversy between research workers. In these circumstances it is the responsibility of the practising engineer to attempt to grasp whatever physical principles may be involved in order to fuel his own intuition, so that he can make reasonable decisions. Absolute certainty is not required here: for example the prudent engineer would note that if he restrained q/p' to less than M, or τ/σ' to less than $\tan\phi'_c$, he should never see a dilating soil failure, irrespective of what the peak strength might actually have been. I have introduced two models of stress–dilatancy, the saw-blades model of section 4.5, which was associated with the shear-box test and the Cam-clay of section 8.3, which was associated with the triaxial test. Before requiring you to surrender to your own intuition I shall attempt to draw these models together somewhat in order to explore the connection between the peak strength and the volumetric expansion of a soil element.

The first step will be to return to the simple stress–dilatancy equation from section 4.5

$$\phi' = \phi'_c + \nu \tag{8.46}$$

where

$$\tan \nu = \frac{dy}{dx} \tag{8.47}$$

referring to figures 4.10 and 4.11. Let us recast this expression in terms of 'pure' rather than 'simple' shear, and express ν in terms of the principal strains. It will be necessary to preserve the plane-strain conditions imposed by the original shear-box mechanism, of course, so that the principal compressive strains will be $\epsilon_1 > 0$, $\epsilon_2 = 0$ and $\epsilon_3 < 0$. The new dilatancy relationship will be useful in the interpretation of plane compression tests, which are easily performed in a standard triaxial machine. Samples which are square in cross section rather than circular are compressed while prevented from expanding in the horizontal 2-direction by greasy walls. This interpretation will, in turn, be useful to the engineer who has a plane-strain field problem such as a long embankment.

With this aim, suppose that the shear box strains were confined to a zone of soil of depth Y. The vertical upward motion of the lid dy would imply a vertical contraction

$$d\epsilon_y = -\frac{dy}{Y} \tag{8.48}$$

Similarly, the increment of horizontal displacement dx would produce an equivalent increment of simple shear angle with respect to xy axes

$$d\gamma_{xy} = \frac{dx}{Y} \tag{8.49}$$

It follows immediately that

$$\tan \nu = \frac{dy}{dx} = -\frac{d\epsilon_y}{d\gamma_{xy}} \tag{8.50}$$

In order to draw the Mohr circle of strain it is necessary to make a further assumption concerning the direct strains in the x-direction. It appears reasonable to assume that the physical imposition of the shear box dictates that

$$d\epsilon_x = 0$$

Following the principles and terminology of section 6.3 the estimated Mohr circle of plane strain for the shear-box element appears in figure 8.32, drawn so

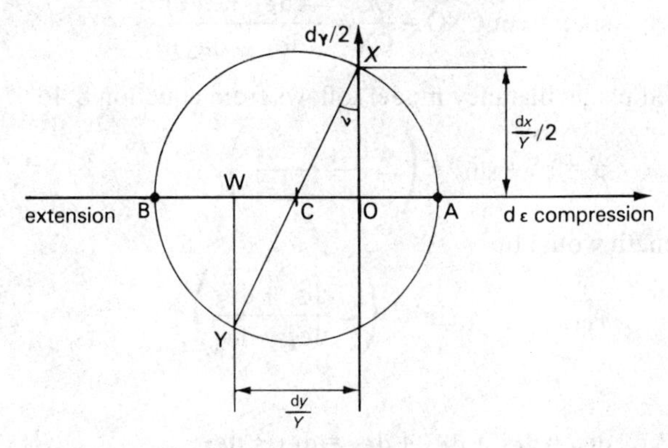

Figure 8.32 *Plane strain increments for dilating soil*

as to imply volumetric expansion. Points X and Y represent the direct and angular strains on the x and y-axes respectively, and they therefore are plotted at opposite ends of a diameter through the centre of the circle C. Point A represents the major principal compressive strain ϵ_1, while point B represents the minor principal compressive strain ϵ_3, which must be strongly negative in order to indicate extension.

The diagram shows that

$$\mathrm{OX} = \frac{\mathrm{d}x}{2Y}$$

while

$$\mathrm{OC} = \frac{\mathrm{OW}}{2} = \frac{\mathrm{d}y}{2Y}$$

An interesting result follows from equation 8.50

$$\tan \widehat{\mathrm{CXO}} = \frac{\mathrm{OC}}{\mathrm{OX}} = \frac{\mathrm{d}y}{\mathrm{d}x} = \tan \nu$$

Therefore

$$\widehat{\mathrm{CXO}} = \nu$$

The angle of dilation of the shear box appears as a strong feature of the corresponding Mohr circle of plane strain.

Now let us assume that the same Mohr circle would have obtained in a pure plane compression test, and therefore that it is meaningful to generalise the saw-blade dilatancy model using the principal strain increments $\mathrm{d}\epsilon_1$ and $\mathrm{d}\epsilon_3$. We must use

$$\mathrm{CX} = \text{radius of circle} = (\mathrm{d}\epsilon_1 - \mathrm{d}\epsilon_3)/2$$

$$\mathrm{OC} = \text{average extension} = -(\mathrm{d}\epsilon_1 + \mathrm{d}\epsilon_3)/2$$

to give

$$\sin \nu = \sin \widehat{CXO} = \frac{OC}{CX} = \frac{-(d\epsilon_1 + d\epsilon_3)}{(d\epsilon_1 - d\epsilon_3)} \tag{8.51}$$

The generalised plane dilatancy model follows from equation 8.46

$$\phi' = \phi'_c + \sin^{-1}\left(-\frac{d\epsilon_1 + d\epsilon_3}{d\epsilon_1 - d\epsilon_3}\right) \tag{8.52}$$

The peak strength would be

$$\phi'_{max} = \phi'_c + \sin^{-1}\left(-\frac{d\epsilon_1 + d\epsilon_3}{d\epsilon_1 - d\epsilon_3}\right)_{max} \tag{8.53}$$

or using

$$d\epsilon_v = d\epsilon_1 + d\epsilon_2 + d\epsilon_3 = d\epsilon_1 + d\epsilon_3$$

in plane strain and

$$d\epsilon_s = d\epsilon_1 - d\epsilon_3$$

$$\phi'_{max} = \phi'_c + \sin^{-1}\left(-\frac{d\epsilon_v}{d\epsilon_s}\right)_{max} \tag{8.54}$$

This is a generalised statement of the very crude saw-blade model of dilatancy. It is quite likely that the form of the new dilatancy model is correct but the detail is in error, and it is sensible to adopt a less specific formula

$$\phi'_{max} = \phi'_c + \phi'_v \tag{8.55}$$

where ϕ'_v is some function of the maximum rate of volumetric expansion, which may be different for triaxial, plane strain or other conditions. In section 4.5 I already indicated that the rate of dilatancy would be expected to increase if there was an increase in the total dilatancy which must take place before the soil reached a critical state. ϕ'_v should increase, therefore, if the soil were made denser or if the effective confining pressure were reduced. You may find it interesting to compare the new model of equation 8.54 with the Cam-clay basic work equation 8.29, which can be rewritten

$$\frac{q}{p'} = M + \left(-\frac{d\epsilon_\kappa}{d\epsilon_q}\right) \tag{8.56}$$

in which you may recall $d\epsilon_\kappa$ was considered to be the plastic component of the volumetric strain and $d\epsilon_q$ a special triaxial shear strain component $\frac{2}{3}\epsilon_s$. Bearing in mind the loose correspondence between q/p' and τ/σ' or between M and ϕ'_c, equations 8.54 and 8.56 can be said to be of the same family.

Not only do they associate peak strength with the maximum rate of dilation, they also imply that any stress ratio greater than critical can only be supported if a certain rate of dilation is taking place. Eventually in a shear test the rate of dilation can increase no more; a rupture will then propagate through the sample at the peak stress, and it will be difficult to prevent the stored energy in the apparatus (or the landslide) from shearing the sliding blocks so far on the dilating rupture plane that the stress ratio drops to its critical value (or beyond, as you will see in the next section). This is a most dramatic event, even in the laboratory: the deviatoric force in a plane compression test on very dense sand may suddenly fall by a factor of 2 at rupture. You should note that both dilatancy models fail to make an absolute prediction of the peak stress ratio. Some other evidence or theory is required which can offer some prediction of the dilatancy rate $-d\epsilon_v/d\epsilon_s$ or $-d\epsilon_\kappa/d\epsilon_q$.

My objective here has been to bolster your grasp of the mechanical consequences of volumetric expansion. If you have absorbed the lesson of the saw-blades and the concepts of the energy equation used in the derivation of Cam-clay then you will have begun to feel certain that any permanent and irrecoverable expansions in volume during shear should be accompanied by evidence of a temporarily enhanced strength. I shall now review some research findings from this new perspective. For this purpose I shall be forced to separate sandy soils from clayey soils because most research workers have assumed that the two were mechanically quite distinct.

8.6.1 Data on Dilating Sand

Cornforth (1973) uses an equation identical to 8.55 in order to interpret the data of both triaxial and plane compression tests on sands of various densities. He does not consider the effect of variations in confining pressure, and refers to ϕ'_v in equation 8.55 as the 'density component' of the strength of the soil (ϕ_{dc} in his own terminology). In this useful paper he shows that in compression tests on sands which had been made as loose as possible ϕ'_v was zero, whereas when the sands were as dense as he could make them ϕ'_v was 12° for triaxial strain and 17° for plane strain in each case. At void ratios intermediate between 'loosest' and 'densest' ϕ'_v could be interpolated roughly linearly. Using Cornforth's empirical relationship an engineer could quickly make an estimate of the peak friction angle of a sand at any desired void ratio. He would firstly tip the dry sand into a loose heap and measure its angle of repose, ϕ'_c. He would then measure the smallest and largest possible void ratios he could attain and calculate the relative void ratio

$$R_e = \frac{e_{max} - e}{e_{max} - e_{min}} \tag{8.57}$$

or with as much empirical justification the relative dry density

$$R_d = \frac{\rho_d - \rho_{d\,min}}{\rho_{d\,max} - \rho_{d\,min}} \tag{8.58}$$

He would proceed to multiply the theoretical maximum dilatancy contribution of 17° by either of the measures of relative compaction

$$\phi'_v \approx R_e\,17° \text{ or } R_d\,17° \tag{8.59}$$

in order to reach the conclusion

$$\phi' = \phi'_c + \phi'_v$$

appropriate to plane-strain conditions.

It happens that the largest ϕ'_v of 17° in plane strain was associated with the greatest rate of dilation of the soil element at failure. Cornforth uses $(-d\epsilon_v/d\epsilon_1)$ to describe the dilatancy rate and reports $\phi'_v = 17°$ at $(-d\epsilon_v/d\epsilon_1)_{max} = 0.9$. Now

$$\frac{-d\epsilon_v}{d\epsilon_1} = \frac{-(d\epsilon_1 + d\epsilon_3)}{d\epsilon_1} = -\left(1 + \frac{d\epsilon_3}{d\epsilon_1}\right)$$

whereas

$$\frac{-d\epsilon_v}{d\epsilon_s} = \frac{-(d\epsilon_1 + d\epsilon_3)}{(d\epsilon_1 - d\epsilon_3)} = -\left(1 + \frac{d\epsilon_3}{d\epsilon_1}\right) \Big/ \left(1 - \frac{d\epsilon_3}{d\epsilon_1}\right)$$

Cornforth's dilatancy rate can therefore be converted easily into the dilatancy rate used in equation 8.54

$$\frac{-d\epsilon_v}{d\epsilon_s} = \frac{-d\epsilon_v}{d\epsilon_1} \Big/ \left(2 - \frac{d\epsilon_v}{d\epsilon_1}\right) \tag{8.60}$$

so that the peak dilatancy rate $(-d\epsilon_v/d\epsilon_1) = 0.9$ leads to $(-d\epsilon_v/d\epsilon_s) = 0.9/2.9 = 0.310$ and to a prediction

$$\phi'_{max} = \phi'_c + \sin^{-1} 0.310$$

or

$$\phi'_{max} = \phi'_c + 18°$$

This is in agreement with Cornforth's data, and lends some support to our line of reasoning.

It is not surprising that Cornforth found that the angle ϕ'_v is dependent on the pattern of strains in three dimensions. The reduced angle of 12° in a triaxial test is no doubt due to the 'wasting' of the available dilation across all the possible shear directions, there being no horizontal restraint. Every different strain pattern could be expected to generate a different dilatancy component ϕ'_v, even when the same sand is tested at the same density. It is interesting to note that by using

$$-\frac{d\epsilon_v}{d\epsilon_q} = -\frac{3}{2}\frac{d\epsilon_v}{d\epsilon_s} = 0.47$$

for Cornforth's peak triaxial dilatancy rate, the Cam-clay work equation 8.56 provides that

$$\left(\frac{q}{p'}\right)_{max} = M + 0.47$$

on the assumption that the elastic component of volume change is negligible so that $\epsilon_v \approx \epsilon_\kappa$. This can be converted back to peak friction angles using equation 8.45, after which it will be discovered that the prediction is

$$\phi'_{max} = \phi'_c + 11°$$

over a wide range of ϕ'_c values. The ability of our plane strain and triaxial dilatancy models to fit Cornforth's data is quite pleasing.

From our present point of view, therefore, the new contribution of Cornforth is his evidence that any sand might be considered to have a zero dilatancy rate when it is very loose and a maximum dilatancy rate

$$\left(-\frac{d\epsilon_v}{d\epsilon_s}\right)_{max} \approx 0.31$$

when it is very dense. This empirical evidence, when used with our own dilatancy models, can be used to predict the peak strength of sand in plane strain or triaxial compression tests.

The earliest and by now the most elaborate and well-tested stress–dilatancy model was developed by Rowe (1962) and was reported by Rowe (1969) as being capable of explaining the differences in behaviour of elements of identical soil tested in triaxial compression, plane compression and direct shear. Although I have in this book relied greatly on Rowe's direct mechanical approach to dilation it would be inappropriate here to explore more fully the diversity of data and interpretations which he has generated. It is interesting, however, to compare the form of his stress–dilatancy theory with that of equations 8.54 and 8.56. Equation 8.61 expresses Rowe's plane-strain theory which is in terms of the ratio of principal stresses

$$\frac{\sigma'_1}{\sigma'_3} = \left(\frac{\sigma'_1}{\sigma'_3}\right)_c \left(1 - \frac{d\epsilon_v}{d\epsilon_1}\right) \tag{8.61}$$

and in which the suffix c is used as before to denote the value at a critical void ratio. He arrives at this by considering all the various possible saw-tooth mechanisms of failure which could offer the measured rate of dilatancy, and chooses that which allows the generation of the intrinsic angle of friction on the sliding surfaces and which causes failure at the lowest principal stress ratio. De Josselin de Jong (1976) provides an elegant proof of the internal consistency of Rowe's approach. Rowe shows that the dilatancy rate $-d\epsilon_v/d\epsilon_1$ approaches, but does not exceed, unity in the densest possible packing of many sands. Like Cornforth he correlates this dilatancy rate, between 1 and 0, with the relative density between 'densest' and 'loosest'. He also demonstrates that ϕ'_v in triaxial strain is roughly two-thirds that in plane strain.

It is interesting to compare, in table 8.1, the two stress–dilatancy rules for plane strain embodied in equations 8.54 and 8.61 using the relationship between strain rates which was established in equation 8.60 and the relationship between ϕ' and the stress ratio which was established following equation 8.43. Considering the good fit to their data obtained by a number of workers using Rowe's more searching model, one must conclude that the generalised saw-blades model over-estimates the dilatancy component ϕ'_v of very dense sand by roughly 20 per cent, this error reducing for looser packings. Either rule is useless as a predictor of strength if the dilatancy rate were unknown.

TABLE 8.1
Plane Strain ϕ' values

$-\frac{d\epsilon_v}{d\epsilon_1}$	$-\frac{d\epsilon_v}{d\epsilon_s}$	$\phi'_c = 20°$ equation 54	61	25° 54	61	30° 54	61	35° 54	61
0	0	20°	20°	25°	25°	30°	30°	35°	35°
0.20	0.091	25°	25°	30°	30°	35°	35°	40°	39°
0.40	0.167	30°	29°	35°	34°	40°	38°	45°	43°
0.60	0.231	33°	32°	38°	37°	43°	41°	48°	45°
0.80	0.286	37°	35°	42°	39°	47°	44°	52°	48°
1.00	0.333	39°	37°	44°	42°	49°	46°	54°	50°

The keystone of Cornforth (1973) and Rowe (1969) is the relationship between the peak dilatancy rate and the relative density of the sand. Clearly the relative density is a fair measure of the remoteness of a sand from a critical state. But what of the confining stress? Ponce and Bell (1971) used a uniform 0.5 mm quartz sand to show that both dilatancy and peak friction angle increased most markedly as the confining stress was reduced from 40 kN/m^2 to 2 kN/m^2. At any particular density the stress ratio at failure doubled over this stress range. With $\phi'_c = 30°$, the densest sand confined by only 2 kN/m^2 in a triaxial test gave $\phi'_v = 22°$ so that $\phi'_{max} = 52°$, while under 40 kN/m^2 these angles were reduced to $\phi'_v = 9°$ and $\phi'_{max} = 39°$. When the Mohr circles were drawn, Ponce and Bell saw that they possessed a common tangent $\tau = c' + \sigma' \tan \phi'$ where c' and ϕ' were constant for a given density. The intercept c' was 1.0 kN/m^2 for loose sand and 1.5 kN/m^2 for dense sand: the authors used the words 'apparent cohesion' to describe the intercept, which they attributed entirely to dilation. It would be surprising if the majority of stress–dilatancy models for sand did not eventually contain some estimate of the likely dilation rate considering the remoteness of some critical state in terms both of stress and density, following Wroth and Bassett (1965).

A recent stress–dilatancy model proposed by Arthur *et al.* (1977) correlates the stress and strain ratios at failure

$$\left(\frac{p'}{\sigma'_3}\right)_f = -H_f \left(\frac{d\epsilon_3}{d\epsilon_1 - d\epsilon_3}\right)_f \tag{8.62}$$

in a manner which resembles Rowe's but which is shown to absorb data from tests with a wide variety of boundary conditions. Their use of X-rays to observe the movement of lead shot within soil samples has persuaded them that the detailed pattern of displacements can be greatly affected by boundary conditions without any corresponding variation in the limiting stresses. In particular they suggest that a simple direct shear-box test (preferably on samples at least 120 mm square) can be taken to offer the same peak angle of shear resistance ϕ'_{max} which would be obtained with a plane-strain compression apparatus. Their complex analysis of microruptures on a variety of possible planes orientated in special directions casts doubt on the validity of the concept of 'strain' just as their dilatancy rule casts doubt on the validity of the concept of 'angle of friction'. No doubt it will become common for research workers to visualise the independent chaotic movement of soil particles in situations where 'stress' and 'strain' are undefinable. This alone will be no more likely to alter the engineer's approach to a heap of soil than it would alter your approach to this book to be told that it is a wildly active collection of atomic particles.

Notwithstanding the evident complexity there are some simple conclusions

(1) There is a lower limit ϕ'_c to the angle of friction of sand no matter what is done to it.

(2) The tendency of dense sands to expand causes an increase in the possible inclination of stress on an imaginary plane through the soil, so that $\phi'_{max} > \phi'_c$.

(3) The dilatant angle of friction ϕ'_{max} in plane strain can be estimated from the inclination of stress across the horizontal rupture plane of a large direct shear-box test on a sand sample at the required density. It is unlikely that ϕ'_{max} will much exceed $(\phi'_c + 17°)$ for any sand, even in its densest possible packing.

(4) The relative size of the intermediate stress σ'_2 between σ'_3 and σ'_1 affects the peak angle of friction on the orthogonal 1–3 plane. In particular the peak angle of friction ϕ'_{max} obtained from triaxial compression tests, in which $\sigma'_2 = \sigma'_3$, is lower than that obtainable from plane-strain tests including direct shear tests in which $\sigma'_2 \approx (\sigma'_1 + \sigma'_3)/2$.

(5) Although the density of initial packing has a dominant effect on the dilatancy component of strength ϕ'_v, the average confining pressure p' also contributes due to the enhanced dilation encouraged by low stress levels. An envelope to Mohr's circles of peak strength will therefore not be a straight line through the origin and may be somewhat curved, after the fashion of figures 8.14 and 8.17.

(6) The situation at the peak strength of a very dense sand in simple direct shear will crudely be represented by figure 8.33. Only when the peak strength is reached will a rupture cleave the sample: from that point onwards the strains are very highly localised. If the whole soil sample is uniform and free from zones of weakness, the initial response of the sample up to inclinations of stress in the region of ϕ'_c will be roughly elastic. From ϕ'_c to ϕ'_{max} dilatancy will occur at an increasing rate. At or slightly after the peak the soil will rupture: from that point onwards the rupture will always be a zone of weakness.

(7) If the sand has a void ratio intermediate between its minimum possible and its maximum possible then the angle of dilatancy ν will roughly be linearly interpolated on a corresponding scale between 17° and 0° in plane strain and between 12° and 0° in triaxial strain.

8.6.2 Data on Dilating Clay

It is understandable, perhaps, that research workers have tended to specialise either on the study of sandy soils or of clayey soils. This, and contrasting behaviour due solely to contrasting pore sizes, has encouraged a needless diversity

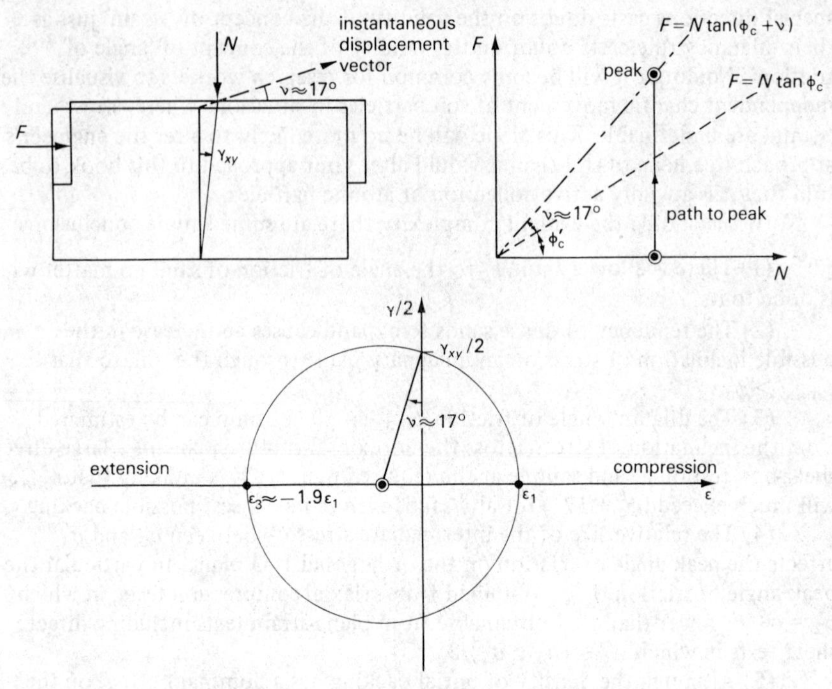

Figure 8.33 *Probable situation in a drained shear-box test on very dense sand at the instant of peak strength*

of notions. I shall attempt here to formulate the hypothesis that the role of dilation in clayey soils is rather analogous to that in sandy soils, and that dilation alone could account for many of the reported cases of 'true cohesion' in dense clays. This argument was first expounded by Rowe *et al.* (1964).

In comparing the behaviour of clay with that of sand three things must be remembered. Firstly, clays can easily be tested drained or undrained and any model of rupture ought to be able to take this into account. Secondly, clays unlike sands possess quite large elastic strains in addition to plastic strains; it may be necessary to consider these separately as we did with Cam-clay. Thirdly, clays unlike sands are easily manufactured at any required density by precompression. They can be mixed into a saturated slurry, plastically compressed to any point such as Q in figure 8.5 and then allowed to expand under a smaller stress to a point such as R. It should be clear that this method of fabrication leaves the tester with a good measure of the potential dilation, namely the overconsolidation ratio p'_Q/p'_R in this case. Overconsolidation ratio, and especially its logarithm, is a good measure of the remoteness of the critical state line: it can replace the 'relative density' of sands in this respect.

I owe most of the data and notions on 'dense' heavily overconsolidated clays to a sequence of publications by Henkel and Parry to which I refer in detail below, some of which were interpreted in Schofield and Wroth's book. Henkel (1959) and Parry (1960) reported the results of triaxial tests on remoulded Weald clay and London clay which had been spherically compressed and then relaxed to achieve overconsolidation ratios up to 24 before deviatoric stresses were applied. A typical τ/σ' diagram for these tests is shown in figure 8.34. The

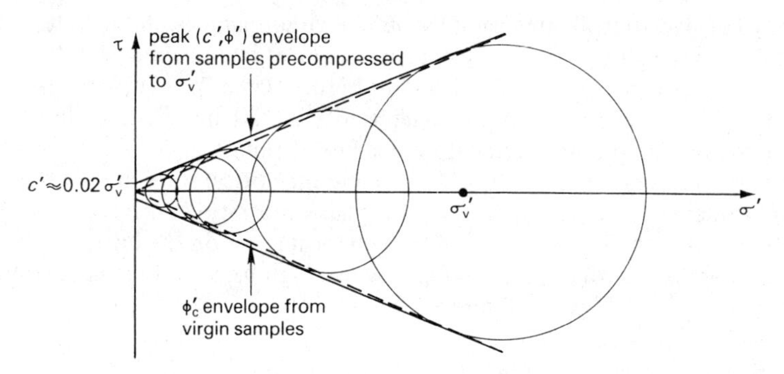

Figure 8.34 *Peak strength of overconsolidated clay*

peak strength envelope for samples which have been precompressed to σ'_v is roughly a straight line $\tau = c' + \sigma' \tan \phi'$ where $c' \propto \sigma'_v$ for a given soil. For Weald clay it appeared that the c' intercept was very roughly $0.02\sigma'_v$, so that the peak strength was only imperceptibly greater than the ultimate strength when the overconsolidation ratio was less than 10.

A different perspective is obtained when the state points of peak strength are plotted in a (q, p', e) diagram such as that of figure 8.35. Using data from

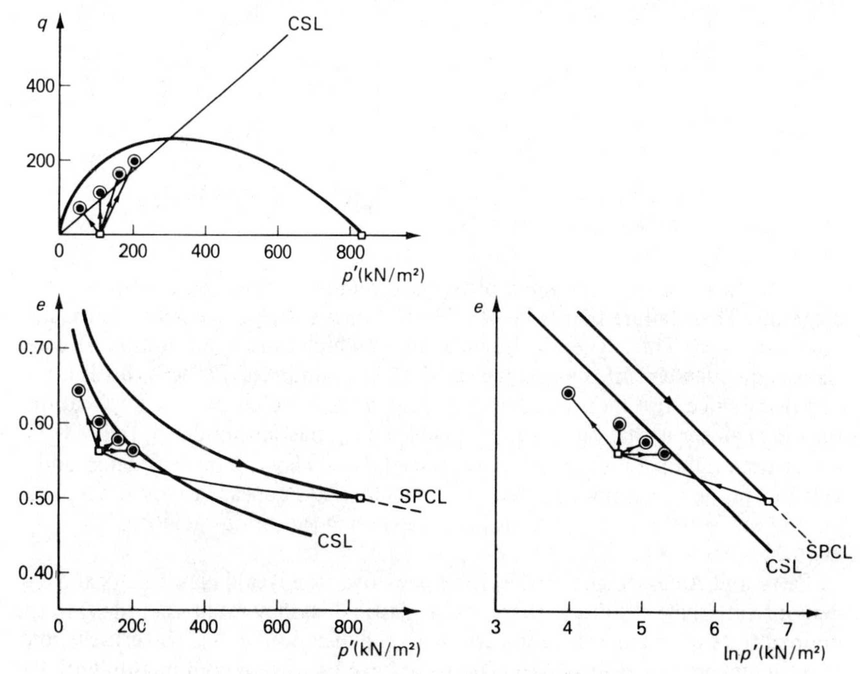

Figure 8.35 *Peak strength of remoulded overconsolidated Weald clay compared with a Cam-clay (from Henkel, 1959)*

Henkel (1959) with a Cam-clay type of framework the following conclusions can be drawn

(1) The overconsolidated samples, unlike virgin samples, did not show any evidence of obeying the Cam-clay yield surface, or of approaching closely to a critical state on both q/p' and e/p' diagrams before they ruptured. Whereas 'loose' samples finished close to the straight critical state line on the $e/\ln p'$ diagram, 'dense' samples fell well below this line at rupture.

(2) Remoteness of the failure state below the critical state line on the e/p' diagram is related to the remoteness of the peak stress ratio above the critical state line on the q/p' diagram. If p'_c is the spherical stress on the critical state line corresponding to the void ratio at failure, then a high p'_c/p'_f ratio is associated with a high q_f/p'_f, as shown in figure 8.36.

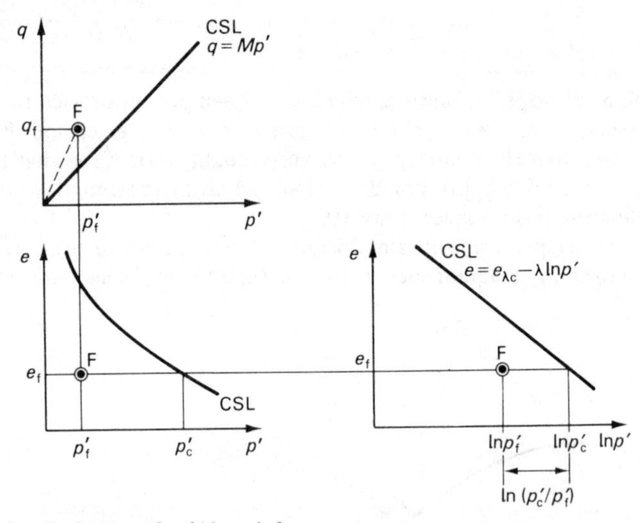

Figure 8.36 *Peak strength of 'dense' clay*

(3) Drained and undrained tests share a unique system on the (q, p', e) diagrams. Their failure points lie on identical lines if they have been identically precompressed. The trigger of the mechanism which causes the rupture of dense clays is independent of the drainage at the soil boundaries. Perhaps this is not surprising since rupture is a sudden and highly localised event. The implication for the engineer is that he can expect an $e/\log c_u$ diagram similar to figure 8.37 rather than a simple unique line. Overconsolidated clay can be very dense and stiff but prone to rupture. Indeed most ancient clays appear heavily fissured and jointed in the field owing to their having experienced various geological disturbances.

Parry and Amerasinghe (1973) look back over the Weald clay results and demonstrate that, at failure, there was a plastic dilatancy rate $(-d\epsilon_\kappa/d\epsilon_q)$ in the drained tests which increased linearly with the distance 'below' the critical void ratio line expressed as $\ln(p'_c/p'_f)$. These data can be used in conjunction with data of stress ratio at failure q_f/p'_f from Parry (1960) to show that a dilatancy rate at

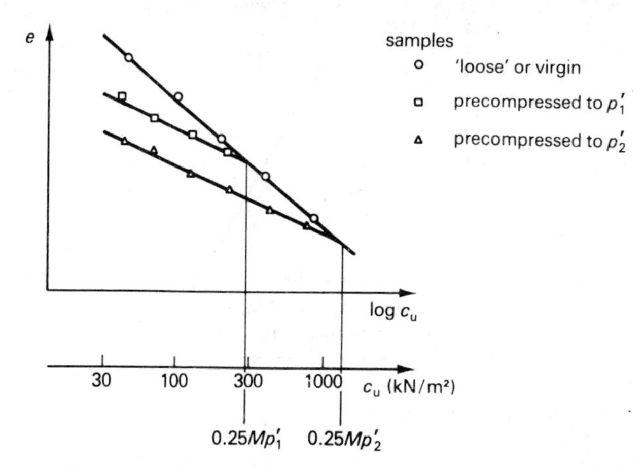

Figure 8.37 *Likely disposition of e/log c_u data for a clay precompressed to various spherical effective stresses*

failure of roughly 0.15 for a p_c'/p_f' ratio of 3.5 accompanied an increase in stress ratio q_f/p_f' to 1.15 from the critical state ratio of 0.95. This accords crudely with the Cam-clay work equation rewritten as a dilatancy rule for triaxial tests in equation 8.56. In the same publication Parry goes on to analyse the results of tests on kaolin clay which had overconsolidation ratios of up to 350: these data are also discussed in Amerasinghe and Parry (1975). Once again he links a large stress ratio at failure ($q_f/p_f' \approx 1.6$) with a remote critical state ($p_c'/p_f' \approx 9$) and a large dilatancy rate at failure ($-d\epsilon_\kappa/d\epsilon_q \approx 0.7$). And once again the Cam-clay dilatancy rule of equation 8.56 would, with $M = 1.02$, have been crudely satisfied. In general terms the results of this investigation correspond rather closely with those reported earlier on sands. One numerical difference is interesting: Rowe reported that the most highly dilatant sand had a rate $(-d\epsilon_v/d\epsilon_1) = 1$ which is $(-d\epsilon_v/d\epsilon_q) = 0.75$ in a triaxial test. Parry reports one test on very highly overconsolidated kaolin in which the critical state was very remote, $p_c'/p_f' = 20$, and the dilatancy rate was $(-d\epsilon_\kappa/d\epsilon_q) = 1.0$. In clay it becomes most important to discriminate that part ϵ_κ of the total volumetric strain ϵ_v which is not recoverable: only this is related to particles sliding. Clearly, clay that is very overconsolidated can be extremely dilatant and therefore very brittle, since the dilatancy contribution to strength is invariably lost as critical states are approached at large strains.

One of the most confusing aspects of the research that I have reported is the uniqueness of the failure criteria and their independence of the drainage of the boundaries. How can an undrained sample dilate? Parry and Amerasinghe (1973) point out that the increasing pore suctions in a dense clay subject to compression cause an increase in the spherical effective stress component p'. The soil can (indeed must) then adopt an elastic volumetric compression with a simultaneous and cancelling plastic dilation, so that its volume remains constant. They show that the rate of plastic dilation $-d\epsilon_\kappa/d\epsilon_q$ in an undrained test at failure was equal to that in a drained test if the samples possessed the same overconsolidation ratio at failure. The cancelling elastic and plastic components of volumetric strain

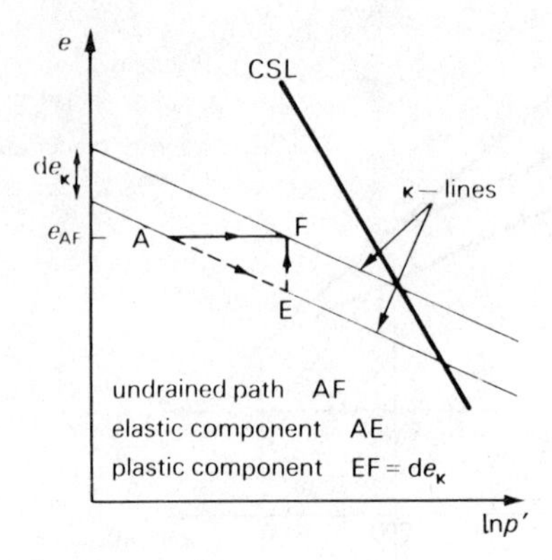

Figure 8.38 *Dilatancy of constant volume*

at constant overall volume are depicted in figure 8.38.

And what happens after the previously intact soil is ruptured? Parry and Amerasinghe (1973) show that the strength of an overconsolidated sample falls acutely after rupture, although more quickly in tests that were drained than in those that were undrained. As far as the rupture itself is concerned, it may be only a few microns thick and it is therefore difficult to measure its internal state. There is some evidence that the rupture zone is able to steal water from its immediate locality to enable a local softening. There will certainly be a strong pore suction in the rupture zone of a 'dense' clay, as figure 8.39 demonstrates. A

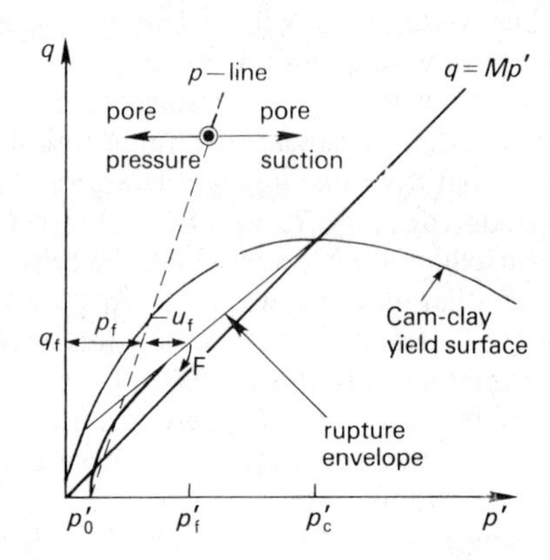

Figure 8.39 *Undrained strength path for compression of 'dense' clay*

further effect may be to align the flaky clay particles along the rupture surface: this is discussed in a later section.

Like the research workers on sand, Amerasinghe and Parry (1975) found that different rates of dilatancy are generated by different strain conditions. They performed triaxial extension tests in which the sample is extended by the reduction in vertical stress while the horizontal stress is kept constant: this requires a mechanical modification to the apparatus. The dilatancy was rather stronger when the ram was squirted upwards from all sides in this way: perhaps you feel intuitively that this is correct.

Out of this complex arise a number of tentative conclusions

(1) Peak strengths of intact 'dense' or heavily overconsolidated clay occur at ratios of q/p' greater than M, which is to say τ/σ' greater than $\tan \phi'_c$.

(2) The peak strength in triaxial strain may be crudely related to the plastic dilatancy rate by

$$\frac{q}{p'} = M + \left(-\frac{d\epsilon_K}{d\epsilon_q}\right)$$

It is impossible for $(-d\epsilon_K/d\epsilon_q)$ to exceed unity without implying a dangerously brittle soil condition, and this will be found to limit peak strengths expressed crudely as friction angles to

$$\phi' = \phi'_c + 24°$$

over a wide range of ϕ'_c values. Such an angle will be observed only when rupture occurs at very large overconsolidation ratios (exceeding 200 perhaps).

(3) The peak strength, due to dilatancy, will depend on the pattern of strains at the boundary of the element. With a given condition, such as triaxial compression which is likely to demonstrate the smallest contribution to strength, the peak strength envelope for samples sharing the same precompression (σ'_v) may be fitted by the tangent

$$\tau = c' + \sigma' \tan \phi' \text{ for } \sigma' < \sigma'_v \text{ where } c' \propto \sigma'_v$$

This tangent probably has no meaning and no application in the region close to the τ-axis, due to (2) above. This is made clear in figure 8.40.

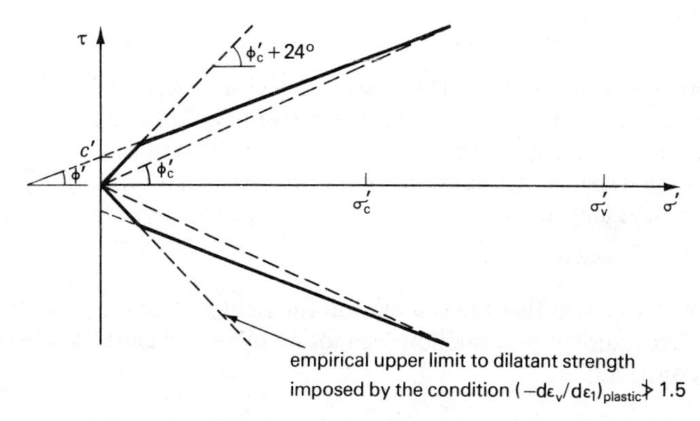

Figure 8.40 *Typical peak strength envelope for triaxial compression tests on dilatant clay*

(4) If the engineer desires to use the strength of heavily overconsolidated clay, he should consider whether the clay will already be ruptured, or whether it might rupture in service. Only when he has satisfied himself that his clay structure in the field will remain a monolith should he use data of peak strength. He should then perform a number of very careful triaxial or plane strain tests,

with a variety of loading conditions. He must not extrapolate beyond the range of his data.

(5) When subject to an undrained test, 'dense' clays rupture after the generation of fairly large transient pore water suctions. Their undrained strength is therefore greater than their drained strength. Their undrained strength is, however, less than that of the identical clay if it had been reconstituted at the same density (void ratio) and tested from a starting condition on the plastic compression line. Indeed the rupture which interrupts the progress of an 'undrained' test may conceivably generate internal drainage, thereby invalidating not only the concept of a unique strength at a unique void ratio for very dense clays, but also the notion that the 'undrained' state exists at all.

(6) (a) It follows that a safe and conservative approach to design in 'dense' clays is to use the prediction

$$\tau = \sigma' \tan \sigma'_c$$

$$\sigma' = \sigma - u$$

where σ is generated by the self-weight of materials and u is estimated from the highest conceivable water table. ϕ'_c can be measured using triaxial tests on saturated but very disturbed (remoulded) samples.

(b) If the designer uses peak strengths

$$\tau = c' + \sigma' \tan \phi'$$

or more properly

$$\tau = \sigma' \tan \phi'_{max}$$

$$\phi'_{max} = \phi'_c + \phi'_v$$

with

$$\sigma' = \sigma - u$$

where u is estimated from the highest conceivable water table, then he must ascertain that the soil mass is effectively unruptured and will remain so. He must measure the strength of undisturbed samples.

(c) If the designer measures the undrained strength of intact nuggets of dense clays and then uses

$$\tau = c_u$$

he must ascertain that the soil mass in the field is effectively unruptured and that transient pore suctions cannot be relieved during the life of his soil construction.

8.7 Difficult Soils

Of course *all* soils are difficult! Their mechanics entails an interaction between seepage, friction, dilation, cohesion, elasticity and plasticity. Their distribution on the face of the earth is erratic and unreliable. In many instances they are varved or layered at small regular intervals. Stiff soils tend to be fissured and fractured so that blocks of them tend to slide into excavations. Virgin soils are highly compressible and cause buildings to crack and tilt due to uneven settlement. These aspects of behaviour and disposition make ground engineering a fascinating career. The soils engineer is called on for a wide range of decisions, and he will be forced to rely heavily on the models that I have outlined so far. He will often take soil samples, so as to create small physical models in permeameters, shear boxes, oedometers and triaxial machines and will employ a selection of mental models to relate their behaviour to the possible behaviour of soil masses in the field. In a large proportion of cases the decisions which he takes – on the necessary capacity of pumps using the seepage model, on the need for densification of the fill for an earth dam using the Cam-clay model, on the required area of a footing on clay using the cohesion model in addition to notions of elasticity – will prove sound. Where the models seem to have led him astray, he will often find that he based his judgement on samples that were not representative of the whole soil mass in the field. But occasionally the engineer will encounter some soil stratum which seems to be haunted. Its behaviour will seem dangerously to disobey the simple models. As most of the mechanics of soil is rather recent, there are a large number of engineers still practising who have encountered puzzling situations which might now be considered perfectly predictable and controllable. This cannot hide the fact that some soil strata behave oddly.

The proper reaction of the engineer is firstly to map out and attempt to correlate strata that behave badly, and secondly to learn how to repair or avoid the bad features of the ground or how to engineer it into a more useful condition. The first function is simply an extension of the art of geological mapping to include geographical or other features. For example, a sensitive soil stratum may be geologically mapped as 'drift, post-glacial marine silt' but it might only give rise to catastrophic landslides in a distinct geographical region and where it is found more than 50 m above sea level on ground rising towards the mountains which is being deeply eroded by young river channels and where the salty water has been washed out of its pores and replaced by fresh water. The second function invites the constructor to learn how to approach the problem of building on the bad soil, and it invites the research worker to adapt or invent a theoretical model which ought to help the constructor construct. Models that are newly invented from scratch are rarely taken with enthusiasm by constructors. But clearly, if some engineering had to be accomplished in frozen Alaskan soil, the models of behaviour in this present textbook would fall very far short of what was needed. In less dramatic circumstances it is more sensible simply to fiddle around with the models I have already described in order to attempt a better fit with reality, or to introduce extra notions which are easy to imagine and to tack on to the broad framework which has already been established. Three very popular extra notions are: the geometrical 'structure' of the assembly of soil particles, the presence of some chemical which acts as a cement between the soil particles, and the presence of interparticle forces. Of these, the third can be more ghostly than the behaviour it is meant to explain: forces are always rather intangible. At least cement can be chemically analysed, dissolved, or remade. And the structure of clay particles can now be observed using scanning electron microscopes.

I shall now proceed to describe four problems: the sensitivity to disturbance of some recent soils, the reduced friction on old landslide surfaces in clay due to mechanical polishing, anisotropy, and creep.

8.7.1 Sensitive Soils

Some soils are sensitive to disturbance at constant void ratio. They typically reach a reasonable peak strength at quite small strains, after which the strength suddenly drops by an order of magnitude while enormous pore-water pressures cause an outflow of fluid corresponding to a dramatic consolidation of the soil skeleton, given sufficient time and drainage. The fall in strength is much greater than that exhibited by heavily overconsolidated soils. Sensitive soils are always recent in the geological sense, being post glacial. Evidence often suggests that they have never suffered erosion, so that they ought to be in a virgin condition. In these circumstances the peak strength is often higher than one might have expected, while the remoulded strength is very much lower. The remoulded strength can be so low that the soil takes on the appearance of muddy water: the sensitive soil is then sometimes described as 'quick'. A typical undrained compression curve is depicted in figure 8.41. The sensitivity ratio S is defined as

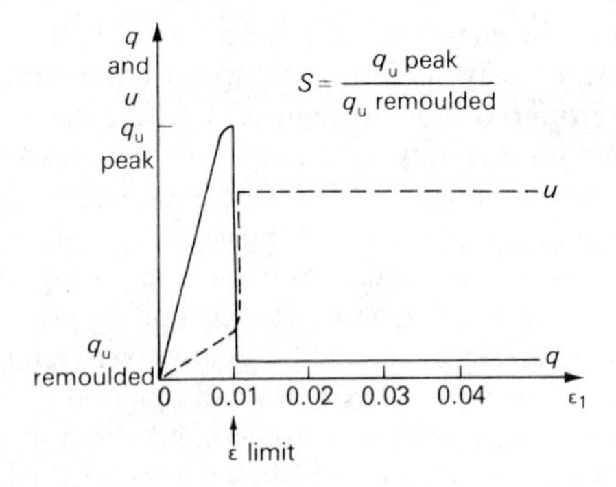

Figure 8.41 *Undrained compression of undisturbed sensitive soil*

the factor between the peak and remoulded undrained strengths: it can exceed 100. If pore-water pressures are being measured then it will be discovered that they rise so as to almost equal the external cell pressure once a very sensitive soil has failed. If the soil sample has not completely slipped off the pedestal, then q and u-values after failure will allow the final effective Mohr's circle to be drawn. As figure 8.42 indicates, it will be consistent with the friction model, but at very small stresses. When the Atterberg limits are determined the high pore-water pressure will be understandable because the soil will be found to have a moisture content not much less than, or perhaps somewhat greater than, its liquid limit.

Sensitivity is important to the engineer because it deals a savage blow to the cohesion model. It is no longer possible to repair accidental wounds in the soil. If strains have been allowed to exceed their limiting values then the soil is ruined for construction purposes. Only if contractors and designers can guard against the peak ever being attained will a soil construction project be viable. But in those regions of Canada and Scandinavia where sensitive clays abound, they are important to whole rural communities who are intermittently assailed by ferocious flowslides of soil slopes which previously had seemed perfectly stable.

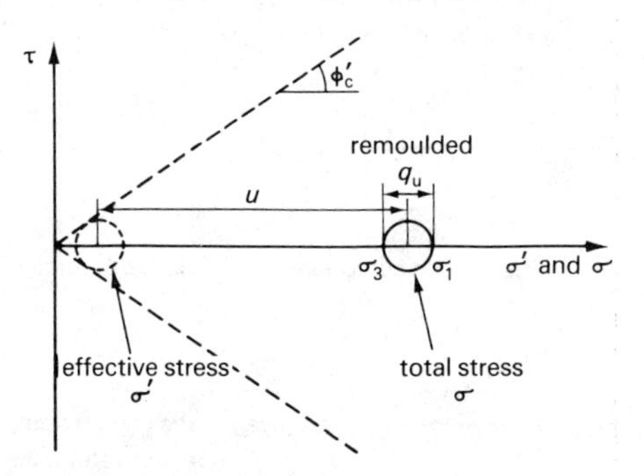

Figure 8.42 *Mohr's stress circles for remoulded sensitive soil*

The regular toll on life and property demands of the research worker some model of behaviour by use of which these soil avalanches could be prevented. Two groups of models have been generated.

Bjerrum (1954) reported that in the majority of Norwegian post-glacial clays, sensitivity was related to the replacement of salt water by fresh water in the pores of the soil. At the close of the glacial period the rushing melt-waters carried out into the fjords a clay/silt/sand mixture which sedimented in the salt water as a fairly homogeneous clayey soil. The whole land mass, recovering from the burden of recently melted ice, then sprang upwards by up to 200 m. This left the newly deposited marine clay above sea level, as a convenient bench for man to settle and farm. In the intervening period of 15 000 years the salt water has been removed in certain locations and elevations either by diffusion or by its direct exchange for rainwater by the action of seepage. Bjerrum was able to correlate the regions of very high soil sensitivity with those where the salt concentration in the pore water had dropped below the original 30 g/l of the native sea water: $S = 5$ at 30 g/l, $S = 15$ at 10 g/l, $S = 100$ at 5 g/l, $S = 500$ at 2 g/l.

This observation led to a model based on the 'structure' of silty clays deposited in salt water. Suppose that the silt and clay particles could adopt some nonrandom open-textured skeletal structure which was fairly unstable even in the presence of ionised sodium chloride ($S = 5$) but which was totally unstable if the salt were removed ($S > 100$), desiring to collapse to a more random assembly. It might be logical to ascribe different compressibilities (λ) and certainly different void ratios ($e_{\lambda v}$) to the two structural arrangements. Variations in friction would be small. The critical state diagram might then be similar to that depicted in figure 8.43. While the unstable structure (1) is subjected only to small quasi-elastic strains, it remains intact. As soon as slippage takes place the skeleton collapses, and some critical state in the random structure (2) is achieved. Figure 8.44 indicates how undrained and drained strain-controlled triaxial tests

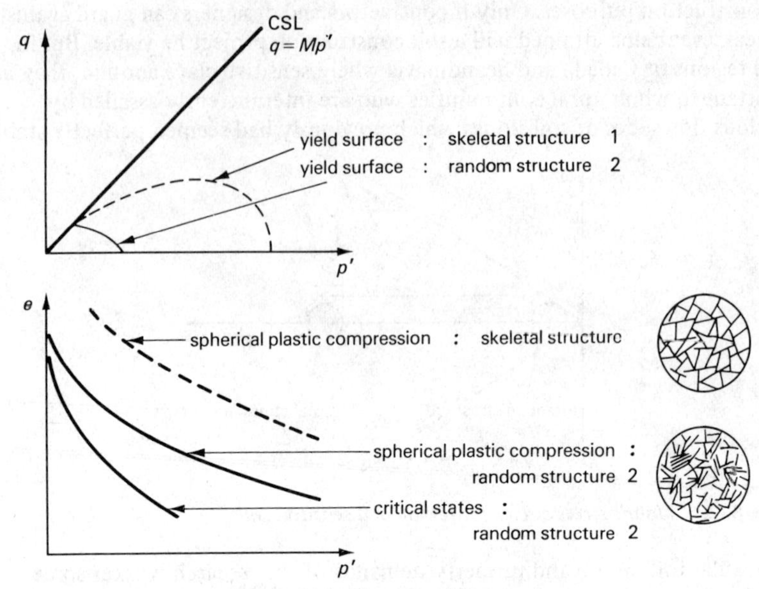

Figure 8.43 *A hypothesis for sensitivity based on 'structure'*

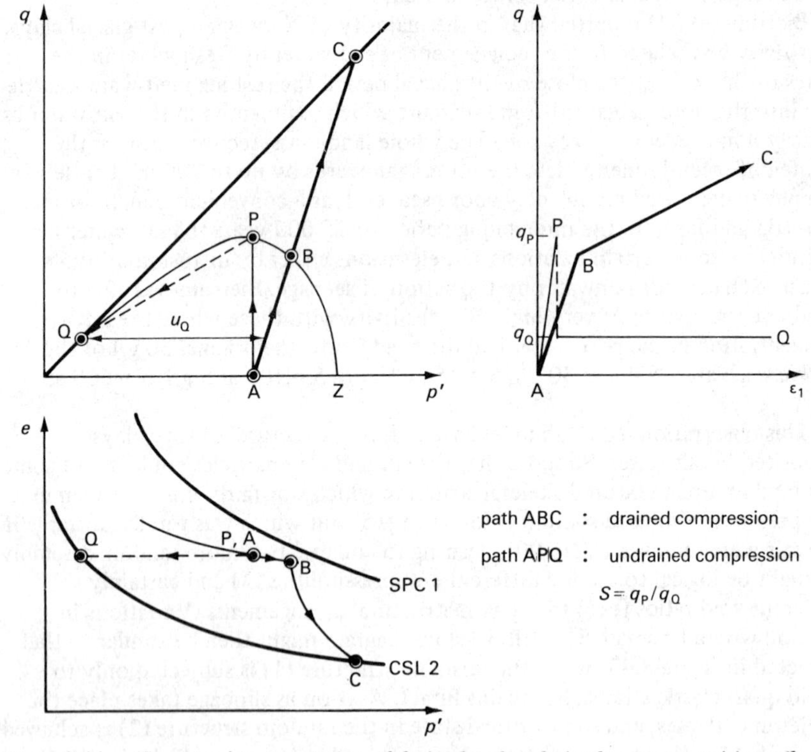

Figure 8.44 *'Structure' interpretation of drained and undrained tests on sensitive soil*

on almost virgin sensitive soil might appear. In a very slow drained compression test the stress vector $dq/dp' = 3$ must be followed. Rising from A, the response is quasi-elastic until the yield surface in the unstable structure is approached at B. There follows an episode of great volumetric compression as the excess pore water which has been trapped within the skeleton is able to escape. Eventually a critical state is reached at C, where the soil is remoulding with a rather random structure. In an undrained compression test the initial elastic response AP must occur at constant void ratio and therefore at constant spherical effective stress. Eventually the yield surface is reached at P. The soil structure collapses and the unstable yield surface evaporates. The soil searches for some equilibrium at constant void ratio and finds point Q which is a critical state in a random structure with a very great deal of water in the way. The strength q_Q may be below that required of soil in a liquid-limit test: of course, the Atterberg limit tests use smashed soil so their positions relate to the random critical state line QC. The pore-water pressure during the undrained test is the horizontal distance from the effective stress path APQ to the total stress path ABC. You will readily see that if p'_A was equal to the cell pressure, having used the drainage lead to remove all internal water pressure before the undrained test began, the final water pressure u_Q may well approach the cell pressure p'_A.

Not only does the model mirror many of the features of the behaviour of sensitive Norwegian clays, but also it offers a few extra tips.

(1) The structure may break down even without the application of deviatoric stresses if the spherical effective pressure is allowed to exceed p'_Z in figure 8.44.

(2) Almost any stress path, including all those for unloading, can lead to a sensitive failure.

(3) After a flowslide pore pressures will be very large if the overburden remains large, and an episode of consolidation will begin.

(4) Perhaps very sensitive soil could be made less sensitive if salt water could be forced back into it!

Sangrey (1972) reported that the majority of Canadian post-glacial clays appeared to be sensitive as a result of having been cemented into a rather open skeletal structure soon after having been deposited. Although the evidence for cementation seemed strong, and a number of possible mechanisms or compounds had been suggested, there was no direct chemical evidence. Sangrey did report on sensitivity in varved clays which had been deposited in fresh water, in addition to deep marine clays, however, so the salt-replacement mechanism is clearly not sufficient in all cases. Strong circumstantial evidence of some form of cementation can be drawn from the volumetric stiffness E'_v of the undisturbed material. For stress changes that did not cause a sensitive collapse, the volumetric modulus was perfectly constant and independent of the existence of strong deviatoric stress components. This rather perfect linear elastic behaviour is in contrast with the normal quasi-elastic behaviour of unconnected soil particles. Furthermore, the spherical effective stress required to cause plastic compression was very much greater than the conceivable geological precompression. Finally, the undisturbed strength of the soil was very much greater than its virgin history should have implied. The sensitivity of the three clays quoted by Sangrey

(Leda clay $S = 780$, Mattagami clay $S = 80$, Labrador clay $S = 400$) certainly class them as very difficult and dangerous materials which reduce to a slurry when they are remoulded.

Sangrey recognised that liquefaction of the clays would always be catastrophic, and concentrated on determining which stresses could cause liquefaction. He expressed his findings as a yield curve in q/p' space, and figure 8.45 represents the results of a variety of tests on samples of Leda clay from

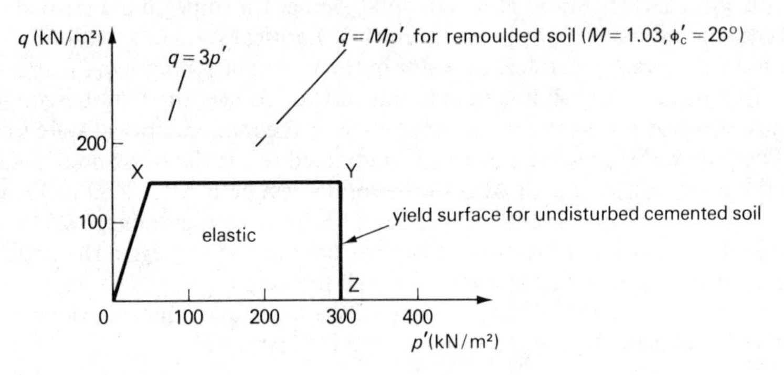

Figure 8.45 *Yield surface for Leda clay (from Sangrey, 1972)*

Ottawa. Although the soil is thought never to have been precompressed, and although the samples were taken from a shallow pit, plastic spherical compression at Z needs 300 kN/m^2, which would hold a rather highly stressed foundation. Small deviatoric stresses q do not appear to alter the limiting spherical pressure p'_Z, so the section YZ of the yield curve is vertical. The roof XY over the stable cemented states is likewise horizontal at $q = 150$ kN/m^2 ($c = 75$ kN/m^2), which would qualify for the description 'firm'.

The failures that occurred in the region OX were observed to be due to vertical splits or cracks in the strongly dilating triaxial test specimens. In order to understand this it is necessary to translate the ratio $q/p' = 3$ (for that was the gradient of OX) back into principal stresses. Using the algebra of equation 8.40, you will find that on line OX $\sigma'_2 = \sigma'_3 = 0$ in a triaxial compression test. Now tensile tests on Leda clay showed that its strength in tension ($\sigma' < 0$) was negligible at 5 kN/m^2: the samples were, in any event, naturally fissured. It follows that line OX represents the onset of lateral effective tension, and therefore of the opening of vertical fissures. Any state of stress within OZYX leaves the soil cemented and almost linear elastic: any attempt to leave these safe confines causes disaster. Sangrey's approach correctly lays emphasis on the determination of the various combinations of stresses at failure by testing undisturbed samples.

In the face of soil that can liquefy, both Bjerrum's and Sangrey's approaches lead to the sensible conclusion: establish what makes it liquefy in the laboratory and then prevent it happening in the field.

8.7.2 Reduced Friction on Old Landslips in Clay

The works of nature or man occasionally reactivate an old landslip in clay within which past movements have created ready-made slip surfaces. The effective angle of shearing resistance on such a surface is sometimes found to be less, and perhaps a good deal less, than the critical state angle ϕ'_c inferred from remoulded 'loose' samples which had not ruptured. Gould (1960) reported that the strength actually mobilised in slides on the Californian coast appeared to be inversely proportional to the amount of displacement that had occurred previously in the shear zone. Bishop *et al.* (1971) reported their design of a ring shear apparatus with which the effective friction across an annular rupture surface could be accurately measured as the relative circular displacement increased very slowly without limit. They confirmed Skempton (1964) in finding that the effective friction, especially of soils with a clay content exceeding 50 per cent, could fall seriously below ϕ'_c to a constant 'residual' angle ϕ'_r after relative displacements of the order of 1 m. Most of the fall in strength occurred in the first few centimetres, as shown in figure 8.46. The strength of undisturbed

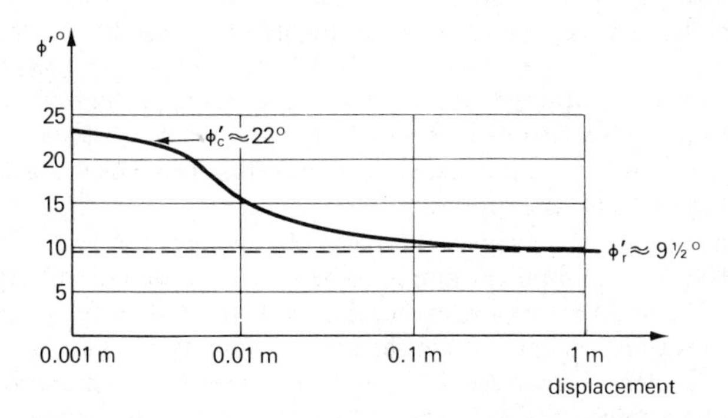

Figure 8.46 *Ring-shear test on blue London clay (from Bishop et al., 1971)*

samples of London clay and Weald clay fell from a peak which is explicable in terms of dilation or slight cementation, through a critical state angle ϕ'_c of the order of 22° and on to a residual angle ϕ'_r of the order of 10°. In common with most soil friction phenomena ϕ'_r was sometimes strongly dependent on the magnitude of the confining stress: for a sample of brown London clay ϕ'_r was 8° at $\sigma' = 200$ kN/m^2, 9° at 100 kN/m^2, 10° at 50 kN/m^2, 11° at 25 kN/m^2, and 13° at 10 kN/m^2.

The residual friction phenomenon is thought to be analogous to any other mechanical polishing process. Once the slip surface is established, the relative displacement of the opposing sides first tends to orientate the platey clay particles to lie parallel to it. Individual particles can then be transported around to create an ever more smooth and highly polished surface.

A good deal of field evidence has now been collected which shows that the residual phenomenon is not as catastrophic a blow to critical state philosophy as may at first have been thought. Natural soils are no more easily brought into a highly polished state than are old bent pennies. A good deal of landslide movement is necessary before individual limited slip zones connect together. Thereafter up to a metre of relative movement is necessary on the unique slip surface before residual friction is generated. Skempton (1970), Chandler (1974) and Chandler and Skempton (1974) provide a sequence of field evidence to show that the residual friction angle ϕ'_r has no relevance to the stability of 'fresh' ground which has not suffered previous landslides. The mobilised strength of first-time slides in a number of ancient and heavily overconsolidated fissured clays which they report agrees very well indeed with the critical state angle ϕ'_c measured after the fashion of Schofield and Wroth (1968) on remoulded 'loose' samples. Chandler (1970) on the other hand provides an interesting close correspondence between the mobilised angle of inclination of stress of a slab slide deduced from the slope angle of $9°$ using equation 4.8 and the effective angle of friction of $18\frac{1}{2}°$ measured on a block sample which contained the rupture surface and which was subjected to a conventional direct shear-box test under the appropriate small confining pressure. This represented a small fall below ϕ'_c on the existing slip surface.

Palladino and Peck (1972) demonstrated that residual angles ϕ'_r become relevant when large historic ground movements have taken place. Whereas the peak dilatant strength of the fissured and highly overconsolidated Seattle clay was fitted by $\tau = c' + \sigma' \tan \phi'$ with $c' = 70$ kN/m^2 and $\phi' = 35°$, the back-analysis of a number of slope failures which reactivated geologically recent but prehistoric slip surfaces offered a good fit to $\tau = \sigma' \tan \phi'_r$ where ϕ'_r was measured to be about $15°$ using a conventional slow shear-box test which was continually reversed so as to polish the rupture plane.

The fact that ϕ'_r can be very small means that no engineer can afford to apply it to every clay design problem simply to be safe: he would needlessly bankrupt his clients. The problem of ϕ'_r becomes the problem of identifying (and ideally avoiding) sites which might contain the hidden wounds of a past landslide, or of locating the ancient slip surface if it exists, and removing a sample containing it in order to determine the friction angle that can be mobilised upon it by conducting a conventional slow drained shear-box test. Only if the slip surface is inaccessible or invisible, or if there are a large number of old slides, would the ring shear test or the cruder reversing direct shear test be an appropriate method of discovering the worst. In identifying ancient landslips the most useful aids are a sound understanding of geology (especially glacial and post-glacial) and a succession of aerial stereo photographs. Hummocky ground, the vertical scars left at the back of old slip circles, and displaced trees or stone walls may each indicate a landslide area, and may each be best observed from the air.

8.7.3 Anisotropy

Due to its original deposition in horizontal beds, ground is often anisotropic in that its horizontal properties are different from its vertical properties. Additionally, imposed strains after deposition can induce anisotropy. One-dimensional vertical compression can leave flaky clay particles predominantly lying in horizontal planes. Shears and displacements can, as you saw in the previous section, cause flat particles to align themselves parallel to the direction of greatest shear strain.

The polished slip surface is one of the clearest exemplars of the problem which anisotropy sets the civil engineer. Think of the engineer who was ignorant of the importance of old landslides but who happened to site a borehole over an ancient slip surface, such that a sample happened to be taken which contained the slip surface as a horizontal wound, which could have allowed the cylindrical sample to fall into halves. If the sample did fall apart, the testing technician in the laboratory would probably curse and throw it away. Otherwise he would set the cylinder up in a triaxial test and compress it, in order to measure the effective angle of friction, perhaps. Now a horizontal plane in the triaxial apparatus has no shear stress on it: it is a principal plane. The very weak polished surface will not be given a chance to display its weakness, and the laboratory will find the friction angle on competent planes other than the ancient slip surface.

Even the provision of triaxial compression tests on vertical cores and direct shear tests on every horizontal plane of soil that looks different and on every suspected shear surface will not provide complete protection for the engineer. Most natural soils are seen to be weaker and more flexible in horizontal cores when these are obtained and their compression compared with that of vertical cores. As an example, the cementation reported in Sangrey (1972) and displayed in figure 8.45 for Leda clay was found to be anisotropic. The limiting deviatoric stress for vertical cores of Labrador clay was 280 kN/m^2, but only 120 kN/m^2 for horizontal cores. In less difficult strata (and excluding old landslide surfaces) there is some evidence that while the horizontal stiffness may be a factor of 2 or 4 less than the vertical stiffness, no plane is so weakened by anisotropy of 'structure' that it cannot at least accept an inclination of stress equal to ϕ_c' measured on random 'structure'. The dilatancy component of strength ϕ_v' may well be expected to show greater variations with respect to particle orientation.

A designer may therefore choose to employ ϕ_c' on all planes, in an anisotropic soil, rather as he might use ϕ_c' throughout all space if he was concerned about the possibility of loose pockets in an otherwise dense isotropic soil. Only if the prediction of *deformations* was vital in a potentially anisotropic soil would it be absolutely necessary to perform the large number of required tests on undisturbed samples of various inclination.

The most important example of anisotropic ground which falls rather outside the scope of this book is rock. Even high-quality intact rock is governed in its performance by its bedding planes and the pattern of its joints and fissures, whicl are probably in preferred directions. It is the disposition in space of these weakness that determines the stability of rock slopes, cuttings and tunnels.

8.7.4 Creep

Creep can be the scourge of the research worker who is trying to validate mechanical theories in great detail. It is rarely a potential source of trouble to engineers in the field except when they encounter highly compressible organic soils or sensitive soils which are not well cemented, or when they might apply vibrating stresses to the ground. I have taken the attitude in this book that only changes in effective stress σ' cause increments of strain in the matrix of soil particles. A corollary is that soil strains should not occur unless there have been changes in effective stress. This is not wholly true. There is evidence that soil creeps at constant effective stress. It is normal for the rate of creep to diminish with time, although creep can accelerate and lead to rupture in soils that are highly stressed and enjoying a brittle peak strength.

Creep is another problem for the hard-pressed soil-testing engineer who would like to report without fuss on the 'modulus' or the 'strength' of a piece of soil. If his testing apparatus is stress controlled, such as an oedometer, he can leave the soil element to creep for as long as his client will continue to hire the apparatus. There is evidence that clayey and especially organic soils continue to compress in an oedometer even when pore-water pressures have dissipated. Very often the continuing secondary compression is linear against a logarithmic scale of time so that equal small increments of creep strain occur between the times 0.1, 1, 10, 100, etc.

Hyde and Brown (1976) showed that a remoulded low plasticity silty clay which was preconsolidated into a firm condition with overconsolidation ratios from 4 to 20 showed axial creep strains which roughly satisfied

$$\epsilon_1 = 10^{-3} \log_{10} t$$

when subjected to undrained constant deviatoric stresses in the normal working range. If an engineer had measured the strain of such a sample after 0.1 days in order to establish a stiffness for the soil so as to calculate the eventual settlement of a foundation after 10 000 days, then he would have missed five cycles of $\log_{10} t$. The underestimated strain due to creep would have been 5×10^{-3}, or very roughly 10 mm settlement for a 2 m square footing. This is on the verge of becoming serious. For this reason soil-testing engineers usually leave clays to settle in oedometers for at least 1 day even though the transient flow of pore water would be mostly completed in 1 hour. A strain *versus* logarithm of time plot in these later stages of the tests can lead to an estimate of strain after the required lifetime. An even more satisfactory creep strain prediction could be made if samples were left to compress for a week.

Creep is difficult to observe directly in a conventional triaxial compression machine, due to its being under strain control. The corollary is that stresses become dependent on strain rates. If the weakest strengths are to be measured, the smallest strain-rates should be used. It is not clear whether fast strain rates have ever led to an overestimate of strength due to insufficient creep, which has then led to a field accident. But it is certainly commonplace for fast strain rates to lead to excess pore-water pressures, positive or negative, in tests which should have been 'drained'. It appears to be sensible practice in soil testing to pay great attention to the curing of leakages and then to run any soil test over as long a period as is economically possible.

8.8 Problems

(1) (a) A cylindrical sample 50 mm in diameter and 100 mm long was prepared from a saturated block of firm clay recovered from a trial pit. The sample was installed in a triaxial testing rig and brought into a known state of equilibrium by leaving it under a cell pressure of 50 kN/m^2 for 1

day with the drain tap open: a burette recorded 1.3 cm^3 of water escaping during this time. Electrical contacts were then used to probe the changes in length and diameter of the sample when the cell pressure was further increased to 75 kN/m^2. The changes recorded after 1 day were $\Delta V = 2.0$ cm^3, $\Delta L = 0.36$ mm, $\Delta D = 0.16$ mm. Show that the three data lead to fairly consistent estimates of the drained bulk modulus of the soil E'_v in the region 2450 kN/m^2.
(b) Following the test described in (a) above, the cell pressure was kept at 75 kN/m^2 while the piston was used to compress the soil with the drain tap closed. When the recorded compression ΔL was 0.40 mm the extra force required was 42.8 N. Show that the undrained Young's modulus E^u_y was 5450 kN/m^2.
(c) Use chapter 6 to show that the following estimates of elastic parameters can be made, applicable in the region $p' = 60$ kN/m^2, $q = 10$ kN/m^2:
$E'_y = 4350$ kN/m^2, $\mu' = 0.20$, $E'_s = 1800$ kN/m^2, $E'_0 = 4850$ kN/m^2
(d) Make the following predictions:
(i) If, following (b), the compressive force was kept constant at 42.8 N and the cell pressure at 75 kN/m^2 while the drain to the sample was opened, what would be the eventual reduction in length of the sample?
(ii) What was the pore pressure following (b)?
(iii) If, following (b), the length of the sample had been fixed while the drain to the sample was opened, to what value would the deviatoric force eventually decrease?

Answers (i) 0.10 mm (ii) 7.3 kN/m^2 (iii) 17.5 kN/m^2.

(e) Comment on the likely problems to be encountered in performing such tests with sufficient accuracy.
(f) Comment on the problems of nonlinearity and use your judgement to predict the likely moduli in the region $p' = 125$ kN/m^2, $q = 10$ kN/m^2.

(2) A sample of soft clay ($G_s = 2.70$) was recovered undisturbed from a depth of 4 m in a borehole through waterlogged ground. After sealing and storage, it was set up as a 100 mm diameter, 200 mm high triaxial sample which weighed 3097 g. The sample was then allowed to drain into equilibrium under an effective cell pressure of 40 kN/m^2 in imitation of its *in-situ* stresses, but the quantity of drainage was immeasurably small. The drain tap was then closed and the following readings were taken during slow axial compression with the drain tap closed and the cell pressure at 40 kN/m^2 as before.

axial compression (mm)	0	0.5	1	2	4	8	16	32
deviatoric force (N)	0	63	123	159	185	213	235	266
pore-water pressure (kN/m^2)	0	3	6	8	11	15	20	22

The moisture content of the sample was then found to be 0.278.
(a) Find the undrained strength of the sample c_u and its associated void ratio e, and show that it was saturated.
(b) Estimate the critical state angle of shearing resistance ϕ'_c.
(c) Estimate the initial shear modulus of the sample E_s and express the

approximate yield point of the sample as both a limiting shear stress τ_e and a limiting mobilised angle of shearing resistance ϕ'_e.
(d) Use your judgement to estimate the overconsolidation ratio of the sample.
(e) Discuss the advantages and disadvantages of this particular type of test procedure.

Answers (a) 14 kN/m^2, 0.75 (b) 27° (c) 1050 kN/m^2, τ_e = 8 kN/m^2, ϕ'_e = 11° (d) roughly 1.1 after considering the q/p' diagram relative to probable *in-situ* stresses: the natural ground must almost be at the point of yielding.

(3) 327.2 g of dry quartz sand was formed into a cylinder of 50 mm diameter and 100 mm length. It was then completely saturated and subjected to a drained triaxial compression test at a cell pressure of 100 kN/m^2. The following readings were taken

axial compression (mm)	0	0.25	0.5	1	2	3	4	5	6	7	8	10	12
deviatoric force (N)	0	412	504	600	656	687	700	683	630	536	504	483	473
net drainage inward (cm^3)	0	–0.1	0.2	0.7	1.8	2.8	4.0	5.2	6.4	7.2	7.7	7.9	8.0

and it was observed that displacement across an inclined ruptured plane was responsible for deformations following the achievement of the peak deviatoric force.

(a) Use your judgement to plot both true deviatoric stress and volumetric strain against axial strain.
(b) Confirm that the initial tangent modulus E'_y was not less than 80 000 kN/m^2.
(c) Infer angles of shearing resistance for the peak and for critical states.
(d) Find the initial void ratio of the sand in the test, and confirm that the extra strength at peak roughly corresponded to what could have been expected in view of the dilatancy rate, and that the sample probably had a relative density of about 65 per cent.

Answers (a) use an area correction up to peak (c) 39°, 32° (d) e_0 = 0.59, $(d\epsilon_v/d\epsilon_1)_{max}$ = 0.6.

(4) An engineer wished to describe the drained strength of a natural grey silt which had very thin black laminae at intervals of about 20 mm. He decided to take cores of the silt at various angles so that the plane of the laminae subtended an angle θ with the horizontal when the core was installed in a triaxial compression machine. The test specimens were 100 mm in diameter and 200 mm long. The deviatoric force causing compression was measured by a load cell. A number of very slow drained tests was conducted at zero pore-water pressure with a 100 kN/m^2 cell pressure, and various inclinations θ of the laminae. When the proving ring reading reached its peak value the dial gauge recording the shortening of the specimen ΔH mm was also read, with the quantity of drainage ΔV cm^3 up to that point. The results are given in table 8.2. Samples 4, 5 and 6

ruptured along a black lamination at peak strength; otherwise the samples remained cylindrical.

TABLE 8.2

Test number	$\theta°$	Peak strength Q(N)	ΔH(mm)	ΔV(cm^3)
1	0	2230	18.5	67.5
2	15	2200	15.0	55.4
3	30	2250	19.5	70.0
4	45	1410	8.2	32.2
5	60	1170	3.0	11.8
6	75	1740	10.0	40.5
7	90	2200	18.0	62.0

(a) Plot Mohr's circles in terms of effective stresses for each of the tests, putting them on one diagram at a scale 1 cm = 20 kN/m^2, and marking the points which indicate the stresses on the black laminations.
(b) Describe the strength of the laminated silt in a concise fashion.

Answer (b) for grey silt $\phi' = 35°$, for black clay $\phi' = 25°$.

(5) (a) A saturated and very soft silty clay is sampled and set up as a triaxial test specimen 50 mm in diameter and 100 mm high. It weighed 381.9 g and was found to have a moisture content of 0.296. Firstly, it was brought to an initial spherical effective pressure of 50 kN/m^2 by the application of cell water pressure: during this time a volume of 5.3 cm^3 of water bled out. Secondly, it was slowly compressed at constant volume by the piston, during which time the readings given in table 8.3 were taken.

TABLE 8.3

Compression (mm)	Deviatoric force (N)	Excess pore-water pressure u(kN/m^2)
0	0	0
0.5	14	4
1	25	10
1.5	33	16
2	42	22
3	48	28
4	53	31
6	55	33
8	58	34

The drain to the sample was then suddenly opened. The deviatoric force began to increase again as the sample was further compressed very slowly, and a volume of 11.1 cm^3 of water drained out before it once again came to a constant value, this time of 202 N at 20 mm total compression. The sample remained cylindrical throughout.
(i) Show that the void ratios before and after the drained phase were 0.751 and 0.649.
(ii) Plot all the information given on a (q, p') diagram, including an estimated path during the whole process on (q, p'), (e, p') and $(e, \ln p')$ diagrams, with the chief points marked.
(iii) Fit a critical state line and determine its equations: find an equivalent angle ϕ'_c.
(iv) Label the 'dense' and 'loose' zones and define the terms.
(v) Make a prediction of the strength of the silty clay if it were saturated at a moisture content of 0.20 after virgin compression.
(vi) Attempt to fit Cam-clay yield surfaces to the data, and confirm that $\kappa = 0.027$ would appear reasonable, if such were the gradients of the elastic swelling lines.
(b) While excavating a city-centre basement behind sheet piles driven through 5 m of a soft silty clay similar to that described in (a) above, a contractor allows a sudden forward movement of 0.2 m to take place owing to insufficient bracing. Although he claims he braced the sheets immediately afterwards, the owner of an adjacent building was appalled to find that the subsidence of his foundations was continuing long after the initial incident. What is the explanation?

Answers (a) (ii) $q = 1.12p'$, $e = 1.033 - 0.088 \ln p'$, $\phi'_c = 28°$ (iv) 155 kN/m^2.
(b) transient outward flow of pore water after shearing 'loose' soil.

(6) If section 2.4 is correct, deduce that $\lambda \approx 0.6PI$.

9

The Collapse of Soil Constructions

9.1 Criteria of Collapse

Reconsider the response of a soil element to a change in stress. If the effective spherical stress component p' is continually increased the element will compress. Although the volumetric strains may become large, especially when any precompression has been exceeded, the element will never 'collapse'. Indeed it will eventually become very stiff and competent. The deviatoric stress component q, however, can only be increased up to some limit q_{max}. This limiting deviatoric stress q_{max} is often called the compressive strength of the soil; the alternative designation being the shear strength τ_{max}, which is the greatest shear stress on any plane through an element which is on the point of collapse: the limiting Mohr circle of stress demonstrates that $\tau_{max} = q_{max}/2$. As this limit is reached the rate of shear strain becomes large: beyond this point the strength of the soil element may well decrease, so that if the applied deviatoric stresses remain at q_{max} the element must collapse rather violently. This behaviour in the laboratory must be witnessed and understood before the engineer will be able to prevent similar catastrophes in the field. In order to relate the collapse of a triaxial soil sample to the collapse of an embankment or the crushing of a concrete cube to the collapse of a floor slab, the engineer must employ some model of structural mechanics. This chapter will be concerned with models of the collapse of whole soil constructions such as slopes, foundations and retaining walls. Their validity will depend not only on their internal logic but also on the careful site investigation, sampling and testing of the soil elements. If all these aspects are carefully considered the soil construction can be made 'safe' and collapse can be prevented. But even if the designer is able to prevent the collapse of soil constructions he may find that finite soil strains cause such large displacements of associated brick or concrete structures that unsightly cracks develop. The problem of damage to structures caused by strains in soil which is itself 'safe' is touched on in a later section on foundation design.

The stress–strain curves obtained from strain-controlled triaxial (or shear-box) tests may not at first sight indicate anything as dramatic as 'collapse'. Figure 9.1

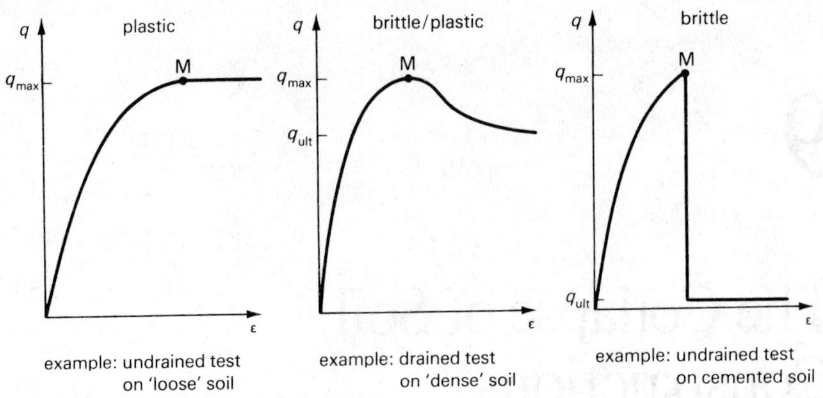

Figure 9.1 *Shear tests on soil elements*

shows three typical curves which might be said to depict the possible range. Of these, only the brittle cemented soil may actually fall off the triaxial pedestal, but this is due entirely to the manner of testing. The normal situation in the field is for soil to be loaded simply by deadweight. If this had been done in the laboratory so that q (or τ) was provided directly by steel weights then collapse would have occurred in each case at point M and the subsequent stress–strain curve would have been inaccessible. When a strain-controlled test reaches $q = q_{max}$ or $\tau = \tau_{max}$, an equivalent stress-controlled sample would collapse. If a sufficient number of elements of soil within a soil construction reach and pass their maximum strength, then the whole construction may collapse.

It is now necessary to choose some specifications for the shear strength of soil which will be useful in the new calculations of the stability of whole constructions. Chiefly on account of their proven ability to generate simple solutions, such as the landslide model of section 4.6 and the foundation collapse model of section 5.2, I shall remain faithful to the simple 'friction' and 'cohesion' models of soil strength presented in chapters 4 and 5. Chapter 8 will have provided further insight into their strengths and limitations.

The updated friction model declares that the maximum shear strength of a soil element is approached when the inclination of the resultant effective stress on any plane through the element approaches a limiting angle ϕ' from the normal to that plane, or

$$\left(\frac{\tau}{\sigma'}\right)_{max} = \tan\phi'$$

The angle ϕ' will be considered to consist of two components

$$\phi' = \phi'_c + \phi'_v \tag{9.1}$$

where ϕ'_c is the angle of friction of the soil when it is in a loose and random state, and ϕ'_v is a component of strength due to volumetric expansion or dilatancy and proportional to the rate of dilatancy against overall deformation. ϕ'_v is zero at a critical state, wherein soil deforms at constant void ratio, and increases as a critical state is made more remote either by a reduction in the void ratio e of a dense soil, or by a reduction in the confining stress p'. The extensive discussions of chapter 8 should have demonstrated that this simple model of behaviour can be used to describe the peak strength of all soils, sandy or clayey, which are not sensitive or cemented. If it be accepted that the rate of dilatancy, and therefore the component ϕ'_v, must be a function of the pattern of strains (triaxial, plane, etc.) allowed by the boundaries of the soil element, then the newly restated friction model can be taken to be operationally equivalent to a variety of stress–dilatancy models. I will presume that the engineer will wish to measure ϕ' and ϕ'_c directly, so that numerical relationships between ϕ'_v and the peak dilatancy rate $(-d\epsilon_v/d\epsilon_s)_{max}$, for example, will be redundant.

This choice of Cornforth's simple friction model for sand to represent the strength of all uncemented soils is not conventional. Many engineers prefer to use

$$\tau_{max} = c' + \sigma' \tan \phi'$$

to represent the strength of a soil at a unique void ratio. The equation fits the data of tests on both sandy and clayey soils. Unfortunately c' varies with the void ratio, which can be expected to vary in the field at random (although within certain limits) in any particular stratum of soil. This makes the good fitting of data at various stresses but constant void ratio an illusory benefit of the (c', ϕ') equation. The density of soil in the field is likely to be variable, so the dilatancy will be variable, and ϕ'_v or c' will be variable. Our simple friction model makes the issue clear: find the least dilatant soil sample (smallest relative density, smallest precompression) if a conservative strength is to be measured. The (c', ϕ') equation obscures this vital relationship. The equation $\tau_{max} = c' + \sigma' \tan \phi'$ must therefore be regarded as inferior owing to the hidden complexity of its interpretation: it also happens to be rather more difficult to use in structural calculations, and I have therefore avoided using it.

Now if we are to use the friction model to predict the strength τ_{max} of soil elements, we must also predict appropriate values of the effective normal stress $\sigma' = \sigma - u$. The total stress σ normal to a plane will invariably be estimated by the resolution of forces which are usually generated by the action of gravity on the overlying surcharge of soil and concrete. The pore-water pressure u may be equated to the 'equilibrium' water pressure in the ground estimated after the fashion of chapter 3, and having sole regard to the pattern of steady drainage, but this is only valid if the engineering construction will not have caused excess pore pressures Δu, that is, only if the work is effectively 'drained'. If transient excess pore pressures Δu are likely to be present at some time when an analysis of the safety of the soil construction is to be undertaken, then they must be added to the 'equilibrium' pore pressures existing before the work was carried out.

We have seen that for a saturated soil element in a triaxial test subject to sudden changes Δp and Δq in the applied stresses at constant volume, the excess pore pressure response will be of the form

$$\Delta u = \Delta p + A \Delta q \tag{9.2}$$

where

$$A = 0 \text{ if the soil is elastic}$$

$A < 0$ if the soil is dilating

$A > 0$ if the soil is contracting

Furthermore, if the soil is not fully saturated the magnitude of Δu will not be as great as that calculated above due to internal drainage. This reduction in Δu is usually expressed by a factor $B < 1$. In this respect partial saturation merely helps in the general process of drainage by which $\Delta u \to 0$ as time passes and as the transient flow of pore fluid ceases. Only very occasionally will it be possible to become so knowledgeable about a body of soil that it will be meaningful to perform undrained triaxial tests to obtain $(\Delta u)_{t=0}$ at constant volume, and then to apply the transient flow model of chapter 7 to assess $(\Delta u)_t$ at some later time. It will be more common to know so little about the state of drainage and uniformity of density of the soil body that simple conservative assumptions will be necessary.

The following steps always lead to the most conservative designs, by employing the most 'positive' conceivable pore-water pressures

(1) Always enhance the pore pressure by the component $\Delta p > 0$ if there is any doubt that drainage might not be good, but ignore any reduction in pore pressure when $\Delta p < 0$ unless it is certain that drainage is impossible.

(2) If the soil is 'loose' it will tend to contract so that pore-pressure parameter A will be positive, and the extra pore pressure due to shear must be taken into account by performing appropriate undrained triaxial tests on the loosest obtainable soil samples. The excess component $A\Delta q$ will only be escaped if the engineer knows that the drainage is good enough to have guaranteed its dissipation.

(3) If the soil is 'dense' it will tend to dilate so that pore-pressure parameter A will be negative. The conservative step is then to ignore the contribution $A\Delta q$ unless the soil body is known to be so impermeable that these excess suctions cannot relieve.

A much simpler way of dealing with the data of undrained triaxial tests on undisturbed samples was introduced as the cohesion model in chapter 4. This involves accepting that the effective stress and void ratio of the soil body would not alter between the measuring of the undrained strength of various samples and the time at which an analysis of distortion at constant volume had to be performed on the whole body. The shear strength c_u measured in one of the undrained tests is simply applied by the engineer to the zone of soil in the field which neighbours the cavity left by the extraction of the sample. The taking of many samples will ensure that the full variation in strength has been observed. This application of raw triaxial data offers an empirical short circuit to the problems of the values of σ', Δu and ϕ' in the field. The cohesion model only leads the designer into gross error when he hasn't taken sufficient undisturbed samples or when he erroneously assumes that the data of undrained tests applies to some field stratum of dilatant soil which will in fact offer a partially drained response to his works. The cohesion model may safely be applied to the problem of the shearing of 'loose' contracting soils, whereas the friction model with zero excess pore pressures may safely be applied to the problem of the shearing of 'dense' dilating soils. While each of these simple techniques may be extended into the opposing range of behaviour, each will need some guarantee; cohesion applied to 'dense' soils will require a guarantee of bad drainage, whereas friction with equilibrium pore pressures applied to 'loose' soils will require a guarantee of good drainage.

The criteria

$$\tau = \sigma' \tan \phi'$$

and

$$\tau = c_u$$

will therefore be used in this chapter to describe the peak strength of all simple soils. It should be clear from section 8.7 that extra care must be taken in the site investigation and testing of very anisotropic soils, existing slip surfaces in clay, or clays that creep a great deal. Sensitive or cemented soils, however, must be placed in a totally different class, for which the 'friction' and 'cohesion' models are inadequate. Section 8.7.1 made clear that such soils collapse when their state of effective stress (q, p') reaches some yield surface which can be determined by performing a number of triaxial tests on undisturbed samples. It is possible to represent Sangrey's simple yield surface in terms of shear stress τ rather than deviatoric stress q, in which case the safe states of stress would be represented by

$$\tau \ngtr \tau_y$$

$$\sigma' \nless 0$$

$$p' \ngtr p'_y$$

where τ_y and p'_y are the shear stress and spherical effective stress which rupture the cement. At first sight, therefore, the analysis of constructions in cemented soil may appear to be similar to a cohesion-type analysis with constant shear strength, accompanied by checks on the effective normal stresses to establish that they do not escape from safe bounds.

In fact, however, the brittle stress–strain curve of cemented soil places it in a totally different class for structural design from that of the simple plastic material, contrasted in figure 9.1. Not only is the criterion of peak strength different, but also the type of calculations in which it is substituted may be entirely different. The extra difficulty of designing safe profiles in brittle materials such as cast iron, glass and ceramics is recognised in every sphere of engineering, and has frequently led to their replacement by ductile alternatives such as mild steel, Perspex, and polythene. Very often the brittle material is stiffer and stronger than its plastic counterpart, but is much more susceptible to sudden overloads that had not been anticipated and allowed for (for example: dropping the tea cup). But the main cause of danger in brittle materials is their 'premature' failure in normal working conditions due to stress concentrations; a striking lesson is afforded by the legendary disasters of the early Comet aircraft, which fell apart due to the propagation of cracks which started at stress concentrations around their rectangular windows. A similar fate befell a number of welded ships, built quickly in the last war to convey supplies, some of which fell to pieces and sank in arctic waters due to the embrittlement of their welds at low temperatures and the excessive stress concentrations induced in the hull of the ship around the large rectangular hole in the deck through which the ships were loaded.

A designer of foundations on, or excavation in, cemented soil will often not be able to avoid encountering brittleness by choosing to build elsewhere: what, then, should he do about stress concentrations? Indeed what are stress concentrations, and how do they cause 'premature' failure? The word 'premature' is based on the engineer's expectation that every piece of a well-behaved plastic construction would be prepared simultaneously to offer to the designer its strength τ_{max} when the construction was on the point of collapse. If the material is plastic in the sense of the shape of the stress–strain curve in shear portrayed in figure 9.1, then the detailed pattern of elastic stresses and strains prior to collapse is irrelevant. If any piece of the construction were subjected to localised forces or were to suffer stress concentrations at sharp corners, the extra local strains would allow the strength τ_{max} to develop at that point first as the whole construction was loaded. The remaining material would then have time to develop greater stresses and to approach the limiting shear stress τ_{max} itself as the zone which had first gone plastic continued to deform at constant limiting stress. The whole construction would continue to carry larger and larger loads until such a large proportion of it had reached its limiting shear stress τ_{max} that the construction could collapse by shearing through a succession of plastic elements which had become joined. Consider now the construction made of brittle material as defined in figure 9.1. An elastic analysis of stress and strain will reveal that some small zone will reach the limiting stress τ_{max} first, as exemplified in section 6.8. As soon as this happens in a brittle solid the small zone of stress concentration softens. In a cemented soil the softening may be so dramatic that the overstressed zone is immediately reduced to the consistency of slurry, which will take no shear stress whatsoever. With the virtual elimination of that most important piece of the construction which had previously taken the greatest burden, there is a danger of progressive collapse. A neighbouring zone which had been close to failure is pushed over the edge and also softens to a slurry: the shear failure propagates through the whole construction, although the average shear stress on the brittle failure surface may have been much less than τ_{max} just prior to failure. The whole construction could fail when any small part of it was overstressed. Now, we shall see that it is very much easier to assess the *average* shear stress on a potential shear surface than it is to predict by elastic analysis the *greatest* shear stress within a construction. An engineer who only calculates average stresses sees the brittle fracture of the construction (and his enforced retirement?) as 'premature'!

The factor by which a brittle construction fails before it has reached its theoretical fully 'plastic' collapse load is increased if

(1) The rate of decline in strength after the peak is sharp and especially if the rate of decline is of the same order of size as the elastic modulus E_s which will be employed by neighbouring elastic elements as they are forced to take up the burden.

(2) The geometry of the construction is such that the load shed by a softened element is unevenly shared by neighbouring elements, such that one of them is forced to carry such an excess that it too fails

(3) the geometry is such that some elements are much more highly stressed than average owing to sharp changes of boundaries or stresses.

The engineer may be able, with practice, to eliminate some unfavourable geometries and to bring the elastic strains and stresses in a brittle construction towards failure in a fairly uniform and simultaneous fashion. If he can do this, then the outstanding problem is that of the design against the *average* shear stress exceeding the limit, which is identical to the plastic design problem discussed in the remainder of this chapter. If stress concentrations are inevitable, as may be the case in soils beneath the edge of loaded foundations, which were the topic of section 6.8, the designer of very brittle soil constructions should refer to whatever elastic stress distributions might offer some hint as to the degree of that concentration. The solutions in Poulos and Davis (1974) would be a useful starting point.

Now consider the dilemma faced by the soils engineer who discovers that the majority of his materials are of the brittle/plastic type characterised in figure 9.1. I have already remarked that undiscoverable variations of the *in-situ* void ratio may force the designer to use critical state strengths ϕ'_c rather than peak strengths ϕ'. A pessimistic designer would similarly invoke only the critical state strength even where he knew that the entire ground would possess a dilatant peak: he would fear that many elements in the collapsing soil construction would have attained a critical state by the time the last plastic link on the failure mechanism had reached its peak strength. Figure 9.2 demonstrates the

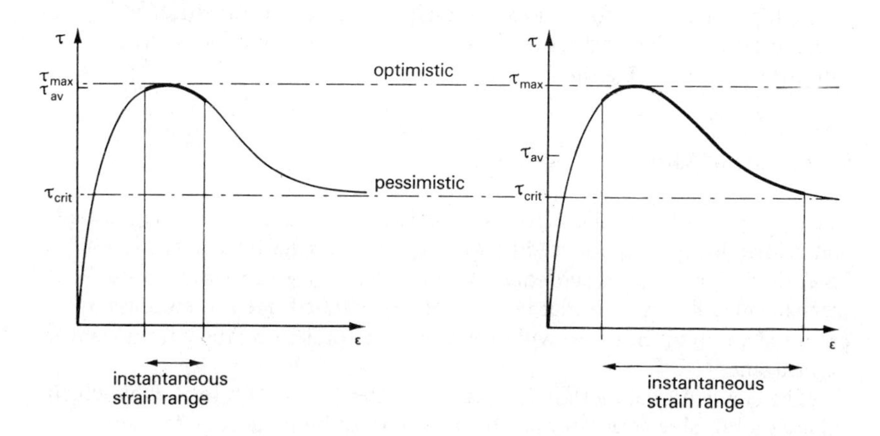

Figure 9.2 *Average shear stress τ_{av} at the collapse of a construction in dilatant soil, which possesses a range of strains at any instant*

optimistic and pessimistic bounds to the choice of an equivalent plastic limit for brittle/plastic soils within constructions which exhibit a large range of strains at different points at any instant. It should be clear that the optimistic choice of the peak strength would only be valid if the construction were stressed in a perfectly uniform manner, with zero range of strains, so that every element of soil reached its peak strength at the same instant. The pessimistic choice of the critical state strength is likely to be quite accurate when the peak is sharp and the strain range is wide at failure. This leaves the designer a safe, pessimistic strategy which he could judiciously depart from in those cases where the peak for the soil is relatively flat and where an elastic analysis predicts that first yield occurs only slightly before the prediction of collapse made by a plastic analysis – in other words when stress concentrations were not too large. The classification of soil constructions with regard to the safety and merit of including for design purposes some fraction of the dilatant peak of strength obtained in tests on elements, is one of the current goals of research workers in soil mechanics.

You will be aware from section 8.7.2 that a more pessimistic strength even than that of the critical state is available for very clayey soils which offer residual strengths $\phi'_r < \phi'_c$ on polished slip surfaces. There is evidence that a clay landform can cultivate such a feature by indulging previously in a very major landslide for which normal soil strengths would have been applicable. Also, Burland *et al.* (1977) show that the extraordinary strain concentrations which develop at the foot of the face of a deep excavation in clay can give rise to a progressive slippage along an induced horizontal slip plane level with the base of the excavation, which can lower the angle of friction in the slip zone to its residual value. There is as yet no firm evidence that residual slip surfaces which were not present at the time of a site investigation, and which were therefore formed for the first time during the course of construction of some engineering works, have ever been responsible for reducing the average mobilised strength of a body of soil below that of the critical states, so as to cause a collapse. The residual strength problem therefore remains principally one of investigation and interpretation of the features of the land in order to determine whether residual slip surfaces are likely to be present.

9.2 Criteria of Safety

The designer must not allow his soil construction to fail, even in the most adverse conditions. If he is experienced he will find that calculations on the stability of his early design sketches subsequently prove that no conceivable failure mechanism will exist and that no zone of soil will be forced to mobilise its potential strength. But how will he discriminate judicious safety from wanton extravagance?

The conventional method is to calculate the 'factor of safety' F by which the likely mobilisable strength τ_{max} of the soil must be reduced before some calculation offers a notional collapse. The designer will attempt to attain what he hopes will be an optimum F-value for safe and economic design, such as 1.50 for important slopes, perhaps. The landslide model of section 4.6 generated just such a factor

$$F = \frac{\tan \phi'}{\tan \beta}\left(1 - \frac{h\gamma_w}{z\gamma \cos^2 \beta}\right) \qquad (4.7)$$

which can be used to demonstrate the problems of this conventional approach. If the designer had a simple, very long, very wide slope on which he had conducted a very expensive site investigation which revealed that each of ϕ', β, γ_w/γ and h/z was accurately constant, surely $F = 1.50$ would be an excessive margin of safety? If the relative water head (h/z) varied from place to place, or was likely to be dependent on the weather, then the designer could take either the likely average or the worst possible value; surely these opposing strategies should not both demand the same safety factor? If the designer takes the worst possible values of ϕ', β, γ_w/γ and h/z surely he needs only a rather small additional safety factor, if any?

The trend in civil engineering design is now away from the calculation of a single safety factor and towards a more intimate assessment of the likely variability of each parameter, leading to a succession of partial safety factors. The code of practice for structural concrete, CP 110, is one such modern code, which deals with the probability of the whole construction being unserviceable due to an accumulation of harmful variations away from the 'likely' value of parameters. If probabilities are to be calculated then sufficient investigation must be carried out to determine the statistical distribution of each parameter. Such a luxury is unfortunately rarely the lot of the soils engineer. Clients (and consulting engineers) have yet to be convinced that a careful site investigation and testing programme is worth while. Yet it appears that the overwhelming majority of insurance claims for new houses are due to foundation failures, and the majority of claims for extra payments to civil engineering contractors are due to unforeseen ground conditions. The fierce criticisms made by Rowe (1972) of site investigation practice still appear to apply. If the soils designer is deprived of probability theory he must rely on common sense and pessimism.

The conventional designer might use the steepest slope section for analysis (worst β) and the most unfavourable foreseeable groundwater location (worst h/z), but he would normally invoke the average density for (γ_w/γ) and the average peak strength for ϕ'. This latter, particularly, would cause him to lose sleep if he hadn't a safety factor F of at least 1.50. Surely it would have been more sensible to continue to take the 'worst possible' values of parameters – the lightest soil and the lowest possible strength – and then to use a lower factor of perhaps 1.20 to cover the inevitable lack of fit between the 'model' and the reality? The lowest likely strength in terms of effective stress, ϕ'_c, is certainly easy to measure. If the soil is sandy, its loose angle of repose measured by a protractor would suffice. Of course, if the designer were absolutely sure of some extra dilatancy component ϕ'_v at every location he could include it: he would then be relying on his excellent site investigation, or site supervision if the soil was being compacted *en masse*. If the designer had been as pessimistic as possible with his choice of every parameter, and if he knew that the theoretical model of the potential collapse was itself either pessimistic or very accurate, then he should be prepared to eliminate the last vestige of 'safety factor' and allow the construction to fail if there was an improbable coincidence of catastrophes which reduced each parameter simultaneously to its worst possible value. Although I shall continue to use the conventional definition for factor of safety, I shall hope to be able to provide 'worst' rather than 'likely' values of the important parameters, which will mean that a desirable value for F would be unity or just a little larger.

Some designers say that they use large factors of safety to protect themselves against unforeseen dangers. Sometimes they mean the danger of not having envisaged the worst value of a parameter in a model of collapse: they could be better advised to apply the factor to the dubious parameter or parameters. Sometimes they mean the danger of not having considered the appropriate collapse mechanism: they could be better advised to spend their client's money on better design rather than more earthworks. Sometimes they mean the danger of having conducted so elementary a site investigation that important strata have been missed, which is rather like putting the boreholes on the wrong side of the boundary fence. Ignorance as comprehensive as this will only lead to safety if the designer chooses the worst values of parameters which have ever been discovered in that part of the country: the resulting constructions are very unlikely to be economic.

Structural designers may be comparing the attention paid here to each parameter to their own use of load factors. It is considered good practice to enhance the permissible live loads on a floor slab for design purposes, due partly to the unforeseeable uses to which the building may be put by future clients: wind pressure poses a similar problem. Most soils engineers have this one advantage over their structural colleagues: live loads are usually a negligible fraction of the dead weight of the soil construction, and the unit weight γ of the earth is almost invariable. Exceptions arise in the design of roads and airfields where the repetition of heavy wheel loads can cause a progressive collapse. Some foundations also suffer from the effects of live load. Live loads in this sense are beyond the scope of this book.

It will often happen with fine-grained soils that both the undrained strength c_u and the effective angle of friction ϕ' must be determined so that the theoretical 'undrained' and 'drained' safety factors can be calculated. If the 'undrained' factor of safety is smaller than the 'drained' the soil must be looser than its eventual critical state: 'loose' soils tend to collapse as soon as loads are applied due to excess pore pressures generated within them. If the 'undrained' factor of safety exceeds the 'drained' value, the soil must be denser than its eventual critical state: 'dense' soils tend to collapse at some time after loading when the transient pore suctions generated within them have had the opportunity to dissipate. If 'loose' soils can be encouraged to drain during construction, the designer might be able to avoid the penalty of the dangerous undrained state. Likewise the evil day of reckoning for a 'dense' soil might be postponed by sealing it against the ingress of water which could allow its suction to dissipate. If the designer is an outrageous pessimist he will assume on the contrary that those soft soils which should have drained will not drain, while those stiff soils which should not drain will drain. He will then proceed to ascribe to each element of clayey ground the lower strength of the extreme alternatives $(c_u)_{t=0}$ and $(\sigma' \tan \phi')_{t=\infty}$.

9.3 Investigating the Collapse of Plastic Soil Constructions by the Methods of Coulomb and Rankine

Coulomb in 1773 introduced the earliest technique for determining the stability of soil constructions. He undertook to check every conceivable mechanism of collapse, which he imagined to take the form of one rigid block of soil sliding against another. He allowed the previously measured strength of the soil to be fully mobilised at every point on the imagined surface of sliding. Every slipping block was bounded either by the atmosphere or by a surface of sliding on which the state of shear stress was known. This was enough to make the block statically determinate. I used Coulomb's method to interpret the data of the shear-box test in chapters 4 and 5, and to analyse the landslide problem in section 4.6 and the strip footing problem in section 5.2. Cross sections of typical mechanisms are depicted in figure 9.3, in which only one member of each family of possible mechanisms is detailed.

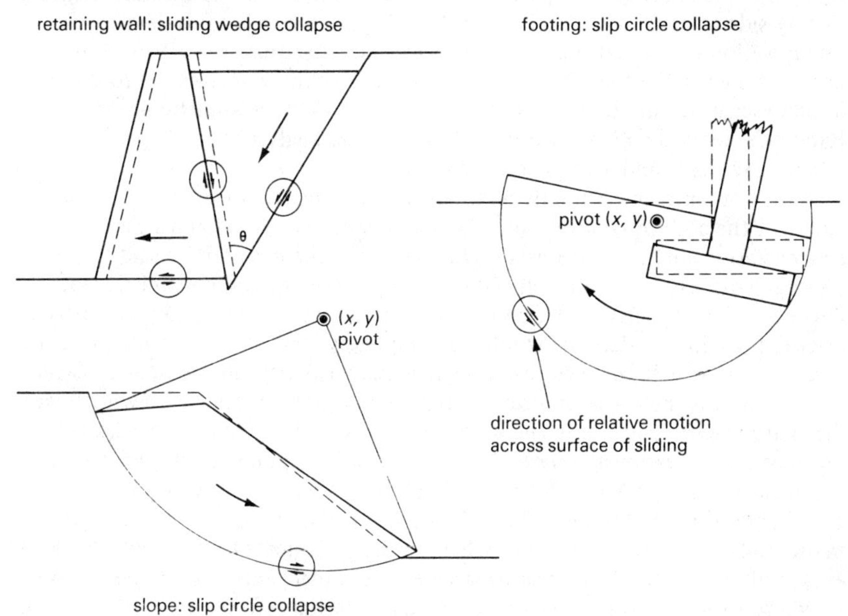

Figure 9.3 *Collapse mechanisms for soil constructions*

A distinguishing operational feature of Coulomb's method is the necessary search for the worst mechanism, being that which mobilises most soil strength and which therefore possesses the smallest factor of safety. The variation in mechanism, whether it be the angle θ of a sliding wedge or the position (x, y) of the centre of a slip circle, can be achieved either by using calculus such as

$$\frac{\partial F}{\partial \theta} = 0$$

and

$$\frac{\partial^2 F}{\partial \theta^2} > 0$$

to determine the mathematical minimum for F as θ is allowed to vary, or by a repeated solution for F at various θ and a graphical interpretation. This search for the most unfavourable mechanism of the family was represented in the discussion of the landslide model by the need to explore sliding surfaces of various depth. It should be intuitively obvious that guessed mechanisms give optimistic or unsafe conclusions: only a fastidious search of mechanisms can lessen the danger.

Rankine in 1857 introduced a contrasting technique in which the soil construction was notionally damaged, usually by interposing imaginary frictionless wounds where none had existed in reality, in such a way that the soil body could be statically analysed as a pile of soil 'bricks', the soil within each 'brick' being just within a limiting state of stress. If Coulomb's method can be thought to be equivalent to placing a number of elementary shear-box elements side by side to generate a mechanism, Rankine's method is equivalent to the juxtaposition of cubical triaxial samples to generate a mass of soil which is approaching the limit at all points. In so far as Rankine damages the soil before he analyses it, his method can be seen to be essentially pessimistic or safe. Rankine's method will be used in later sections to analyse the collapse of retaining walls, foundations and cavities.

Both Coulomb and Rankine methods require a plastic shear stress–strain curve, or the additional attention to the elimination of stress and strain concentrations in brittle materials. Only then can the methods be said to possess internal logic which is optimistic (unsafe, upper bound) and pessimistic (safe, lower bound) respectively. Of course, a potentially safe and pessimistic analysis after the fashion of Rankine would be completely spoiled by an inadequate site investigation which failed to reveal a particular weak spot in the field construction which was therefore omitted from its representation on the drawing board. Plasticity theory, laid out in the useful textbook by Calladine, embraces the optimistic and pessimistic methods of Coulomb and Rankine within a powerful and logical framework which leads ultimately to 'exact' solutions for the point of collapse of plastic bodies. These theoretical collapses occur when every point in the body is in equilibrium, but when so many elements have reached (but not exceeded, of course) their limiting stresses that a mechanism of failure can be shown to exist which satisfies the normality criterion referred to in figure 8.20. If you ever discover that the pessimistic method of Rankine and the optimistic method of Coulomb offer the same answer for the collapse condition of a plastic soil construction, then you may have discovered one of these 'exact' solutions by accident. Such an 'exact' answer merely excuses you from further paper searches, whether of Rankine zones of stress or of Coulomb's mechanisms: it says nothing of the validity of your initial assumptions or of the quality of your site investigation. A course on plasticity theory would be a useful adjunct to the education of any civil engineer, but since it is not universally offered I have avoided using rigorous plasticity theory.

9.4 Coulomb's Slip Plane Mechanism

Slip planes can be applied with ease to most problems of the collapse of soil constructions in order to generate optimistic solutions. Occasionally they can be proved to offer solutions which are not so optimistic as to be a hazard. The landslide model depicted in figure 4.15 is an example: the slope is supposed to be infinite and uniform and the potential landslide relatively shallow, so that any mechanism other than of a parallel flake slide would appear to create unwanted asymmetries or singularities. You should now work through the whole landslide model again, and take note of the method of attack.

9.4.1 Landslide Analysis using the Cohesion Model

The landslide model could just as easily have used the cohesion criterion

$$\tau_{max} = c_u$$

rather than the effective friction criterion. The analysis would have reached a very quick conclusion with

$$\tau = W \sin \beta \tag{4.3}$$

Let the conventional failure occur at

$$\tau = \frac{\tau_{max}}{F} = \frac{c_u}{F} \tag{9.3}$$

so that

$$F = \frac{c_u}{W \sin \beta} = \frac{c_u}{\gamma z \cos \beta \sin \beta} \tag{9.4}$$

or when $F = 1$

$$\sin 2\beta_{max} = \frac{2c_u}{\gamma z} \tag{9.5}$$

For the parallel flake landslide to be reasonable c_u must not vary along the slope, but it may vary with depth. The most unfavourable possible slip plane is that on which the factor of safety is least, which is that possessing the smallest ratio $c_u/\gamma z$. It is this critical plane that determines the steepest slope angle β_{max}. If the strength c_u were constant the slip plane would be as deep as possible.

Natural clay slopes usually have a dry shallow crust of strong clay masking the general tendency, in the soil beneath, for the undrained strength c_u to increase with depth. As the depth increases the overconsolidation ratio decreases towards unity as the inevitable depth of historical erosion becomes smaller in proportion to the depth of soil that still remains; the soil therefore approaches a virgin condition. Virgin soil that has not been precompressed usually displays a strength c_u proportional to depth, as mentioned in section 8.5.1, so that the factor of safety given by equation 9.4 would remain constant in virgin soil. Slope failures that occur due to some sudden cause, such as river erosion at their toe during a rare flood, tend to search out a relatively deep slip surface upon which the $c_u/\gamma z$ ratio approaches its lowest virgin value. Clay slopes, like sand slopes, tend to search out deep slip surfaces. Since no slope is infinite in extent this can lead to the critical mechanism being a deep slip circle rather than a parallel flake. Slip circles are dealt with shortly.

9.4.2 *Thrust on Retaining Walls: Active and Passive*

The classical use of the slip plane, however, is to estimate the thrust of earth against retaining walls. Two limiting cases exist, termed active and passive. In the active case the wall is allowed to move away from the earth so that the soil must actually support itself. In the passive case the wall is allowed to move into the soil mass which presents a passive obstacle to motion. The active and passive conditions are limiting states in that the soil generates its full strength τ_{max} either to support itself or to prevent the wall from pushing it aside. Assumed wedge failures are sketched in figure 9.4 with the directions of motion and of

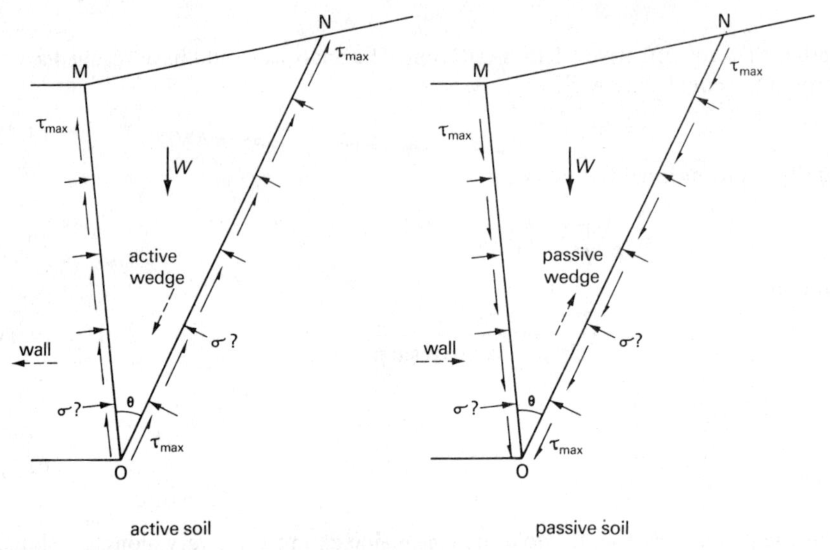

Figure 9.4 *Active and passive wedge collapses*

the shear stress τ_{max} acting on the soil wedge, included. I presumed that in each case the wall slid horizontally. In the active case the soil wedge slid downwards and was therefore dragged back by upward shear stresses against the wall and the underlying soil. In the passive case the failing wedge of soil slid upwards so that boundary friction, in opposing motion, acted downwards. The remaining problem is to determine the most critical wedge angle θ in each case, by considering a variety of angles and choosing that which mobilises most strength or which has the smallest factor of safety.

9.4.3 *Thrust on Retaining Walls: Friction Model: Analytical Method*

Elementary analytical solutions are obtained in the case of a frictionless wall with an assumption of the validity of the friction model in the retained soil, with zero pore-water pressures and a horizontal surface. The wedges are drawn in figure 9.5 where W is the weight of the wedge, P is the unknown reaction between the wall

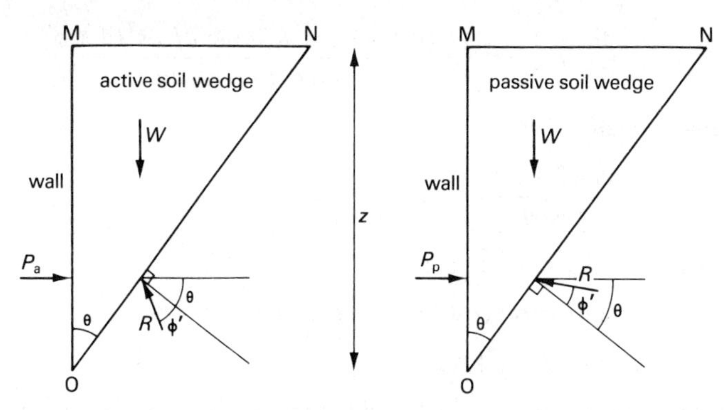

Figure 9.5 *Forces acting on active and passive soil wedges: soil obeying friction model, smooth wall*

and the soil with suffixes a for active and p for passive, and R is the unknown reaction between the wedge and the underlying soil. The forces P are unknown in magnitude and position but known to lie normal to the smooth wall. The forces R are likewise unknown in every respect except that they must incline at angle ϕ' to normal to the slip plane ON in each case. This knowledge is, however, sufficient to fix the two unknowns by simple resolution of forces. The wall force P can be found directly by resolving perpendicular to R. In the active case

$$P_a \sin(\theta + \phi') = W \cos(\theta + \phi')$$

In the passive case

$$P_p \sin(\theta - \phi') = W \cos(\theta - \phi')$$

Taking the unit weight γ of the soil to be constant, the height of the wall to be z, and the wall to be long so that a section of unit length can be considered

$$W = \gamma \times 1 \times \text{area OMN}$$

$$W = \gamma \times \tfrac{1}{2} z^2 \tan\theta$$

Interpreting P_a and P_p as forces per unit length of the wall we now obtain

$$P_a = \frac{1}{2}\gamma z^2 \frac{\tan\theta}{\tan(\theta + \phi')} \tag{9.6}$$

$$P_p = \frac{1}{2}\gamma z^2 \frac{\tan\theta}{\tan(\theta - \phi')} \tag{9.7}$$

It is now necessary to mitigate the accepted optimism of Coulomb's method by choosing the most unfavourable wedge angle θ in each case.

In the active case the soil is supposed to be pushing the wall over. In the most unfavourable condition it will support itself least and lean most heavily on the wall. The least optimistic active thrust will be that which is largest, so it is necessary to determine the maximum of P_a as θ varies. Now by differentiating equation 9.6

$$\frac{\partial P_a}{\partial \theta} = \frac{1}{2}\gamma z^2 \frac{\tan(\theta + \phi')/\cos^2\theta - \tan\theta/\cos^2(\theta + \phi')}{\tan^2(\theta + \phi')}$$

So $\partial P_a/\partial\theta = 0$ when

$$\frac{\tan(\theta + \phi')}{\cos^2\theta} = \frac{\tan\theta}{\cos^2(\theta + \phi')}$$

This is easily solved by multiplying out

$$\sin(\theta + \phi')\cos(\theta + \phi') = \sin\theta\cos\theta$$

$$\sin 2(\theta + \phi') = \sin 2\theta$$

which has the useless solution $\phi' = 0$ and the more interesting condition

$$2(\theta + \phi') = 180° - 2\theta$$

$$\theta = 45° - \frac{\phi'}{2}$$

It is straightforward to check that this value of θ offers $\partial^2 P_a/\partial\theta^2 < 0$ which therefore means that we have discovered a mathematical maximum for P_a. Further checks reveal that this critical wedge angle has generated the largest active thrust of all. From equation 9.6

$$P_a = \frac{1}{2}\gamma z^2 \frac{\tan\left(45° - \frac{\phi'}{2}\right)}{\tan\left(45° + \frac{\phi'}{2}\right)}$$

$$P_a = \frac{1}{2}\gamma z^2 \tan^2\left(45 - \frac{\phi'}{2}\right) \tag{9.8}$$

In the passive case the soil is supposed to be preventing the wall from pushing into it. The most unfavourable condition is when it does this least, so we must search for the least optimistic mechanism by finding the minimum value of P_p offered by equation 9.7. On differentiating

$$\frac{\partial P_p}{\partial \theta} = \frac{1}{2}\gamma z^2 \frac{\tan(\theta - \phi')/\cos^2\theta - \tan\theta/\cos^2(\theta - \phi')}{\tan^2(\theta - \phi')}$$

So $\partial P_p/\partial\theta = 0$ when

$$\frac{\tan(\theta - \phi')}{\cos^2\theta} = \frac{\tan\theta}{\cos^2(\theta - \phi')}$$

or when

$$\sin 2(\theta - \phi') = \sin 2\theta$$

which has the interesting solution

$$\theta = 45° + \frac{\phi'}{2}$$

This can be proved to represent the minimum passive thrust, which is given by substituting back into equation 9.7.

$$P_p = \frac{1}{2}\gamma z^2 \frac{\tan\left(45^\circ + \frac{\phi'}{2}\right)}{\tan\left(45^\circ - \frac{\phi'}{2}\right)}$$

$$P_p = \frac{1}{2}\gamma z^2 \tan^2\left(45^\circ + \frac{\phi'}{2}\right) \tag{9.9}$$

The use of the notion of least optimism to create the appropriate solutions for maximum P_a and minimum P_p deserve your closest possible scrutiny. Note also that if $\phi' = 30^\circ$ $P_a = (1/6)\gamma z^2$ while $P_p = (3/2)\gamma z^2$, which is nine times larger. If $P \leqslant P_a$ the wall falls outwards. If $P_p > P > P_a$ the wall is stable. If $P \geqslant P_p$ the wall pushes inwards. The central range of wall thrusts between P_a and P_p is enormously wide.

Although we have checked all possible slip *planes*, might the soil not discover some more favourable mechanism of failure? If it could do so, P_p would be less than our least estimate while P_a would be greater than our greatest. In order to protect himself against such an eventuality the analyst may like to increase the calculated estimate of P_a, and to decrease P_p, by a factor of safety, in addition to using prudent pessimism in the choice of ϕ'.

9.4.4 Thrust on Retaining Walls: Friction Model: Graphical Method

The analytical approach very soon gets out of hand if the wall is to be considered inclined or rough, or if the earth is not level. A simple graphical technique enables the appropriate polygons of force for each of a number of trial wedges to be superimposed as shown in figure 9.6 for an active case. The weights of the various wedges OMN_1, OMN_2, etc. are calculated to be W_1, W_2, etc. It is convenient if MN_1, N_1N_2, N_2N_3, etc. mark off equal lengths along the surface of the soil so that the weight increments are equal. Each wedge offers a different value for the thrust P on the wall, which is supposed to act at an angle δ to its normal. The angle of friction δ of soil against a surface is usually somewhat less than the peak friction angle ϕ' of the soil alone. Drained shear-box tests should be performed with the lower half of the box filled flush with the material of the wall and the upper half filled with soil at the appropriate density, compressed to the appropriate effective direct stress. All the forces P are parallel and at an inclination to the horizontal of $\alpha + \delta$. The various triangles of force are completed by knowing the various lines of action of the resistances R_1, R_2, etc. which are at ϕ' to the normals to the various slip planes. The locus of P can then be sketched and its maximum value determined. This is our best estimate of P_a and the corresponding plane is the plane of potential sliding.

This graphical technique easily absorbs the problem of line loads running on the earth parallel to the wall. Suppose that the wall represented by OM in figure 9.6 were supporting a railway track and that a heavy train could be idealised as a line load of ΔW kN/m length acting above point N_2 as shown in figure 9.7.

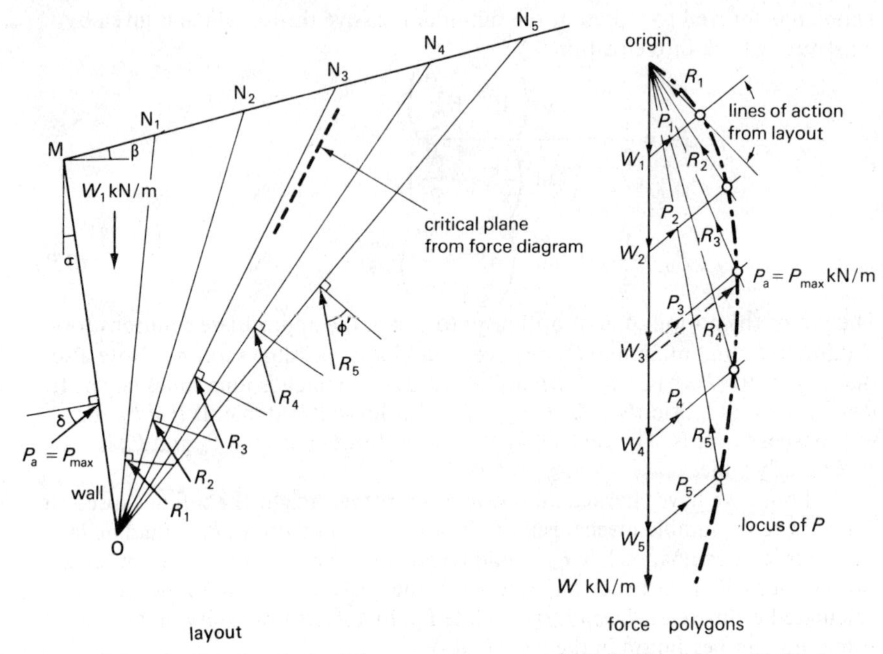

Figure 9.6 *Graphical analysis of forces on a sequence of Coulomb wedges with friction*

The corresponding force polygons are also shown in figure 9.7 in which W_1, W_2, etc. represent the weights of the trial wedges of soil as before. Wedge OMN_1 did not contain the train and is therefore unchanged. The weight of wedge OMN_2 considered an infinitesimal distance before the line load is also unchanged at W_2. An infinitesimal distance beyond, however, and the extra load ΔW must be added without any other changes. Wall thrust P_2 thereby changes to P_2^+. Thereafter the weight of the railway train must be included and the diagram continued as before.

You must guard against thinking that thrusts P on either side of the maximum mean very much. The soil won't fail other than on the critical wedge: only the greatest thrust and its associated slip plane are significant. You must imagine the wall OM well supported by car jacks so that $P > P_a$. If the jacks are slowly taken out the soil will mobilise its own strength to replace that lost by the jacks. Eventually P will drop to P_a, which occurs at the maximum on the locus of P on the force diagram. The wall then fails, and the noncritical values of P for other wedges never get an opportunity to develop. Don't forget that although you may have been painstaking in your search for large active thrusts, the soil will be more so. What we have achieved is the least optimistic slip plane.

This graphical technique can also handle the extra forces arising from water pressures, which must be taken into account in the friction model, if they exist. In such cases it is necessary to split the total forces P and R into their effective and fluid components P', U_w on the wall and R', U_θ on the trial slip plane. The thrusts due to the water must usually be calculated by integrating or averaging the pore-water pressures over the relevant areas. The pore pressures may have to

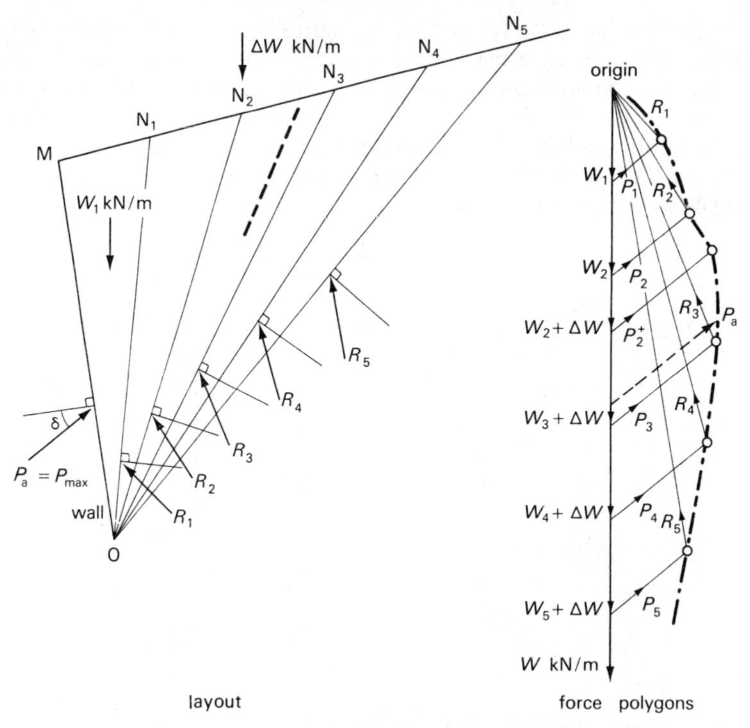

Figure 9.7 *Graphical wedge analysis including a line load ΔW for soil obeying friction model*

be derived from a level groundwater table, or from a flownet if the groundwater is flowing steadily; and in the case of clayey soils transient pore pressures Δu may need to be added. Whatever the case, the magnitudes of U_w which acts normal to the wall and U_θ which acts normal to the slip plane must be determined before their vectors can be included in the force diagrams. Figure 9.8 depicts the wall of figure 9.6 with the fill submerged to the level of the top of the wall. In such a simple case it is easy to see for the first wedge, for example, that the average water pressure on the wall OM and the submerged part of the plane OT_1 is one-half the water pressure at O. Multiplying by the relevant lengths OM and OT_1 we obtain the thrusts due to the water U_w and U_1. The diagram of force polygons can now be begun. From the origin o the weight W_1 is scaled by ow using the average unit weight γ of the soil in the first wedge, which in this case would probably be saturated. The two water thrusts can now be added to the diagram, zo corresponding to U_1 and wx corresponding to U_w. Finally, the R_1' line drawn from z and the P_1' line drawn from x cross at point y, which must close the polygon of forces. Check that the arrows on the forces rotate in the same sense. When there is hydrostatics the horizontal components of the water thrusts are equal and opposite, as here. Furthermore their net upward thrust would exactly carry a weight of water which could fill triangle OMT_1, this being the difference in length between ow and zx. A short-circuit to the answer when there is a level groundwater table is therefore simply to concentrate on the *effective* forces P_1', R_1' and W_1', the latter being the effective weight of the wedge calculated by using γ above the water table and $(\gamma - \gamma_w)$ below it. Whichever method is adopted, the various wedges must be considered, as before, to determine which has the greatest thrust on the wall. You must not of course, forget to add U_w to the final effective active thrust P_a' in order to arrive at the total thrust P_a. Nor must you forget that the addition is vectorial, after the fashion of wx + xy → wy in figure 9.8.

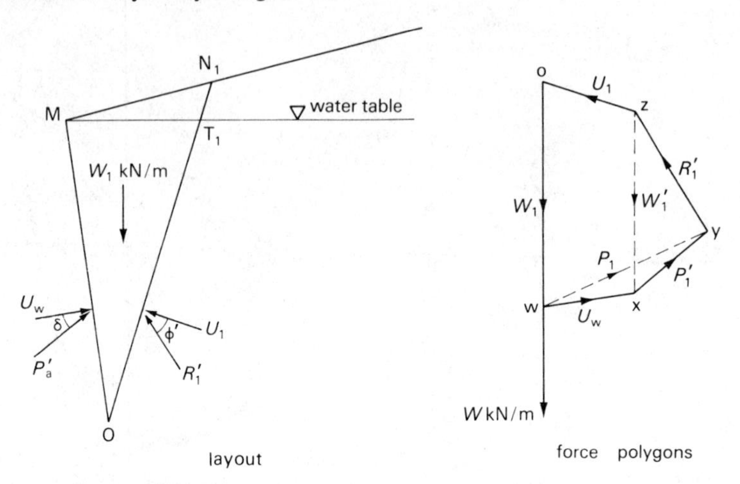

Figure 9.8 *Graphical wedge analysis including hydrostatic water for soil obeying friction model*

Passive thrust can be estimated in a similar way by reversing the shear stresses and choosing the *minimum* thrust offered by the locus of P as a number of wedges are calculated.

9.4.5 Thrust on Retaining Walls: Cohesion Model: Analytical Method

Simple solutions are obtained in the case of a frictionless wall with an assumption of the validity of the cohesion model in the soil, using c_u = constant on trial slip planes. Figure 9.9 demonstrates the forces to be considered in both an active and passive analysis of a rectangular block of earth of depth z. Once more the soil is assumed to be plastic and to develop its maximum shear strength along the whole slip plane.

It is necessary only to resolve the forces on the wedge parallel to the slip plane. In the active case

$$P_a \sin\theta = W \cos\theta - F$$

$$P_a = \frac{1}{2}\gamma z^2 - \frac{c_u z}{\sin\theta\cos\theta}$$

$$P_a = \frac{1}{2}\gamma z^2 - \frac{2c_u z}{\sin 2\theta} \qquad (9.10)$$

In the passive case

$$P_p \sin\theta = W\cos\theta + F$$

$$P_p = \frac{1}{2}\gamma z^2 + \frac{2c_u z}{\sin 2\theta} \tag{9.11}$$

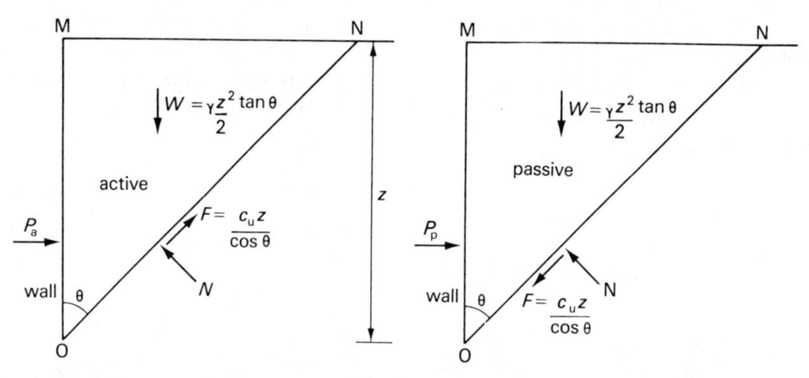

Figure 9.9 *Forces acting on active and passive soil wedges: soil obeying cohesion model, smooth wall*

Now the true active thrust is that which will be generated when the first mechanism is activated as P falls and the wall moves outwards. It is necessary, as before, to search for the largest P_a. This can be done by inspection: looking at equation 9.10, P_a is largest when the term $-2c_u z/\sin 2\theta$ gives the smallest decrement, which occurs when $\sin 2\theta$ is largest, which is when $2\theta = 90°$ so that $\theta = 45°$. The least optimistic estimate of the active thrust on the critical 45° wedge is

$$P_a = \frac{1}{2}\gamma z^2 - 2c_u z \tag{9.12}$$

Likewise, the true passive thrust is that which will first cause a mechanism to slide, as P is increased so much as to push the wall into the soil. It is necessary to find the smallest P_p. Once again, inspection of equation 9.11 reveals that P_p is the smallest when $2c_u z/\sin 2\theta$ is smallest, which occurs when $\sin 2\theta$ is largest, at $\theta = 45°$. The least optimistic passive thrust is

$$P_p = \frac{1}{2}\gamma z^2 + 2c_u z \tag{9.13}$$

9.4.6 Thrust on Retaining Walls: Cohesion Model: Graphical Method

The graphical method can easily be adapted to suit the use of the cohesion model on the slip plane and, if the analyst desires it, on the wall. One might intuitively say that soft sticky clay could adhere to a wall and mobilise cohesion, whereas stiff clay would tend merely to offer a straightforward coefficient of friction. Just as δ on the surface may be less than ϕ' in the body of the soil, so might c_u^w against the wall be less than c_u in the mass. Figure 9.10 demonstrates the method of estimating the active thrust on a wall when the cohesion model is used throughout. For clarity only the force polygon for the first wedge is shown. The forces W, S and F must first be calculated directly. The vector W can then be set away from the origin. The cohesive shear forces F and S can then be entered at opposite ends of the W-vector in directions taken from the layout sketch. The lines of action of the unknown normal forces N and Q are likewise known: their intercept closes the force polygon. The 'maximum' of the locus of the intercept must be determined by solving a number of wedges so that the least optimistic estimate of the active thrust P_a can be determined: it is the vector sum of S and Q_a as shown.

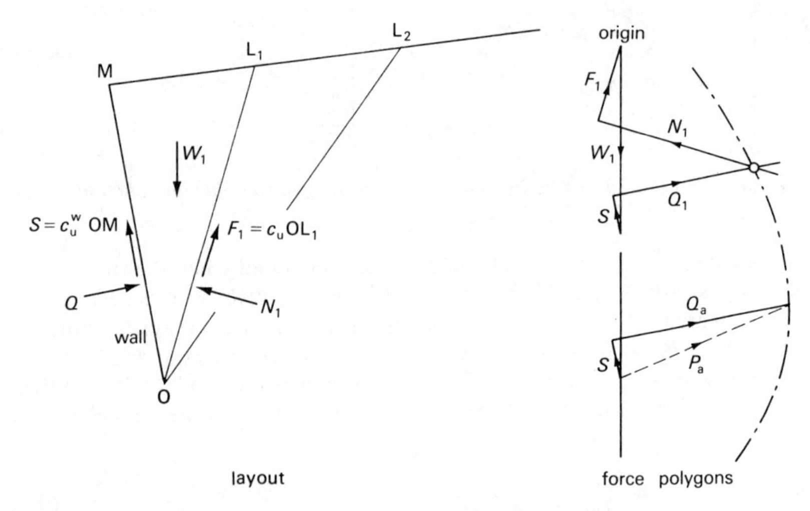

Figure 9.10 *Graphical analysis of forces on Coulomb wedges for soil obeying cohesion model*

If the least optimistic estimate of the passive thrust is to be determined the cohesive forces in figure 9.10 must be reversed and the *minimum* thrust P_p determined by solving a number of wedges and drawing the locus of P.

Note that both the analytical equation 9.12 and the graphical method demonstrated in figure 9.10 can imply a negative active thrust if the soil is very strong, so that if the wedge is to slide down, the wall must actually pull it forwards. This is unlikely to be the most critical condition in the possible life of the structure: if it were, there might be no need for the wall! Although negative thrusts are not of great practical significance they must not be confused with positive thrusts. This can happen if the analyst forgets to put arrows on the force diagram, forgets that the arrows should be cyclic, or forgets that the wall thrust P which appears on the force diagram is the reaction of the wall on the soil so that the action of the soil on the wall is equal and opposite.

9.5 The Slip-circle Mechanism

Slip circles which allow one rigid body of soil to rotate against another are widely used to investigate the stability of soil constructions. Slip planes might be considered a special subset of slip circles which have infinite radii. As with slip planes, the analyst must check as wide a variety of possible slip circles as he can in order that he may choose the most unfavourable of them. This will be the one that is least optimistic, since it will have the smallest factor of safety and will mobilise the greatest proportion of the soil strength.

9.5.1 The Long Strip Footing: Cohesion Model: Analytical Method

In section 5.2 I introduced the semicircular mechanism for the failure of a long strip footing shown in figure 5.3. The concrete strip was considered to rotate about one of its lower edges, taking a semicircle of potent soil with it. The soil above the founding level KCJ was considered to be overburden with no strength. The method employed the conventional factor of safety F by supposing that slip failure took place at a reduced shear strength c_u/F rather than c_u, which was presumed to have been measured and found constant in the region. You should read the section through again.

The simple geometry of the strip footing allows a mathematical search for other slip circles. You were invited to do some of this in problem 3 at the end of chapter 5. Almost all other conceivable slip-circle problems are most easily tackled in a trial-and-error fashion. A fairly rigorous solution to the problem of the least optimistic slip circle for a very long strip footing is set out below.

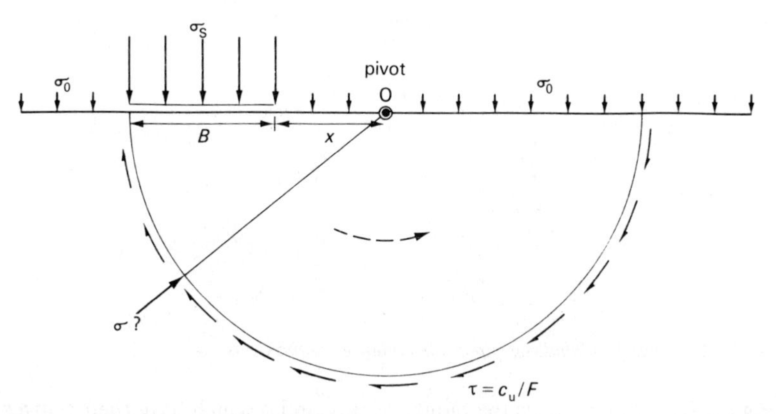

Figure 9.11 *Slip circle collapse of strip footing on uniform soil obeying* $\tau_{max} = c_u$: *section of unit length*

Figure 9.11 shows the stylised founding level with a surcharge pressure σ_0 replacing the overburden whose strength is ignored, and a 'safe' contact pressure σ_s due to the weight of the foundation and the vertical structural load. A family of slip circles has been assumed which have their centre O on the founding level and at a distance x from the edge of the footing. The shear stress preventing the circle from being driven round anticlockwise is set at c_u/F where c_u is the likely mobilisable undrained shear strength of the soil in this vicinity and F is the factor that renders the foundation pressure 'safe'. The semicircle of soil beneath the foundation has no inherent tendency to rotate about the pivot O.

Allowing slip failure to take place at the reduced shear strength and taking moments about O

$$\sigma_s B\left(\frac{B}{2}+x\right)+\sigma_0 x\frac{x}{2}-\sigma_0(B+x)\left(\frac{B+x}{2}\right)=\frac{c_u}{F}\pi(B+x)(B+x)$$

$$\sigma_s=\frac{\sigma_0(B^2+2Bx)+2\pi c_u(B^2+2Bx+x^2)/F}{(B^2+2Bx)}$$

$$\sigma_s=\sigma_0+\frac{2\pi c_u}{F}\left(1+\frac{x^2}{B^2+2Bx}\right)$$

Now the least optimistic slip circle is that which decrees that safe contact pressure σ_s is smallest. Since $x^2/(B^2+2Bx)$ is positive when x is positive, the smallest σ_s coincides with $x = 0$ to repeat the solution

$$\sigma_s=\sigma_0+\frac{2\pi c_u}{F} \tag{9.14}$$

for the simple semicircular mechanism considered before. Although moving the centre of the circle to the right increased the turning moment from the foundation it also increased both the magnitude and lever arm of the restraining shear force.

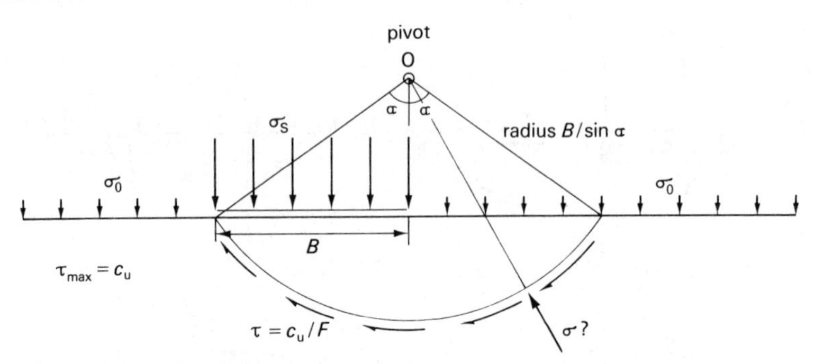

Figure 9.12 *Search of shallow slip circle collapse mechanisms*

Figure 9.12 accompanies the family of slip circles which have their centres above one side of the footing, using the semi-angle α as the parameter. Allowing the soil to fail at a factored strength c_u/F so that the contact pressure σ_s can be referred to as 'safe', it is possible once again to short-circuit the unknown direct stresses σ on the slip surface by taking moments about O.

$$(\sigma_s-\sigma_0)B\frac{B}{2}=\frac{c_u}{F}2\alpha\frac{B}{\sin\alpha}\frac{B}{\sin\alpha}$$

$$\sigma_s = \sigma_0 + \frac{4\alpha}{\sin^2\alpha}\frac{c_u}{F} \tag{9.15}$$

The worst slip circle is that which offers the minimum safe stress σ_s. Now by differentiation

$$\frac{\partial \sigma_s}{\partial \alpha} = \frac{4c_u}{F}\,\frac{\sin^2\alpha - \alpha\, 2\sin\alpha\cos\alpha}{\sin^4\alpha}$$

$$= \frac{4c_u}{F}\,\frac{1 - 2\alpha/\tan\alpha}{\sin^2\alpha}$$

This is zero when $\tan\alpha = 2\alpha$. It is not difficult to differentiate again in order to prove that this solution gives $\partial^2\sigma_s/\partial\alpha^2 > 0$ so that it represents a minimum value of σ_s and it is then easy to show that this is the lowest value of σ_s as α is allowed to vary between 0 and $-\pi$. By looking up tangent tables you will discover that $\tan\alpha = 2\alpha$ occurs in our range at $\alpha = 67^\circ$, which on substitution in equation 9.15 gives

$$\sigma_s = \sigma_0 + 5.5\,\frac{c_u}{F} \tag{9.16}$$

As expected, the successful search for a less optimistic mechanism has produced a slightly lower estimate of the safe foundation pressure than that first deduced in equation 9.14. Although we have not exhaustively searched every slip circle, and notwithstanding that the soil in practice may command some slightly noncircular slip surface, the search has been sufficient to demonstrate that the calculated safe pressure is not very sensitive to changes in the mechanism: this is encouraging.

9.5.2 The General slip Circle for Prismatic Soil Constructions: Cohesion Model

Figure 9.13 depicts the general slip circle which is used so frequently by analysts who must report on the stability of heaps of earth. Once more the analysis is performed for a section of unit length taken out of a long prismatic heap. The neighbouring sections (above and below the page) are considered to be identical, so that they can offer no support without contravening symmetry. Real constructions and real slip surfaces have ends, of course, and do derive some benefit from the extra friction developed at them. By considering only the soil strength which is mobilised on the cylindrical part of the slip surface, the analyst makes a small contribution to pessimism. This may or may not compensate for his optimism in not considering an infinite variety of failure mechanisms.

It is necessary only to have the overturning moments balanced by the restraining moments. The lever arm of the slipping mass W in figure 9.13 is x_W, while those of the general surface forces P and Q are x_P and y_Q respectively. Equating moments about the pivot O

$$Wx_W + Px_P - Qy_Q = \Sigma\, \tau l R \tag{9.17}$$

where the working shear stresses τ are summed over small lengths of arc l and

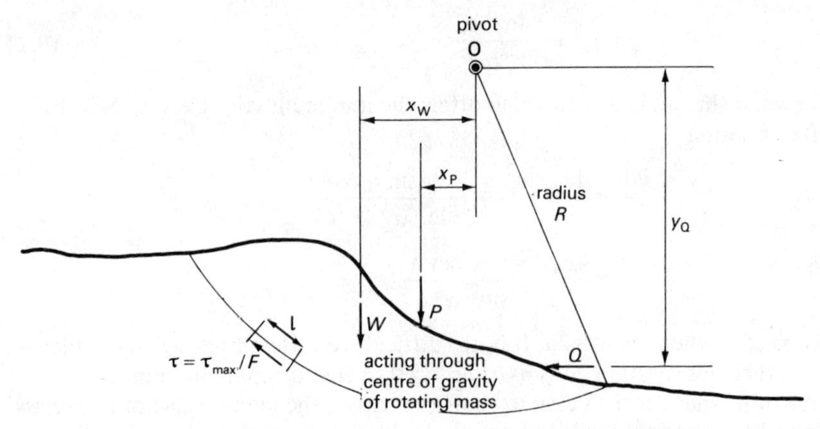

Figure 9.13 *A slip circle through a unit section of a long landform suffering gravity and the general line loads P and Q*

provided with their lever arm R. This is the basic slip circle equation. If the analyst can ascribe a mobilisable strength τ_{max} to every point and is prepared to assume that as failure approaches the soil elements on the slip plane all mobilise the same proportions of their strength (whatever it might be) at a given instant, then he can write

$$\tau = \frac{\tau_{max}}{F}$$

where a unique F applies to the whole mechanism. Equation 9.17 can then be rewritten

$$Wx_W + Px_P - Qy_Q = \sum \frac{\tau_{max} l R}{F} = \frac{R}{F} \Sigma\, \tau_{max} l \qquad (9.18)$$

and F can be calculated using

$$F = \frac{R\, \Sigma\, \tau_{max} l}{Wx_W + Px_P - Qy_Q} \qquad (9.19)$$

or ignoring surface forces P and Q

$$F = \frac{R\, \Sigma\, \tau_{max} l}{Wx_W} \qquad (9.20)$$

The most convenient way of calculating the moment Wx_W is to split the slipping mass into vertical parallel slices as shown in figure 9.14. It is permissible to sum the moments of each slice to obtain

$$Wx_W = \Sigma\, wx$$

and it is accurate enough to characterise each slice as a trapezium if there are sufficient (usually 10 or more). In that case the weight of a slice is

$$w = \frac{b(z_1 + z_2) \times 1 \times \gamma}{2}$$

where z_1, z_2 are the heights of the sides and γ is the average unit weight within. The centre of gravity of the slice, needed to fix x, can be estimated between the likely limits of $b/3$ from either side.

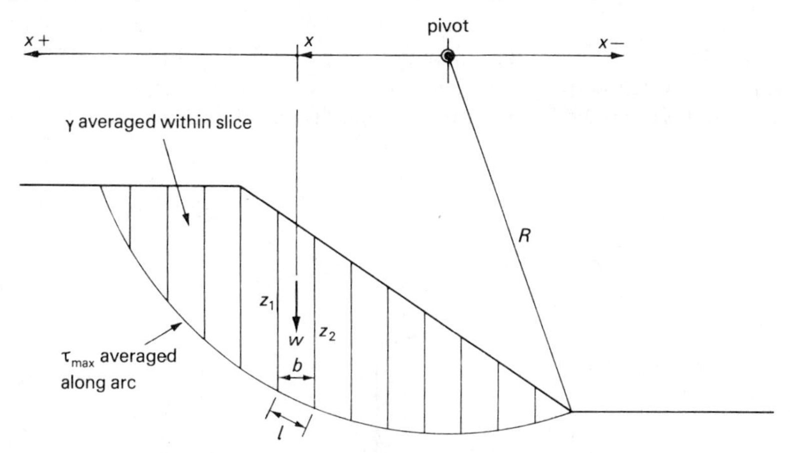

Figure 9.14 *Subdivision of rotating mass into vertical slices*

Now that the slip circle is split into lengths by the intersecting slices, the analyst may as well use these subdivisions for his summing of $\tau_{max} l$. It will be necessary to have a scatter of data of τ_{max} so that values close to the base of each slice can be selected and individually averaged. Equation 9.20 can then be rewritten

$$F = \frac{R \, \Sigma \, \tau_{max} l}{\Sigma \, wx} \tag{9.21}$$

If the cohesion criterion $\tau_{max} = c_u$ is to be used then

$$F = \frac{R \, \Sigma \, c_u l}{\Sigma \, wx} \tag{9.22}$$

provides a simple arithmetic formula which is easily evaluated using a pocket calculator. The most common mistake made by the unwary is to forget that x is algebraic; slices to the downhill side of the pivot have a negative lever arm x and their restraining moment must indeed be deducted from the string of positive moments wx of those slices that are tending to cause failure.

9.5.3 The General Slip Circle for Prismatic Soil Constructions: The Friction Model

Equation 9.17 can be treated equally well using an effective friction analysis. Ignoring surface forces P and Q as before, and replacing Wx_W by $\Sigma \, wx$ using the slices of figure 9.14, we obtain

$$\Sigma\, wx = R\, \Sigma\, \tau l \tag{9.23}$$

If the analyst wishes to use

$$\tau = \frac{\tau_{max}}{F} = \frac{\sigma' \tan \phi'}{F}$$

he is faced with the task of determining σ' under each slice. Bishop (1955) has provided an elegant approximate solution. Consider all the forces that can act on a single slice withdrawn from figure 9.14 and displayed in figure 9.15. Surface

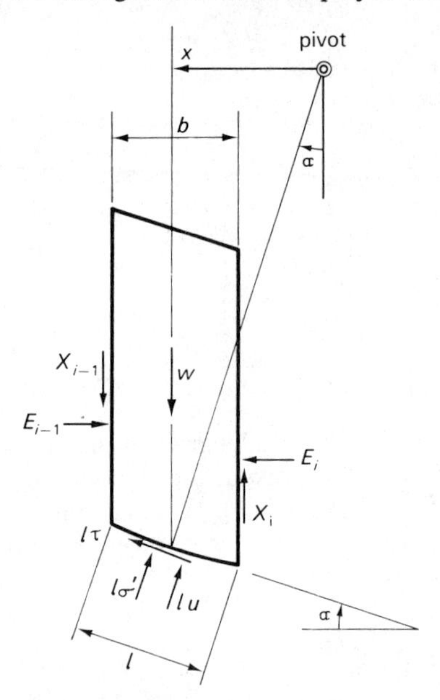

Figure 9.15 *Forces acting on the ith slice within a unit section*

forces have been excluded, but general forces on the three buried sides of the strip have been included. The convention has been that forces E act perpendicularly between strips and that shear forces X act downwards on the uphill side of each element. Since action and reaction are equal and opposite X must also act upwards on the downhill side of neighbouring elements. Strictly, I should have used l_i, τ_i, u_i, etc. to designate the values of these parameters on the base of the ith slice, but I have omitted the i for clarity. Only the side forces X_i, E_i, etc. have been so designated. The uphill inclination α of the base of the slice relative to the horizontal, which is equal to the uphill inclination of the generating radius to the vertical, fixes the geometry of the base of the slice which is of length l, as before.

The analyst has little chance of an exact solution with so many unknowns! He can calculate w, take steps to measure u in the field, and he can designate a safe value for τ/σ': but what of X_{i-1}, X_i, E_{i-1}, E_i? These side forces never arose in the cohesion analysis, and only arise here because the friction criterion demands an intimate knowledge of the stresses on the base. Bishop resolved all the forces vertically, so that the lateral thrusts E could remain unknown, and obtained for the slice

$$l\sigma' \cos\alpha + lu \cos\alpha + l\tau \sin\alpha = w + X_{i-1} - X_i \tag{9.24}$$

Now assuming conventionally that

$$\frac{\tau}{\sigma'} = \frac{\tan\phi'}{F}$$

σ' can be eliminated to give

$$\frac{l\tau F}{\tan\phi'} \cos\alpha + lu \cos\alpha + l\tau \sin\alpha = w + X_{i-1} - X_i$$

$$\tau = \frac{(w + X_{i-1} - X_i - lu \cos\alpha)\tan\phi'}{Fl \cos\alpha + l \sin\alpha \tan\phi'}$$

This can be substituted into equation 9.23 to obtain

$$\Sigma\, wx = R \sum \left[\frac{(w + X_{i-1} - X_i - lu \cos\alpha)\tan\phi'}{F \cos\alpha + \sin\alpha \tan\phi'} \right] \tag{9.25}$$

Now assume that the factor of safety F of each slice will be the same, or at least that its average value is of great interest to us. If we attempt to take it out of the summation we only obtain

$$\Sigma\, wx = \frac{R}{F} \sum \left[\frac{(w + X_{i-1} - X_i - lu \cos\alpha)\tan\phi'}{(\cos\alpha + (\sin\alpha \tan\phi')/F)} \right] \tag{9.26}$$

Let us tidy this up using $l \cos\alpha = b$ and $x/R = \sin\alpha$ to obtain

$$F = \frac{\sum \left[\dfrac{(w - bu + X_{i-1} - X_i)\tan\phi'}{(\cos\alpha + (\sin\alpha \tan\phi')/F)} \right]}{\Sigma\, w \sin\alpha} \tag{9.27}$$

At least an answer for F can now be obtained by iteration, putting a value on the right-hand side and getting a better answer on the left. Unfortunately, however, forces X are unknown and must be ignored so that the remaining formula is approximate.

$$F \approx \frac{\sum \left[\dfrac{(w - bu)\tan\phi'}{(\cos\alpha + (\sin\alpha \tan\phi')/F)} \right]}{\Sigma\, w \sin\alpha} \tag{9.28}$$

If many slices are taken each of width b, b can be eliminated. If at the same time the average height of slice $z = (z_1 + z_2)/2$ is used with unit weight γ to replace w/b, and if u is replaced by $\gamma_w h$, equation 9.28 becomes

$$F \approx \frac{\sum \left[\dfrac{(\gamma z - \gamma_w h)\tan\phi'}{\cos\alpha + (\sin\alpha \tan\phi')/F} \right]}{\Sigma\, \gamma z \sin\alpha} \tag{9.29}$$

Equation 9.29 can easily be shown to reduce to the 'infinite slope' equation 4.7, when the following constraints are applied: z, h, ϕ' equal for all strips, α constant for all strips (equivalent to the constant slope angle β). How serious was the omission of the side shear forces X? The term which has been missed from equation 9.28 is

$$\frac{\Sigma\left[\dfrac{(X_{i-1} - X_i)\tan\phi'}{(\cos\alpha + (\sin\alpha\tan\phi')/F)}\right]}{\Sigma\, w \sin\alpha}$$

Now of course α certainly varies from slice to slice and so the term $[\cos\alpha + (\sin\alpha\tan\phi')/F]$ cannot be withdrawn from the upper summation. But if it *could*, and if $\tan\phi'$ could also, we should have been left with

$$\Sigma\,(X_{i-1} - X_i) = (0 - X_1) + (X_1 - X_2) + (X_2 - X_3) + \cdots$$

which is zero. This may explain why the omission of the differences in X from equation 9.28 is not thought to cause more than a 2 per cent error, which is negligible.

Equation 9.28 is used widely in the estimation of the stability of soil constructions. It converges uniformly and very rapidly, so that by starting with a guess of $F = 1$ for each slice on the right-hand side, a sufficiently accurate new value is obtained after the summations have been carried out. A typical sequence of iterations might be: take $F_0 = 1$ to get $F_1 = 1.35$, take $F_1 = 1.35$ to get $F_2 = 1.39$, take $F_2 = 1.39$ to get $F_3 = 1.39$. If the factor starts rising it will keep rising, if it starts to fall it will keep falling.

Errors are often made by taking liberties with the method of summation. Remember that

$$\frac{\Sigma\left(\dfrac{a}{b+c}\right)}{\Sigma\, d} \equiv \frac{\dfrac{a_1}{b_1 + c_1} + \dfrac{a_2}{(b_2 + c_2)} + \cdots}{(d_1 + d_2 + \cdots)}$$

so that $\Sigma\,[a/(b+c)]\,/\Sigma\, d$ is *not* equal to $\Sigma(a/b + a/c)/\Sigma\, d$ or to $\Sigma\,[a/d(b+c)]$ or tc $\Sigma\, a/\Sigma(b+c)\,\Sigma\, d$.

9.5.4 Searching for the Least Favourable Slip Circle

You must check a wide diversity of slip circles before assuming that you have a sufficiently accurate answer. You will no doubt be guided that unfavourable circles are likely to pass through the weakest ground (high u, low c_u), have the shortest arc length, and the largest lever arm for their weight. Nevertheless, the safe design of an important earth dam may require the analysis of 100 slip circles. It is therefore vital for the designer to possess a computer program which can accept any ground profile, split it into slices, and compute the factor of safety of any required slip circle according to equations 9.22 or 9.28. Many such programs exist, and they are not very difficult to write. But how should the task of searching an infinity of possible slip circles be attempted?

Figure 9.16 shows the profile of a benched slope in which at least four families of circles need attention. Family A refers to a shallow slide emerging at the toe of the upper slope. Family B is concerned with a similar shallow slide in the lower slope. Family C investigates a composite shallow slide, whereas family D is concerned with a deep slip circle which emerges well beyond the lower toe rather after the fashion of a foundation failure. Circles in A and B have a clear advantage to the analyst who is looking for low safety factors, since they clearly have a very big lever arm to length ratio. If A and B separately might be guilty so might C. And D offers the landslide the chance to delve deep into the earth where the strength of the soil is likely to be iowest in proportion to the weight of overburden.

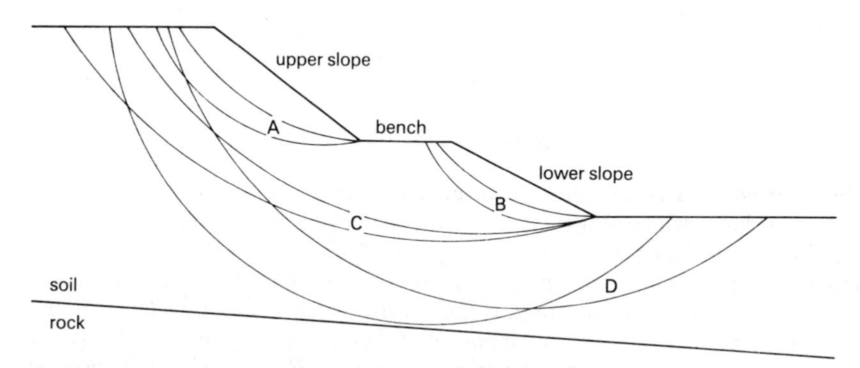

Figure 9.16 *Searching for the worst slip circle mechanism*

In each family there will be a number of members each with a different centre. It will be possible to write their factors of safety against each of their centres on a scale drawing in order that contours can be drawn through them to determine the location and value of their minimum.

Certain simple and useful problems have been 'solved' by repeated application of these numerical methods, and published in the form of charts. Taylor (1948) studied the simple slope sketched in figure 9.17 using equation 9.22. He first took account of the constancy of c_u and γ by rewriting equation 9.22 as

$$F = \frac{Rc_uL}{Wx_W} = \frac{Rc_uL}{\gamma Ax_W}$$

and again as

$$\frac{c_u}{F\gamma H} = \frac{Ax_W}{RLH} \tag{9.30}$$

The term Ax_W/RLH is a purely geometrical and dimensionless group of parameters. Every circle which could be drawn in figure 9.17 could possess a different magnitude of Ax_W/RLH and therefore a different magnitude of $c_u/F\gamma H$. Once Taylor had found, for a particular slope angle β and bedrock depth ratio Y/H, the circle with the ***greatest*** ratio of Ax_W/RLH he had found the least favourable circular slip mechanism which would indicate the *smallest*

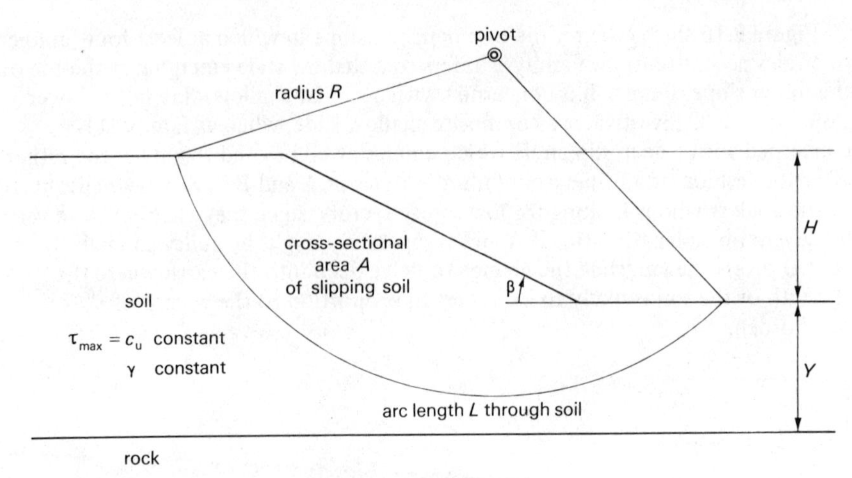

Figure 9.17 *Taylor's simple uniform slope problem*

factor of safety. Having thereby determined the critical magnitude of $c_u/F\gamma H$ from the slope geometry it is clear that the factor of safety F of a slope with that geometry is then proportional to $c_u/\gamma H$. Figure 9.18 is taken from Taylor's charts. The two most common uses for the chart are to determine safe short-term slope angles and factors of safety. If the site investigation has provided Y, γ and c_u and the engineer has a requirement for H then he can impose a certain safety factor, enter the chart at appropriate values of $c_u/F\gamma H$ and Y/H and deduce the safe slope angle. Alternatively he can find the existing factor of safety by entering at Y/H and travelling upwards until he intersects the present slope angle, reading off $c_u/F\gamma H$, from which F can be deduced. The accuracy of the technique depends mainly on the accuracy of the engineer's site investigation, by which he will judge whether the undrained strength of the soil is relevant, and what conservative average value it may be given.

Although similar presentations have been made for the same prototype slope based on the friction model and equation 9.28, the charts are not so useful. By the time the slope conditions, and in particular the groundwater pressures, have been sufficiently homogenised, the simpler 'infinite slope' model described in chapter 4 offers as likely a solution. If effective friction analyses based on slip circles are worth doing they are usually worth doing well, with water pressure and friction variations accounted for on the appropriate slope profile, and a rigorous search for the critical circle.

9.6 Rankine's Active and Passive Zones of Collapse

9.6.1 Elements in Limiting Equilibrium

The fundamental piece in Rankine's jigsaw puzzle is the element of soil in equilibrium, but on the verge of plastic collapse. Rankine's criterion for plastic collapse is therefore best symbolised by the drawing of the largest Mohr circle for a particular element, which will lie by definition in the principal plane of

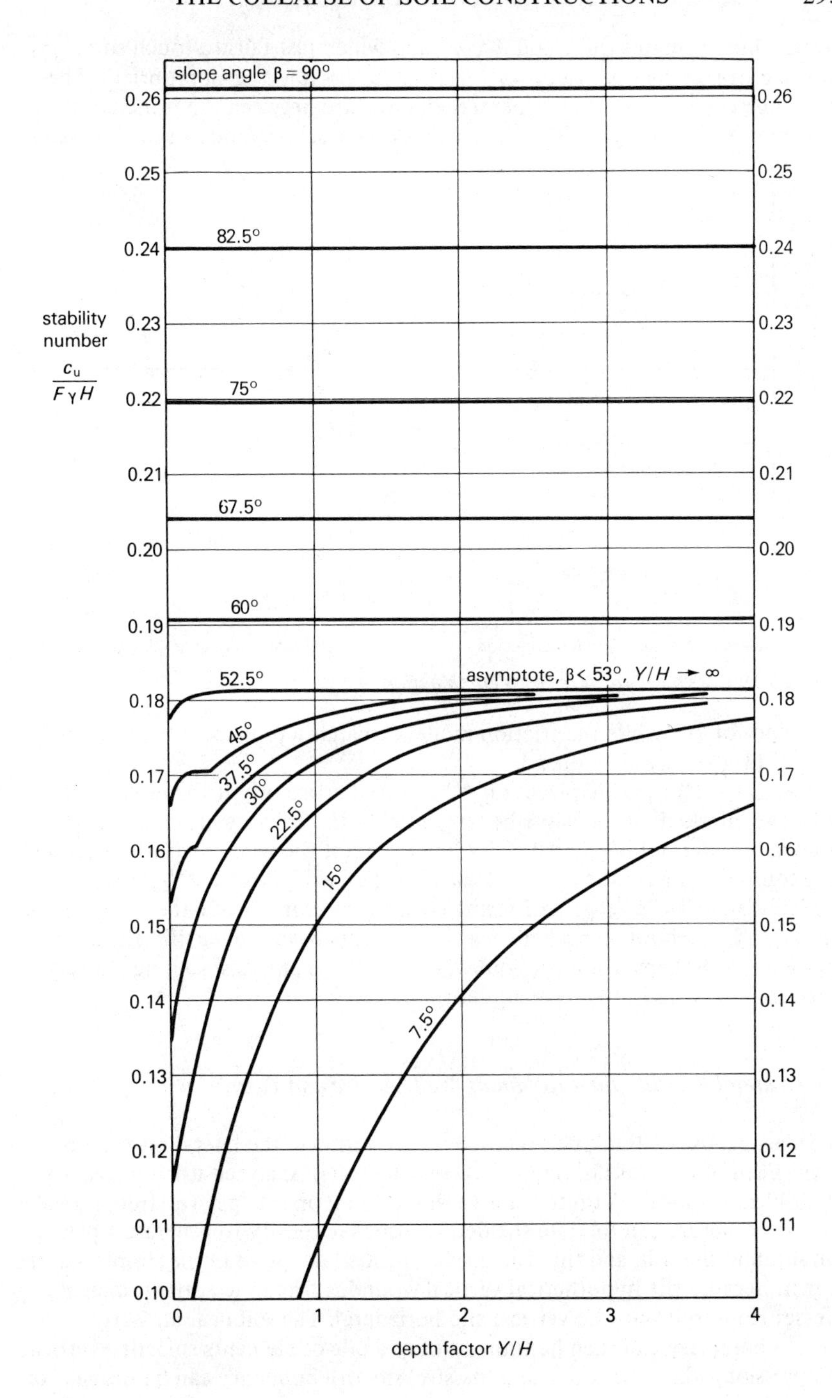

Figure 9.18 *Taylor's stability chart (taken from Taylor, 1948)*

stress which contains the 1 and 3 axes, and which just fails to touch some limiting stress envelope, $\tau_{max} = c_u$ or $(\tau/\sigma')_{max} = \tan \phi'$ as appropriate. These two independent rules then generate relationships between the principal stresses, as demonstrated in figure 9.19. The cohesion model demands a principal stress

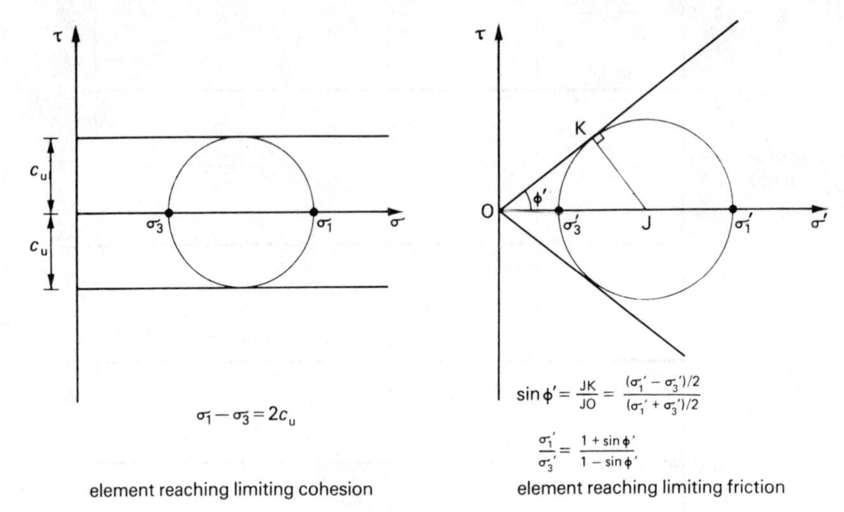

Figure 9.19 *Typical just-safe stress combinations*

difference of $2c_u$ while the friction model demands a principal effective stress ratio of $(1 + \sin \phi')/(1 - \sin \phi')$.

The jigsaw (for plastic pieces only!) is solved efficiently if the Mohr's circles of all the interlocking elements lie tangential to the collapse envelopes, if their boundaries have not been artificially 'oiled', and if the whole pattern of pieces is in equilibrium with the applied loads. The jigsaw is solved *safely*, if some of the Mohr's circles lie tangential to the collapse envelopes while the remainder lie within, or if the boundaries between the elements have been artificially weakened – perhaps made frictionless – to ease the analysis, so long as every piece and every boundary is in equilibrium.

9.6.2 Lateral Pressure on a Retaining Wall: Active and Passive

The classical use of Rankine's method was to estimate the lateral pressure of earth. Each of the walls in figure 9.20 may be *safely* analysed using imaginary, frictionless wounds AA on the 'active' side and PP on the 'passive' side, provided that the structure can tolerate the deformations necessary to generate a plastic condition in the soil, and that the displacements take place in the simple manner shown. Because the hypothetical vertical wound is frictionless, the principal stresses close to it must be vertical and horizontal. The soil near an 'active' smooth boundary can then be thought of as a pile of elements suffering vertical compression, while the soil near a 'passive' smooth boundary can be thought of as a similar stack of elements suffering horizontal compression. Since there are no side shear forces on the stacks, the vertical stress at any depth z in the stack

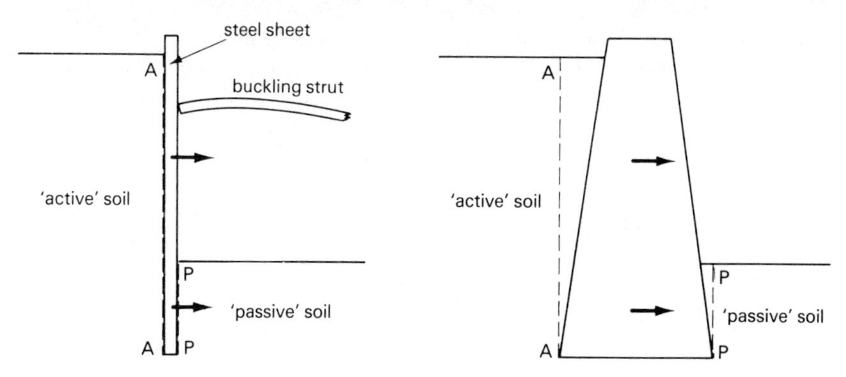

Figure 9.20 *Rankine's analysis of collapsing retaining walls by imposing frictionless surfaces AA, PP*

must be γz if gravity is the only action. Figure 9.21 must then represent the limiting equilibrium of cohesive level ground in the neighbourhood of a smooth vertical plane. If the horizontal stress approaches its 'active' value of $\gamma z - 2c_u$ the soil will be on the point of unlimited vertical compression. If the horizontal stress approaches its 'passive' value of $\gamma z + 2c_u$ the soil will be on the point of horizontal compression. At intermediate horizontal stresses the soil will be in nonlimiting equilibrium.

Likewise, figure 9.22 represents the limiting equilibrium of level ground to which the friction model is to be applied close to a smooth vertical wound. If the vertical plane is 'active' it must retreat away from the soil element and allow the horizontal stress to drop to σ_a. This can only be calculated if the water pressure is known, since the friction model demands a certain ratio between principal *effective* stresses, as shown in figure 9.19. The same is true if the passive pressure is to be calculated, when the smooth boundary moves into the soil element to generate a limiting lateral soil stress σ_p. Figure 9.22 demonstrates the appropriate routines, and defines the so-called active and passive earth pressure coefficients K_a' and K_p'. Although the only concept here is the ratio of principal effective stresses, it is conventional to define and use the two parameters notwithstanding that they are simply the inverse of each other. You should study figure 9.22 until some of the apparent complexity drops away to reveal the very simple skeleton.

As always, safety flows from pessimism: the designer should choose 'worst' values for c_u, ϕ', u and any other parameter that is likely to be variable.

9.6.3 Comparing Rankine with Coulomb for Simple Smooth Vertical Walls: Friction Model

Rankine's stresses should be pessimistic when used with rough walls. Real active pressures would be lower, and real passive pressures higher than the σ_a and σ_p estimates made above. But shouldn't the frictionless vertical wall (if one were to exist) be exactly solved? Consider a fill against such a wall which possesses constant γ and ϕ' and zero pore-water pressures ($u = h = 0$). Taking the active

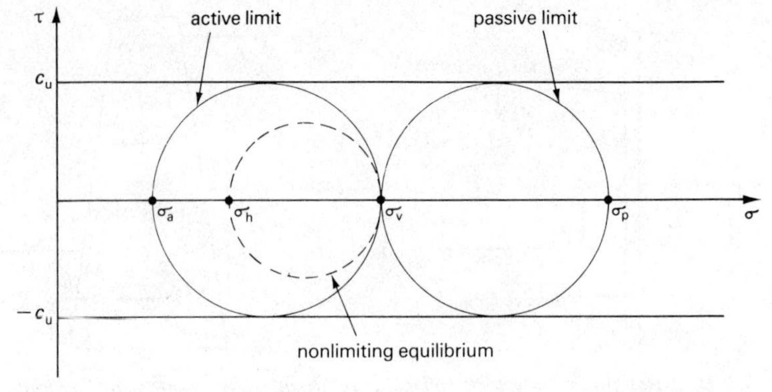

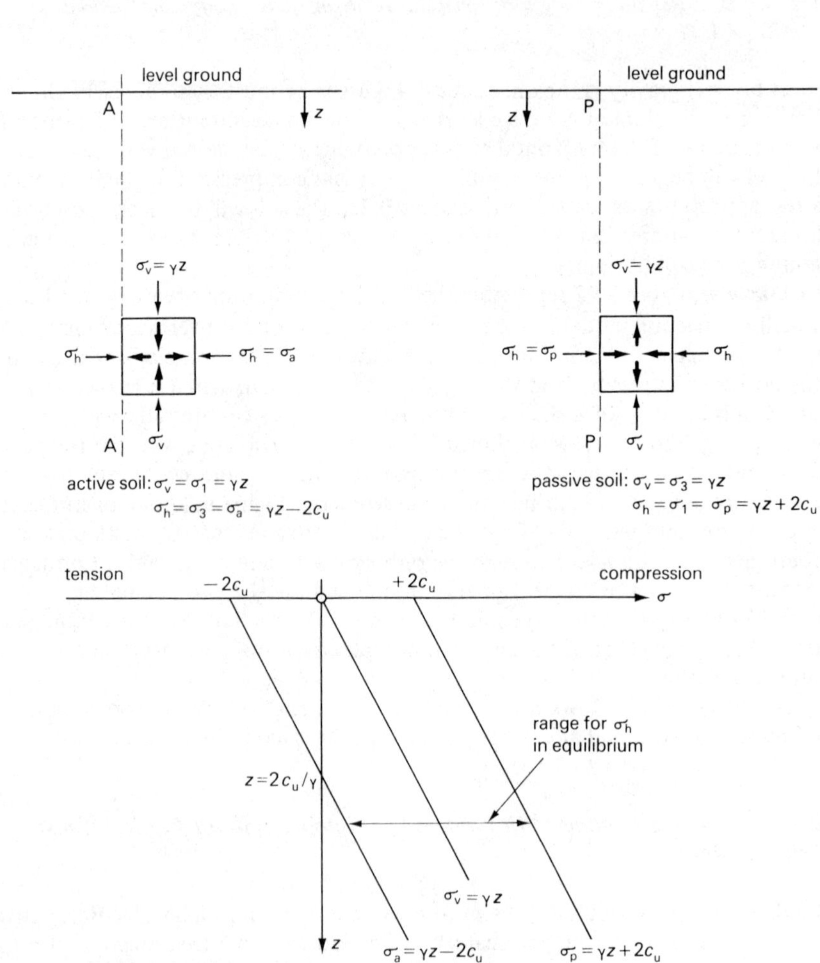

Figure 9.21 *Active and passive conditions in ideal cohesive ground*

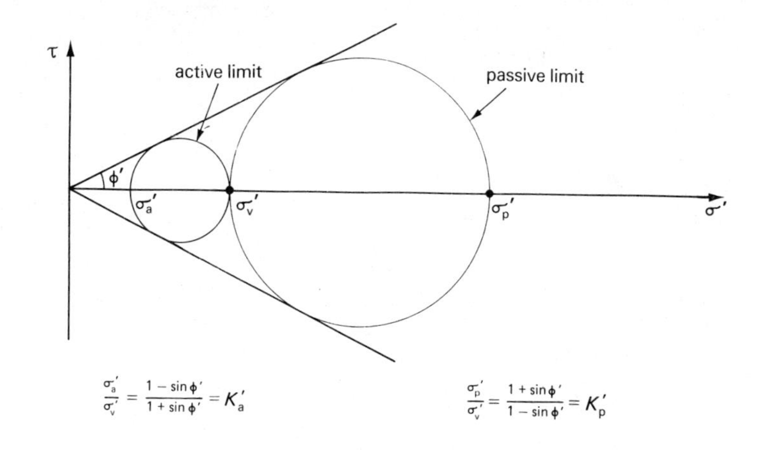

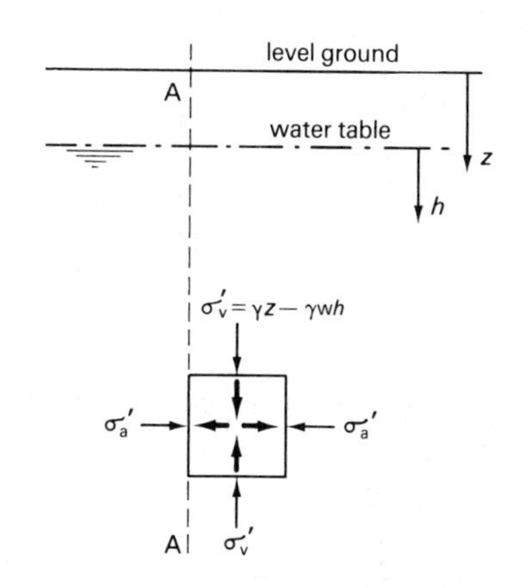

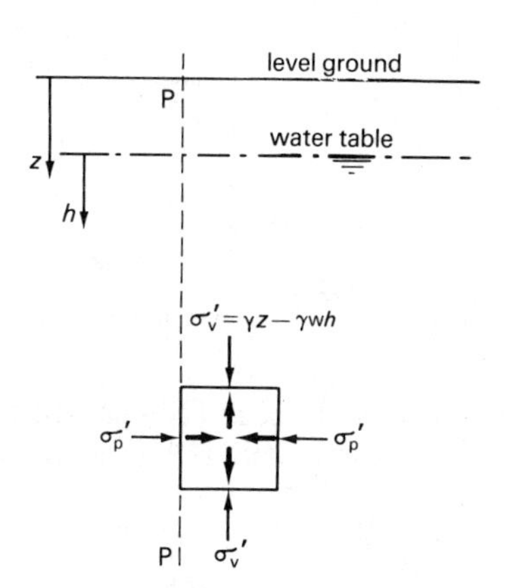

active soil: $\sigma_1 = \sigma_v = \gamma z$
$\sigma_1' = \sigma_v' = \gamma z - \gamma w h$
$\sigma_3' = \sigma_a' = K_a'(\gamma z - \gamma w h)$
$\sigma_3 = \sigma_a = K_a'(\gamma z - \gamma w h) + \gamma w h$

passive soil: $\sigma_3 = \sigma_v = \gamma z$
$\sigma_3' = \sigma_v' = \gamma z - \gamma w h$
$\sigma_1' = \sigma_p' = K_p'(\gamma z - \gamma w h)$
$\sigma_1 = \sigma_p = K_p'(\gamma z - \gamma w h) + \gamma w h$

Figure 9.22 *Active and passive conditions in ideal frictional ground*

pressures of figure 9.22

$$\sigma_a = \sigma_a' = K_a'(\gamma z - \gamma_w h) = K_a'\gamma z \tag{9.31}$$

while the passive pressures will be

$$\sigma_p = \sigma_p' = K_p'(\gamma z - \gamma_w h) = K_p'\gamma z \tag{9.32}$$

These lateral pressure distributions are shown in figure 9.23. The lateral thrust per unit length of wall is equal to the area of the pressure diagram ($dP = \sigma dA = \sigma \times 1 \times dz$) so that

$$P_a = \frac{1}{2}\sigma_a z = \frac{1}{2}K'_a \gamma z^2 \tag{9.33}$$

and

$$P_p = \frac{1}{2}\sigma_p z = \frac{1}{2}K'_p \gamma z^2 \tag{9.34}$$

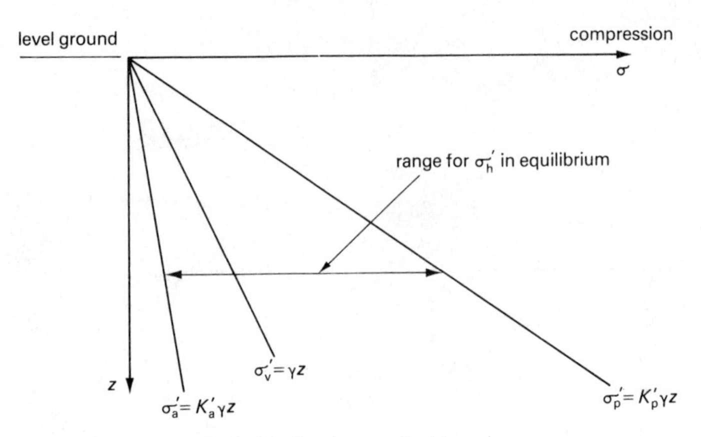

Figure 9.23 *Limiting stresses in frictional ground with zero pore pressures*

Now go back and compare these forces with those obtained from Coulomb's method in equations 9.8 and 9.9. Look closely at

$$\tan\left(45° - \frac{\phi'}{2}\right) = \frac{1 - \tan\dfrac{\phi'}{2}}{1 + \tan\dfrac{\phi'}{2}} = \frac{\cos\dfrac{\phi'}{2} - \sin\dfrac{\phi'}{2}}{\cos\dfrac{\phi'}{2} + \sin\dfrac{\phi'}{2}}$$

$$\tan^2\left(45° - \frac{\phi'}{2}\right) = \frac{\cos^2\dfrac{\phi'}{2} + \sin^2\dfrac{\phi'}{2} - 2\cos\dfrac{\phi'}{2}\sin\dfrac{\phi'}{2}}{\cos^2\dfrac{\phi'}{2} + \sin^2\dfrac{\phi'}{2} + 2\cos\dfrac{\phi'}{2}\sin\dfrac{\phi'}{2}}$$

$$\tan^2\left(45° - \frac{\phi'}{2}\right) = \frac{1 - \sin\phi'}{1 + \sin\phi'} = K'_a$$

Clearly equation 9.33 is identical to 9.8 while equation 9.34 is identical to 9.9! The internal logic of the methods combines to give the whole solutions which are depicted in figure 9.24. This correspondence lends credence to both techniques. Rankine's method is more useful in this very simple case since it makes clear that the line of action of the thrust, passing as it does through the centre of area of the triangular pressure diagram, is at one-third the height of the wall above its base. This offers $P_a z/3$ and $P_p z/3$ as the overturning moment and restraining moment respectively, of smooth vertical walls on the point of causing the neighbouring soil to collapse.

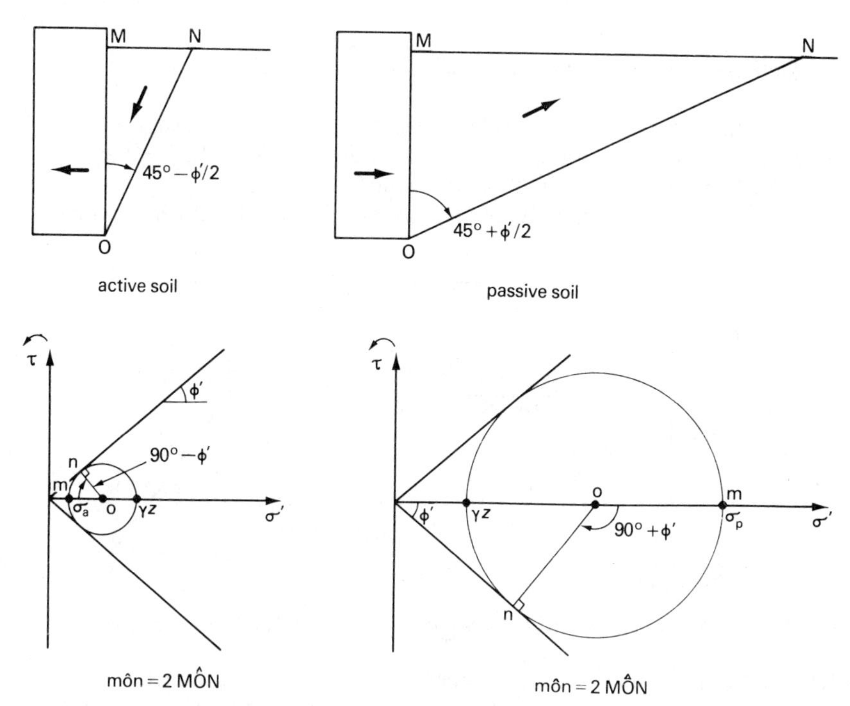

Figure 9.24 *Relationship between Rankine's and Coulomb's techniques for smooth walls retaining frictional soil*

Sections 4.5 and 8.6 demonstrated that the peak angle of friction of a dense frictional soil could be associated with dilatancy, which was shown in figures 4.10 and 8.33 as causing the vertical loading platen of a direct shear box to displace it an inclination ν to the imposed horizontal surface of gross sliding. You should now notice that neither of the derivations of Rankine's and Coulomb's simplified solutions to the lateral earth pressure problem dictated the direction of motion of the collapsing fill. It is, of course, accepted that the active Coulomb wedge is falling and the passive wedge is rising, just as it is accepted that the active lateral stress is as small as it may be while the passive stress is at its largest. However, no constraint has been placed on the *precise* nature of the ensuing strains. The wedges are therefore free to slide and dilate at an angle ν to the gross surface ON if they wish: the Mohr's circle of strain can similarly achieve any internally consistent configuration such as in figure 8.33. In no sense, therefore, does the existence of dilatancy invalidate these classical predictive techniques. The contribution of twentieth century soils engineers has been to clarify those situations in which the *peak* angle of shearing resistance of a soil fill may not be mobilised due to the lateral strain being either too small so that the interaction remains elastic or too large so that the dilatant peak is lost as the soil approaches a looser critical state. Dilatancy makes it more difficult to choose an appropriate value for ϕ'.

9.6.4 Rankine's General Method for Walls: Friction Model

While Coulomb's mechanisms provided straightforward graphical solutions to problems with peculiar geometries or line loads, Rankine's method of limiting stresses easily accommodates hydrostatic groundwater.

Consider the wall shown in figure 9.25; let us calculate the distribution of

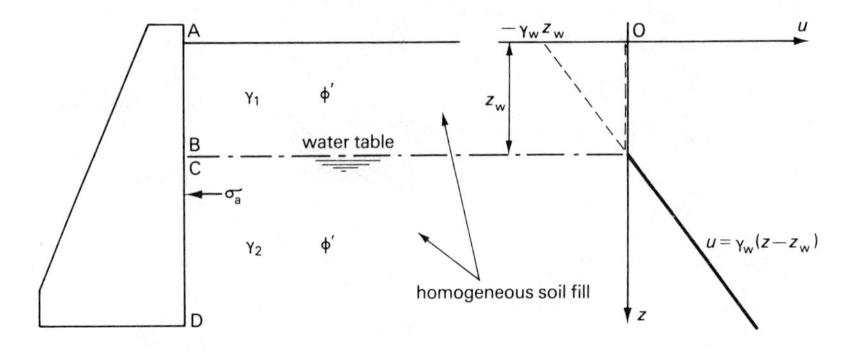

Figure 9.25 *Retention of partially submerged fill*

active pressure, rehearsing all the arguments as an example. Although the wall may not be very long we must isolate a section of unit length by imagining the insertion of two plane frictionless wounds 1 m apart. Although the wall cannot possibly be frictionless we imagine a further plane frictionless wound coinciding with the vertical back face of the wall. These pessimistic steps isolate a smooth rectangular block of soil in which active earth pressures must be calculated. The water table is at a depth z_w and in general the bulk density of the soil above the water table (γ_1) may be a little less than that beneath (γ_2) owing to partial saturation. As we saw in section 3.1, the typical dry and saturated bulk densities of a soil are 18 and 21 kN/m^3 respectively, so the difference between γ_1 and γ_2 is likely to be less than 15 per cent. In many cases it will be sensible to assume that both zones are saturated and possess identical unit weights γ. The major effect of the water is exerted by its pressure, not by its extra weight. Although it is clear that the hydrostatic pore-water pressure increases linearly below the water table, there is some uncertainty in the upper zone. Figure 3.4 indicated some of the complexity of the zone of soil above the water table, in which pore pressures are influenced by the pore size of soil controlling the height of capillary rise, and by the weather. Consider the extremes: if the soil concerned were gravel, the pore pressure above the water table would remain zero. If the soil were homogeneous clay, the pore pressure would probably drop on average to $-\gamma_w z_w$ at the surface, more in winter, and less in summer. If the analyst wants a pessimistic answer he should take the highest possible water pressures; zero above the water table, in this case.

The calculation of lateral pressure should be made at every depth at which the soil parameters change. Table 9.1 sets out the steps. In this case there was clearly no need to calculate the lateral pressures at both B and C, just above and below the water table. This is only necessary if the angle of friction changes across a bedding plane, when the different K'_a values cause the lateral pressure

TABLE 9.1

Location	A	B	C	D
Depth z	0	z_w	z_w	z
Vertical stress σ_v	0	$\gamma_1 z_w$	$\gamma_1 z_w$	$\gamma_1 z_w + \gamma_2(z - z_w)$
Water pressure u	0	0	0	$\gamma_w(z - z_w)$
σ'_v	0	$\gamma_1 z_w$	$\gamma_1 z_w$	$\gamma_1 z_w + (\gamma_2 - \gamma_w)(z - z_w)$
σ'_a	0	$K'_a \gamma_1 z_w$	$K'_a \gamma_1 z_w$	$K'_a[\gamma_1 z_w + (\gamma_2 - \gamma_w)(z - z_w)]$
σ_a	0	$K'_a \gamma_1 z_w$	$K'_a \gamma_1 z_w$	$\gamma_w(z - z_w) + K'_a[\gamma_1 z_w + (\gamma_2 - \gamma_w)(z - z_w)]$

to jump. The components of the active pressure diagram are depicted in figure 9.26. They can be integrated to give the total thrust per metre length; the lever

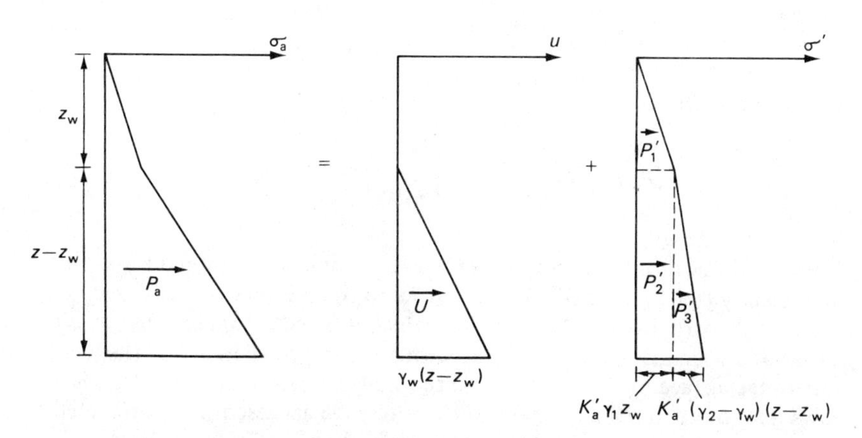

Figure 9.26 *Active stresses in partially submerged fill*

arms of each component of force are also demonstrated, being located at the centroids of their respective areas.

$$U = \frac{1}{2}\gamma_w(z - z_w)^2$$

$$P'_1 = \frac{1}{2}K'_a\gamma_1 z_w^2$$

$$P'_2 = K'_a\gamma_1 z_w(z - z_w)$$

$$P'_3 = \frac{1}{2}K'_a(\gamma_2 - \gamma_w)(z - z_w)^2$$

and

$$P_a = U + P'_1 + P'_2 + P'_3$$

So much for the general method, but what about the order of size? Take the special case of fill which is totally submerged, with $z_w = 0$ and $\gamma_1 = \gamma_2 = \gamma$.

$$P_a = \frac{1}{2}\gamma_w z^2 + \frac{1}{2}K'_a(\gamma - \gamma_w)z^2$$

$$P_a = \frac{1}{2}z^2[\gamma_w + K'_a(\gamma - \gamma_w)] \qquad (9.35)$$

Compare this with equation 9.33 for perfectly drained fill with zero water pressures. Take $\gamma = 20$ kN/m^2, $\gamma_w = 10$ kN/m^2 and $\phi' = 30°$ so that $K'_a = (1 - \frac{1}{2})/(1 + \frac{1}{2}) = 1/3$. With no water pressure in the fill

$$P_a = \frac{1}{2}\frac{1}{3}20z^2 = 3.33z^2$$

With the fill submerged

$$P_a = \frac{1}{2}z^2\left(10 + \frac{10}{3}\right) = 6.67z^2$$

The hypothetical rise of groundwater level has doubled the lateral force on the wall: clearly the drainage of water is crucial to the stability of walls.

The only form of superficial loading which it is easy to include in the analysis is that of a wide blanket of surcharge with a vertical pressure σ_0 on the whole horizontal surface of the earth behind the wall. Supposing that any transient flows due to the initial placement of the surcharge are essentially completed so that water pressures have become hydrostatic, the effect of σ_0 is simply to increase the total vertical stress σ_v at every horizon. The calculated horizontal stress at every depth is therefore increased by $K'_a\sigma_0$ where K'_a is the local active earth pressure coefficient. This is easily absorbed in a table such as 9.1. If the transient flow due to a uniform surcharge σ_0 on the earth has not *begun*, then it is easy to see that the increment in active pressure at every depth is simply σ_0, which is entirely due to the increase σ_0 in the pore-water pressure of the soil.

Any lateral nonuniformity of the conditions near the wall – hummocky ground, horizontal pore pressure gradients, non-homogeneity – must cause the analyst some concern since they may tilt the principal stresses away from the vertical and horizontal. A safe response is to analyse a 'shaft' through the 'worst' ground in the vicinity of the wall, protecting it from any possible support at its edges by the pessimistic provision of frictionless wounds.

If safe passive pressures σ_p were to be estimated, rather than active pressures σ_a, then all the foregoing would be altered only by the rigorous substitution of the subscripts p for a, and in particular K'_p for K'_a. Equation 9.35 would then be rewritten

$$P_p = \frac{1}{2} z^2 [\gamma_w + K'_p(\gamma - \gamma_w)] \tag{9.36}$$

to obtain the passive thrust of submerged soil. Using $\gamma = 20$ kN/m^2, $\gamma_w = 10$ kN/m^2, and $\phi' = 30°$, as before, $K'_p = 3$ so that, with no water pressure

$$P_p = \frac{1}{2} \times 3 \times 20z^2 = 30z^2$$

with the fill submerged

$$P_p = \frac{1}{2} z^2(10 + 30) = 20z^2$$

The hypothetical rise in groundwater level has reduced the maximum lateral restraint of the ground by a factor of 2/3.

9.6.5 Rankine's General Method for Walls: Cohesion Model

At first sight, the method described in figure 9.21 allows a simple calculation to be performed for the undrained active or passive pressure in soil at any depth z at which the undrained shear strength c_u has already been measured, namely

$$\sigma_a = \gamma z - 2c_u \tag{9.37}$$

and

$$\sigma_p = \gamma z + 2c_u \tag{9.38}$$

By integrating these stresses with depth, the lateral forces per unit length

$$P_a = \int_0^z \sigma_a \, dz = \frac{\gamma z^2}{2} - 2c_u z \tag{9.39}$$

and

$$P_p = \int_0^z \sigma_p \, dz = \frac{\gamma z^2}{2} + 2c_u z \tag{9.40}$$

are easily obtained and seen to be equal to Coulomb's estimates in equations 9.12 and 9.13. The internal logic points to a very satisfactory plastic solution to the collapse of soil against a frictionless vertical wall, using the cohesion model for the soil. The coincidence of Coulomb's and Rankine's estimates only leads to a useful solution if all the assumptions are valid, however. In this case the tensile zone ($\sigma_a < 0$) between the surface of the soil and depth $z = 2c_u/\gamma$ in the active state, marked in figure 9.21, is of some concern.

It is clear that uncemented soils can deliver 'tension' only by virtue of suction in the pore fluid. If the undrained shear strength of a firm to stiff clay were $c_u = 100$ kN/m^2, then at the surface of an active zone the total stresses would be $\sigma_v = 0$, $\sigma_h = -200$ kN/m^2, according to equation 9.37. This might be understandable if the pore pressure u were -300 kN/m^2, because the effective stresses would then be $\sigma'_v = 300$ kN/m^2, $\sigma'_h = 100$ kN/m^2 which would be an easily understood state of 'active' collapse based on effective stresses with $K'_a = 1/3$ and therefore $\phi' = 30°$. But can the negative pore pressure be relied on to provide overall tension like this? It is difficult to research this question, because the standard triaxial apparatus automatically provides that $\sigma_3 \geqslant 0$. In the absence of evidence the analyst must be pessimistic, and he normally chooses to supplement the shear stress limit $\tau \not> c_u$ with a direct tension limit $\sigma \not< 0$. This further limits the permissible Mohr circles, as shown in figure 9.27, and leads to the following 'active' stresses

$$\begin{aligned} \sigma_a &= 0 &&: 0 < z < 2c_u/\gamma \\ \sigma_a &= \gamma z - 2c_u &&: \quad z > 2c_u/\gamma \end{aligned} \tag{9.41}$$

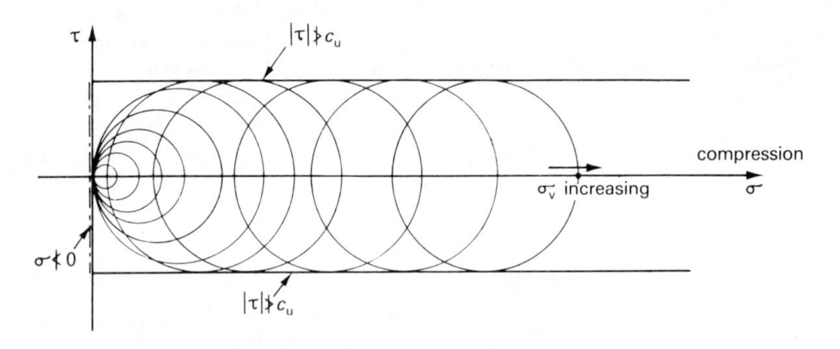

Figure 9.27 *'Limiting cohesion' with 'limiting tension' criteria for active stress conditions*

The tensile horizontal stresses, which would have tended to stop the wall falling outwards, have been ignored. In this way the analyst has taken into account the possible lack of suction between the wall and the clay, and the possible pre-existence of vertical fissures which could gape open rather than deliver tension. He may indicate the nature of his assumption by drawing vertical cracks on the cross-section of the soil up to a depth $2c_u/\gamma$. This leads to the pessimistic calculation of the lateral thrust, which is shown in figure 9.28.

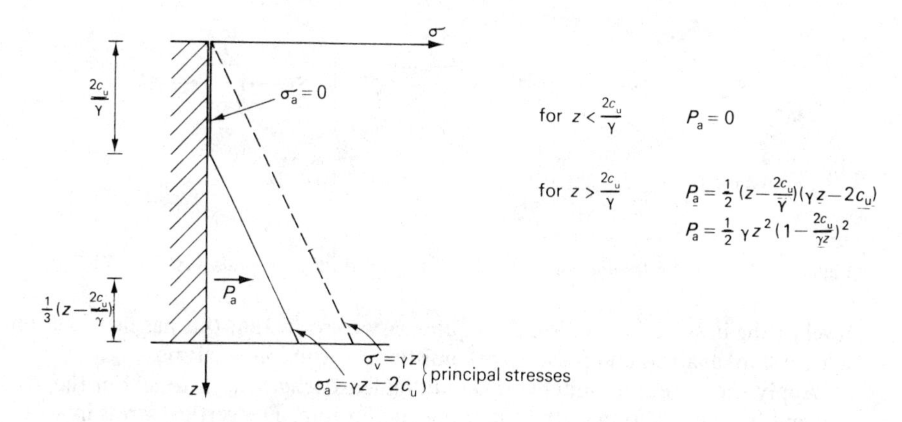

Figure 9.28 *Pessimistic lateral thrust of active cohesive soil*

The application of a blanket surcharge σ_0 merely increases all the stresses and reduces the depth of the tension zone, thus

$$\sigma_a = \sigma_0 + \gamma z - 2c_u \text{ for } z > (2c_u - \sigma_0)/\gamma \tag{9.42}$$

and

$$\sigma_p = \sigma_0 + \gamma z + 2c_u \tag{9.43}$$

9.6.6 Pessimistic Analysis of the Collapse of Strip Footings

The Coulomb-type analysis in sections 5.2 and 9.5.1 of the stability of a strip footing actually started with two pessimistic assumptions: it neglected the strength of the soil above the foundation bed, and considered a section 1 m long which received no support at its ends. At this point it is possible to adopt Rankine's technique and to continue being pessimistic. Consider the insertion of two deep, frictionless, vertical wounds VW and XY on either side of the footing which will also be made frictionless on its lower surface VX as shown in figure 9.29. As the foundation contact stress is increased up to its value σ_f at failure, the zone A will tend to compress vertically and bulge out sideways while zones P will therefore be compressed horizontally and will bulge upwards. In order to simplify the analysis let us ignore the effects of gravity on the soil beneath the

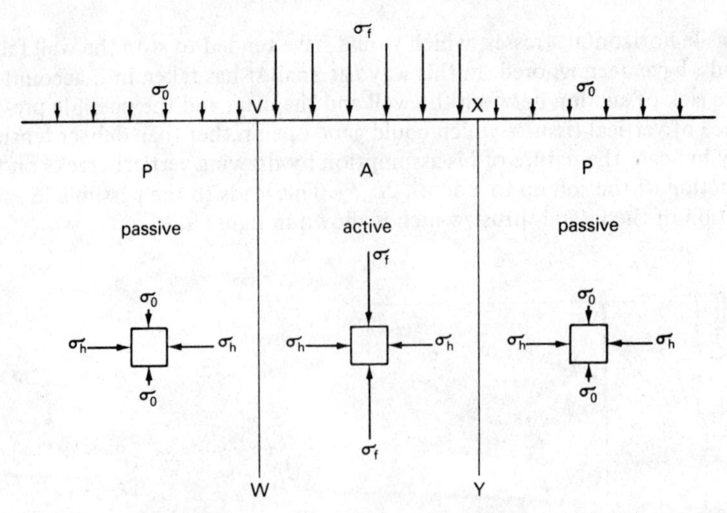

Figure 9.29 *Simple foundation stress field using frictionless wounds VX, VW, XY*

level of the base of foundation. It is quite easy to show that this has no effect on a 'cohesion' analysis and is safely pessimistic in a 'friction' analysis.

Apply the cohesion model with an undrained strength c_u throughout the soil and let us look for a state of limiting equilibrium. The vertical stress in block A is σ_f, while that in blocks P is σ_0. The horizontal stress across the wounds which separate the blocks is some unknown σ_h which must be equal in both blocks A and P. Since VW, VX and XY are frictionless, the stresses in these regions must adopt vertical and horizontal principal axes. If A is to compress vertically then $\sigma_f > \sigma_h$, while if P is to compress horizontally $\sigma_h > \sigma_f$. If all the stress differences are at their limit, then the Mohr circles of total stress for all points in zones A or P are those depicted in figure 9.30. The circles being of radius c_u it is clear that the pessimistic estimate is

$$\sigma_f = \sigma_0 + 4c_u \tag{9.44}$$

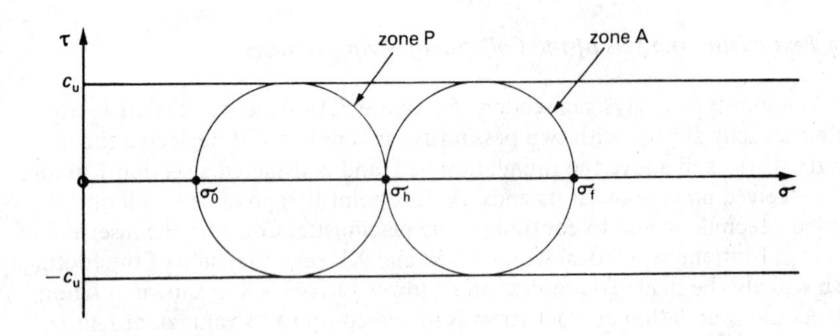

Figure 9.30 *States of stress using limiting cohesion within the wounded foundation subgrade*

If the conventional view of safety were adopted in addition to the pessimistic analysis then some factor of safety F could be used to generate a foundation stress σ_s which was 'safe' against collapse due to the weakness of the soil; thus

$$\sigma_s = \sigma_0 + \frac{4c_u}{F} \tag{9.45}$$

You will see that the pessimism leading to equation 9.45 has indeed offered the analyst a lower safe foundation stress than that of the optimistic, though careful, analysis which led to equation 9.16. Irrespective of the numerical value of equation 9.45, there is great value in the concept of a soil foundation bed comprising an active zone beneath the footing which can deform rather like a triaxial compression sample confined by soil rather than cell pressure, while these neighbouring passive zones deform rather like triaxial compression samples on their sides and are confined in their turn by the pressure of overburden σ_0 above the founding plane.

Figure 9.31 demonstrates a similar technique using the friction model, in

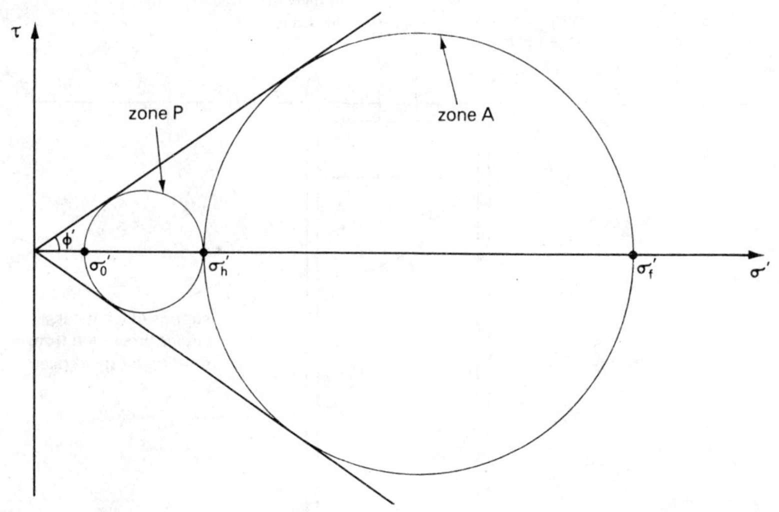

Figure 9.31 *States of stress using limiting friction within the wounded foundation subgrade*

which the two Mohr's circles representing zones A and P are just tangential to the envelope $(\tau/\sigma') = \tan \phi'$, and to each other. Once again the common point of touching represents the state of stress on the frictionless vertical wounds. Since $\sigma'_f/\sigma'_h = \sigma'_h/\sigma'_0 = K'_p$ it follows that the pessimistic estimate of the effective contact stress is

$$\sigma'_f = \sigma'_0 K'^2_p \tag{9.46}$$

which is the first estimate we have made of the foundation capacity of soil which obeys the friction model. If the analyst desires a 'safe' effective contact stress σ'_s then he should choose a safe angle of friction ϕ'_s for his estimation of a safe value of K'_p. If he stuck rigidly to convention then he would choose

$\phi'_s = \tan^{-1}(\tan \phi'/F)$, but he might otherwise choose $\phi'_s = \phi'_c$ or some other pessimistic angle of shearing resistance.

Let me choose an example to demonstrate how these pessimistic analyses of wounded soil foundation beds may be of value. Suppose that a foundation on clay was evidently in danger of collapse and that the responsible engineer was a slip-circle man. He would probably believe that the bearing capacity which was being approached was, putting $F = 1$ in equation 9.16, and taking the average undrained strength as c_u

$$\sigma_f = \sigma_0 + 5.5c_u$$

and that the critical mechanism was that of figure 9.12. He might then propose to drive steel sheets at the edges of the foundation in order to prevent the critical circles from sliding, as shown in figure 9.32. He would naturally specify new and lightly oiled sheet piles fresh from the store, so that their driving would not drag down the foundation. In so doing he damages the soil exactly after the fashion

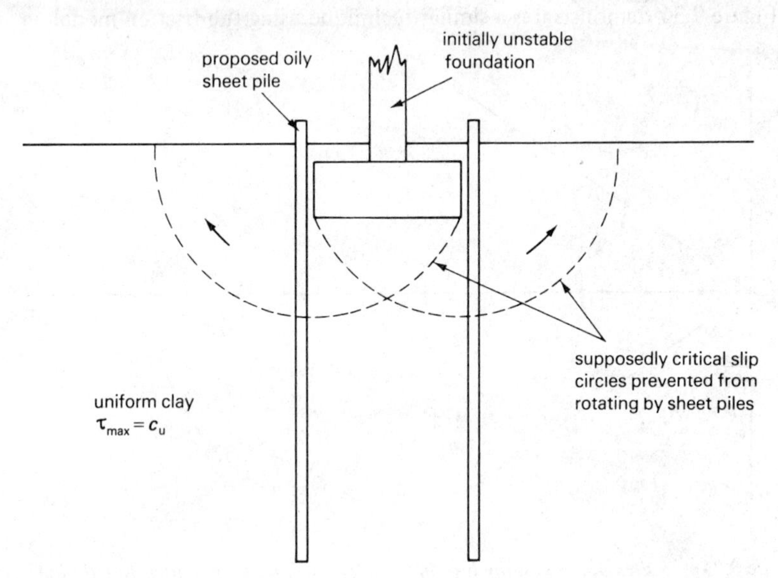

Figure 9.32 *The road to hell is paved with good intentions*

of figure 9.29 and guarantees that the lower bearing capacity of equation 9.44 would apply

$$\sigma_f = \sigma_0 + 4c_u$$

Rankine's active and passive zones could save him an embarrassment.

9.7 The Collapse of Soil Arches

It is possible to adopt Rankine's technique of allowing Mohr's circles to touch the envelopes of limiting strength, without adopting rectangular elements.

Consider the collapse of the crown of an underground cylindrical cavity – perhaps part of a tunnel heading, a plastic sewer pipe, a void in some earthwork which must be collapsed by surface compaction, or a plastic bottle deep in a city dump which is to be levelled and developed. Figure 9.33 demonstrates that

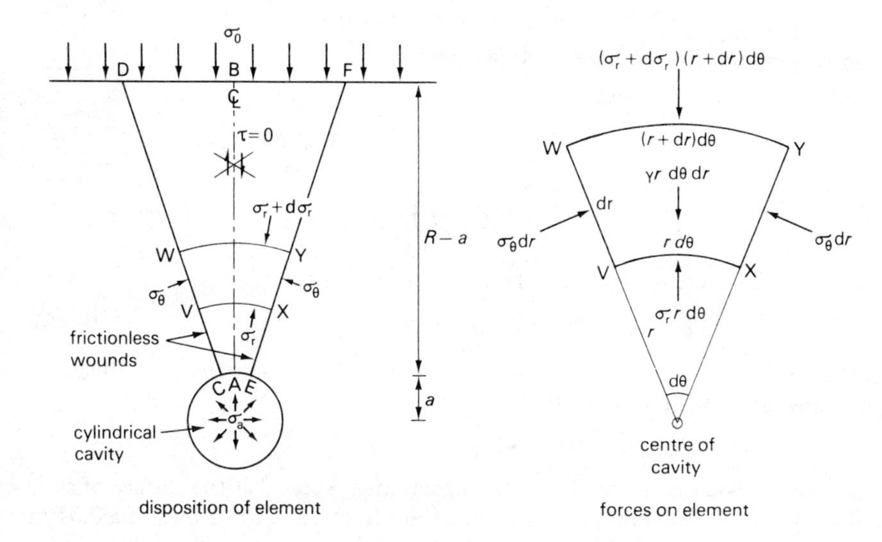

Figure 9.33 *Cross section through overburden above the crown of a cylindrical cavity*

if the soil conditions and the surcharge pressure σ_0 are uniform there can be no shear stress across the vertical plane AB, by the requirement of symmetry. Interposing frictionless radial wounds CD and EF on either side of the vertical is therefore unlikely to be outrageously pessimistic, if the included angle $d\theta$ is small. It is then possible to analyse pessimistically the collapse of the wedge CDFE into the crown of the cavity.

Since the radii CD and EF are frictionless, and both the surcharge σ_0 and the internal cavity pressure σ_a are perpendicular to the circumferences DF (which is indistinguishable from an infinitesimal arc) and CE respectively, it follows that the principal stresses in the wedge must be radial and circumferential. If the cavity is to collapse, then DC and FE must extend. It follows that the radial stress σ_r must be so small at every radius that the circumferential stress σ_θ is able to cause something akin to a passive collapse of each of the circular wedge elements represented by VW YX in figure 9.33. The stresses must be allowed to change, and they are depicted as σ_r and σ_θ in the figure, which goes on to calculate the forces on a wedge element by summing the stresses over the relevant length of their side. The weight of the element is included, being the unit weight of soil γ multiplied by the volume of the element. Resolving these forces vertically

$$(\sigma_r + d\sigma_r)(r + dr)\, d\theta - \sigma_r r\, d\theta - 2\sigma_\theta\, dr \frac{d\theta}{2} - \gamma r\, d\theta\, dr = 0$$

$$\sigma_r\, dr\, d\theta - \sigma_\theta\, dr\, d\theta + r\, d\sigma_r\, d\theta - \gamma r\, d\theta\, dr = 0$$

$$\frac{d\sigma_r}{dr} = \frac{(\sigma_\theta - \sigma_r)}{r} - \gamma \qquad (9.47)$$

Now if the cohesion model is applied with $\sigma_\theta > \sigma_r$ it follows that, in the limit

$$\sigma_\theta = \sigma_r + 2c_u$$

Substituting this into equation 9.47 we obtain

$$\frac{d\sigma_r}{dr} = \frac{2c_u}{r} - \gamma$$

$$\int_{\sigma_a}^{\sigma_0} d\sigma_r = \int_a^R \left(\frac{2c_u}{r} - \gamma\right) dr$$

where the limits are σ_a at $r = a$ and σ_0 at $r = R$. On integration

$$\sigma_0 - \sigma_a = 2c_u \ln R/a - \gamma(R - a) \qquad (9.48)$$

If the soil were a heavy liquid ($c_u = 0$) then equation 9.48 satisfyingly reduces to hydrostatics, with

$$\sigma_a = \sigma_0 + \gamma(R - a)$$

But consider what pressure of static compaction σ_0 applied at the surface of a stiff clay, $c_u = 100$ kN/m^2, would be necessary to collapse the crown of a 0.01 m radius cavity at 0.2 m depth, assuming atmospheric pressure in the cavity ($\sigma_a = 0$),

$$\sigma_0 = 2 \times 100 \ln 20 - 20 \times 0.99 \text{ kN/m}^2$$

$$\sigma_0 = 600 - 20 = 580 \text{ kN/m}^2$$

Perhaps it is no coincidence that this is the order of size of the tyre pressure of pneumatic-tyred compaction plant, used to compact loose clayey fill in layers up to 0.2 m thick. This result can be generalised by observing that no load could apply a pressure much greater than the $5.5c_u$ of equation 9.16 which generates a bearing failure for long strip footings on undrained clay. Ignoring the weight of the soil and equating the cavity pressure σ_a to zero in equation 9.48, it follows that the greatest possible surcharge pressure is

$$\sigma_0 \approx 5.5c_u \approx 2c_u \ln R/a$$

so that largest possible ratio of R/a at which collapse occurs is roughly 16.

Now consider what compressed air pressure σ_a would be required in a tunnel heading to prevent the collapse of an unlined 2 m diameter tunnel with a cover of 7 m of firm clay with $c_u = 50$ kN/m^2. Ignoring surcharge

$$\sigma_a = 20 \times 7 - 100 \ln 4$$

$$\sigma_a = 140 - 139 = 1 \text{ kN/m}^2$$

The tunnel is almost safe (temporarily – remember c_u must be undrained) without being lined or inflated. Clearly equation 9.48 has wide and interesting applications in 'cohesive' ground.

If, on the other hand, the friction model were used with zero pore-water pressures, the criterion of collapse for the element would be

$$\sigma'_\theta = K'_p \sigma'_r$$

or, since $\sigma'_\theta = \sigma_\theta$ and $\sigma'_r = \sigma_r$

$$\sigma_\theta = K'_p \sigma_r$$

Substituting this into equation 9.47, we obtain

$$\frac{d\sigma_r}{dr} = \frac{(K'_p - 1)\sigma_r}{r} - \gamma \tag{9.49}$$

This is most easily integrated by substituting $\psi = \sigma_r/r$ from which

$$d\sigma_r = r\,d\psi + \psi\,dr$$

so that by equation 9.49

$$r\frac{d\psi}{dr} + \psi = (K'_p - 1)\psi - \gamma$$

$$r\frac{d\psi}{dr} = (K'_p - 2)\psi - \gamma$$

$$\int_{\sigma_a/a}^{\sigma_o/R} \frac{d\psi}{(K'_p - 2)\psi - \gamma} = \int_a^R \frac{dr}{r}$$

$$\frac{1}{(K'_p - 2)} \ln \left[\frac{(K'_p - 2)\dfrac{\sigma_0}{R} - \gamma}{(K'_p - 2)\dfrac{\sigma_a}{a} - \gamma} \right] = \ln \frac{R}{a}$$

or

$$\frac{\dfrac{\sigma_0}{\gamma R}(K'_p - 2) - 1}{\dfrac{\sigma_a}{\gamma a}(K'_p - 2) - 1} = \left(\frac{R}{a}\right)^{K'_p - 2} \tag{9.50}$$

Consider what internal supporting stress $\sigma'_a = \sigma_a$ would be required in an otherwise unlined tunnel through dry sand with zero surcharge. From equation 9.50 with $\sigma_0 = 0$

$$\sigma_a = \frac{\gamma a}{(K'_p - 2)} \left[1 - \left(\frac{a}{R}\right)^{K'_p - 2} \right] \tag{9.51}$$

which is remarkable in that the effect of increasing depth R is swiftly reduced to zero so that

$$\sigma_a \to \frac{\gamma a}{(K'_p - 2)} \text{ as } \frac{a}{R} \to 0 \tag{9.52}$$

It is instructive to substitute some typical numbers for a large tunnel, $\gamma = 20$ kN/m^3, $a = 3$ m, $R = 12$ m and $\phi' = 42°$ so that $K'_p = 5.0$. From equation 9.51

$$\sigma_a = \frac{20 \times 3}{3}\left[1 - \left(\frac{1}{4}\right)^3\right]$$

so that the (a/R) term is readily seen to approach zero and the required supporting pressure in the tunnel is only 20 kN/m^2, notwithstanding that the average vertical stress at 9 m depth must be 180 kN/m^2. Clearly the stress has 'arched' around the cavity. The requirement for internal support to bolster the friction in the ground is surprisingly little. It should be clear, however, that whereas compressed air is able temporarily to support a tunnel through clay, it would be useless in dry sand since it would enhance total stresses and pore pressures equally and leave unchanged the effective stress of the soil in the roof. The effective supporting stress σ'_a in dry sand must therefore be provided by such structural components as sheets of corrugated steel carried over arched ribs. Rabbits, on the other hand, use both the capillary suction of damp earth and the tensile strength of plant roots to provide the effective stress required in the soil forming the roof of their burrows.

Now let us use equation 9.50 to calculate what surcharge pressure would be needed on the surface of a similar sand fill in order to collapse the crown of a cylindrical cavity 0.10 m in radius and 1 m deep which was so soft that it could only offer a token resistance of 1 kN/m^2 (perhaps a plastic bottle)

$$\frac{\dfrac{\sigma_0 \times 3}{20 \times 1} - 1}{\dfrac{1 \times 3}{20 \times 0.1} - 1} = 10^3$$

$$0.15\sigma_0 - 1 = 500$$

$$\sigma_0 = 3\ 300 \text{ kN/m}^2$$

It would require the equivalent of a column of steel 40 m high to crush the 0.2 m diameter cavity only 1 m below its base. The static compaction of sand requires enormous stresses if loose pockets are to be collapsed and densified.

Although I have not considered the stability of the whole construction, it seems reasonable to suppose that wedges other than at the crown might actually require less support than that calculated pessimistically for the crown. Atkinson and Potts (1977) present a more rigorous analysis of the whole cavity, and report data which fit our less sophisticated treatment quite well.

Arching zones cause great difficulties for research workers who want to measure soil stresses. If they introduce any sort of flexible diaphragm into a soil in an attempt to relate its deflection to the pressure acting on it, they may well create a soil arch around the 'soft' diaphragm which could cause a reduction of pressure upon it by a factor of 10 or 100. Consider, for example, the problem of a flexible plate or trapdoor underneath a bed of sand with $\phi' = 42°$ and $K'_p = 5$. Taking equation 9.51 to refer approximately to a long plate of width $2a$, the vertical pressure on the plate would be roughly $\gamma a/3$ whatever its depth. The equivalent depth $a/3$ of sand 'resting' on the plate could be a tiny fraction of the depth of overburden.

9.8 Limiting Stresses in an Infinite Slope

It is sometimes necessary to estimate the lateral thrust on walls which retain steep slopes, or the tangential stresses on floors which support heaps of granular material – coal, flour, sugar, iron ore, etc. In these cases an estimate of the state of stress on the appropriate vertical, horizontal or inclined surfaces may be obtained on the assumption that the presence of the structure does not influence the state of stress in the 'infinite' slope.

Figure 9.34 shows an infinite slope in which it is required to find the state of

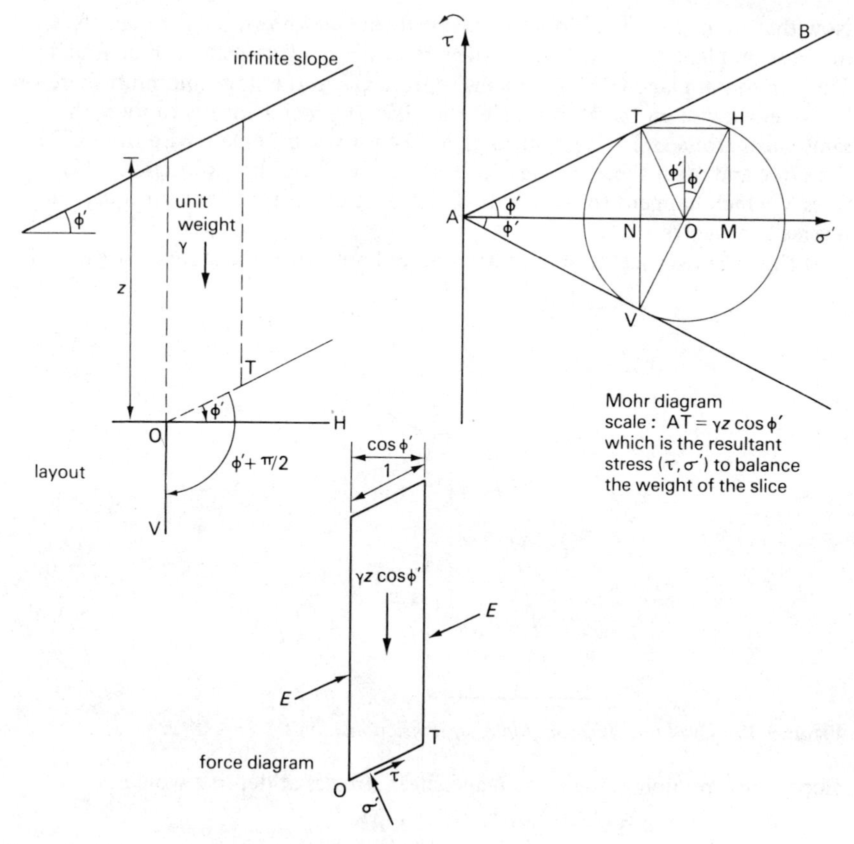

Figure 9.34 *Stresses within an infinite slope at its frictional limit*

stress on every possible plane passing through point O which is z deep. In order to make the analysis easy I have chosen to make the slope as steep as it can be, ϕ', where the friction model $\tau = \sigma' \tan \phi'$ is applicable and there are no pore-water pressures so $\sigma' = \sigma$. It is clear that the limiting inclination of stress, ϕ', occurs on plane OT which is tangential to the slope. It follows that point T representing plane OT can be placed on the friction line AB on the Mohr diagram, but at what distance AT from the origin? Consider a vertical slice of soil above the element of plane OT which will be considered to be 1 unit long. The width of the slice is therefore 1 cos ϕ' units so that the weight of the slice is $\gamma \times 1 \times 1 \cos \phi'$ per unit length. The force diagram displays all the forces on the element. The side forces E must be equal and opposite if the conditions on the slope are indistinguishable at points O and T. Since the base of the element OT is of unit area, the stresses τ and σ' upon it are numerically equal to forces. By resolving the weight parallel and perpendicular to the plane OT, therefore, we obtain

$$\tau = \gamma z \cos \phi' \sin \phi' = \text{NT on Mohr diagram}$$

$$\sigma' = \sigma = \gamma z \cos^2 \phi' \qquad = \text{AN on Mohr diagram}$$

Now that point T on the Mohr diagram is fixed and known to represent the state of stress on plane OT in the slope, the stress on any other plane can be found. The horizontal plane OH is ϕ' clockwise from OT in the slope and must therefore be $2\phi'$ clockwise on the Mohr circle: this fixes H which happens to share the same shear stress as T. The vertical plane OV is $(\phi' + \pi/2)$ clockwise from OT in the slope and must therefore be $(2\phi' + \pi)$ clockwise on the Mohr circle: this fixes V which happens to share with T the same direct stress and an equal but reversed shear stress.

If the rough vertical wall in figure 9.35 did not affect the state of stress in the

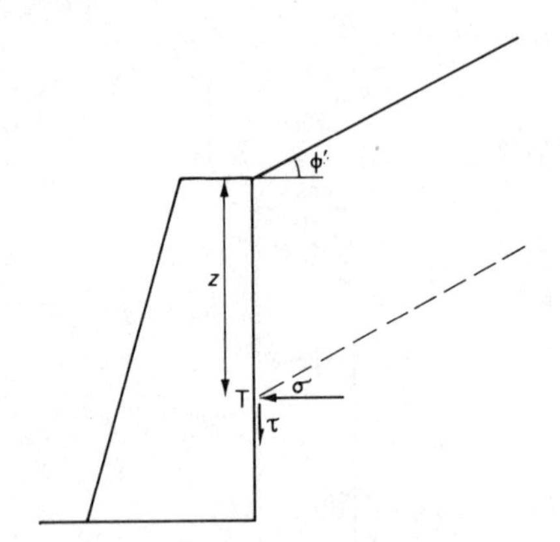

Figure 9.35 *Wall retaining soil sloping at its frictional limit*

slope it was retaining, the normal and shear stresses at depth z would be

$$\sigma = \sigma' = \gamma z \cos^2 \phi' \qquad = \text{AN}$$

$$\tau = \gamma z \cos \phi' \sin \phi' = \text{NV}$$

This is very much larger than the normal active stress $K'_a \gamma z$ for smooth walls retaining level earth. If, for example, $\phi' = 45°$ then $\cos^2 \phi' = 0.5$ while $K'_a = 0.17$. If the rough horizontal floor in figure 9.36 did not affect the state of stress in

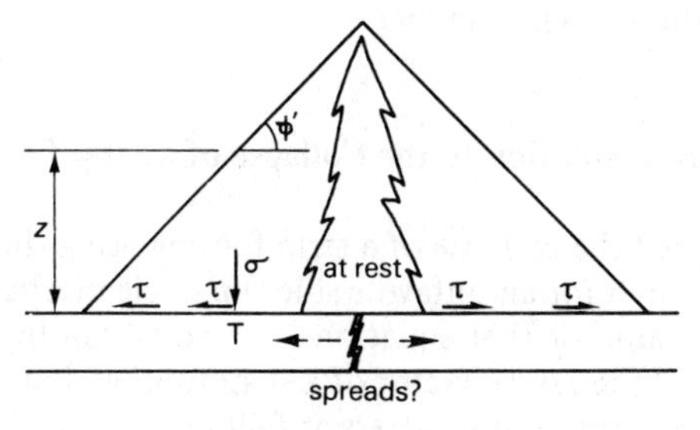

Figure 9.36 *Stresses beneath a steep earth bank*

the slope it was retaining, the normal and shear stress at depth z would be

$$\sigma = \sigma' = \gamma z(1 + \sin^2 \phi') = \text{AM}$$

$$\tau = \gamma z \cos \phi' \sin \phi' = \text{HM}$$

If $\phi' = 45°$ then $\sigma = 1.5\ \gamma z$ and $\tau = 0.5\ \gamma z$. The shear stresses caused by granular heaps have been known to cause tensile cracks in concrete floors, and earth dam failures by spreading on soft clay layers in the subgrade soils.

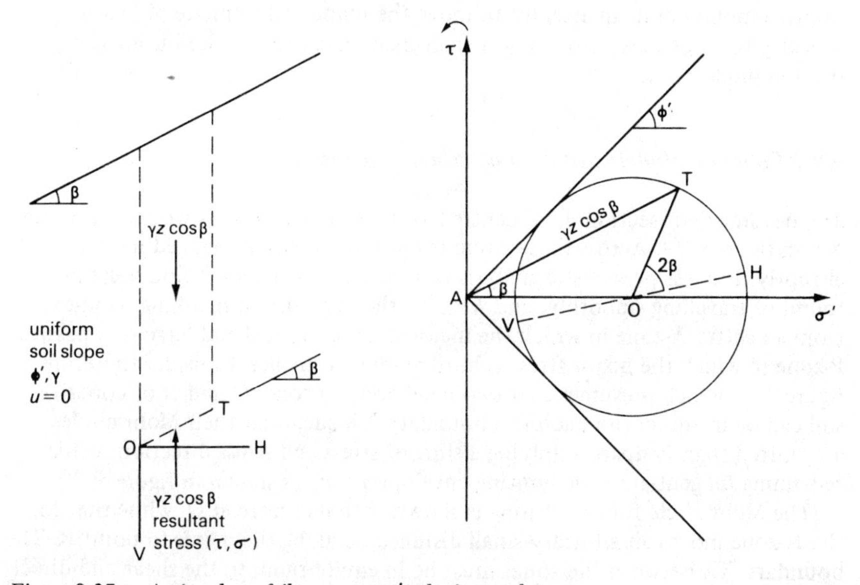

Figure 9.37 *Active slope failure at $\beta < \phi'$: the internal stresses*

Figure 9.37 demonstrates that a simple graphical technique will be successful in determining the active stresses within a long uniform slope with zero pore pressures collapsing at a slope angle $\beta < \phi'$ due perhaps to the failure of a retaining wall or of sliding upon some weak subgrade. Once again the resultant stress on a plane OT parallel to the slope is simply $\gamma z \cos \beta$ which can be set out as vector AT on the Mohr diagram. The limiting active circle can then be drawn tangential to the limiting envelope. The state of stress on any plane can then be deduced using OT as the known reference.

9.9 Approaching a Plastic Solution to the Collapse of a Strip Foundation

We have already explored the collapse of a strip footing using the cohesion model. Our previous search for an unfavourable slip circle mechanism in sections 5.2 and 9.5.1 led us to suppose that equation 9.16 would not be an over-optimistic assessment of the safety factor of a strip footing. Putting $F = 1$ this reduced to an estimate of the contact stress at failure

$$\sigma_f = \sigma_0 + 5.5c_u$$

The rather pessimistic assumption of deep frictionless wounds separating active and passive zones under a strip footing led to an estimate in equation 9.44

$$\sigma_f = \sigma_0 + 4c_u$$

which accompanied an interesting portrayal of the stresses to be carried at every point. In the present section I shall continue to use pessimistic techniques after the fashion of Rankine but shall attempt to reduce my analytical damage of the soil to a minimum in an attempt to move the numerical estimate of bearing capacity towards $5.5c_u$ from $4c_u$. I shall then repeat the method using the friction model.

9.9.1 Cohesion Model: Rotation of Principal Stresses

The pessimism of section 9.6.6 centred around the frictionless wounds VW and XY in figure 9.29. Across these wounds the state of stress changed most abruptly. If a less pessimistic analysis is to be performed, a method must be found of travelling smoothly, and without the necessity of invoking wounds, from an active A-zone in which the major stress is vertical and large to a passive P-zone in which the major stress is horizontal and smaller. Consider therefore figure 9.38 which presumes that two neighbouring zones Q and R of cohesive soil can be in equilibrium across a boundary XX such that their Mohr circles are shifted slightly apart, implying a shift of stress and stress direction, while remaining tangential to the limiting envelope $\tau = c_u$ as shown in figure 9.39.

The Mohr circle for the Q-zone is shown with its centre at Q, while that for the R-zone moves an arbitrary small distance $d\sigma$ along the σ-axis to point R. The boundary XX between the zones must be in equilibrium, so the shear and direct stresses σ_x and τ_x must be the same in each zone. This means that the common point X shared by the two Mohr circles must represent the boundary between the zones. Now, the direction of the major principal stress σ_{1Q} in the Q-zone is angle $\widehat{XQ\sigma}$ clockwise from X, whereas the direction of the major principal stress σ_{1R} in the R-zone is angle $\widehat{XR\sigma}$ clockwise from X. But X represents a fixed

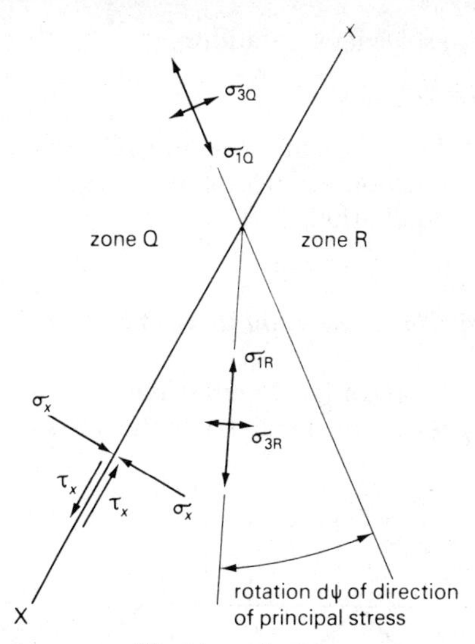

Figure 9.38 *Neighbouring zones of limiting cohesion*

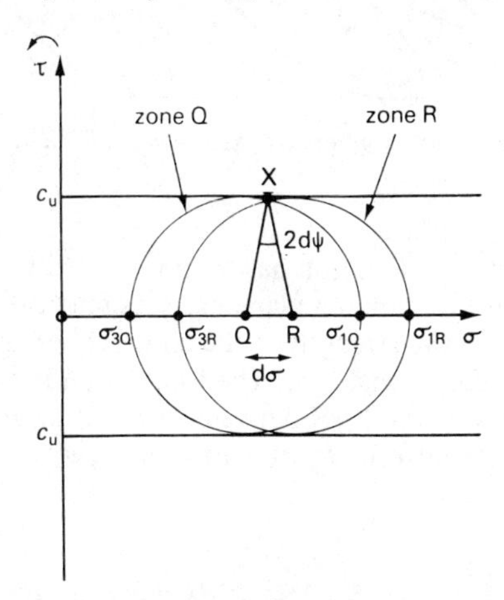

Figure 9.39 *Limiting stresses in equilibrium across boundary XX*

direction. So while the major principal stress was increasing by $d\sigma$, its direction was rotating clockwise by $X\hat{R}\sigma - X\hat{Q}\sigma = Q\hat{X}R$ on the Mohr diagram. If $Q\hat{X}R$ is a small angle it can be written as $d\sigma/c_u$. Directions on Mohr circles rotate twice as fast as their corresponding directions in the material. The rotation $d\psi$ of the principal stress direction in the plastic material must therefore be $Q\hat{X}R/2$, or

$$d\psi = \frac{d\sigma}{2c_u}$$

which can be written

$$d\sigma = 2c_u d\psi$$

and easily integrated to solve large rotations

$$\Delta\sigma = 2c_u\,\Delta\psi \tag{9.53}$$

This very simple result links the shift of Mohr circle very closely to the rotation of the principal stress direction in a plastic 'cohesive' material which is everywhere in limiting equilibrium.

9.9.2 Cohesion Model: Plastic Solution to Simple Strip Footings

Figure 9.40 depicts our conventional representation of a 1 m section taken out of a long strip footing resting on a material with constant strength $\tau_{max} = c_u$.

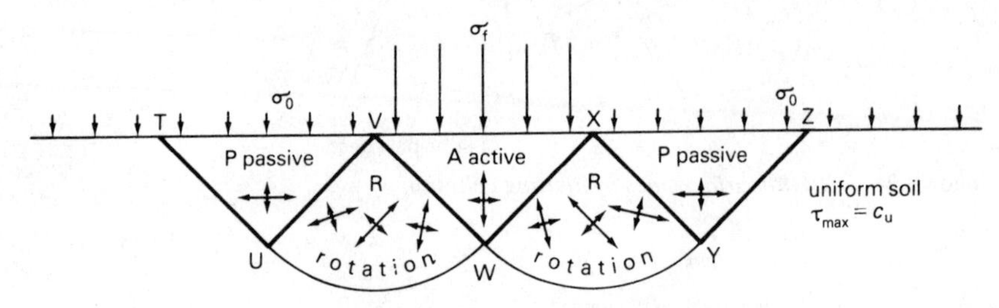

Figure 9.40 *Active and passive zones must exist beneath TVXZ: they may be separated by zones R as shown*

The contact stress on the foundation has been increased to σ_f at which collapse occurs. The soil above the founding plane has pessimistically been assumed to make no contribution to the strength of the construction and it has been replaced simply by its dead weight σ_0. The foundation has been 'damaged' by interposing a frictionless wound VX between the footing and the soil. The effect of gravity on the soil beneath the footing has been ignored.

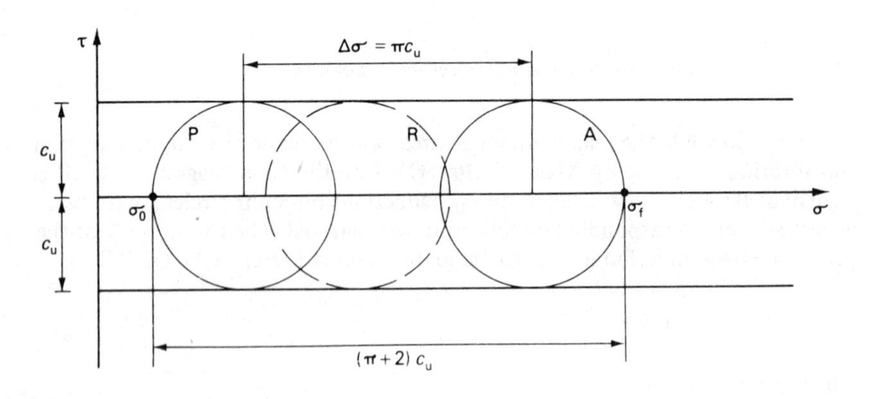

Figure 9.41 *Mohr circles for zones P, A and R (typical)*

Suppose that some active zone A exists under the footing, and that passive zones P exist on either side. The principal stress direction therefore rotates through an angle of $\pi/2$ from zone A in which it is vertical, to zone P in which it is horizontal. Using equation 9.53 therefore, $\Delta\sigma = 2c_u\Delta\psi = \pi c_u$ is the limiting shift in Mohr circles from A to P. But the limiting Mohr circle for zone P was already known. This enables the desired limiting Mohr circle for zone A to be fixed, as shown in figure 9.41. Also shown is one of the spectrum of circles between P and A which is applicable to the zone R of rotating principal stresses, marked as a segment on figure 9.40.

The new pessimistic estimate of strength is clearly

$$\sigma_f = \sigma_0 + (\pi + 2)c_u$$

or

$$\sigma_f = \sigma_0 + 5.14c_u \tag{9.54}$$

with a conventional factor of safety F this would be written

$$\sigma = \sigma_0 + 5.14c_u/F \tag{9.55}$$

which has almost reached the optimistic estimate of equation 9.16. Before moving on I should remark on the shape of the zones P, R and A in figure 9.40. Their shape did not enter into the argument above and might therefore be considered irrelevant. So would it be, were it not for the fact that a coincidental mechanism based on the triangular active and passive zones with the intervening shear fan offers an answer identical to equation 9.54. The mechanism is represented in figure 9.42 and is seen to be a combination of Coulomb's slip planes and slip circles. I have arbitrarily allowed the footing to slip to the right instead of the left.

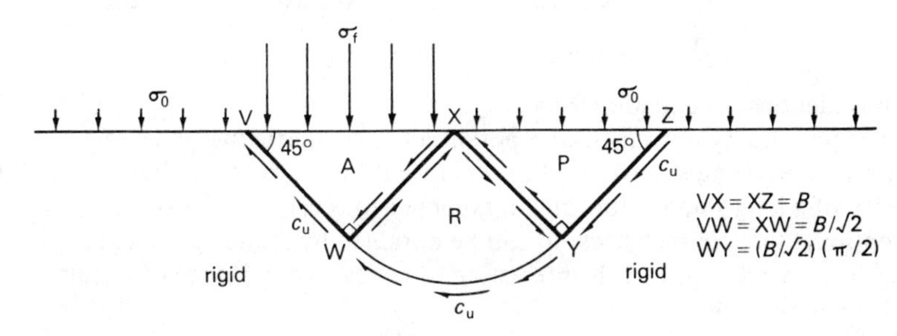

Figure 9.42 *Foundation failure mechanism in cohesive material*

The analysis of such a mechanism is only a little tedious. Figure 9.43 establishes estimates of the known forces on the boundaries of the zones. Zone A is being pushed down into the earth and will therefore be resisted by upward shear stresses c_u on its boundaries VW and XW. The normal stress σ_W on these boundaries is initially unknown, but it must be the same on each side otherwise there would be an unbalanced horizontal force. It is reasonable to infer that the stress distribution for σ_W is uniform, so that its integrated force is $\sigma_W B/\sqrt{2}$ acting at each of the mid-points of VW and WX. Resolving vertically for zone A

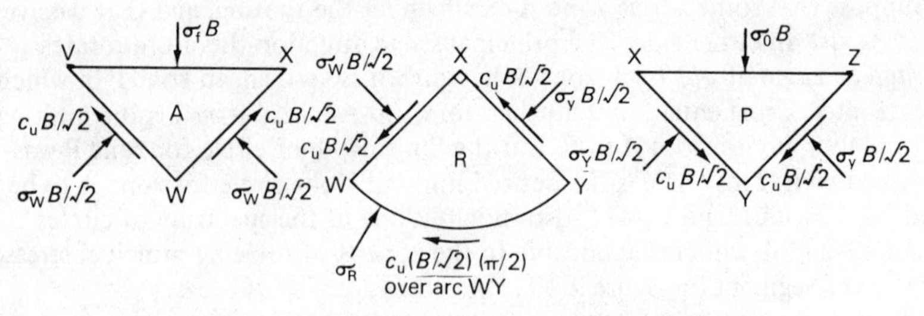

Figure 9.43 *Forces acting on elements of failure mechanism*

$$\sigma_f B = 2c_u \frac{B}{\sqrt{2}} \frac{1}{\sqrt{2}} + 2\sigma_W \frac{B}{\sqrt{2}} \frac{1}{\sqrt{2}}$$

$$\sigma_W = \sigma_f - c_u$$

Zone P, on the other hand, is being pushed into the air, so it will be restrained by downward shear stresses c_u acting on sides XY and YZ. An argument similar to that used for the A-zone can be used to deduce that the normal stresses σ_Y on XY and YZ are equal and that

$$\sigma_Y = \sigma_0 + c_u$$

It is finally possible to take moments about X for the sector WXY containing the R-zone. The normal forces on the straight sides are $\sigma_W B/\sqrt{2}$ acting at the mid-point of WX and $\sigma_Y B/\sqrt{2}$ acting at the mid point of XY. The tangential stress on the circular arc length $(B/\sqrt{2})(\pi/2)$ is c_u and has a lever arm of $B/\sqrt{2}$. Therefore

$$(\sigma_f - c_u) \frac{B}{\sqrt{2}} \frac{B}{2\sqrt{2}} - (\sigma_0 + c_u) \frac{B}{\sqrt{2}} \frac{B}{2\sqrt{2}} = c_u \frac{B}{\sqrt{2}} \frac{\pi}{2} \frac{B}{\sqrt{2}}$$

$$\sigma_f = \sigma_0 + (\pi + 2)c_u$$

which is identical to equation 9.54.

Since potentially optimistic and potentially pessimistic methods lead to a unique answer we must have reached a plastic solution to the undrained bearing capacity of a long smooth footing on homogeneous clay. As always, a conventional safe contact stress σ_s can be obtained by reducing the expected strength c_u by a factor F. It is interesting to review the sequence of estimates which we have made

optimistic : $\sigma_f = \sigma_0 + 6.3c_u$: semicircle

optimistic : $\sigma_f = \sigma_0 + 5.5c_u$: worst slip circle

optimistic, smooth : $\sigma_f = \sigma_0 + 5.1c_u$: wedges + sectors

pessimistic, smooth : $\sigma_f = \sigma_0 + 5.1c_u$: active, rotating stress, passive

pessimistic, smooth : $\sigma_f = \sigma_0 + 4c_u$: deep frictionless wounds

Although the method of accounting for the rotation of principal plastic stresses must have smacked a little of magic, perhaps you will be tempted by its relative swiftness and elegance to study plasticity in a little more depth.

Some engineers like to use the fact that foundations are likely to be rough rather than frictionless by allowing full adhesion to develop notionally between footing and soil. They go on to assume that self-equilibrating shear stresses c_u exist on either side of the footing, as shown in figure 9.44, through an assumed

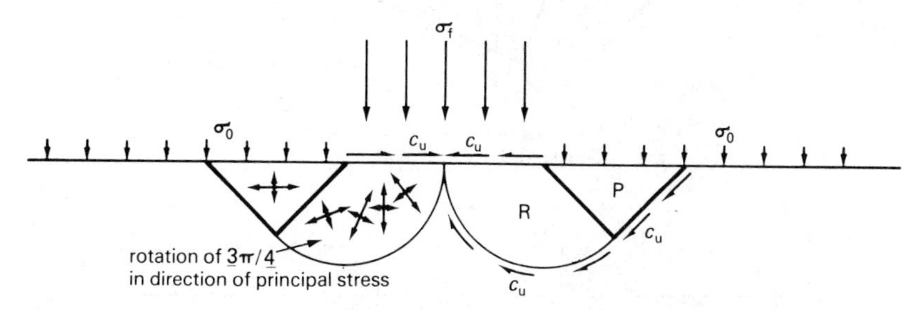

Figure 9.44 *Hypothetical collapse of rough footing at $\sigma_f = (1 + 3\pi/2)c_u$*

desire of zone A to expand sideways. This sequence of assumptions, together with a revised application of the stress zone method, leads to an estimate $\sigma_f = \sigma_0 + 5.7c_u$. This is suspect for the following reasons

(1) the mechanism of figure 9.42 does not demand that zone A expand sideways

(2) $5.7c_u$ exceeds the $5.5c_u$ derived from a slip circle, for which roughness or smoothness is irrelevant

A much more important correction is involved if the foundation is forced to carry an inclined load. Figure 9.45 is applicable to a foundation on cohesive ground which is forced to carry a shear stress τ_f in addition to a normal stress σ_f. The magnitude of τ_f marked on a Mohr circle fixes the angle ψ_0 by which the principal stress direction is forced to rotate in the active zone beneath the footing. From the Mohr circle

$$\sin 2\psi_0 = \frac{\tau_f}{c_u}$$

The overall rotation of principal stress in zone R is thereby reduced to $(\pi/2 - \psi_0)$, so that the separation between the extreme circles of stress is reduced to $\Delta\sigma = 2c_u(\pi/2 - \psi_0)$ or $c_u(\pi - \sin^{-1}\tau_f/c_u)$. It is then clear that

$$\sigma_f - \sigma_0 = c_u + c_u\left(\pi - \sin^{-1}\frac{\tau_f}{c_u}\right) + (c_u^2 - \tau_f^2)^{1/2} \tag{9.56}$$

The greatest possible shear stress is $\tau_f = c_u$ which generates $\psi_0 = 45°$ and a much simpler bearing formula

$$\sigma_f - \sigma_0 = c_u(1 + \pi/2) \tag{9.57}$$

which happens to have halved the shear strength component compared with equation 9.54. The capacity of a foundation on cohesive material to resist inclined loads is summarised in table 9.2, in which the relatively fast decline in potency as τ_f approaches c_u is made quite clear.

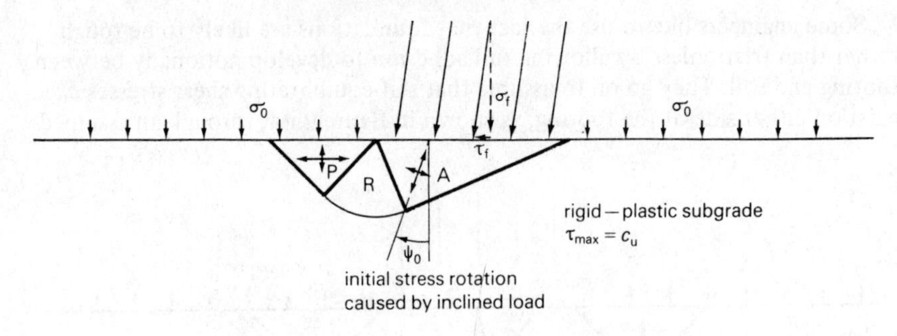

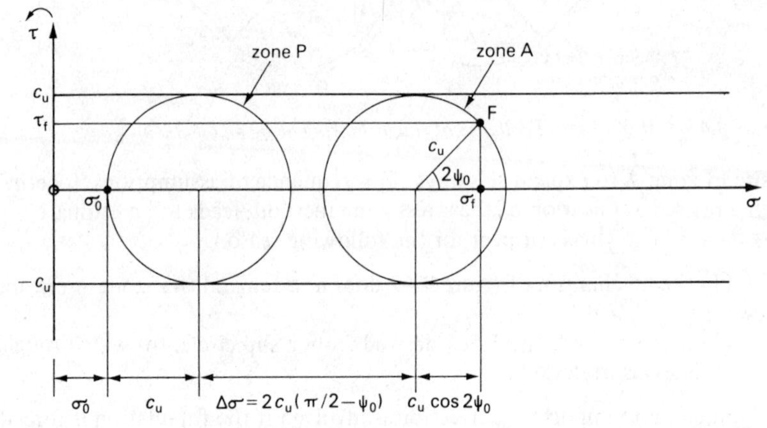

Figure 9.45 *Unit strip foundation over cohesive ground collapsing owing to inclined load*

TABLE 9.2
Inclined Cohesive Bearing Capacities

τ_f/c_u	0	0.2	0.4	0.6	0.8	1.0
$(\sigma_f - \sigma_0)/c_u$	5.14	4.92	4.65	4.30	3.81	2.57
$\tan^{-1}\left(\dfrac{\tau_f}{\sigma_f - \sigma_0}\right)$	0	2°	5°	8°	12°	21°

9.9.3 Friction Model: Rotation of Principal Stresses

An improvement in the estimate of bearing capacity was made possible above by considering the impact on stresses of a rotation in the principal stress direction. Let us repeat the treatment using the friction model, in order to improve on equation 9.46 which represented the bearing capacity of a smooth strip footing with two deep frictionless wounds separating simple active and passive zones.

Figure 9.46 depicts a stable boundary XX between two slightly different

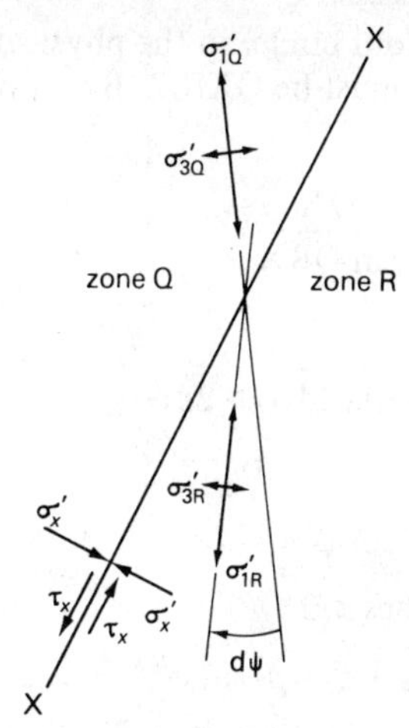

Figure 9.46 *Neighbouring zones of limiting friction*

zones Q and R. There is an infinitesimal rotation and shift of principal effective stresses across XX, but the boundary would not be stable unless the two zones shared a common boundary stress (σ'_x, τ_x) which must be a common point on their Mohr circles, as shown in figure 9.47. Q and R mark the centres of the two

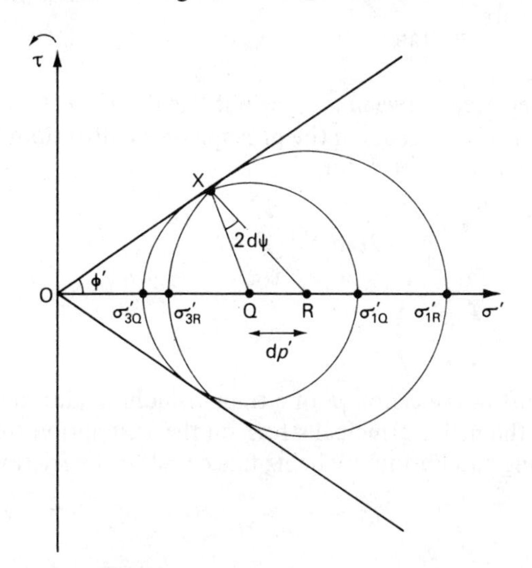

Figure 9.47 *Stresses in equilibrium across boundary XX*

Mohr circles, and the shift QR is the increase dp' in the mean effective stress p'. The principal stress direction in zone Q has a stress σ'_{1Q} which is angle $X\widehat{Q}\sigma'$ clockwise from X on the Mohr circle for Q. The principal stress direction in zone R has a stress σ'_{1R} which is angle $X\widehat{R}\sigma'$ clockwise from X on the Mohr circle for R. But X is a fixed direction. The principal stress direction must itself have rotated through an angle $X\widehat{R}\sigma' - X\widehat{Q}\sigma' = Q\widehat{X}R$ clockwise on the Mohr circle.

Angles are doubled on the Mohr circle, so the physical angle of rotation $d\psi$ of the principal stress direction must be $\widehat{QXR}/2$. But applying the sine rule to triangle QXR

$$\frac{QR}{\sin Q\hat{X}R} = \frac{QX}{\sin Q\hat{R}X}$$

But if $Q\hat{X}R$ is small

$$\sin Q\hat{X}R = \sin(2d\psi) = 2d\psi$$

and since OXR approaches $\pi/2$

$$\sin Q\hat{R}X = \cos \phi'$$

and since $O\hat{X}Q$ also approaches $\pi/2$

$$QX = OQ \sin \phi' = p' \sin \phi'$$

Making these substitutions

$$\frac{dp'}{2d\psi} = \frac{p' \sin \phi'}{\cos \phi'}$$

so

$$\frac{dp'}{p'} = 2\tan \phi' \, d\psi \tag{9.58}$$

This is easily integrated between two regions S and T between which a large rotation $\Delta\psi = \psi_T - \psi_S$ occurs in the principal stress direction, all the intervening states being tangential to the ϕ'-line.

$$\int_{p'_S}^{p'_T} \frac{dp'}{p'} = 2\tan \phi' \int_{\psi_S}^{\psi_T} d\psi$$

$$\ln\left(\frac{p'_T}{p'_S}\right) = 2\tan \phi' \, \Delta\psi \tag{9.59}$$

This links the shift in the centre p' of a train of Mohr circles to the rotation of the direction of the major principal stress, on the assumption that each Mohr circle is in limiting equilibrium by being tangential to the friction line $\tau = \sigma' \tan \phi'$.

9.9.4 Friction Model: Plastic Solution to Simple Strip Footings

Figure 9.48 depicts our conventional representation of a 1 m length of strip footing which is supposed to receive no benefit from adjoining sections. It might therefore be taken out of an endless footing, or pessimistically out of a real footing which has received two parallel frictionless wounds 1 m apart along its length. The effective stress between footing and soil has been increased to σ'_f, at which point the effective overburden pressure σ'_0 is pushed into the air and the footing collapses. Not only has the strength of the overburden been ignored, but also the stabilising effect of gravity on the founding layer has been ignored. The foundation has been slightly damaged for analytical purposes by passing a frictionless sheet VX between the footing and the soil. Some simple active zone A

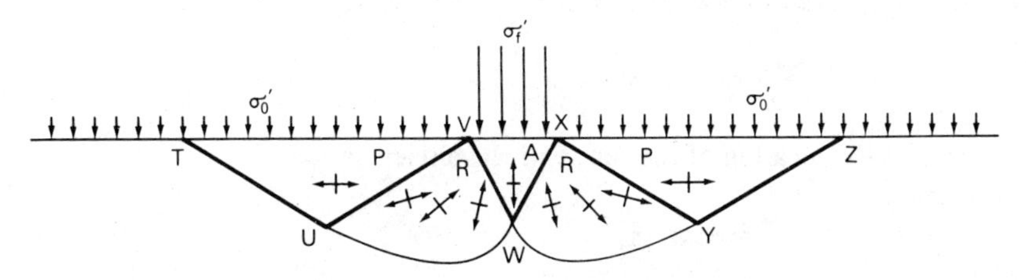

Figure 9.48 *Principal effective stress directions in active (A), passive (P) and rotation (R) zones if frictional soil beneath a collapsing strip footing*

must exist under the smooth footing, together with passive zones P under the overburden. The principal stress direction rotates through an angle $\pi/2$, from vertical in A to horizontal in P. The Mohr diagram for effective stresses can then be drawn, starting with P, then using the shift in centre over region R predicted by equation 9.59, and deducing the circle for zone A: it is shown in figure 9.49.

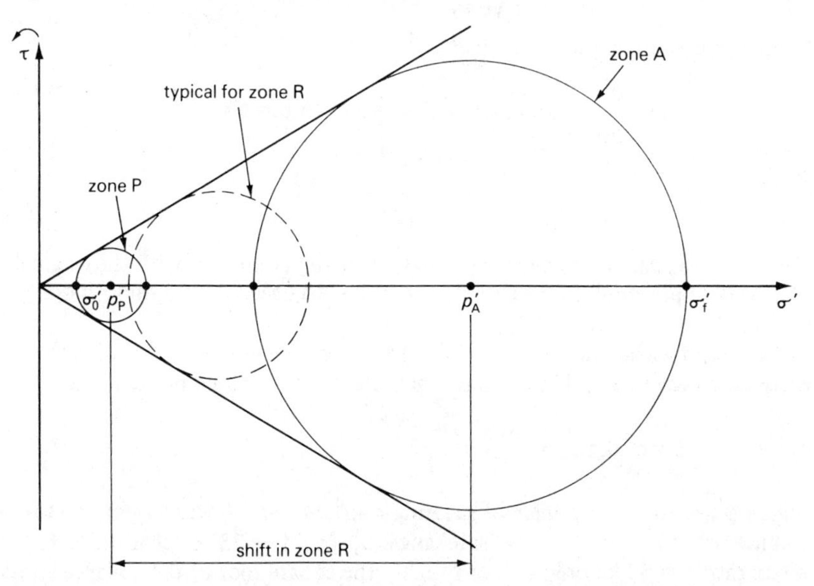

Figure 9.49 *Mohr circles of effective stress in frictional soil beneath a collapsing strip footing*

Clearly

$$\frac{\sigma_f'}{\sigma_0'} = \left(\frac{\sigma_f'}{p_A'}\right)\left(\frac{p_A'}{p_P'}\right)\left(\frac{p_P'}{\sigma_0'}\right) \tag{9.60}$$

But just as

$$\frac{\sigma_1'}{\sigma_3'} = \frac{1 + \sin\phi'}{1 - \sin\phi'} = K_p' = \text{constant}$$

for any circle drawn between ϕ'-lines, so

$$\frac{\sigma_1'}{p'} = \text{constant}$$

where p' is the location of the centre of the circle and

$$\frac{\sigma_3'}{p'} = \text{constant}$$

It follows that

$$\left(\frac{\sigma_f'}{p_A'}\right)\left(\frac{p_P'}{\sigma_0'}\right) = \left(\frac{\sigma_1'}{p'}\right)\left(\frac{p'}{\sigma_3'}\right) = \frac{\sigma_1'}{\sigma_3'} = K_p'$$

which can then be used with equation 9.60 to obtain

$$\sigma_f' = \sigma_0' K_p' \left(\frac{p_A'}{p_P'}\right)$$

Using the rotation equation 9.59

$$\frac{p_A'}{p_P'} = \exp\,(2\tan\phi'\,\Delta\psi) = \exp\,(\pi\tan\phi')$$

so that

$$\sigma_f' = \sigma_0' K_p' \exp\,(\pi\tan\phi') \tag{9.61}$$

Table 9.3 lists values of the new estimate of σ_f'/σ_0' compared with the original and crudely pessimistic estimate of equation 9.46 based on the smooth deep wounds.

The exponential increase of σ_f'/σ_0' with $\tan\phi'$ makes equation 9.61 rather difficult to use safely. The conventional safe angle ϕ_s' would be such that

$$\tan\phi_s' = \frac{\tan\phi'}{F}$$

where ϕ' was the likely angle of shearing resistance and F was a safety factor. Taking $\phi' = 40°$ and $F = 2$ the 'safe' angle ϕ_s' would be 23° so that the 'safe' stress ratio would be only 8.5, or roughly the square root of the collapse stress ratio of 64 for $\phi' = 40°$. A safety factor of 2 against $\tan\phi'$ has turned out to require a factor of 8 against the foundation collapse pressure. If the designer was incautious enough to believe that he only needed a factor of 2 against the collapse pressure then, if he thought ϕ' were 40°, he would believe that σ_f'/σ_0' was 64 so he would choose 32 for a 'safe' design. You will see that an error of only 5° in his estimate of ϕ' would bring disaster. The collapse pressure is so sensitive to ϕ' that it is probably unwise to use any value greater than ϕ_c', the loose, remoulded, angle of repose. Any dilatancy component ϕ'_v would then be ignored for safety's sake.

TABLE 9.3
Estimates of σ'_f/σ'_0 for Smooth Foundation on Weightless Frictional Soil

ϕ'	K'^2_p	$K'_p \exp(\pi \tan \phi') \rightarrow N_q$
20°	4.1	6.4
25°	6.1	10.7
30°	9.0	18.4
35°	13.5	33
40°	21	64
45°	34	135
50°	57	319

The wounded footing offers another route to safety in design, since it gives the designer some feeling for the consequences of having missed weak pockets of soil in the site investigation. Perhaps the use of K'^2_p based on an estimate of the likely angle of shearing resistance ϕ', together with the secondary use of $K'_p \exp(\pi \tan \phi')$ based on the critical state angle ϕ'_c, would set the scene for a sensible decision. Suppose that the designer believes that $\phi' = 40°$ comprising $\phi'_c = 30°$ and $\phi'_v = 10°$. Here are his options

(1) likely ϕ'; (σ'_f/σ'_0) at collapse $= 64$

(2) smallest $\phi' = \phi'_c$; (σ'_f/σ'_0) at collapse $= 18$

(3) wounded soil; likely ϕ'; (σ'_f/σ'_0) at collapse $= 21$

He might be persuaded that a safe ratio (σ'_s/σ'_0) was in the vicinity of 20. Of course, he might prefer to use even less than that if he feared the effects of settlement.

As with the previous cohesion solution, the new friction solution did not require the exact shape of the various zones in figure 9.48 to be known. Only the rotation of principal stress direction really mattered. But once again, a mechanism can be shown to provide the same answer, therefore proving that a perfect plastic solution to the smooth footing has been determined. The mechanism employs a logarithmic spiral for curves WU and WY.

Terzaghi went on to alter the zones of stress so as to take advantage of the supposed roughness of the footing. The net effect was to increase the ratio σ'_f/σ'_0 by roughly 20 per cent, which is the sort of improvement which can be wrought by increasing ϕ' by 2° overall. Considering the uncertainty of the existence of lateral expansion under the footing, and in the value of the coefficient of friction between the footing and the soil, this offer of an extra 2°

maximum makes little contribution to the engineer's difficult decision on what 'safe' ϕ' value to use in the analysis.

Of much greater significance is the reduction in capacity caused by an inclination in the load. Figure 9.50 is analogous to figures 9.48 and 9.49 except

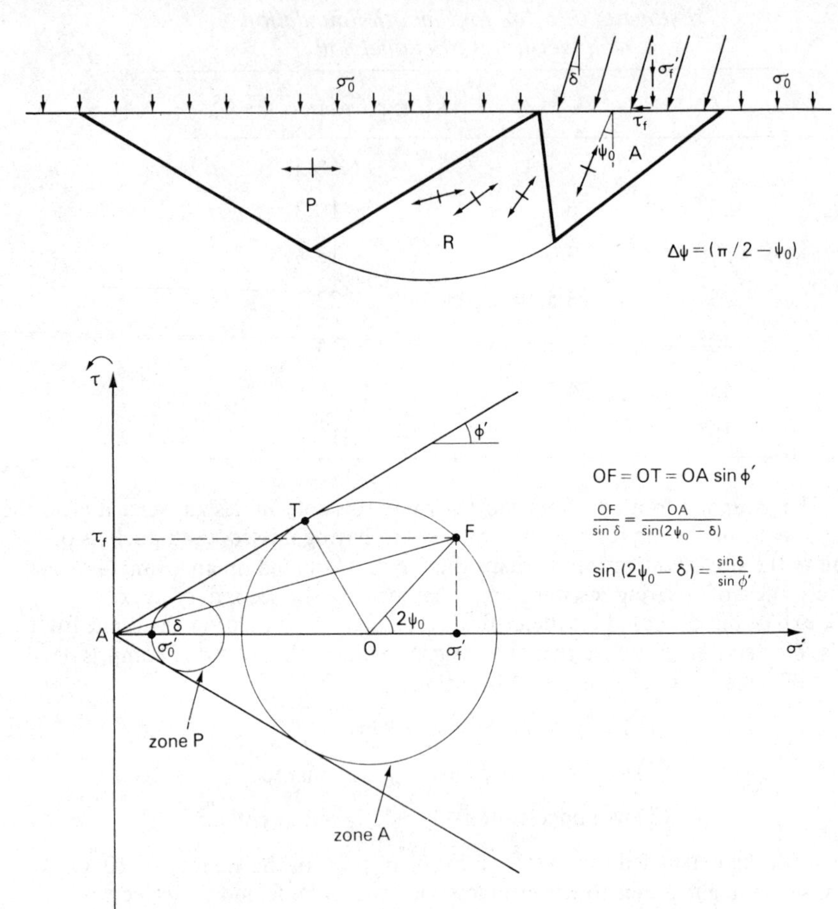

Figure 9.50 *Unit strip foundation over frictional ground collapsing owing to inclined load*

that some inclination δ in the applied load generates failure stresses σ_f' and τ_f at point F on the Mohr circle and thereby causes an initial rotation ψ_0 of the major principal effective stress directly under the base. By trigonometry on triangle AFO

$$2\psi_0 = \delta + \sin^{-1}\left(\frac{\sin\delta}{\sin\phi'}\right) \tag{9.62}$$

This initial rotation ψ_0 reduces the overall rotation of stress direction in zone R to $(\pi/2 - \psi_0)$ and this has an exponential effect on the bearing capacity via equation 9.59. Logic identical to that used previously leads to a much reduced value of

$$\sigma_f' = \sigma_0' \frac{(1 + \sin \phi' \cos 2\psi_0)}{(1 - \sin \phi')} \exp \left[2\tan \phi' \left(\frac{\pi}{2} - \psi_0 \right) \right] \quad (9.63)$$

If the base were rough enough to allow $\delta = \phi'$, and if the designer allowed such a large inclination of stress to develop, the rotation ψ_0 would simply be $(\pi/4 + \phi'/2)$ and the bearing capacity would reduce to

$$\sigma_f' = \sigma_0' (1 + \sin \phi') \exp [\tan \phi' (\pi/2 - \phi')] \quad (9.64)$$

which is an order of magnitude less than its perpendicular value, as table 9.4 makes clear. If a given horizontal thrust must be accepted it is counterproductive to incline the resultant force at greater than 20°. Indeed, the optimum inclination of the force is roughly 15° whatever the friction properties of the soil may be: the perpendicular stress component at failure then equals about one-half its normal value.

9.9.5 The Effect of Gravity on the Subgrade

Our previous analyses have taken into account the surcharge pressure above the founding level which will be caused by the weight of soil above. They have ignored the weight of soil subgrade beneath the plane of the foundation interface.

This hardly matters when using the cohesion model. The strength is not affected by the stresses, so the only effect of gravity is the self-weight experienced by the subgrade stratum. Taking a mechanistic view, this self-weight will cancel out in such as figure 9.12 because there will be no tendency for the heavy subgrade to pivot. Taking a stressing point of view, the train of Mohr circles in figure 9.30 or 9.41 will simply move as a unit a distance γz to the right if an analysis were to be performed on a horizon z deep. Each layer, at whatever depth, would reach failure at the same instant.

The neglect of gravity on the subgrade is definitely pessimistic when the friction model is being employed, however. The only stress in figure 9.48 which is responsible for the enormous foundation strengths generated by equation 9.61 is the effective overburden stress σ_0' on the founding plane. You may like to imagine the shear stresses on each of the potential surfaces of sliding TU, UW, WX, etc. as being proportional to σ_0'. But clearly the effective weight of the ground in zones TUV, etc. also rests on these slip surfaces, and must also generate friction. It transpires that the extra depth of the zones of shear can be treated like extra overburden. Figure 9.51 demonstrates an empirical correction to the overburden pressure which replaces σ_0' by σ_E', the effective vertical stress at a depth $B/2$ below the founding plane. The actual founding plane OO has been moved analytically downwards by $B/2$ to position EE which is roughly on the same horizon as the average element of soil in the zone of deformation. Terzaghi and Peck (1967) chose to express Terzaghi's own calculations and experimental results in a more complicated and less accessible fashion, but there is very close numerical agreement between his own method and that set out below.

TABLE 9.4
Bearing capacity factors σ'_f/σ'_0 and τ_f/σ'_0

ϕ' \ δ		0°	5°	10°	15°	20°	25°	30°	35°	40°	45°	50°
20°	σ'_f/σ'_0	6.4	5.6	4.6	3.7	2.1						
	τ_f/σ'_0	0	0.5	0.8	1.0	0.8						
25°	σ'_f/σ'_0	10.7	9.2	7.7	6.1	4.6	2.4					
	τ_f/σ'_0	0	0.8	1.3	1.6	1.7	1.1					
30°	σ'_f/σ'_0	18.4	15.6	12.9	10.4	8.0	5.7	2.7				
	τ_f/σ'_0	0	1.4	2.3	2.8	2.9	2.6	1.6				
35°	σ'_f/σ'_0	33	28	23	18.1	13.9	10.2	6.9	3.1			
	τ_f/σ'_0	0	2.4	4.0	4.9	5.1	4.8	4.0	2.2			
40°	σ'_f/σ'_0	64	53	42	33	25	18.7	13.1	8.4	3.4		
	τ_f/σ'_0	0	4.6	7.5	8.9	9.2	8.7	7.5	5.9	2.8		
45°	σ'_f/σ'_0	135	108	85	66	49	36	25	16.8	10.2	3.7	
	τ_f/σ'_0	0	9.5	15.0	17.6	17.9	16.8	14.6	11.8	8.6	3.7	
50°	σ'_f/σ'_0	319	249	190	143	105	75	52	35	22	12.4	4.1
	τ_f/σ'_0	0	22	34	38	38	35	30	24	18	12.4	4.8

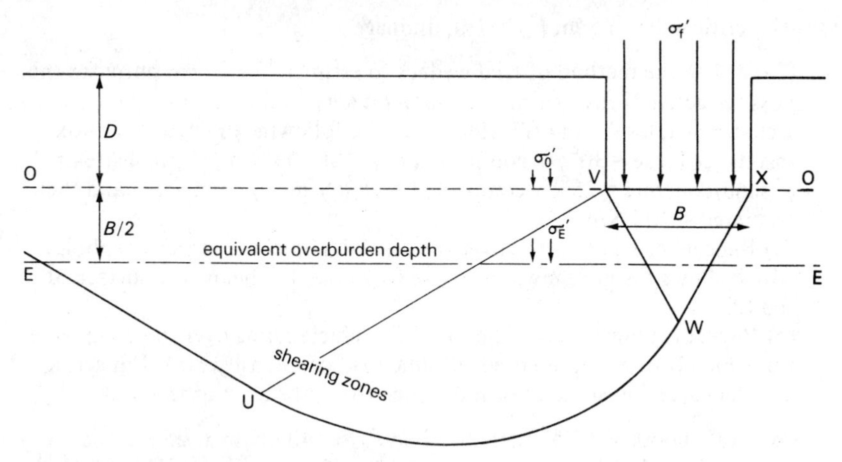

Figure 9.51 *Empirical determination of equivalent depth of overburden (D + B/2) for a strip footing of width B and depth D on a frictional subgrade*

Let the average effective weight of the soil between OO and EE in figure 9.51 be γ^E. In other words let

$$\sigma_E' = \sigma_0' + \gamma^E \frac{B}{2}$$

Then transferring the foundation notionally to horizon EE we obtain an equivalent failure stress of $(\sigma_f' + \gamma^E B/2)$ and an equivalent effective overburden pressure of $(\sigma_0' + \gamma^E B/2)$. Using these in equation 9.61 we obtain

$$\sigma_f' + \gamma^E \frac{B}{2} = \left(\sigma_0' + \gamma^E \frac{B}{2}\right) K_p' \exp(\pi \tan \phi')$$

or alternatively

$$\sigma_f' = \sigma_E' K_p' \exp(\pi \tan \phi') - (\sigma_E' - \sigma_0') \tag{9.65}$$

It is therefore crudely possible to imagine an equivalent depth of a strip foundation below ground surface to be its true depth plus half its width. Its strength from friction is approximately proportional to its equivalent depth.

9.10 Problems

(1) An extensive valley slope of 12° average inclination is deeply scarred by landslips and mudflows. A site investigation at one location revealed 7 m of clay overlying weathered granite. The undrained strength of the clay was measured on site at various horizons using a portable triaxial machine: 1 m, 45 kN/m^2; 2 m, 40 kN/m^2; 3 m, 30 kN/m^2; 4 m, 20 kN/m^2; 5 m, 20 kN/m^2; 6 m, 25 kN/m^2; 7 m, 30 kN/m^2. Perform slope stability calculations which might clarify the problem; suggest remedial measures.

Answer critical depth 5 m, $F_u \approx 1.0$, drainage.

(2) (a) Use the method of trial wedges to estimate for design purposes the possible active thrust per metre length on a vertical wall 8 m high retaining well-drained, level, sand fill which gave the following drained shear box results: ϕ' loose = 30°; ϕ' compacted = 43° at 50 kN/m^2 normal stress; ϕ' sand/concrete = 25°; γ compacted = 18 kN/m^3; γ compacted and then saturated = 20 kN/m^3.
(b) Repeat the solution to cover the case of a blockage in the weepholes which allows the groundwater to rise to a level 4 m below the surface of the fill.
(c) Repeat (a) but to cover the case of a vehicle riding over the completed fill, which is to be represented as a line load of intensity 30 kN/m acting parallel to, and at some critical distance from, the back of the wall.

Answers (a) choose $\phi' = 35°$ perhaps, then $P_a \approx 140$ kN/m at 25° to the horizontal (b) $P_a \approx 200$ kN/m at 16° comprising $P'_a \approx 124$ kN/m, $U_w \approx 80$ kN/m (c) $P_a \approx 152$ kN/m, at about 4 m separation

(3) Repeat the previous question, but employ Rankine's pessimistic technique throughout.

Answers (a) if $\phi' = 35°$, $P_a = 156$ kN/m (b) $P_a = 219$ kN/m (c) $P_a \approx 178$ kN/m if the load is arbitrarily represented as an alternative 10 kN/m^2 over the whole area above the fill.

(4) If the designer of the previous wall decides to resist 200 kN/m of horizontal active thrust solely by generating passive pressures in a horizontal layer of identical fill compacted in front of the toe of the wall, and if he fears that this passive zone may be inundated with water at some time and that excessive strains may develop if the average mobilised angle of shearing resistance in the passive zone exceeds 30°, how deep should this zone be?

Answer 3.2 m

(5) Take figure 3.11 to represent a dam constructed of a glacial sandy clay compacted to an undrained strength in the region 90–110 kN/m^2 and a density of 20 kN/m^3, and suppose that triaxial tests provided an angle of shearing resistance no lower than $\phi'_c = 38°$. Perform both undrained stability calculations relevant to the state of the dam immediately after construction and drained calculations relevant to the long term seepage depicted in figure 3.11 on a potential slip circle which is tangential to the upper surface of the drainage blanket DCE at a point 7 m inside the toe of the slope and which has its centre located 42 m above that surface. Use Taylor's chart in figure 9.18 as an approximate check on the undrained safety factor of the downstream slope.

Answers $F_u \approx 1.9$, $F_d \approx 1.6$ neglecting any suction above the water table.

(6) Confirm that a shallow flake slide on the downstream slope referred to in the previous question is prevented by a safety factor of only 1.35 if potential suctions in clayey fill are once again ignored, but note that suctions are much more likely in this region. What further collapse mechanisms should be checked?

(7) An engineer in the process of designing a sheet pile wall as protection against the erosion of a river bank decides to calculate some active and some passive pressure distributions. A typical finished profile would involve sheet piles driven deeply below a wide, horizontal river bank comprising 5 m of soft clayey sandy silt overlying a further 3 m of dense gravelly sand upon a deep stratum of firm to stiff clay. The river channel could be considered to be dredged to a depth of 3 m below the bank. Establish pessimistic profiles of pressure on each side of the wall, up to the base of the gravelly sand, on the supposition that at least the top 8 m of the sheet piles lean sufficiently far into the river to generate limiting soil pressure on either side, and in harmony with these further assumptions

(i) the river level and the ground water level are likely to remain locked together at 1 m below the bank level

(ii) the average undrained strength of the silt is 15 kN/m^2 at a unit weight of 18 kN/m^3 while its effective angle of shearing resistance is not less than $\phi'_c = 25^\circ$

(iii) the sand is likely to have an *in-situ* peak strength corresponding to no less than $\phi' = 45^\circ$ at a unit weight of 20 kN/m^3 and a remoulded strength corresponding to no less than $\phi'_c = 33^\circ$.

Amalgamate the pressures down to a depth of 8 m below the bank, in order t estimate

(a) total active thrust, assuming silt undrained and sand drained

(b) total active thrust, assuming that both silt and sand are drained

(c) total passive thrust (including river water), assuming silt undrained and sand drained

(d) total passive thrust, assuming both silt and sand drained.

Write an account of how these estimates, and the further consideration of variations in each of the previous assumptions should influence the designer.

Answers if the design value of ϕ' for the sand is taken to be 33°, and for the silt 25°, (a) 324 kN/m (b) 359 kN/m (c) 630 kN/m (d) 594 kN/m

(8) What minimum width of strip footing, constructed with its base at a depth of 1 m in a deep homogeneous soil stratum below a level ground surface, might be considered to be safe under a vertical applied load of 100 kN/m at the surface, in the following circumstances

(a) soil is compact well-graded sand with $\gamma \nless 20$ kN/m^3, $\phi'_{max} \nless 43^\circ$, $\phi'_c = 33^\circ$ and with a deep water table

(b) as (a) but soil is medium dense with $\phi'_{max} \nless 35^\circ$

(c) as (b) but water table may rise to ground level

(d) as (c) but 30 kN/m of lateral load also acts

(e) soil is stiff clay with $\gamma \nless 20$ kN/m^3, $c_u \nless 150$ kN/m^2, $\phi'_c = 28^\circ$ and with a water table at 1.5 m depth

(f) as (e) but clay is softer with $c_u \nless 50$ kN/m^2

(g) as (f) but 30 kN/m of lateral load also acts.

Answers (a) 0.2 m using safe $\phi' = 33^\circ$ (b) 0.4 m using safe $\phi' = 25^\circ$ (c) 0.7 m (d) 1.2 m (e) 0.7 m controlled by the drained strength if a safe ϕ' of 20° is used, in comparison with a safety factor of 3 on undrained strength (f) 1.2 m controlled by the undrained strength with a safety factor of 3 (g) 1.9 m.

10

Towards Design

10.1 Objective and Method

Unless he is requested to act antisocially, the objective of the designer should be to please his client. Clients normally seek constructions of individuality, economy, efficiency, beauty and safety. All these attributes are understood by the client only in comparison with constructions which belong to other people. A bridge which fell apart after 20 years might well have lasted longer than a good motor car and might have been strikingly beautiful and very cheap to build, but it would be unlikely to satisfy its owners in the long run. Part of their concern would have been for the safety of the people crossing it as it fell, but even if there were ample warning of imminent danger the owners of a 20-year-old structure would almost certainly feel cheated if it collapsed. People do not expect civil engineering constructions to collapse at all, and certainly not if they were only built in living memory. And although the number of yearly road deaths (*circa* 7000 in Great Britain) would hardly be affected if some bridge or other collapsed every week, the newspapers will proclaim every single instance of structural collapse as though it were a threat of mutual annihilation. This public demand for the relative safety of structures in comparison to motor cars must obviously be satisfied by the engineer, who may also like to draw society's attention to the 'illogical' nature of its demands. Society may well wish to retain a feeling of relative security in its buildings and bridges, of course. The relativity of first-rate structures, second-rate sewage treatment facilities and third-rate public recreation facilities is a matter for political debate and not for a declaration of unilateral independence by the engineer.

Safety is a relative concept, but so also are economy, efficiency, beauty and individuality. It follows that engineers must become experienced in what exists already before they can add meaningfully to it. While a price is fairly easy to establish, efficiency, beauty and individuality are less tangible. Intangibles are only mastered by those who are sensitive to the contrasting nature of things, and the various reactions of people to each other and to their environment. Designers will need to make and test models of their clients' behaviour so that they can attempt the better to satisfy them! Those engineers who desire to improve the quality of the life and environment of their clients set themselves a prolonged task which can scarcely have begun by the time they graduate. Although his education should offer him some appropriate avenues for the exploration of quality in life, the graduate engineer will probably measure its success – and that of his teachers and textbooks – in a more technical dimension.

I shall therefore concentrate on the very practical issue of designing soil constructions – cuttings, trenches, banks, dams, retaining walls, foundations – which do not deform excessively in use. To ask that they do not deform excessively is clearly to require that they do not collapse, or slump into unsightly heaps, or settle so as to cause associated brittle materials such as brick or concrete to crack, or tilt so as to unnerve the client or jam his machinery. The designer often operates in this characteristically negative fashion: he imagines as many evil influence as he can and then attempts to prevent them. To do this he will work on three levels: creating the outline of a totally new solution or option amongst contrasting alternative schemes which have previously been invented, proportioning the chosen alternative, and detailing the jobs which craftsmen and labourers must undertake. These three functions are often thought to reside with different people, senior engineer, junior engineer and technician respectively, but most good designers are successful at each function. Raw graduates clearly must not expect to be able to pass drawings to the workforce since they will be ignorant of convention. Likewise they will find it difficult to take the initial formative decisions: they will find it relatively easy to calculate the width of a drain or an anchor and relatively difficult to decide whether one is necessary at all! One hallmark of the experienced engineer is his ability to make strong initial decisions which clarify and simplify the later more technical phase of calculation. Especially when faced with an uncertainty the engineer should lean away from exhaustive analyses and towards bold and simple physical solutions. Complex analyses are useful in problems of great simplicity, and vice versa.

In the design of a 'pure' soil construction, such as a cutting slope or an earth embankment for a new road, it is common to ignore strains, and to provide simply for a margin of safety against total collapse, after the fashion of the previous chapter. The desire for structural stability usually requires the control of groundwater, and indeed surface water, lest the soil construction erode away. This leaves the designer cycling between slope profiles with and without drains or seals in order to find a suitable noncollapsible cross-section. Only in exotic circumstances may he need to recognise actions other than that of gravity: earthquakes, vandals and burrowing animals are examples.

New considerations arise in the case of 'composite' soil constructions, such as foundations for buildings, retaining walls and braced trench cuttings. Of course the first step might be to choose a design which does not cause the soil to collapse. The second step must be to check the integrity of other composite materials whether brick, concrete, steel, wood or other kind. Small soil strains might damage any brittle building materials, and it may therefore be necessary to make rough calculations of the likely strains at working load after the fashion of chapter 6. It is also necessary to consider whether the composite materials could be damaged other than by the action of gravity. Sulphates in groundwater attack cement; anaerobic bacteria in saturated soil can attack steel; termites eat wood; frost can burst pipes and can similarly cause saturated soil to heave under shallow foundations; traffic or machine vibrations can shake down loose sands;

badly ventilated boilers can cause clay subsoils to shrink by evaporation; porous building materials can offer groundwater the chance of capillary rise to cause unsightly damp patches on walls which have not been sealed at their base. This list of environmental vicissitudes is not meant to be exhaustive, but it should be enough to convince you of the importance of statutory Standards and Codes of Practice, whose chief function is to jog the memory of the designer so that crucial criteria are not left unconsidered. I shall make reference to British Codes as I consider a variety of soil constructions in turn, but I shall not attempt to cover them in any detail. In particular I shall forsake environmental enemies in favour of gravity-induced failures, simply in order to bring this elementary textbook to a halt. Reconsider what I will almost totally omit: beauty, individuality, economy, good planning of services and functions, and environmental attack on materials. Neither will I have space to consider the enormous range of inventions which engineers have used in the design and construction of foundations, walls, roads, dams, trenches, tunnels and slopes over the thousands of years of their construction. I will dwell on the mechanics of just a handful of types of soil construction. You must therefore appreciate that design is a much larger topic than soil mechanics, and much more complex.

The complexity of his task usually forces the designer to adopt a deceptively simple style: guess the whole solution and then check all the mechanisms of failure to see that they could not occur. Into the 'guess' goes his accumulated wisdom, intuition and experience, and into the 'check' goes whatever mechanical or other models of behaviour he may command. It should have become clear that in this textbook I aim to serve the checking phase in some detail while reserving a place of honour for the initial act of creation.

10.2 Cuttings

The designer of a cutting has two important options, where to put it and whether to install drains in it. The most important calculation which he must then make is that of the allowable slope, based on either the flake or circular slide models. He will also be forced to consider the possible transportation of soil fines by the seepage of groundwater, which could lead to internal erosion followed by a major piping failure or by subsidence.

He will naturally choose to site the cutting, if possible, through well-graded soils with a high angle of friction and above the water table. There will be a limit to the designer's freedom, however, and road alignment (as an example of a type of scheme which frequently demands cuttings) is beset by problems other than those of mechanics. Volumes of cut and fill might need to be locally equalised to avoid the excessive transportation of earth. Environmental considerations may require that the road be sunk in deeper cuttings than strictly necessary simply to hide it and obviate the nuisance of noise. Steep inclines must be avoided on a section of road where traffic congestion would ensue: this may entail cutting into unavoidable hills. These further considerations may leave the designer little choice but to submit to cuttings at or below the previous water table: he must then consider the installation of drains. A typical soil slope streaming with water collapses, as you have repeatedly observed, at roughly one-half its friction angle. Deep buttress drains running down the slope and leading to a trench or pipe drain running along the toe, may draw the water table down so low that the slab-slide and slip-circle calculations reveal a collapse slope angle approaching the angle of friction. If the use of drains can nearly halve the required purchase of land required for a road in cutting, by doubling the side slope angles, they may well be economic.

The intelligent choice of site clearly requires early exploration of the various possibilities. The objective of the exploration is to reach a simple decision, and it would be wasteful to recover so much information that the cutting could be designed in detail in a handful of locations when only one will be used. The engineer should therefore initially recover only sufficient information to make a sensible and practical decision on the location of a site. This would certainly include geological maps, aerial photographs and a visit to any locations on the photographs which appear to have significance. Existing landslides can be most useful in determining the likelihood of future landslides. Steep ground or quarries can bear testimony to the strength of the ground. Streams, springs and wells offer evidence about the groundwater levels.

Once the location of the cutting has been fixed, a detailed site investigation should be undertaken with the objective of discovering the present and possible future groundwater regime, and the effective angle of shearing resistance ϕ' of the soils involved. Grain size classification of the soils will also be necessary in order to decide upon, and if necessary design, a system of drains. The investigation should probably be carried below the toe of the proposed cutting to a depth equal to its height or to intact rock if that is shallow. Boreholes can be shelled out in various cross-sectional groups of three or four so as to obtain a three-dimensional picture of the ground variations. Some relatively undisturbed samples of soil should be taken for insertion in a triaxial cell or shear box so that their strength can be determined. Sands are difficult to recover in sample tubes, especially from below the water table. Various penetration tests, such as the SPT mentioned in section 6.6, may be used to classify the relative density of the sand. It may then be possible to employ a crudely equivalent compaction on disturbed samples in order to estimate the dilatant strength ϕ' ($= \phi'_c + \phi'_v$) in the ground. The critical state component of strength ϕ'_c can be found fairly reliably, of course, since it refers to completely shattered remoulded and dilated soil samples.

Once the borehole has been excavated it can be turned into an observation well for groundwater by the simple expedient of installing a standpipe in it. The hole can be backfilled with tamped clay up to the point at which the water pressure u is to be measured. A bag of sand can then be emptied into the borehole and a plastic pipe inserted down the hole and into the sand. The annular space outside the pipe can then be grouted or filled with clay. If the selected zone of earth is itself clayey, there would be a very long delay before water would percolate into the pipe to reach an equilibrium head. It would be necessary in such ground to install a piezometer with a response at small volumetric displacement: pressures would then be properly recorded with little delay, assuming that the cavity in which the device was placed was entirely saturated with water.

If he can obtain pessimistic estimates of the angle of shearing resistance and the highest likely groundwater levels, the designer can follow the method used in the case study of section 4.6. If he has the choice he must assess the relative cost of sloping the cutting at some angle such as $0.8\phi'_c$, which could be safe if the groundwater were deeply drained away from the face, or at a much lower angle such as $0.4\phi'_c$, which might be safe regardless of drainage. If he decides that drains will be required, the designer should probably make full use of them so that the slopes can be made as steep as possible. If the consequences of slope failures are not catastrophic, the designer may attempt to employ soil shear angles somewhat in excess of ϕ'_c if he believes that there is a good average dilatancy contribution ϕ'_v. Occasional soft spots would then slide and require local repairs.

There is little merit in performing calculations based on the initial undrained strength of clayey ground for the purpose of designing permanent cuttings. Clay is almost certain to soften by transient flow when vegetation is removed, or overburden is removed, and when shear stresses are increased due to steepening a slope in 'dense' clay. Undrained strength calculations might offer a guide to the temporary safety of a foundation excavation, but are bound to be increasingly dangerous as time passes. The consequences of failure of a temporary excavation can be dreadfully costly; in lives if the sides were so steep that workmen could be crushed, in time and money if concrete foundations are displaced. Engineers who are aware that their cuttings could fail in the long term but will stand temporarily according to their undrained strength c_u inserted into equation 9.22 or its derivative figure 9.18, must attempt to make time stand still until they can backfill their excavations! This means ensuring that surface water and groundwater are not available to participate in the softening process. Surface water can be drained quickly away and not allowed to pond above the slope. Groundwater can also be removed by installing a pattern of wells containing submersible pumps around the excavation. The objective here is to reduce pore-water pressures: a temporary cutting might be saved if a few litres per day of water are removed from a thin silty band which would otherwise have allowed a neighbouring clay band to soften in response to the relief of overburden and increase of shear stress.

The major obstacle to the rational and economic design of permanent cuttings is the uncertainty over present and future groundwater pressures. It is rare to have good standpipe or pressure transducer data over representative dry and wet seasons. Only when he has observed the response of groundwater to rainfall can the designer use his hydrological intuition to foretell the most serious chronic wet season or catastrophic storm and therefore the highest groundwater levels which will occur in the useful life of the cutting. In the absence of such information the designer must fall back on some pessimistic assessment, such as an assumption that only his positive provision of drains will prevent groundwater levels rising everywhere to ground level in some future wet season. Occasionally this latter assumption can be shown to be grossly pessimistic, where the inferred volume of groundwater flow would greatly exceed the potential rate of rainfall over the appropriate hinterland. Consider the groundwater regime of figure 10.1: you will see that I have chosen a particular system of hydrogeology which makes the flownet easy! The precipitation of rain and snow over the mountainous

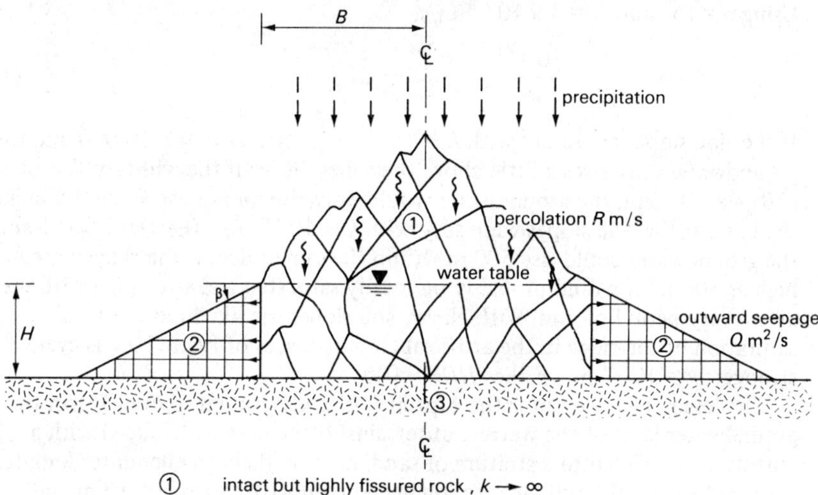

Figure 10.1 *Hypothetical landform for study of groundwater*

hinterland causes some fraction to percolate downwards into the hypothetical zone 1 of highly fissured rock. A groundwater table is established at height H above the valley floor. This coincides with the height of the heaps of soil, zone 2 on the flanks. The flownet through these soil flanks is simply rectangular if zone 1 is relatively very permeable and zone 3 is relatively very impermeable. In these circumstances the soil in the flanks would stand at angle $\beta \approx \phi' (\gamma - \gamma_w)/\gamma \approx \phi'/2$ which would probably be 10° to 20°, with a typical value of 15°.

The equipotentials in the soil flanks are vertical so it is clear that the hydraulic gradient, being the rate of drop of head along the flowline, is simply $\tan \beta$. The rate of flow of water entering each flank

$$Q = Aki = Hk \tan \beta$$

per unit length of the ridge. This must be provided by the average rate of percolation, neglecting storage, so that

$$Q = BR$$

entering each flank from above. The average percolation rate R must be somewhat less than the average rainfall, due to surface runoff and evaporation. If the yearly rainfall were 2 m in a wet temperate climate, the percolation might be in the region of 1 m per year or 3×10^{-8} m/s. Equating supply to discharge

$$Hk \tan \beta = BR$$

Therefore

$$\frac{H}{B} = \frac{R}{k \tan \beta} \tag{10.1}$$

Using $\beta = 15°$ and $R = 3 \times 10^{-8}$ m/s

$$\frac{H}{B} \approx \frac{10^{-7}}{k} \tag{10.2}$$

If the side slopes are sandy with $k \approx 10^{-3}$ m/s, then $H/B \approx 1/10\,000$ and the groundwater only rises a little above the valley floor: if the width of the ridge ($2B$) were 10 km, the groundwater would only rise on average some 0.5 m above the base. If the side slopes were silty with $k \approx 10^{-6}$ m/s, then $H/B \approx 0.1$ and the groundwater could rise 500 m within the same ridge: if the ridge were not as high as 500 m, it would probably be totally saturated and streaming with water after every period of rain. With clayey soil slopes the likelihood of total saturation of the ridge in the artificial circumstances of figure 10.1 is even stronger: $k = 10^{-9}$ m/s makes $H/B \approx 100$.

The lesson to be learnt is the capacity of sandy ground to draw down the groundwater level of the wettest of terrains. If the designer is faced with a cutting 10 m deep into a stratum of sand, he is unlikely to encounter long-term seepage through the full face: following equation 10.2 with $H = 10$ m and $k = 10^{-3}$ m/s would necessitate $B = 100$ km. It is, of course, possible that water would tend to drain into the cutting from a swathe of land 100 km wide, but much more likely that other sinks would be operating. This implies that a new cutting below the previous groundwater table may well stream with water temporarily as it empties the pores of neighbouring zones of permeable soil, but may then settle down with a trickle of seepage representing the average percolation of rainwater over its hinterland. The soil above the final groundwater level could well be capable of standing very steeply, perhaps vertically, due to capillary suction in its pores.

Many a coastline abounds with near-vertical cliffs of soil perhaps 50 m high. Typical cross-sections include alternating clayey and sandy strata, some seepage occurring on the face of the cliff at the base of each sandy stratum and especially at the base of the lowest sandy stratum. Frequent cliff landslides may be caused either by tidal erosion at the base or by prolonged or intensive rainfall causing groundwater levels to rise and suctions to dissipate. Engineers may occasionally feel free to emulate nature and allow earth to be piled up at angles steeper even than the angle of shearing resistance, by taking advantage of pore suction. They must only recall that such suction could be wiped out by a strong downward percolation or by the rise of groundwater in some future deluge: they must be very sure of their surface water and groundwater drainage! Evidence of the historic stability of steep local landforms may bolster their courage. You may find it instructive to devise a variety of regimes of sand, silt and clay layers which could offer a variety of unstable ground profiles under the influence of rainfall. You may also like to consider the relative susceptibility to a sudden rainstorm of rock such as jointed granite which may possess very little volume of groundwater storage.

The designer must also consider whether groundwater seepage will be capable of transporting the smaller soil grains. If they are washed out on to the surface they may leave subterranean pits or tunnels which threaten the slope. If they are washed into drains, they can progressively clog them and cause back-up of the groundwater. Internal erosion such as this can be prevented at the potential exit by the use of a graded filter which employs layers of successively more permeable and large-grained soils such that the fine constituents of each will just be unable to wash into the voids of the succeeding layer. The final line of defence against the disturbance of soil grains on the surface of a slope is a good mat of grass roots.

The final steps in the design of an important cutting are likely to include a careful assessment of all potential slip-circle mechanisms after the fashion of sections 9.5.3 and 9.5.4, employing the philosophy of reasonable pessimism in the choice of friction angles and water pressures. Intelligent use of the simpler flake-side model almost always precludes the necessity for adjustments at this late stage.

10.3 Compaction

When a soil construction such as an embankment is to be built, the earth must be removed from pits and mechanically reconstituted to the desired shape. The designer has a wide range of options concerning both the final condition of the soil and the method by which it is attained. He can request that the soil be stacked in heaps to dry, though this is likely to be counterproductive in a wet and temperate climate like that of Britain. He can certainly request that the soil be sprayed with water from hose pipes if, for some reason, he required it. He can also have the soil compacted as it is finally placed: this is usually accomplished by running heavy wheels or rollers over it. Water can be expelled almost immediately from saturated sands and gravels due to their high permeability but it is very rare for their compaction to take place under water and their pores are usually filled with air. Water certainly cannot be expelled immediately from silts and clays below a rolling load: their times for consolidation are too great. The immediate effect of such compaction is therefore likely to be restricted to the expulsion of air due to the collapse of internal cavities left when the soil was dumped. In the long term however, the artificially induced pore-air and pore-water pressures must establish an equilibrium with themselves and with the natural soils at the boundaries of the soil construction: the resulting transient flows can alter the nature and even the shape and size of the compacted material. If the compacted soil was left almost saturated and with excess positive pore-water pressures, then outward transient flow would probably take place and result in some settlement and compression. If the compacted soil was left with its pore water in suction, then inward transient flow would probably soften the compacted soil and cause heave.

The designer of a new soil construction must therefore choose the void ratio and degree of saturation of his material according to its desired properties. He or his contractor must also choose the plant and the method by which the compaction will be achieved. Table 4 in CP 2003 Earthworks (1959) sets out a variety of recommendations concerning the method of compaction of loose materials and it is summarised below. The loose earth is usually spread in layers not exceeding 0.2 m in thickness. Heavy plant is then driven over it a number of times determined by experience early in the job; four to eight passes is the normal range. The compaction machine may simply be a piece of general earth-moving equipment, but it is more likely to be either a smooth steel roller or a pneumatic-tyred roller towed by a tractor. The mass of the compaction vehicle will probably be in the range 2 to 50 tonnes. Smooth rollers can apply very large local pressures to the soil and are particularly useful in producing a smooth weather-resistant surface to clayey fills. The contact pressure of pneumatic tyres equals their inflation pressure, which is normally in the range 200 to 800 kN/m^2. The analysis in section 9.7 of the collapse of subterranean cavities demonstrated the importance of not burying cavities too deeply before attempting to squash them. Following the cohesion analysis of equation 9.48 we showed that it may not be possible to compact cylindrical voids smaller than 10 mm radius under a 0.2 m deep layer of clay, and that the size of a stable cavity would be proportional to its depth of cover. The strict control of layer thickness for compaction should come as no surprise, therefore. The friction model of equation 9.51 gave even less hope for the eventual static compaction of fully drained soils. In these circumstances the kneading and shearing of the soil in the high pressure zone under the moving wheel or roller may be the source of the compaction of coarse-grained soils. The further provision of an engine-induced vibration at about 20 Hz with a 10 to 20 tonne smooth steel roller will dramatically improve the compaction potential of fairly dry sands and gravels beyond that of a simple rolling load.

Heavy compaction vehicles may well damage structures if they are allowed to pass too close to them. The compaction of fill against retaining walls, bridge abutments, pipes and culverts is therefore usually effected either by small smooth wheeled rollers up to 1 tonne in mass, or by mechanical punners and rammers which use an internal petrol explosion to lift their 100 kg to 500 kg mass repeatedly by up to 0.5 m in a hopping action.

Compaction of the natural ground without its being removed is also an option, sometimes neglected, for the improvement of soil upon which foundations are to be established. In order to compact up to substantial depths in loose sands, waste tips, or uncontrolled fill, it is necessary to adopt some form of ground penetration. Large vibrating pokers are used in the vibroflotation method, the resulting sink holes being filled with crushed rock which is itself compacted by the vibrator. Another technique is to allow a very heavy reinforced concrete block of perhaps 10 tonne mass to fall freely from as great a height as possible, perhaps 15 m, in successive locations so as to produce a checkerboard pattern of craters which can then be filled with compacted rock. Although these deep compaction techniques are expensive, they may be justified where their use can obviate the need for piled foundations.

The success of compaction is conventionally measured in terms of the increased number of soil particles which can be jammed into a given volume. The associated conventional definition is the dry density of the soil ρ_d equal to the mass of solids contained in a unit volume of soil, which by reference back to table 1.1 can readily be seen to be given by

$$\rho_d = G_s/(1 + e) \tag{10.3}$$

Since the only fundamental variable at stake here is the familiar void ratio e, I shall use both the practical conventional parameter ρ_d and the fundamental parameter e in the descriptions of compaction performance which follow. The reason that engineers in the field will continue to use ρ_d is that they can readily compact and weigh damp soil into a cylinder of known volume to obtain the bulk density ρ and can then measure the moisture content m of representative samples and deduce the dry density directly

$$\rho_d = \frac{\rho}{(1 + m)} \tag{10.4}$$

The void ratio can only be deduced by the further step of employing 10.3 to obtain

$$e = \frac{G_s}{\rho}(1 + m) - 1 \tag{10.5}$$

which demands the extra measurement of the specific gravity of soil solids G_s with little or no extra benefit to the practical man in a hurry.

Whichever method is chosen to represent the packing tight of soil particles, it is arguable that engineers have become unnecessarily obsessed with achieving it. The conventional ideal is sometimes presented as compaction which expels all air and which leaves the void ratio as low, and the dry density as high as possible. This could well be undesirable. Very dense clayey soils tend to be brittle and subject to cracking, which can be catastrophic in compacted dams. They also tend to swell and soften on contact with water. Even the exclusion of air may not be wholly desirable in all circumstances if it entailed the application of large stresses which could be diverted to cause deformation in neighbouring structures. Nothwithstanding these concerns, that compaction which induces the greatest possible dry density and smallest possible void ratio in a given set of circumstances is invariably and therefore incorrectly described as the 'optimum' compaction. Three variables control the final void ratio of a compacted soil, the proportion of water present, the type and weight of the 'blow' or 'pass' which is used to achieve densification, and the number of its applications. Every blow from the compacting device tends to drive air out of the soil beneath but there is an eventual limit to the degree of saturation achievable, however many blows are delivered. A heavier device might be capable of offering a slightly greater degree of saturation but however heavy the device it is unlikely to be able to achieve void ratios below 0.30 or dry densities above 2100 kg/m^3 simply because of the inevitable voids between the irregularly shaped soil particles. Neither will it be able to drive the initial water out of fine-grained soils; their moisture content will remain constant during the compaction process.

The influences of moisture content and compactive effort can be demonstrated with respect to the standard laboratory compaction test described in BS 1377, which is used as a crude guide to the likely compaction performance of a soil in the field. Soil is rammed into a cylindrical brass mould of volume 9.44×10^{-4} m^3 by either a 'standard' action of a 2.5 kg hammer dropping vertically 0.305 m or a 'heavy' action 4.5 kg hammer dropping 0.457 m. The soil is placed in three consecutive layers, each layer being subjected to 25 blows from the hammer. The tests are normally conducted over a range of moisture contents, and results for a typical glacial sandy clay of low plasticity are presented in figure 10.2 as a chart of void ratio against moisture content. The diagonal fan of lines generated

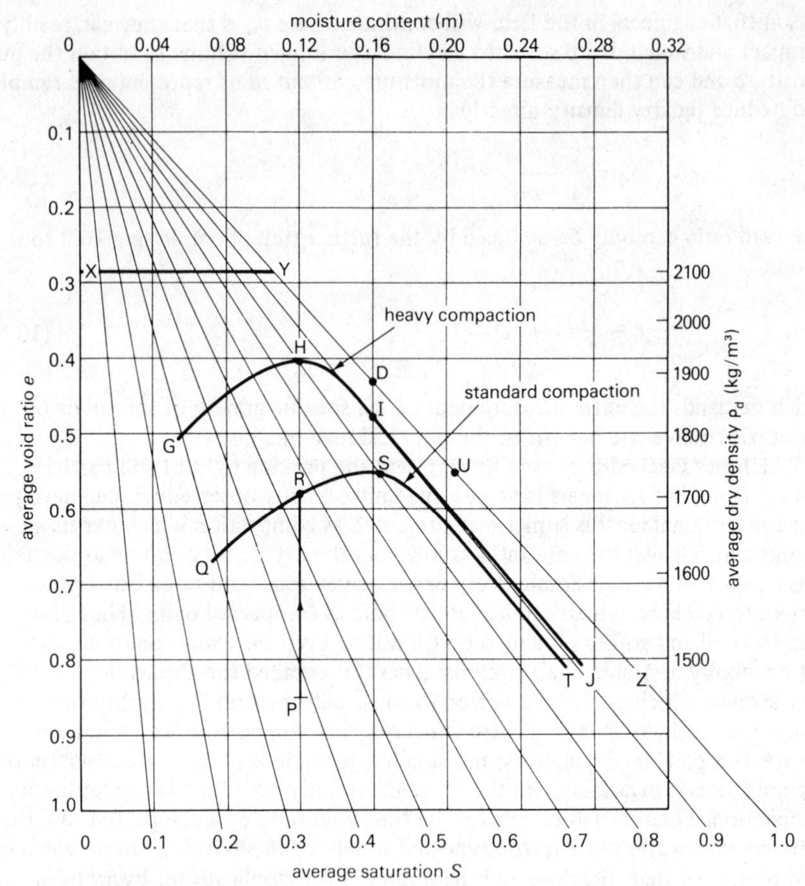

Figure 10.2 *Compaction diagram*

from the origin in the top left corner of the chart represent different degrees of saturation of the voids, using the simple relationship $m = eS/G_s$ derived in section 1.1.3. The magnitudes of the dry density ρ_d are included for interest on a nonlinear scale parallel to the void ratio scale: a value of 2.70 has been used for G_s.

Consider that the clayey soil is initially thrown into the cylinder at a moisture content and average void ratio indicated by point P in figure 10.2. If it is rammed by the standard hammer its state will begin to journey towards R, which it will eventually reach at 75 blows. More blows would only have reduced the final void ratio very marginally. If the heavy hammer had been used, however, the state would eventually have reached point H. The standard compaction curve QRST is only determined when a number of moisture contents have been attempted by mixing water with the original soil or perhaps by drying it. Likewise the heavy compaction curve GHIJ requires the fresh compaction of the soil at perhaps five different moisture contents. It should be clear that no amount of compaction can lift the soil higher than the full saturation line YZ, or above some irreducible minimum void ratio X.

Points S and H are said to represent the 'optimum' conditions for standard and heavy compaction respectively. It is self-evident that the 'optimum' moisture content reduces as the compactive effort increases. In the high moisture content ranges ST and HJ, there is a maximum achievable saturation of the voids of about 90 per cent; these states correspond to the complete encirclement by soil and water of small air bubbles which are almost impossible to move by compaction. In the low moisture content range SQ and HG the voids are becoming increasingly filled with air: the surface tension of menisci between the clay particles holds the cavernous soil structure together so strongly that very large forces are required to break it down. The tripling of compactive energy implied by the use of heavy rather than standard compaction would achieve state H rather than state R which was on the dry side of 'optimum', the degree of saturation increasing from 0.56 to 0.81 due to the reduction of 0.17 in the void ratio. The actual moisture content which will be used in the field compaction will depend on the natural moisture content of the soil. If this were 0.18 in our example, the engineer would probably decide that heavy compaction was valueless, but that some standard compaction would create a fairly dense and well-saturated clay, albeit slightly wetter than its 'optimum' moisture content. If on the other hand its natural moisture content were 0.10, the engineer might either use heavy compaction and risk producing a rather brittle unsaturated structure or add 5 per cent water to it in order to achieve a more saturated plastic structure after standard compaction. One danger of leaving a compacted clayey soil with a state such as Q is that rain or groundwater may be able to percolate rapidly through the air-filled voids, destroying the capillary suction which is binding the soil together. The suctions in sandy soils are less significant.

The principal concerns of the designer of a compacted soil construction will be its strength, both short term and long term, and its susceptibility for settlement or heave. The designer of a coarse-grained fill will be prepared to adopt simple groundwater pressures and use the friction model to assess the stability of the fill at all times: the high permeability of the soil will preclude the existence of transient excess pore pressures. Neither will he be disturbed about the likelihood of settlement if he can compact the sand or gravel so that its void ratio approaches the smallest achievable void ratio. He may be prepared to use Cornforth's correlation of equation 8.59 to estimate the dilatancy contribution to strength ϕ'_v from the relative density of the sand. In each case he will want to compare the void ratio or density achieved in the field compaction with the greatest state of compaction available in the laboratory, such as that of point H in figure 10.2 perhaps, and with the loosest state achievable in the laboratory such as by quickly inverting a measuring cylinder three-quarters full of a measured mass of the dry sand.

The designer of a clayey fill may well be prepared and able to guess the highest possible long-term pore pressures, so that the long-term strength problem can be solved using the friction model if he has performed a few very slow drained shear box or triaxial tests on the compacted soil in order to find ϕ'. The remaining problems of short-term stability and volumetric strain are crudely soluble if the post-compaction pore pressures can be estimated. Imagine that the soil were saturated. While soil is being remoulded and compacted it should be at some critical state such as Y or Z in figure 10.3. In particular the pore pressures

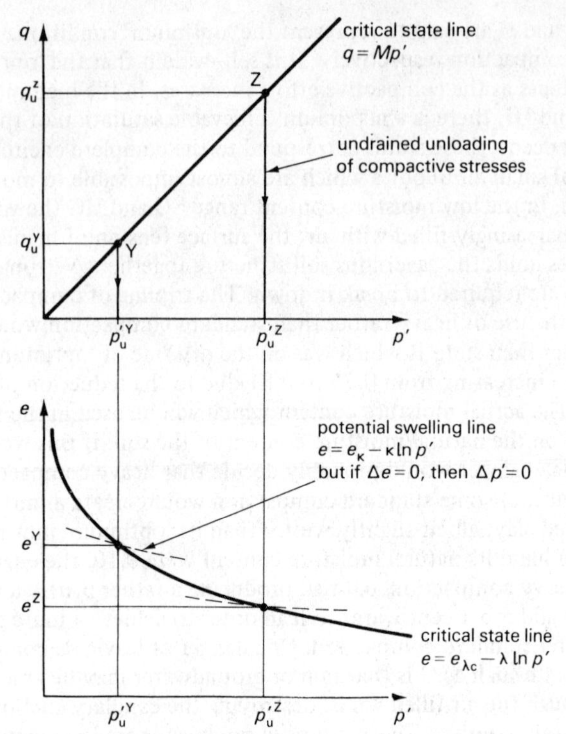

Figure 10.3 *Fixed positions on critical state line for soil compacted at a unique void ratio*

of the soil will have been altered by the tendency of loose soils to contract and dense soils to dilate. When the remoulding process has finished and the soil lies in its final position, there will be an immediate undrained elastic response to the removal of the direct and shear stresses under the machine which will generate an increment of pore suction to add to the previous critical state pore pressures. The effective pressure p' of the soil is locked at its critical state value. The sudden burial of the compacted soil element by the remaining fill would finally generate an increment of positive pore-water pressure due to the undrained elastic compression. Once again the requirement for zero drainage locks the average effective stresses, which remain at their critical values $p_u'^Y$ or $p_u'^Z$.

The assumption made above that the soil remained undrained and saturated throughout the compaction and burial process is a familiar one. The undrained strength of the soil $q_u^Y (= 2c_u^Y)$ and $q_u^Z (= 2c_u^Z)$ remained constant, and in conformity with the cohesion model. The prediction is for an exponential drop in strength with an increase in void ratio, which would be evident at the time of compaction. It would, if the assumptions held, be possible to check the design of the construction using slip circles in the fashion of section 9.5 after determining the undrained strength of well-compacted soil in a field trial. The inspection of the final construction works could then consist of check measurements of the undrained strength of recently compacted soil to make sure that it did not fall below a safe value.

The assumption of no change in volume during construction produces an equally simple rationale for the calculation of the eventual settlement or heave of a clayey fill. Suppose that some drained triaxial tests on the saturated fill material had provided a value for M. The control and measurement of q_u in the field is then equivalent to the control and measurement of p'_u in the field, using $p'_u = q_u/M$. While the fill remains undrained the effective spherical stress remains at p'_u. The fully drained effective stresses should be crudely guessable: if the fill were to be submerged the vertical effective stress of $(\gamma - \gamma_w)z$ would be a fair estimate for the average effective stress, which would be seen to increase linearly from zero at the surface. An upper zone of clay would therefore expand as its effective spherical stress fell from p'_u to $(\gamma - \gamma_w)z$, while a deeper zone would consolidate as its stress rose from p'_u to $(\gamma - \gamma_w)z$. Some soil modulus from chapter 6 would be needed to achieve the required strains, of course, and these would then need to be integrated before the whole composite displacements emerged. Large settlements may be expected if the void ratio of the fill were large so that p'_u is small, which will be experienced as a small undrained strength q_u or c_u, or if the fill was very deep implying that the final stresses were large. Severe heave may be expected if p'_u is large due to a large undrained strength implying that the void ratio was small and the soil very dense, and especially if the fill was shallow. All movements due to transient drainage are inversely proportional to the stiffness of the soil. Very clayey plastic soils are most susceptible.

If there is drainage of the soil to its boundaries during construction it is usually presumed that its state is moving towards the ultimate condition which the designer has allowed for. If the designer can guarantee the safety of his works on both the assumptions of zero drainage and completed drainage, he will feel aggrieved if the structure suffers a collapse at some interim condition. Likewise he will presume that the volumetric strain of the construction after completion must be less than that calculated as above if some movement due to transient flow was already taking place during construction.

It is, unfortunately, much more difficult to be properly pessimistic about the 'undrained' strength and capacity for volumetric strain of partially saturated soils. States with a saturation less than 0.7, such as Q or G in figure 10.2, are potentially unstable due to their capacity to soften quickly and collapse on contact with water. States such as S may be viable, but their 'undrained' strength is difficult to interpret as you saw in figure 8.13 and its accompanying discussion. The measurable strength of a sealed sample with state S might be equal to that of a saturated sample with any void ratio between D and U depending on the speed of test and the confining pressure. It could be sensibly pessimistic to attempt to replace entrained air with water, so as to change the state from S to U, before testing the 'undrained' strength of a sample of fill. The alternative of confining the sample at such high pressure that the air voids disappear while the moisture content is constant leads to optimistic state D. Many engineers go to the other extreme and perform a quick triaxial compression test without membrane or cell pressure and hope that the unconfined compression strength which they measure will be inferior to all confined strengths in the bank.

A failure to correctly predict the future groundwater regime in the compacted fill leads to errors in the analysis of both long-term stability and long-term deformation. If the designer assumes that a gravel-filled trench drain beneath a clayey fill will draw down the water table but forgets to secure the exit to the drain, he may find that the fill expands and softens as the groundwater rises. If the designer assumes that a clayey embankment dam in a tropical country will possess continuous seepage from its reservoir and therefore moderate pore-water pressures, he may find that the dry season evaporates the pond and promotes deep shrinkage cracks due to the increase in pore suction, and that the wet season arrives so quickly that floodwater can seep through the cracks to erode the bank away. Clay fills should ideally be protected against any rise in the groundwater table at their base and against harsh evaporation at their surface.

10.4 Embankments

An earth embankment can serve a number of functions, such as the provision of an elevated highway or canal, the prevention against flooding from a river or the sea of low-lying land, and the storage of water in a permanent reservoir for the purposes of irrigation, flood control, water supply or electricity generation. In each case it is in competition with other structural solutions such as viaducts, walls and concrete dams. One frequent cause for its selection rests with its relative ability compared with reinforced concrete structures to deform without damage when built over compressible soils. Likewise, the purposes to which it is put frequently require an earth embankment to be sited on low-lying waterlogged ground or across river channels and estuaries. Good subgrades for earth banks are rarely the lot of the hard-pressed designer; indeed, the foundation materials are likely to cause him more anxiety than the bank itself.

Four types of failure mechanism dominate the designer's calculations: slope failure of the bank alone, foundation failure mainly within the subgrade, internal erosion in or under water-retaining banks, and functional failure due to the settlement, heave or cracking of the bank caused by long-term transient flow. An initial exploration of sites must reveal the most favourable location both for the embankment and for the borrow pits which will yield the material for its construction. Once the sites have been fixed the designer must decide whether to treat or remove the subgrade beneath the bank. Fissured rock beneath an earth dam would probably require the injection of a cement grout to reduce the quantity of seepage. Peat or very soft clay beneath a proposed road embankment might, if it were shallow, be removed by dragline or displaced by dumped rock fill. Each operation on the subgrade would have to be preceded by a specific investigation, such as a pumping test to determine the permeability of the dam foundation and a simple sounding test with rods to determine the extent of very soft material beneath the road embankment. This second round of site investigation only needs to be accurate enough to allow the designer to take simple yes/no decisions. At the start of the third stage of the site investigation, the designer should have a fairly clear impression of the situation and materials of construction of the embankment, and an explicit grasp of the likely and less likely hazards which may afflict the project. The remaining investigation must accurately recover those properties of the fill and subgrade which will allow the detailed design of compaction technique, construction sequence, slope angles,

Veronica
Knight

23'4271
84.32
47.46
74.31

Pippin's Farm

Pembury (089 282)
4624

On Maidstone Road
(B 2015)

Go in ent to Farm Shop
(Big hoarding)
Out turn left to Pembury
then 1st left by bus
stop & telephone box
180° turn.
Down narrow lane.
→ Farm ent.
& house.

TAKE
WELLIES

seepage cut-offs, and drains, to offset the anticipated hazards.

The extremes of embankment design can be represented by road embankments and earth dams. Road embankments are commonly designed to be homogeneous and constructed of any soil which can be compacted into a dense and strong state. It is only necessary to condemn fine-grained materials which are so soft that machines would be bogged down and undrained failures be unavoidable. Earth dams, on the other hand, must specificially be designed to be anhomogeneous due to the requirement for dealing safely with the inevitable seepage. Figure 10.4 depicts just two of the safe solutions that may be adopted,

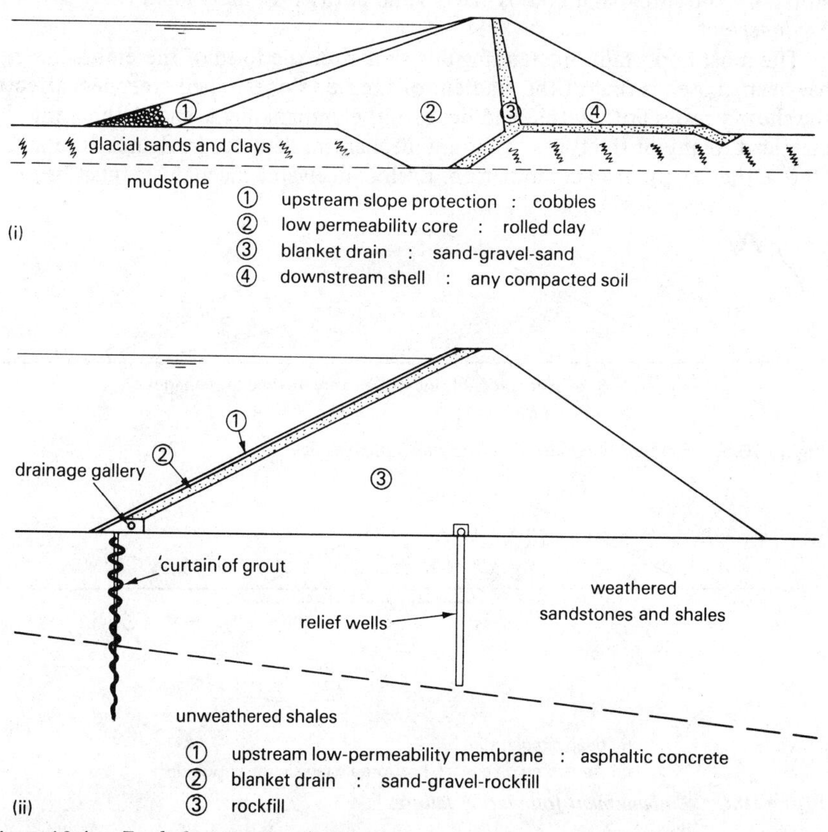

Figure 10.4 *Earth dam cross sections*

the choice being dominated by the properties and abundance of the soils in the neighbourhood of the project. Type (i) features in zone 2 a rolled clay core sloping upstream which acts as the seepage inhibitor. The core is continued downwards as a cut-off trench through the erratic glacial drift. The clay is protected against wave action on its upstream face by a zone 1 of dumped rock or gravel. The inevitable seepage through the core is intercepted by a blanket drain in zone 3 which prevents water pressures rising in the downstream shell, zone 4. The interface between zones 2 and 3 must be a suitable graded filter to prevent fines washing into and blocking the drain. Dam type (ii) depicts a rockfill dam which is kept entirely free of seepage by an upstream membrane of asphaltic concrete in zone 1. Any damage to the membrane would allow water to enter the graded drain zone 2 beneath, thence to the inspection gallery. A 'curtain' of cement grout has been used beneath the upstream toe to reduce the seepage through a sequence of interbedded sandstones and shales. In addition a line of relief wells with filter casing has been provided to minimise the risk of piping or heave due to upward seepage beyond the downstream toe. The range and diversity of problems outside the scope of this book which must be faced in the design and construction of earth dams is reflected in the collection of papers edited by Hirschfeld and Poulos (1973) and entitled *Embankment Dam Engineering.*

The most important outstanding decision after the form of the embankment has been chosen is that of the gradient of the side slopes. Their steepness affects the shear stresses both within and beneath the embankment, and in the limit may lead to any of the types of failure mechanism depicted in figures 10.5 and 10.6, either singly or in combination. Each conceivable mechanism must be

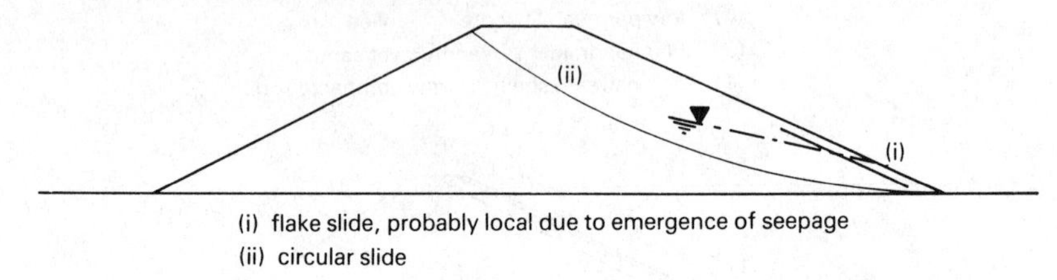

Figure 10.5 *Embankment slope failure mechanisms*

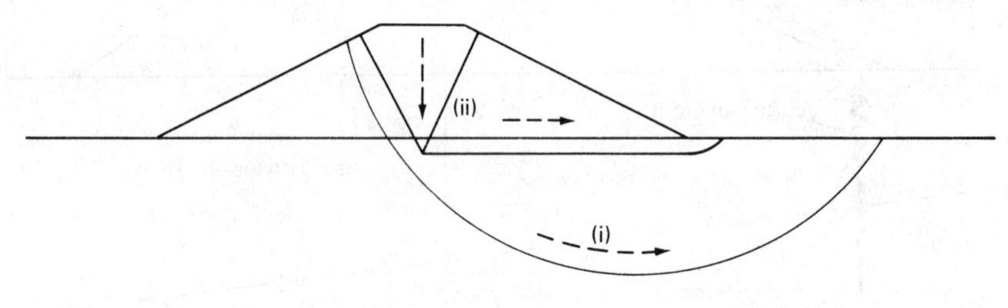

Figure 10.6 *Embankment foundation failures*

proved to be inoperative at the end of construction, and in the normal long-term condition, and during any possible but extraordinary event such as an earthquake. If the embankment fails by such a shear mechanism the cause will usually be that pore-water pressures are higher than the designer had allowed. An earthquake-induced slippage might be said to be partially caused by the designer incorrectly estimating the body forces on the soil and hence the total stresses. Some slope failures may be due to a careless estimation of the angle of shearing resistance of the fill. But it is the water pressure u in the effective strength equation $\tau = (\sigma - u) \tan \phi'$ which contributes most failures; during construction due to soft impermeable zones of clayey soil in or under the bank, in the long term due to unforeseen high groundwater levels following the cracking of clay cores in dams or the blocking of drains, after gentle earthquakes if the effect has been to shake down some zone of saturated loose sand or silt which liquefies until the excess water can escape.

In the previous section 10.3 I discussed the danger of leaving clayey fills in a dry condition, and the possibility of choosing a pessimistic equivalent 'undrained' strength c_u by saturating the air voids or by performing an unconfined compression test, which might then be applied with safety to the condition of clayey fill immediately after construction. In addition the second round of site investigations should have revealed the nature of the subgrade soils and in particular the presence of any potentially dangerous zones of weak and poorly draining fine-grained soils. These materials must be sampled in the third round of investigations so that a pessimistic value can be chosen for their undrained strength. Since the client will demand that his embankment be built whatever are the data of undrained strength of the subgrade, and since it is perfectly possible to build the bank slowly or to install vertical subgrade drains in boreholes so that the foundation consolidates towards its fully drained strength as the bank is being built, it is usually necessary to attempt a judgement of the likely speed of drainage of the subgrade soils. Aspects of this difficult decision were covered in chapter 7. The assumption of zero volume change in soft-grained soils, both in the bank and its foundation soils, will lead to a pessimistic assessment of the stability of the bank immediately after construction, and therefore of the necessity or otherwise of assessing and improving their drainage. If the bank appears perfectly stable using the undrained strength of clayey soils and the drained strength of sandy soils in the range of mechanisms depicted in figures 10.5 and 10.6, it follows that drainage of the clayey soils during construction is not a vital matter.

The parameters for the investigation of the long-term stability of the bank are simply the lowest conceivable effective angles of shearing resistance ϕ' of the materials and the highest conceivable water pressures u at every point. The designer will have more opportunity to control the quality of the fill than he will to inspect the quality of the subgrade, and he will therefore be able to employ less pessimism in the choices of parameters in the bank than he will in the subgrade. The discussions of chapters 8 and 9 should have made clear that pessimistic angles ϕ' emanate from samples which are as loose and as heavily stressed as any element in the field. The correct prediction of the greatest long-term pore-water pressures rests chiefly with the correct prediction of the permeabilities of the field materials, leading to the correct estimation of the worst flownet. Drainage material which is not as permeable as it should have been due to poor selection, placement or subsequent clogging has often been responsible for the failure of earth dams. Similarly, clayey fill which is not as uniformly impermeable as was thought can lead to the embarrassing result of the seepage being carried through an earth dam without entering the drains. Consider a borrow pit from which the clayey fill was being taken to construct an earth dam designed after the fashion of figure 3.11; if this was to encounter a pocket of silty sand which was placed unnoticed as a uniform horizontal layer across the whole bank at mid-height, the reservoir would wash through it and then erode away the whole construction, avoiding the filter drains at the base entirely. Experiences such as this have led the designers of large earth dams to the cross sections of figure 10.4 within which no rogue seepage path can avoid the carefully graded drains: the loss of water itself is not of prime concern but rather the erosion and local failure which can accompany seepage when it is allowed to approach a free surface. Seepage must be attracted to buried drains which are sufficiently heavily stressed to ensure that soil particles are not washed around by the inevitable concentration of hydraulic gradients into the drain.

The parameters for the investigation of rare events, such as an earthquake or the sudden drawing down of an impounded reservoir, require special stress-path triaxial tests to be performed on soils whose time for transient drainage may be longer than the process of loading. The designer must estimate the likely changes in total stress on a variety of field elements which will have achieved some estimated equilibrium, and must then attempt to perform equivalent triaxial tests with a similar sudden change in stress superimposed over a similar equilibrium. If he is correct in replicating the stress path of a number of field elements in a triaxial machine he will be able to treat the strengths he achieved as undrained strengths q_u^* or c_u^* which will be operative just before the future event. I showed in section 8.5.3 that soils which were 'loose' and which would tend to contract or consolidate under a load increment, were highly susceptible to sudden shear stresses. This may give the designer some cause to use heavier compaction at the base of the bank in order to achieve smaller void ratios and a tendency to dilate rather than contract upon the application of shear stress, in spite of the consolidating effect of the earth above. In other words the designer may decide to zone his compaction, very heavy at the centre of the base, heavy at the mid-depth, and light at the top and immediately under the slopes so that every element of soil would be a little 'denser' that its critical state at its own effective pressure of confinement in the bank. If calculation shows that the sudden event will cause a failure, the designer must choose between warning his client of the risk of future repair bills and third-party insurance claims, attempting to improve the properties of the soil by compaction, reducing the slope angles thereby increasing the volume and cost of the earthworks, or installing such measures of drainage as will allow the anticipated event to be drained rather than undrained.

The designer will be guided in his preliminary choice of slope angles by a simple slope friction formula such as equation 4.6. If a slope were to be well drained with no possibility of water pressure within it, the designer would desire it to stand at some angle such as $0.8\phi'$. This might apply to such examples as road embankments, to the downstream slope of dam (i) and to both slopes of dam (ii) in figure 10.4. If the slope were to be allowed at any time to be water-logged, with groundwater seeping out of its face, then it should stand no steeper than $0.4\phi'$ following the reasoning of section 4.6. This might apply to a poorly drained and cheaply built flood protection embankment but not to the carefully drained profiles of figure 10.4. Indeed the upstream slope of dam (i) in figure 10.4 is subjected to inward seepage which can easily be shown not to disturb it all but to clamp it more securely in position, and to generate safe slope angles steeper than the angle of shearing resistance ϕ'. This fact has little bearing on the design, however, since the designer will presumably wish the slope to stand on some future occasion when the reservoir has been drawn down and the inward seepage is absent. If the drawdown was more rapid than the sympathetic transient flow of water out of a clay core such as zone 2 in dam (i), the resulting condition of *outward* flow from the core may resemble figure 4.17, or with the emplacement of gravel in zone 1 the designed slope might be based roughly on the analysis of figure 4.21. The upstream slope of a dam, unless it is well drained in its entirety, is therefore unlikely to be placed steeper than $0.5\phi'$ due to its potential instability following 'rapid drawdown' of the reservoir.

The designer will employ Taylor's curves of figure 9.18 to determine whether the drainage of clayey fills will be a problem. Figure 9.18, for example, would show him that a 40 m high dam constructed on a good subgrade but with a compacted sandy clay slope of equivalent undrained strength 100 kN/m^2 and unit weight 20 kN/m^3, would just collapse at a slope angle of about 26°, and would have a conventional safety factor of 1.5 at a slope angle of 13°. If the clay fill had a friction angle of 30° and was to be used as a core sloping upstream, the previous friction calculation for the case of rapid drawdown might have already convinced the designer that 15° was the maximum slope angle. His later undrained strength calculation would merely support a conclusion that the soil could be compacted sufficiently strong to stand without the requirement of drainage. If the dam had been higher, say 80 m, and the 'undrained' strength remained 100 kN/m^2, the desirable factor of safety of 1.5 would actually demand a slope angle of 5°! The response of the designer would then be to discover what effect the provision of a reasonable delay between two or three stages of construction would have on the dissipation of the positive excess pore pressures in the compacted fill. He would almost certainly not entail his client in the extreme expense of a flat upstream slope merely to protect against an undrained failure when he could see that a drained calculation offered stability. If he could not promote drainage he might, in a case such as dam (i) in figure 10.4, increase the width of zone 1 and reduce the extent of zone 2 in order to trap the relatively dangerous clay between stable shells of good granular material. Or he might choose another design entirely.

The undrained strength of the subgrade would be treated by equation 9.54 in the same spirit in which the bank slopes had been checked by Taylor's curves. The designer might notionally replace a triangular earth dam of height H and total width B by a crude 'equivalent' uniformly rectangular strip of width $3B/4$ and height $2H/3$. If the average strength of the subgrade to a depth of the same order as the height of the bank was c_u, the designer would conclude that in the absence of drainage there was a likelihood of a foundation failure of the type (i) in figure 10.6 when the 'equivalent' foundation pressure $2\gamma H/3$ equalled the bearing capacity $5.14c_u$ or in other words when H rose to roughly $7.7c_u/\gamma$. As it happens, the 40 m high dam would just require a subgrade with c_u = 100 kN/m^2 in order to be on the verge of a foundation failure on completion of the embankment, neglecting consolidation of the subgrade. If the designer was not totally convinced by evidence of the existence of sand lenses and partings that the subgrade clay could drain during construction, he might consider the installation of piezometers so that construction could be halted if they registered excessive pore-water pressures, and only resumed when a safe proportion of dissipation had taken place.

The designer is not at liberty to ignore soft badly drained strata of clay or silt even where they may be rather thin and where sound rock at shallow depth may appear to preclude a deep slip circle. Mechanism (ii) of figure 10.6 demonstrates the ability of a bank to spread on just such a soft shallow surface. At first sight chapter 9 may appear to have left you without any apparatus for investigating such a mechanism, but figure 10.7 shows how simple such a calculation can be

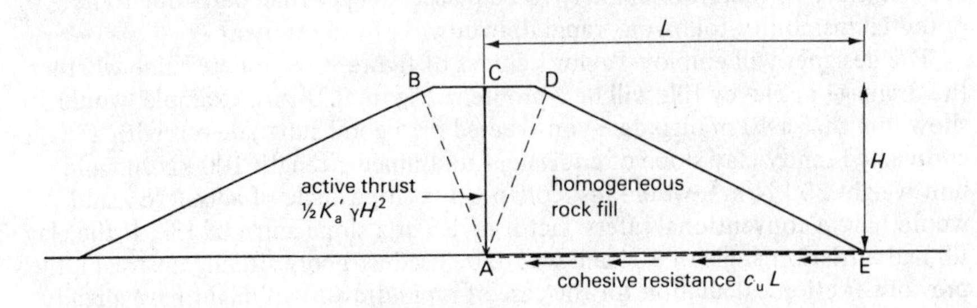

Figure 10.7 *Spreading of a granular heap on a cohesive layer*

once the imagination has been provoked into providing the sketch. The figure depicts the potential spreading of an embankment of dry rock fill with an angle of friction ϕ', on a very shallow sliding surface AE with an average strength c_u developed on it. An 'active' zone of fill ABD sinks and spreads as the shell ADE is pushed out sideways. It is reasonable to assume that the lateral pressure against the vertical place AC is simply $K_a'\gamma z$ where z is the depth beneath the crest BD, following equation 9.31. The total lateral thrust promoting the spreading mechanism would therefore be $K_a'\gamma H^2/2$ following equation 9.33. It immediately follows that the strength of the layer should certainly not drop below

$$c_u = \frac{1}{2}\frac{K_a'\gamma H^2}{L}$$

if spreading is to be avoided. A safety factor of 1.5 would be some comfort. If the fill were rather poorly compacted with $\phi' = 35°$ then K_a' would be 0.27. The designer would presumably like to achieve a slope angle in the region of 30° so that H/L becomes roughly 0.55 and the desirable minimum shear strength of the shallow layer would be 1.5 x 0.27 x γH x 0.55 x 0.5 or $0.11\gamma H$. Such a rockfill dam 40 m high might be susceptible to spreading on a thin stratum weaker than 90 kN/m^2 and certainly susceptible if less than 60 kN/m^2. In this particular geometry spreading is actually almost as critical as the conventional deep slipping mode as you will appreciate if you look back a paragraph. Once again the attitude of the designer will be to attempt the drainage of the suspect material. In all these cases of insufficient undrained strength there is usually a parallel calculation to show that the long-term strength is adequate. The designer will appreciate, therefore, that the short-term strength c_u could alternatively be expressed in friction terms as $(\sigma - u_u)\tan\phi'$ where the short-term pore pressure u_u must be large to explain the relatively small strength. The provision of borehole drains or of slow construction monitored by pore-pressure gauges in the suspect material may offer the designer an escape from his constraint.

Now that the designer has supposedly chosen his preliminary slope angles, and has decided whether or not the drainage of any relatively weak clay zones is likely to be necessary, he must carefully analyse the detailed proposal with the full panoply of stability techniques. In particular, he must use a wide diversity of slip circle and other mechanisms to assess the stability of the dam under all the detailed loading and seepage cases which may appertain. Any evidence of lack of stability must be reflected by appropriate measures, such as better drainage or flatter slopes. Only the most experienced of engineers will achieve a solution which is both safe and economical. The final mode of failure, mentioned at the outset and since ignored, was that due to transient flow which might simply cause the deterioration of the riding quality of the road over a settling embankment, or more seriously the cracking of clay cores in dams due to differential settlement or to the influx of stored water. The problems of settlement of completed structures and the low undrained strength of their subgrades arise from the same cause and have the same solutions, drainage during or before construction. A little more will be said about the effects of settlement on building foundations in section 10.6.

The cracking of clay cores or cut-offs when subjected on one side to the high pressure of the stored reservoir is a complex topic currently receiving attention following the collapse of Teton Dam in Idaho in 1976. It may be that a sudden rise in pore-water pressure on one face of a clayey or silty cut-off can decrease the local effective stresses so that the upstream face attempts to heave, thereby stealing total vertical stress from the downstream face of the cut-off, which may then suffer a horizontal crack which may then be enlarged by erosion due to the concentration of seepage through it. If such a scenario were realistic, the reservoirs behind such dams should be filled slowly enough to allow a step-by-step acclimatisation of the whole cut-off to the higher ambient water pressures.

10.5 Trenches

A trench is usually a means to an end, a steep temporary cutting which will be backfilled as soon as a drain, pipeline, sewer or foundation has been constructed. The structural design of a trench is usually the responsibility of the construction company that has contracted to do the work, although the client's engineer will wish to veto any temporary works that are unsafe. The siting of a trench is normally entirely constrained by its function: if a housing estate needs a new sewer there is little scope for flair in choosing its location. If any degree of choice exists, such as in the routing of a cross-country pipeline, the client's engineer must take his decision partly in recognition of the difficulty and expense of buying or hiring land and partly in recognition of the difficulty of constructing pipelines in rock or in waterlogged or compressible ground. The cheapest land to acquire often affords the most expensive construction.

The chief options in the design of a trench concern the slopes of its sides, the degree of structural support of its sides, the length opened at any one time, and the removal of groundwater. The most frequent sources of trouble are depicted in figure 10.8. (i) the collapse of the vertical sides of poorly supported trenches, which can kill workers in the trench and cause neighbouring ground to subside, (ii) the heaving or softening of the base of the excavation due to high

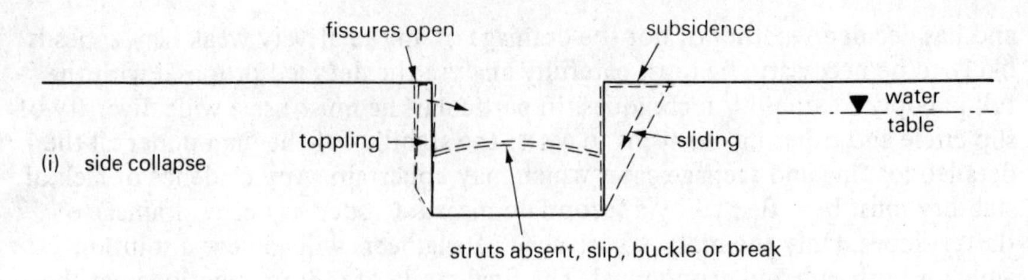

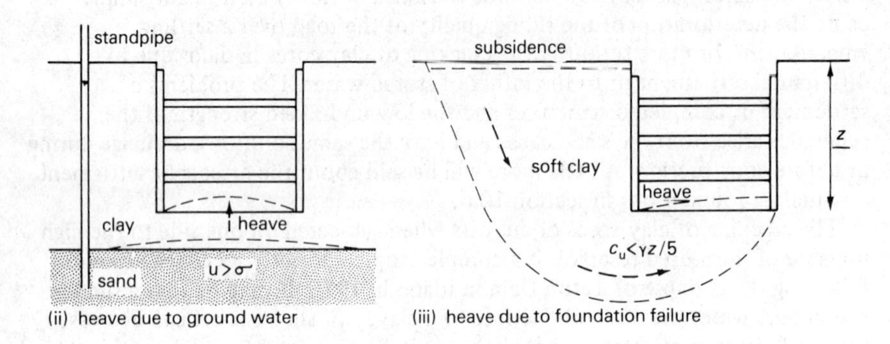

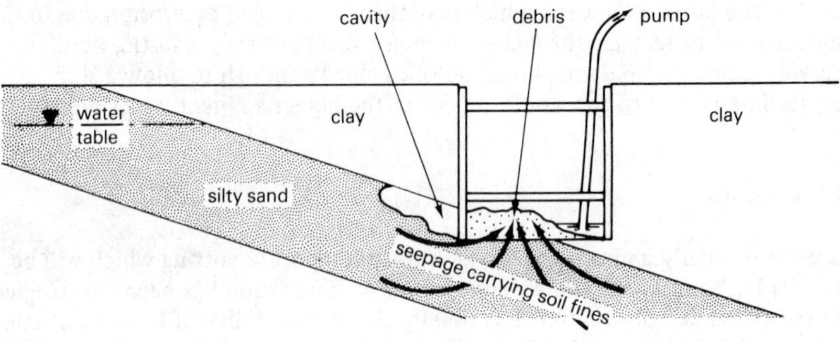

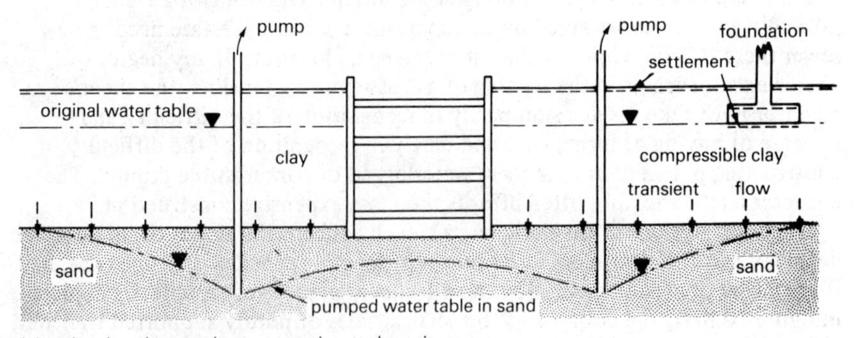

Figure 10.8 *Modes of trench failure*

groundwater pressures, which spoils the foundation for the pipe, etc. (iii) the heaving of the base of the excavation due to shear failure of the soil, causing local subsidence outside the trench and spoiling the levels inside, (iv) the erosive action of the groundwater, washing sand or silt into the trench and causing local subsidence, and (v) the consolidation of neighbouring compressible soils due to the local reduction in groundwater pressures which causes settlement of neighbouring foundations or the opening of sewer joints, etc.

The pressure of groundwater in the vicinity of the proposed trench is the key to each of the modes of failure depicted in figure 10.8. I showed in section 9.6.4 that the presence of static groundwater behind the face of a retaining wall could double or treble the thrust on it, perhaps causing the inadequate strutting to break or buckle as a wedge of soil slides into the trench as shown in (i). If the ground is sandy or silty, high groundwater levels will be immediately obvious due to the influx of water into the trench, which will either flood it or be successful in drawing down the water table. Cedergren (1967) depicts the successive lowering of the water table beside a trench in sandy ground. The final equilibrium water table must offer a regime with a quantity of seepage which corresponds to the rainfall over the appropriate catchment area. Unless he is certain of the high permeability of all the ground, the engineer will often assume, for the purpose of undertaking stability calculations, that groundwater lowering is too slow to be of any help. In section 9.4.6 I pointed out that clay could often be relied on to stand unsupported if it were undrained, and in section 9.6.5 I argued that not only trenches but also vertical gaping cracks might exist to at least a depth $z = 2c_u/\gamma$. Such cracks, if they followed natural joints and fissures, might allow the toppling of narrow columns of even the stiffest clays if trench supports had been omitted. Even if a clay were initially strong enough to resist any mass slide, surface water or groundwater could well soften it by transient flow to satisfy the suction caused by the removal of horizontal stress, and by the generation of shear stresses in 'dense' clay. It is often difficult to assess the variation in groundwater levels along the route of a trench, especially when the soil is predominantly clayey. It is equally likely that vital silty or sandy layers may be missed which alter completely the rate of transient flow. One solution is simply to design trench supports which can separately withstand not only the possible undrained thrust arising from a c_u calculation but also the drained pressure of the notionally submerged ground according to a ϕ' calculation similar to equation 9.35. Another is to take positive measures to pump down the groundwater table around the trench and design the supports to withstand either the immediate undrained stresses or the long-term stresses based on ϕ' and zero pore pressures. It is probably unsafe to allow the design of supports based solely on the large undrained strength of a soil, since the pore suction will be being relied on. Only calculations on the most uniform and predictable firm or stiff clays might, if surface water were well drained and the groundwater table were deep, be reliably based on the undrained strength of the clay as it was excavated: experience would prove whether the time interval before the pore suction was relieved to a dangerous degree would be sufficient to allow completion of the works.

A variety of methods can be used to support the vertical walls of a narrow trench. The traditional method is well illustrated in CP 2003 Earthworks (1959)

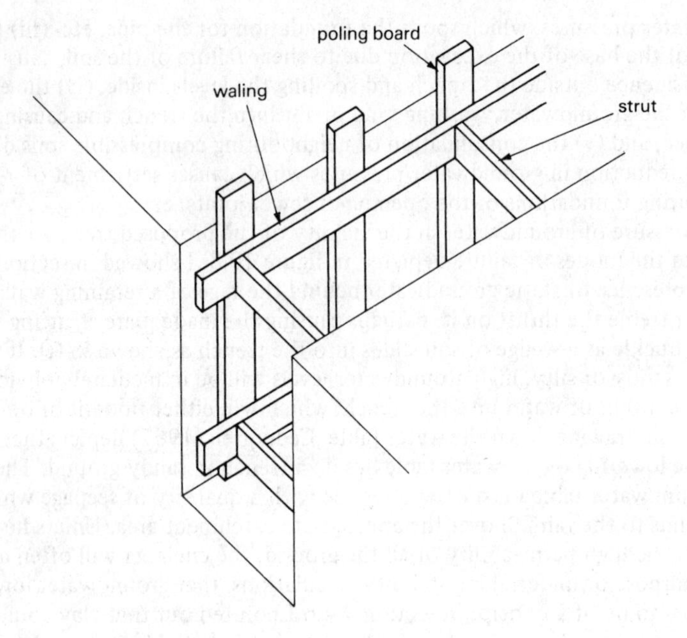

Figure 10.9 *Timbered trench in firm ground*

and used timber planks in a variety of ingenious systems which allowed the work to go ahead in a trench which was always well propped: figure 10.9 depicts a typical scheme for firm ground. The poling boards would be set at some safe interval, restrained by the horizontal walings which would be propped apart by struts. The required spacing and tightness of timbers would be achieved by further props and wedges. The poling boards could be set closely in 'running' ground in an attempt to prevent erosion. It is much more common now to use interlocking steel sheet piles, which reduce seepage from submerged ground to a trickle and which can be predriven to the required depth and then propped apart at various spacings by adjustable steel struts acting on wooden walings. The trend towards mechanisation of narrow trenches is likely to continue, with the development of trenching machines which either obviate the necessity for men to work below ground, or which provide them with a mobile trench support operated by hydraulic rams. Wide trenches will probably be supported by deeply driven sheet piles which act as embedded cantilevers: the bending strength of the sheets may be supplemented by temporary anchors as shown in figure 10.10. If there are no space restrictions the engineer should also consider whether the sides can simply be cut back to some slope angle which will temporarily remain stable with perhaps some elementary drainage provision at the toe.

Section 3.6 dwelt on the heave which is bound to occur if water pressures exceed the weight of overburden. This is exemplified by mode (ii) failure of the base of the trench in figure 10.8, which would be particularly dangerous in clay since there would be no obvious vertical seepage into the trench to warn of the imminent danger. Avoidance demands firstly the recognition that a high groundwater pressure exists beneath the proposed trench, and secondly its

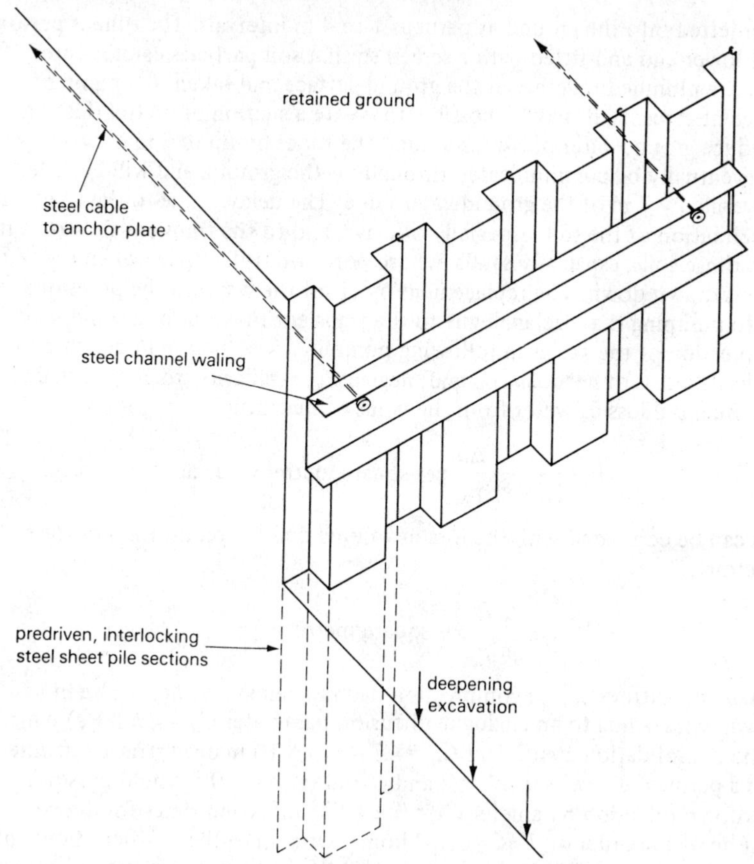

Figure 10.10 *Wide trench supported by anchored sheet piles*

pumping down to some safe level. If the trench were to intersect a large permeable aquifer, as in mode (iv), not only would the quantity of seepage into the trench be troublesome, but also the likely washing-in of the soil particles would ultimately lead to the collapse of cavities in the soil behind the supports. It is easy to sketch a crude flownet for such a profile in order to prove to yourself that the hydraulic gradients adjacent to the trench are so large as to generate the quicksand condition. The designer must constantly remember the concepts of section 3.6, and that seepage with a hydraulic gradient i affects the soil skeleton in a similar fashion to gravity and buoyancy so that the gravitational body force γ downwards is to be compared with the buoyancy body force γ_w upwards and the seepage force $i\gamma_w$ in the direction of seepage. The soil skeleton will stand most stably if it is stacked behind a surface slope which is perpendicular to the line of action of the resultant body force, although slopes up to $\pm\phi'$ on either side will also be stable by virtue of friction. If the resultant body force near the surface has an upward component, the soil is reduced to quicksand. The only economic remedy to instability caused by seepage into the trench is to remove the seepage by a line of wellpoints on one or both sides of the trench. A typical wellpoint installation would consist of 50 mm diameter steel tubes which can be water-jetted into the ground at perhaps 1 to 4 m intervals. The tube is perforated at its lower end and fitted with a screen so that soil particles cannot enter. If the tubes are plumbed together at the ground surface and taken to a pump of sufficient capacity it may be possible to create a suction of up to 70 kN/m^2 so as to reduce groundwater pressures around the tubes by up to 7 m of water. This will eventually be communicated throughout the ground, and will be reflected as a lowering by 7 m of the groundwater table. The delay is due to the transient consolidation of the soil, especially of clays, and to the transient flow of water from those soils, especially sands, whose pores will fill with air when the water table is drawn down. The replacement by air of the water in the pores of a soil due to pumping is a true analogue to the transient flow out of a soil due to compression of the skeleton following pumping. A reduction in pressure by Δu will lead to loss of head $\Delta u/\gamma_w$ and, neglecting capillarity, to an eventual consequential loss of water from the voids of magnitude

$$V = \frac{e}{(1+e)} \frac{\Delta u}{\gamma_w} \text{ per square metre of surface}$$

This can be compared with the loss in volume due to compression of the soil skeleton

$$V = \rho = \frac{\Delta u d}{E'_0} \text{ per square metre}$$

so that the stiffness E'_0 in compression has an analogue $\gamma_w d(1+e)/e$ in drawdown, leading to an analogue diffusion parameter $C_d = kd(1+e)/e$ instead of the consolidation coefficient $C_0 = kE'_0/\gamma_w$. A 10 m deep stratum of fine sand with a permeability $k = 10^{-4}$ m/s and a void ratio $e = 0.5$ would possess a drawdown diffusion parameter $C_d = 3 \times 10^{-3}$ m^2/s and times for drawdown must be of the order of d^2/C_d or 10 hours. Of course, the transient flow only takes place if air actually enters the ground. If all the water in the soil above the draw-down water table were to remain in place by capillary suction, and if the soil were incompressible, the attempt to reduce water pressures would be almost immediately successful: pressures would drop with very little flow. Sands do not possess a deep zone of capillary saturation and therefore participate in transient flow due to air entry. Clays have a highly compressible soil skeleton and participate in transient flow due to compression. Some silts have a relatively deep zone of capillary saturation and yet are quite stiff: the effects of pumping in such soils can spread very rapidly and quite widely.

Trench failure mode (v) in figure 10.8 is a reminder that 'failure' must be interpreted broadly. If concern about the effects of seepage had caused the engineer to reduce the groundwater pressures by pumping, he is vulnerable to the claim by the owners of adjoining land that he had encouraged the settlement of their foundations. The effects are likely to be particularly serious if there are clayey or organic soils in the vicinity, which must compress as their effective stresses increase in response to the reduction in water pressure. The responsible engineer must either convince himself that the ultimate settlement is small or that existing foundations, gas pipes, sewers, etc. are sufficiently remote to be out of the reach of transient consolidation for the duration of the pumping, or he must find some way of achieving his objectives without pumping. Perhaps it would be possible to keep the trench well braced but flooded so that there was no tendency for inward seepage. A technique has been evolved of flooding a trench with bentonite slurry which has the multiple effects of supporting the walls and base with its hydrostatic pressure (γ_s = 10 to 13 kN/m^3) and reversing the trend of seepage in such a way that water leaks out of the slurry into the neighbouring ground. The softening of clayey soils neighbouring the slurry-filled trench is restricted to a few centimetres due to the tendency of the bentonite to form a highly impermeable filter cake on the walls ($k \approx 2 \times 10^{-11}$ m/s), so that the effects of the slurry are entirely beneficial. If a Mohr circle would show that the walls of an unsupported trench were safe up to a height $z = 2c_u/\gamma$ it would show that the walls of a slurry-filled trench would be safe from sliding up to $z_s = 2c_u/(\gamma - \gamma_s)$ while the strength remained constant at c_u. Alternatively he could drive sheet piles as the side supports, and drive them so deep that they cut off and isolated the aquifer beneath the trench so that the effects of pumping from a row of well points in the trench itself would be confined to the foundation bed of the trench.

Mode (iii) in figure 10.8 depicts the heaving of the trench base which can occur as a large-scale mechanism rather than as the piecemeal aggregation of eroded soil particles of mode (iv). If the soil in the vicinity of the trench is entirely clayey, mode (iii) can occur with very little warning, there being a negligible seepage quantity. Mechanism (iii) can be thought of as a foundation failure of the earth flanks with their vertical stress γz being resisted solely by the soil's strength. Applying figure 9.12 with equation 9.16 or figure 9.40 with equation 9.54 would lead to the concept of their being a maximum depth for a well-propped trench in clay with an undrained strength c_u at which heave would occur

$$\gamma z_h \approx 5c_u$$

$$z_h \approx 5\frac{c_u}{\gamma}$$

Trenches deeper than this would simply suffer bottom-heave to such an extent that z_h remained constant while the trencher removed the heaving clay, causing a corresponding subsidence of the neighbouring land. If the engineer attempted to prevent the subsidence by driving the sheet pile trench supports deeper into the clay, he would simply achieve a situation akin to figure 9.29 and would reduce the critical depth to

$$z_h \approx 4\frac{c_u}{\gamma}$$

If the trench were flooded with water or slurry not only would its walls be well supported but also its critical depth for heaving would be increased owing to the restraint offered by its pressure on the base. Application of equation 9.54 for example would then offer

$$\gamma z_h - \gamma_s z_h \approx 5c_u$$

$$z_h \approx \frac{5c_u}{(\gamma - \gamma_s)}$$

which is roughly double the depth previously obtained. If the construction could not be placed in a trench flooded with water or bentonite mud, or if bottom heave remained a threat, the engineer might be forced to adopt the very expensive technique of freezing the ground prior to construction so that the trench could be carved from within a solid block.

When he has sketched out a scheme which he supposes to be viable, concentrating on the benefits and dangers of groundwater lowering, the designer must finally choose the dimensions of any sheet piles, struts, etc. If he has specified steel sheet piles with adjustable steel struts, the engineer is in the enviable position of being able to insert more struts and perhaps even to drive a second wall of piles in an emergency, without suffering enormous costs. As the collapse of a flexible steel and soil construction approaches, large deflections will almost certainly provide an unmistakable alarm. He should nevertheless attempt to be pessimistic, not only about the strength of the soil to be retained and the level of the groundwater table, but also about the likelihood of struts being accidentally knocked away, for example. A useful guide will be obtained if the designer can discover a system of support which could just conceivably collapse at the worst section along the route: he can then double or treble the number of struts. His next step might be to confirm that the provisional design would not collapse if any one strut were removed.

Rankine's active and passive earth pressures calculated after the format of figures 9.21 and 9.22 will probably offer the greatest insight into collapse modes. Consider, for example, the possible positioning of a single line of supposedly immovable struts or anchors to support a supposedly rigid sheet which retains dry sand, as shown in figure 10.11. If the support is placed too high, the sand

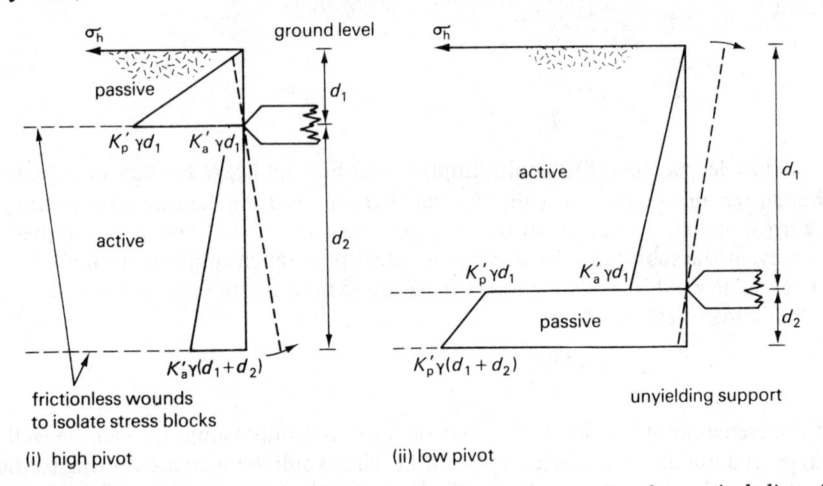

Figure 10.11 *Lateral stresses on a rigid sheet collapsing by rotation about a single line of supports*

can spill out underneath as shown in mode (i), whereas if the support is placed too low, the sand can topple outwards as shown in mode (ii). Assuming pessimistically that the sheet is smooth and that horizontal wounds separate the stress blocks, and assuming optimistically that the deflections of the sheet will be sufficient to mobilise the entire soil strength at every point in the retained soil, the pressures will be passive where the sheet moves into the soil and active where the sheet moves away. The ruling condition for the limiting equilibrium of modes (i) and (ii) is that the moment of the active soil forces just balances that of the passive soil forces. For mode (i), the moments about the support are

$$\frac{1}{6}K'_p\gamma\, d_1^3 = \frac{1}{3}K'_a\gamma\, d_2^3 + \frac{1}{2}K'_a\gamma\, d_1 d_2^2$$

so that

$$2\left(\frac{d_2}{d_1}\right)^3 + 3\left(\frac{d_2}{d_1}\right)^2 - K_p'^2 = 0$$

by using $K'_a = 1/K'_p$. It is very easy to show that the solution is roughly

$$\frac{d_2}{d_1} = 0.4K'_p$$

in the normal range of K'_p values, so that (d_2/d_1) for collapse by mode (i) will vary from 1.2 for $\phi' = 30°$ to 3.0 for $\phi' = 50°$. For mode (ii)

$$\frac{1}{6}K'_a\gamma\, d_1^3 = \frac{1}{3}K'_p\gamma\, d_2^3 + \frac{1}{2}K'_p\gamma\, d_1 d_2^2$$

$$2\left(\frac{d_2}{d_1}\right)^3 + 3\left(\frac{d_2}{d_1}\right)^2 - K_a'^2 = 0$$

which has the rough solution

$$\frac{d_2}{d_1} = 0.55K'_a$$

so that (d_2/d_1) for collapse by mode (ii) will vary from 0.18 for $\phi' = 30°$ to 0.07 for $\phi' = 50°$.

Because very large passive pressures have been inferred to act close to a supposedly unyielding support, and since the soil away from the support might suffer strains larger than those required to generate the peak soil strength ϕ'_{max}, it would be folly to employ K' values based on a higher estimate of strength than that of the critical state ϕ'_c of the poorest conceivable soils to be encountered in the trench wall. In sands and silts such an angle might be 30°, so that the collapsing of a dry sand face of height $D = (d_1 + d_2)$ supported by a sheet and a single line of supports could possibly occur when the depth ratio of the supports (d_1/D) lay outside the bounds 0.45 to 0.85. If three lines of supports were placed at ratios d/D of 0.25, 0.50 and 0.75, therefore, either the bottom or the top two supports could be removed without generating a rigid-body collapse. It is an easy matter to calculate the prop force F and bending moment M in the sheet at a single support just as the sheet is about to collapse.

For mode (i) with $(d_2/d_1) = 0.4K'_p$

$$F_{(i)} = \frac{1}{2}K'_p\gamma\, d_1^2 + K'_a\gamma\left(d_1 + \frac{d_2}{2}\right)d_2$$

$$M_{(i)} = \frac{1}{6}K'_p\gamma\, d_1^3$$

For mode (ii) with $(d_2/d_1) = 0.55K'_a$

$$F_{(ii)} = \frac{1}{2} K'_a \gamma d_1^2 + K'_p \gamma \left(d_1 + \frac{d_2}{2}\right) d_2$$

$$M_{(ii)} = \frac{1}{6} K'_a \gamma d_1^3$$

In the case of a sheet of height D supporting soil of friction angle $\phi' = 30°$ these reduce to

$$F_{(i)} = 0.44\gamma D^2$$

$$M_{(i)} = 0.047\gamma D^3$$

$$F_{(ii)} = 0.54\gamma D^2$$

$$M_{(ii)} = 0.034\gamma D^3$$

If for some reason it was not possible to employ supports which could withstand reactions as large as $0.5\gamma D^2$, it would be necessary to base the design on a sheet supported either by more anchors, or by being driven below the trench base. Serious analytical problems then arise, due partly to the redundancy of the structure, and partly to the deflection-controlled nature of the lateral earth pressures. As an example, the tendency for elastic movement of a single support is harmful since the effective pivot point would be moved further towards the extremities of the sheet and would cause premature collapse: the bounds on the anchor depth ratio would be tighter. If there are many supports, however, and they all yield a little, the *total* thrust upon them will reduce to $K'_a\gamma D^2/2$ since the whole soil profile will probably be in an active state. Comparing the total active thrust for $\phi' = 30°$ which is $0.17\gamma D^2$, with $F_{(i)}$ and $F_{(ii)}$ calculated on the basis that some soil is in a passive condition, it is clear that anchor or prop forces will drop dramatically if the designer can safely eliminate the single support condition and can safely allow sufficient yielding of the supports to generate active soil pressures throughout. It should be clear, however, that the stiffness of the sheet will also have a strong influence on the pattern of deflection, and therefore on the lateral pressures, forces and bending moments.

There may be some merit in using the methods of plastic frame analysis in order to choose a reasonable set of spacings for the support of a braced sheet. Suppose that it were required to use a sheet pile which could develop plastic hinges with a plastic moment of resistance M_p (kNm/m). Any textbook on structural analysis will prove that the uniform pressure required to collapse such a member acting as a beam is $16\,M_p/d^2$ where d is the span between simple supports and where plastic hinges of capacity M_p can be inferred over both the supports and at the mid-point of the span. Figure 10.12 lays out an approximate calculation for spacings to support a face of dry sand, starting with the top span as a simple unpropped cantilever whose applied moment is known to be $K_a\gamma d_1^3/6$. Once the critical spacings have been calculated, the structure can be made safe by the emplacement of additional intermediate supports. Due to the acknowledged redundancy of the 'safe' structure, *each* of the supports actually provided in the region of point 2 in figure 10.12, for example, should itself be

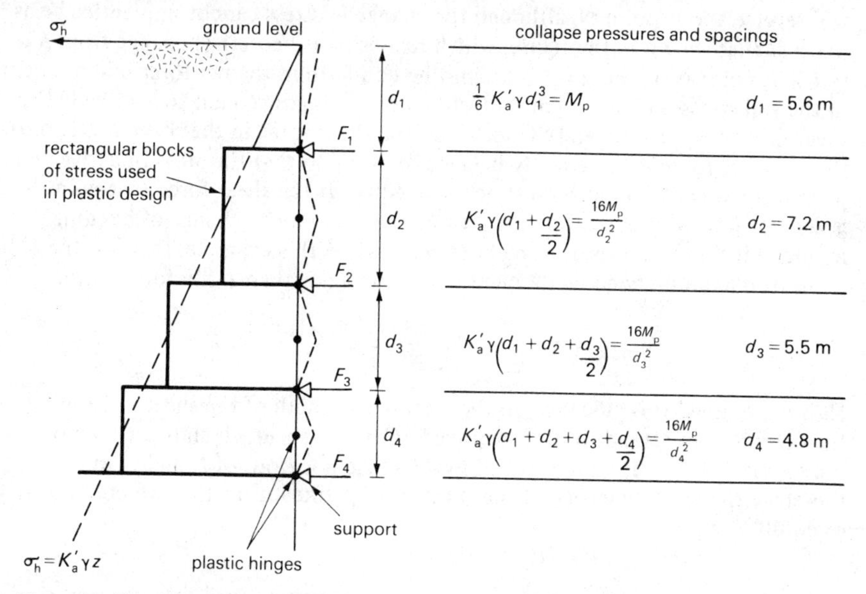

Figure 10.12 *Plastic design of support spacings to generate collapse with example for* $M_p = 200$ *kNm/m* $K'_a = 1/3$, $\gamma = 20$ *kN/m³*, $u = 0$

able to withstand the force F_2 which corresponds to the collapse of the construction due to the accidental removal of its redundant neighbours.

If the engineer has been forced to drive sheet piles well below the trench base in order to reduce the upward hydraulic gradient within the trench, he may face an even more complex analytical problem. Figure 10.13 portrays the crude active and passive soil zones which might protect an embedded sheet AC against monolithic rotation about some point N. The active and passive soil conditions

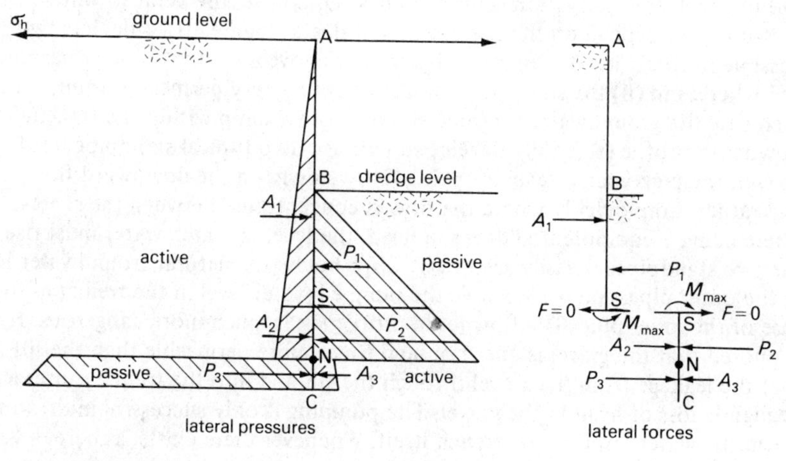

Figure 10.13 *Limiting equilibrium of an embedded sheet*

will reverse about point N, although the change in stress cannot in practice be as sharp as that shown in the figure, which relates to a theoretical change from K'_a to K'_p at point N. If the sheet is in limiting equilibrium the net force and moment of the soil stresses acting upon it must be zero. It is convenient to subdivide the sheet into two parts AS and SC such that the shear force in the sheet at S is zero: this requires force A_1 to exactly balance force P_1 so that the areas of active and passive stress above point S must be made equal. If the shear force F at point S is zero, any structural textbook will show that the rate of change of bending moment at S must be zero. It ought to be clear, in this example, that S is the point of maximum bending moment M_{max}, which is then easily found using

$$M_{max} = A_1 \frac{AS}{3} - P_1 \frac{BS}{3} = A_1 \frac{AB}{3}$$

Having obtained this guidance on the required strength of the sheet in bending, the designer can attend to the lower part SC of the sheet which is carrying only the moment M_{max} at its top. It follows that the positions of N and C and therefore the limiting length of the sheet, can be fixed using the two conditions of equilibrium

$$A_2 + P_3 = P_2 + A_3$$

and, ignoring the slightly uneven nature of the pressure distributions

$$M = (P_2 + P_3 - A_2 - A_3) \frac{SC}{2}$$

The active and passive pressures on the sheet must be estimated after the fashion of figure 9.21 or 9.22, and following the general methods laid out in section 9.6. The principal difficulty will be the assessment of water pressures for the purpose of a friction calculation. In order to be properly pessimistic it is necessary for the designer to imagine the highest possible unbalanced pore-water pressures. This will usually correspond to a flownet after the fashion of figure 3.10 in which the ground is supposed to be inundated up to ground level outside the excavation, and in which the excavation is pumped dry. Of course, the detailed water pressures will depend on the pattern of soil strata. Figure 10.14 depicts two possible profiles: in (i) the ground is uniform above a relatively impermeable bed whereas in (ii) the sheet piles are driven into a very permeable aquifer. In each case the groundwater is removed only from a sump within the trench. The flownet in profile (i) is fully developed with the two typical standpipe heads as shown, the piezometric level dropping fairly slowly on the downward flowpath and rather more quickly where the flow is concentrated between the sheets. There being 8 equipotential drops in head, the level to which water must rise in the two standpipes is easily seen to be $2H/8$ below the natural groundwater level in the outer pipe, and $4H/8$ above the pumped water level in the trench in the case of the inner pipe. The flownet in profile (ii) is much more dangerous. It is supposed that the gravel is one thousand times more permeable than the silt, so that the leakage from the gravel through the silt and into the trench causes a negligible loss of head in the gravel. The pumping is only successful in drawing down the water levels in the trench itself. Whenever there exists, as here, a very permeable buried aquifer charged with high pressure water, it is necessary to

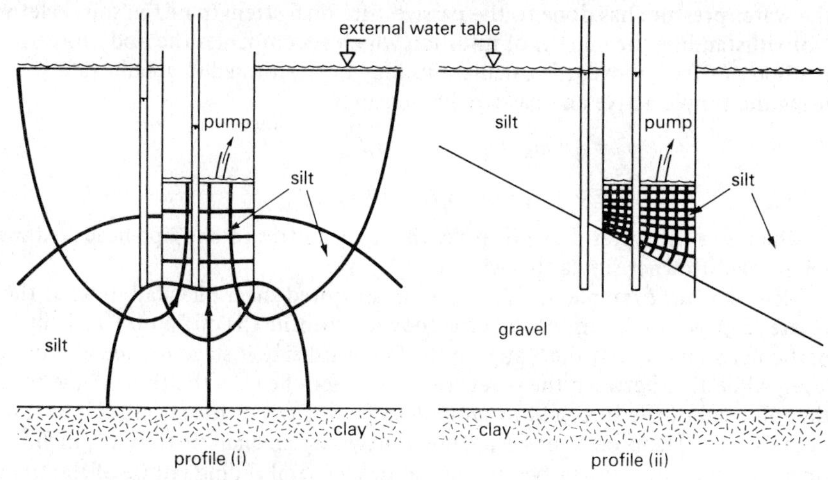

Figure 10.14 *Seepage into sheeted trench*

puncture it with wells if overlying zones of less permeable soil are to be stabilised. The flownet drawn in figure 10.14 for profile (ii) makes clear that the external water pressure remains high while the internal hydraulic gradients are made much more severe than was the case in profile (i). If the engineer expects some dangerous circumstance such as profile (ii) to occur only infrequently along the route for a pipeline, for example, he may adopt the design-as-you-go philosophy expounded by Peck (1969). In this instance he might have a wellpoint system permanently on standby and then design his trench supports on the basis that unfavourable seepage will not occur. If the trench base subsequently starts to heave at one location, or if the supports begin to displace badly, he can then fall back on his prepared remedial plan which would begin with a dewatering operation.

Once he has decided on the water pressures against which to base his initial plan of action, the engineer can go ahead with his calculation of active and passive pressures, by substituting the various standpipe heads h and depths z into the formulae of figure 9.22. The water pressures are bound to form a significant part of the total stresses on the sheet. If the designer intended to drive the sheets only as deep as was necessary just to avoid the quicksand condition in the base, then it would be quite illogical to ascribe any passive resistance in front of the embedded zone, and the sheet would have to be designed propped above the base. Even if the sheet were driven so deeply that the upward hydraulic gradient was perfectly safe, its effect on passive resistance is still very important. If the upward hydraulic gradient were 0.4 then the upward body force due to seepage and buoyancy would be $1.4\gamma_w$. The downward body force due to gravity is γ, so in such a case the net effective downward force would be $(\gamma - 1.4\gamma_w)$ or perhaps 6 kN/m^3 compared with the 20 kN/m^3 that the same soil would possess if it were dry. At some depth x beneath the trench base the effective passive soil stress would then be $6xK'_p$ rather than $20xK'_p$, with only a very small benefit in terms of the extra water pressure to be added to the effective stress, which would be numerically insignificant compared with the enormous damage which the water pressure has done to the passive frictional strength of the soil skeleton. Notwithstanding the benefit of intuition which accompanies the body force method outlined above, the calculations may be performed in greater safety using the simple active and passive relationships

$$\sigma_a = u + \sigma'_a = u + (\gamma z - u)K'_a$$

$$\sigma_p = u + \sigma'_p = u + (\gamma z - u)K'_p$$

in which u is the water pressure at depth z derived from standpipe heads h drawn on a scaled flownet similar to figure 10.14.

Consider, for example, the design of an anchored sheet pile wall to resist the severe seepage conditions of the type shown in profile (ii) of figure 10.14. In particular suppose that the water on the landward side is stagnant and at ground level, while that between the sheets rises from zero head at the trench base to the full head of the surrounding ground at the bottom of the sheet, as shown in figure 10.15. The process of design might begin by an analysis to find which small depth of embedment would just allow a general sliding failure of the trench wall, with the line of anchors yielding at some force T per unit length as the soil

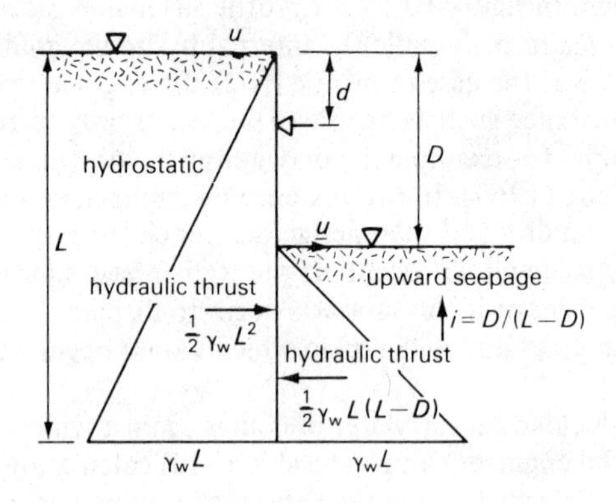

Figure 10.15 *Water pressures*

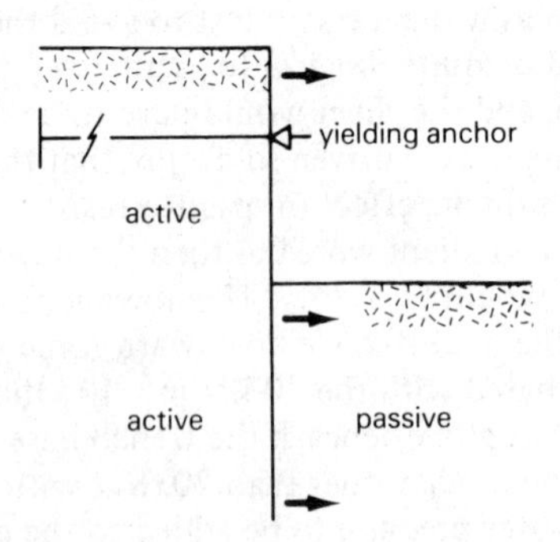

Figure 10.16 *Deformations*

in the trench base is forced into a passive condition by the large active pressures from the outside: the hypothetical deformation diagram leading to the clarification of the active and passive soil zones appears in figure 10.16. The effective lateral stresses of figure 10.17 are then easy to derive on the assumption

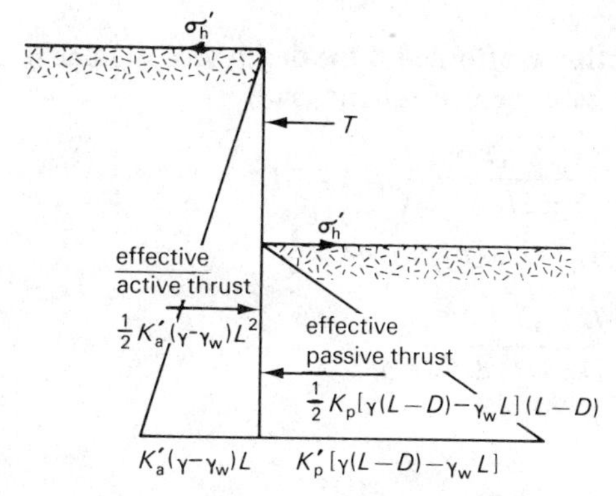

Figure 10.17 *Effective stresses*

that the surface of the sheet is frictionless. The unknown sheet length L and anchor force T which just produce this limiting state of equilibrium can then be discovered by resolving forces and taking moments.

Figure 10.18 displays the resultant forces per unit length found by summing

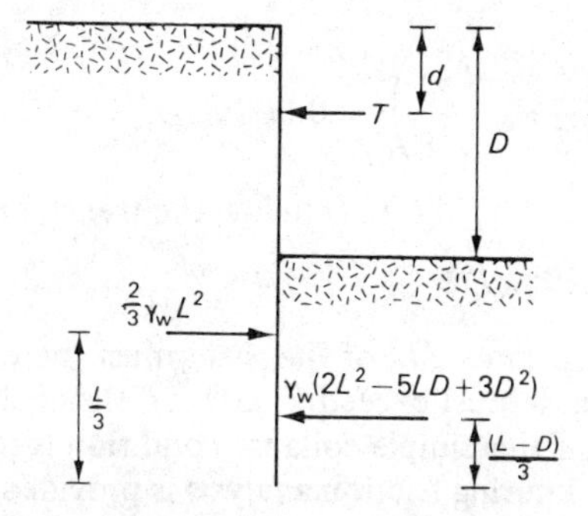

Figure 10.18 *Resultant forces (for case $\gamma = 2\gamma_w$, $\phi' = 30°$*

the pressures of figures 10.15 and 10.17 in the special case of soil with $\gamma = 2\gamma_w$ and $\phi' = 30°$ so that $K'_a = 1/3$ and $K'_p = 3$. Resolving horizontally

$$T = \frac{2}{3}\gamma_w L^2 - \gamma_w(2L^2 - 5LD + 3D^2)$$

$$T = \gamma_w\left(5LD - 3D^2 - \frac{4}{3}L^2\right)$$

Taking moments about the line of action of the passive thrust

$$T\left[L - \frac{(L-D)}{3} - d\right] = \frac{2}{3}\gamma_w L^2 \left[\frac{L}{3} - \frac{(L-D)}{3}\right]$$

$$T = \frac{2\gamma_w L^2 D}{3(2L + D - 3d)}$$

A rather easy solution is afforded if the depth d of the line of anchors is chosen to be $D/3$. In that case the two equations offer

$$\frac{2\gamma_w L^2 D}{3 \times 2L} = \gamma_w\left(5LD - 3D^2 - \frac{4}{3}L^2\right)$$

$$4L^2 - 14LD + 9D^2 = 0$$

$$\frac{L}{D} = \frac{14 \pm \sqrt{52}}{8}$$

$$\frac{L}{D} = 2.65 \text{ and } 0.85$$

Ignoring the impossible value of 0.85 the method has shown that a sheet of length $L = 2.65D$ supporting a trench of depth D will just collapse forward against a yielding support which offers $T = 0.88\gamma_w D^2$ per unit length at a depth of $D/3$ below the crest, when the soil can generate $\phi' = 30°$ and when the seepage system is very harsh. The maximum bending moments in the sheet must occur when its shear force passes through zero. This happens firstly at the anchor where

$$M = \frac{1}{3}\frac{2}{3}\gamma_w\left(\frac{D}{3}\right)^3 = 0.008\gamma_w D^3$$

and secondly at a depth of roughly $0.2D$ below the trench base, where

$$M = 0.39\gamma_w D^3$$

The plastic moment of resistance M_p of the sheet must therefore exceed $0.39\gamma_w D^3$, just as its length must exceed $2.65D$ and the anchor yield strength must exceed $0.88\gamma_w D^2$, if this simple collapse condition is to be avoided.

A further example of limiting friction analysis is provided by the design of a sheet anchor to withstand some lateral force T. Once again it is important to base the design on some angle ϕ' which can be mobilised without undue strains in ground which is quite likely to be disturbed during construction; triaxial tests may confirm that $(\phi'_c - 10°)$ is a safe choice for loose soils and that ϕ'_c is safe for dense soils. The second assumption must be of a groundwater table which is as high as might reasonably be expected. The third step is to imagine a reasonable pattern of deformations, and to apply Rankine's active and passive zones where appropriate in order to make the soil structure statically determinate and in a condition of limiting equilibrium. Figure 10.19 depicts the limiting equilibrium of a vertical sheet anchor of unit length buried up to its top surface in dry sand with the anchor cable attached at one-third height. Since the anchor force is applied at its centre of pressure, the sheet simply tends to slide forward. It is

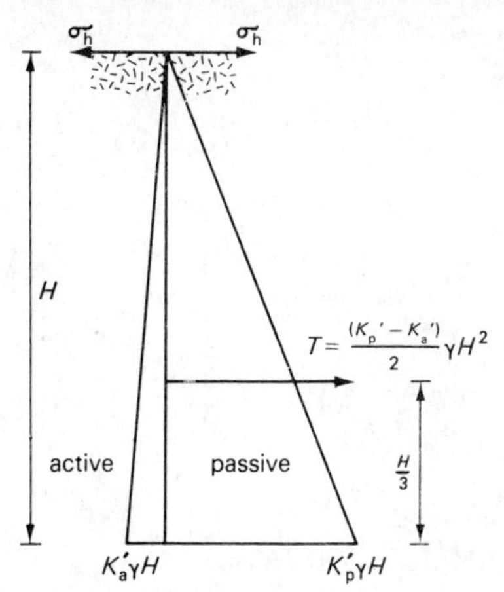

Figure 10.19 *Anchor at centre of pressure of vertical sheet buried in dry sand*

clear that the anchor resistance is the sum of the passive pressures minus the sum of the active pressures

$$T = 0.5\,(K_p' - K_a')\gamma H^2$$

If the sand had been totally submerged, the water pressures on the two sides would have cancelled out, but the effective stresses would all have been reduced to give

$$T = 0.5\,(K_p' - K_a')(\gamma - \gamma_w)H^2$$

or roughly one-half of its previous value. If on the other hand the cable were attached to the top of the sheet, there would be a tendency to rotate, and figure 10.20 shows the sheet in dry sand and pivoting about N with the appropriate zones of active and passive pressure. The unknowns are T and H_1/H_2, and they are found by considering the limiting equilibrium of the sheet. Taking moments

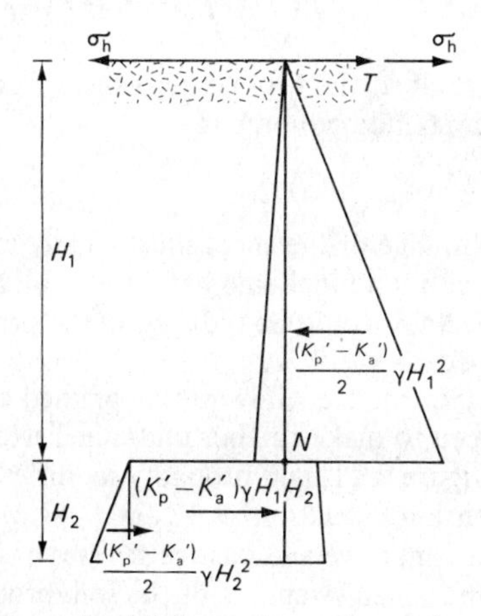

Figure 10.20 *Net forces on buried sheet with anchor at the top*

about the top

$$(K'_p - K'_a)\gamma \frac{H_1^2}{2}\frac{2H_1}{3} = (K'_p - K'_a)\gamma H_1 H_2 \left(H_1 + \frac{H_2}{2}\right) + (K'_p - K'_a)\gamma \frac{H_2^2}{2}\left(H_1 + \frac{2H_2}{3}\right)$$

On rationalising this becomes

$$\left(\frac{H_2}{H_1}\right)^3 + 3\left(\frac{H_2}{H_1}\right)^2 + 3\left(\frac{H_2}{H_1}\right) - 1 = 0$$

which has the solution

$$\frac{H_2}{H_1} = 2^{1/3} - 1 = 0.26$$

It is then easy to resolve forces horizontally in order to show that

$$T = 0.13\,(K'_p - K'_a)\gamma H^2$$

If the sand had been submerged, the effective stresses would have been proportional to $(\gamma - \gamma_w)$ instead of to γ, so that the anchor force would be

$$T = 0.13\,(K'_p - K'_a)(\gamma - \gamma_w)H^2$$

Rankine's analysis has been useful in demonstrating a tenfold reduction in anchor capacity if the soil is flooded and the cable attached to the top of the sheet instead of at the centre of pressure. Of course, the sheet must be strong enough to withstand the bending moments implied by figures 10.19 and 10.20. The greatest moments occur where the shear force is zero. The sheet in dry sand with its anchor cable at the centre of pressure will possess a simple bending moment diagram with its maximum at the point of connection

$$M = (K'_p - K'_a)\frac{1}{2}\frac{\gamma}{3}\left(\frac{2}{3}H\right)^3 = 0.049\,(K'_p - K'_a)\gamma H^3$$

The sheet with its anchor cable at the top has zero shear force roughly at its mid-point, with a maximum bending moment

$$M = 0.044\,(K'_p - K'_a)\gamma H^3$$

The plastic moment of resistance of the sheet must clearly exceed these values if the sheet is not to bend about a hinge line before the full anchor load can be applied. Once again these values would be reduced in the proportion $(\gamma - \gamma_w)/\gamma$ if the sheets were submerged.

If the soil in the vicinity of the trench were fine grained and compressible, the engineer would be forced to make further undrained strength calculations based on the concepts of figure 9.21, and probably modified after the fashion of figure 9.27 to eliminate tension. Equations 9.42 and 9.43 would then represent a good working hypothesis for active and passive stresses based on some pessimistic value for the undrained strengths of the soil, probably obtained by undrained triaxial tests. These propositions lead to the conclusion that trenches shallower than $2c_u/\gamma$ should stand unsupported, that props or some overdriving of sheets is necessary between depths of $2c_u/\gamma$ and $4c_u/\gamma$, and that trenches deeper than $5c_u/\gamma$ can never stand however deep the sheets may be driven into the clay of strength c_u and however well propped the sides may be, because of the impossibility of preventing a passive failure of clay beneath the trench base. It is, of course, possible to perform the same sort of structural calculations as those previously introduced which were based on the friction model, and to replace some or all of the limiting pressures by their cohesive alternates. It is certainly necessary to use the cohesion calculations as a guide to whether trenching is likely to be impossible, and to take the decision to avoid the open trench problem if $D > 4c_u/\gamma$ in a very deep stratum of clay. Alternatives depend on the original purpose of the trench and include tunnelling in compressed air, pipe jacking, slurry trench construction and ground freezing. Preconsolidation by groundwater lowering is an option in marginal ground if it can be sufficiently strengthened to make the trench stable. If the weak cohesive soil of strength c_u is not itself too deep, it is possible to overcome the problem of passive heave of the trench base by driving the sheet piles deep into some more competent soil beneath, which can provide the necessary passive restraint. Sheets which were well supported in the trench and of great stiffness would be required if the clay were both weak and of great depth, since the large active pressures from the clay must then be carried over a long span between the props at the top and the passive bed at the base.

I have not included any surcharge on the ground close to the trench in any of the foregoing structural calculations. It should be clear however, that an extra pressure σ_0 on the ground surface would raise the lateral stress down the whole active zone by $K_a'\sigma_0$ in a friction calculation, and by σ_0 in a cohesion calculation. The effects of placing excavated soil close to the open trench can be dramatic, therefore, and the engineer must correctly anticipate any such necessity if the strip of land available during construction is only narrow. Any heavy machines which will work close to the trench must likewise be included in the designer's calculations, possibly being converted notionally into an equivalent surcharge applied over their plan area. The technique of trial wedges introduced in section 9.4 might alternatively be used to adjudicate on the influence of line loads or an irregular heap of surcharge which might be considered to be in the vicinity of the trench. The qualitative importance of surcharge σ_0 can be appreciated by considering that the stability of the trench is almost equivalent to that of a second trench without surcharge but σ_0/γ deeper than the first. A surcharge of 40 kN/m^2 is then roughly equivalent to deepening the trench by 2 m since driving lengths are proportional to D, support forces to D^2, and bending moments in the sheets to D^3, such an increment may well be fatal if it were unforeseen.

The designer must be sensitive to the purpose of his construction, and must be certain that his calculations have guaranteed that it will fulfil its designated purpose adequately. Many temporary supports for trenches can perfectly well be designed using active and passive pressures to guard against a gross collapse. Others may be designed safely by choosing a very conservative value for the strength of the soil. Some trenches, however, will be so close to existing buildings or services that the collapse calculations must be complemented by predictions of the deflections of the structure and the adjoining ground. Rowe (1951) and (1955) has proposed a quasi-elastic analysis of the sheet pile wall problem which is normalised against the results of a large number of model tests, and which offers a route to the design of such a structure in relation to its stress–strain behaviour rather than to its limiting state of collapse. Bransby and Milligan (1975) present model test results which support some fairly simple propositions regarding the strains and displacements behind a sheet pile cantilever wall. Their paper offers an excellent introduction to the exploration of strain fields based on a Mohr circle of strain for a dilatant material. Milligan and Bransby (1976) go on to analyse in detail the fields of stress and strain in the vicinity of a simple sheet-pile wall retaining dry sand in terms of the peak angle of shearing resistance and angle of dilation of the soil. An elementary treatment of the elastic–plastic interaction between a wall and its retained soil was included in section 6.10. Finally, some appreciation can be gained of the variety of responses elicited from soil mechanicians by the soil structure interaction problem, on reading the proceedings of the 5th European Conference on Soil Mechanics and Foundation Engineering, Madrid 1972, which was devoted to that topic.

10.6 Foundation Design

10.6.1 Economy

Rigidity and cheapness are the desired attributes of a foundation: they are in conflict. If a foundation is not to move, it must either spread the weight of its structure over a large area so as to reduce stresses or it must seek out some deep stratum of very stiff soil or rock. The most rigid foundations therefore require the greatest excavation of materials, and the greatest volume of concrete in replacement. The difficult task of the foundation engineer is to discover some compromise which is relatively cheap and which is rigid enough to reduce the risk of structural damage to within acceptable limits. The options open to the designer include shallow foundations (pad footings probably in the form of rectangular pits filled with plain concrete, or strip footings and rafts in reinforced concrete), basements which complement the action of a raft with that of buoyancy, and piles which are long slender members searching out deeper, more competent, strata of soil or rock.

Economy in the design of foundations is achieved less by reducing expenditure on materials than by reducing the time and labour necessary to construct them. For example, it becomes exponentially rather than arithmetically more expensive to dig a square pit to a greater depth due to the extra precautions which must be taken to keep the pit free of water and its sides safely supported. Likewise it may cost much more to install reinforced concrete than it does to place plain concrete, due to the difficulty of fixing the steel rather than the cost of buying it. While shallow holes will remain cheaper than deep holes, and while simplicity will always rule in a designer's favour, the relative merits of the various options are bound to change as the relative costs of labour, fuel, concrete, steel and land also change. It is also clear, however, that a designer who uses double the necessary number of piles is wasting half his client's money.

One of the most difficult aspects of the designer's struggle to save his client unnecessary expense concerns the initial investigation of the site. If no investigation is carried out at all, the designer will be forced into an ultra-pessimistic stance and may feel obliged to specify, for example, a close matrix of piles driven until they can be driven no more. Not only might such a plan be grossly expensive, it might even be futile. If there were an erratic distribution of boulders in a glacial clay, for example, piles may be driven to refusal on top of a boulder, only to settle under the structural loads owing to the eventual consolidation of the clay beneath. If on the other hand the designer requests knowledge of every cubic metre of ground, perhaps he could take precise advantage of every strong zone and avoid every soft spot. This plan, like that of the other extreme, would be grossly expensive and similarly runs the risk of futility. When the engineer investigates he disturbs: the boring of holes and the sinking of pits offers information about the pre-existing ground but it also creates new 'ground' when the holes are backfilled. Heisenberg would have enjoyed this particular uncertainty, which is yet another example of an observation changing that which is observed. The extremes having been shown to be useless, the designer must adopt a reasonable compromise in which information is sought until the additional cost of supplying it outweighs the likely benefit of having it. The key word is 'likely'. The danger in stopping the investigation too soon is that some soft spot may have escaped attention, perhaps a bowl of peat in a glacial sand, perhaps an abandoned mine shaft which has been capped beneath ground and covered over. It is the risk of having left such a possible feature undiscovered which must be offset against the money saved in a minimal investigation. It is at this juncture that the experienced professional engineer must part company with a client who will not accept his judgement, and who refuses to release sufficient money to conduct a reasonable investigation.

10.6.2 Fixity

The designer must provide foundations of sufficient fixity. Excessive movements lead to the cracking of walls and the distortion of frames, or to the disruption of machinery. The designer must therefore be able to link his expectation of the movement of foundations to the damage which might follow. In estimating this movement he must consider the settlement of the ground due to the loads he is about to apply, the reciprocal settlement of an existing building due to a new building, the settlement of compressible soils which may follow any lowering of groundwater level, the similar movement due to whatever cycles of temperature or wetness are likely to afflict the ground surface, the exacerbation of suction caused by vegetation in a drought, the shakedown of loose sands by vibration, the slow ground motions which must accompany most forms of mining, and the fast ground motions which are referred to as earthquakes. And if he satisfies all these criteria but forgets to protect his foundations from chemical or biological attack, his work may be in vain. The designer's initial reaction to a threat should be to attempt to forestall it: only when he has taken strong avoiding action should he proceed to compute the residual risk. If elementary settlement calculations based on the methods of chapter 6 seem to offer the likelihood that some part of a building will settle by more than 50 mm if it is supported on conventional shallow footings, the designer will begin to consider either the use of piles or a buoyancy basement. If seasonal shrinking and heaving appear to have cracked the façades of neighbouring buildings founded on compressible clay, the designer may consider using deeper footings. He may alternatively bore wells into some underlying aquifer, and fill them with fine sand, through which the water could rise, so that the local groundwater pressures in the clay are stabilised at the roughly constant reference of the head in the aquifer. If the ground floor or basement of the building is to house a boiler or a refrigerator the designer will pay great attention to insulating the ground and air-conditioning the plant. If the site is underlain by pockets of fill or loose natural sands the designer might consider compacting the entire area rather than bothering to locate the weak spots, and in preference to piling. If the site of his prospective structure has been undermined, the designer will have to take steps to discover whether any cavities remain and to fill those which may collapse and cause damage: cement grout can be injected through boreholes. If the site is to be undermined in the future the designer must attempt to minimise future damage. By subdividing his structure into strong monolithic compartments separated by wide compressible packing pieces, and by founding each compartment on a stiff raft underlain by sheets with a small coefficient of friction, the designer is providing the greatest opportunity for his structure to 'ride' the waves of subsidence which will cause first a spreading, then a distortion and finally a recompression of the ground as they pass beneath. A similar philosophy may be required if a structure is to be located in an earthquake zone. Stiff 'boxes' founded on rafts have the best chance of resisting the strong horizontal accelerations which are the most destructive component of earthquakes. Finally, the designer should sample the chemical and biological environment in which his foundation is to be constructed in order to avoid using materials which would degrade, or to choose some resistant coating.

10.6.3 Preventing Collapse

Having tentatively chosen a scheme for the structure and its foundations, the designer's second step is to assess the largest forces and moments that may be delivered to the foundations by the structure. For this purpose the foundations are often assumed to be rigidily fixed into the ground, although there is increasing recognition that the pattern of forces and moments in the composite structure of ground plus columns plus beams may be altered quite strikingly by incorporating such equations as 6.21 to 6.23 into the standard slope-deflection analysis of the structural components above ground. In section 6.9 we discussed the stiffness of a shallow footing on its supposedly elastic bed relative to that of a supposedly elastic cantilever column, and deduced simple rules from equation 6.32 with regard to the necessity of disregarding the fixity of bases which are likely to prove too flexible. In section 9.9 I pointed out that the bearing capacity of a shallow footing was drastically cut if it were forced to deliver an inclined load through its horizontal bearing surface. Moments will likewise be shown to reduce the load-carrying capacity of a shallow footing of a given size. These considerations may either provoke a designer into providing very large or deep foundations which will be capable of offering the degree of elastic and plastic fixity that his flexible structure demands, or they may induce him to reduce or at least equalise the actions on his light foundations as much as possible by stiffening the structure or by providing a substructure of groundbeams to tie the footings together.

When the designer has chosen a scheme and calculated the forces which will be delivered to the foundations in the full variety of loading cases appropriate to the working life of the structure, his third step might be to so proportion the foundations that they will never collapse according to the concepts laid out for long strip footings in chapter 9, modified to include foundations of any shape or depth. The designer must firstly assess the state of the soil and attribute a safe value to its angle of shearing resistance ϕ'. In addition to ϕ' the mobilisable undrained strength c_u of fine-grained soils must be estimated if there is any doubt about their capacity to consolidate quickly when the structural load is applied. Soils with a transient response to stress which is much slower than the rate of application of the load will force the designer to duplicate his calculations, so that his construction may be proved capable of standing safely in the short term when the subgrade is undrained and its strength is c_u, and in the long term when it is drained and its strength is $(\sigma - u) \tan \phi'$ with the pore pressure u determined from some steady long-term groundwater regime. It is necessary for the designer to choose the lowest conceivable values of c_u or ϕ' and the highest possible future groundwater pressures u which may apply to his foundation. Chapters 4 and 8 dealt in some detail with the determination of the shear strength of soils using direct shear and triaxial compression. If the designer wants to use pessimistic remoulded nondilatant strengths relevant to the critical states of the soil, he will be prepared to accept disturbed samples of the soil from a site investigation. These should not be too difficult to supply. Clays and silts can be recovered in an almost undisturbed condition using standard tubes 100 mm in diameter jacked into the natural ground beneath a bored hole. Highly fissured clays should be completely remoulded in order to find their critical effective angle of shearing resistance ϕ'_c. The ideal for fissured soils is otherwise to test a sample very much larger than the fissure-spacing, so that it can fail along just the sort of interconnection of fissures which the full-scale construction would also choose. There is no alternative to large diameter sampling and testing if their operational 'undrained' strength is required, although different degrees of internal drainage are inevitable and the meaning of c_u will therefore be shrouded. Sands can be recovered in a similar fashion if the sampling tube has a nonreturn flap at its base: their speed of consolidation or dilation will inevitably have allowed the density to change, but the soil particles ought to be representative. Gravels, and any soil containing stones or boulders, can only be recovered haphazardly. If the designer wants to use the *peak* strength of naturally dilatant, cemented or sensitive soils, he faces a rather more difficult problem. Soft fine-grained soils without stones may be recovered in excellent condition using thin walled tubes, but the recovery of other soils in an undisturbed condition poses great practical difficulties which can be overcome only with an expensive provision of equipment and skilled hands.

10.6.4 In-situ Tests

The expense and difficulty of recovering good soil samples has led many foundation engineers to question their necessity. In particular, the use of a quick exploratory boring survey followed by an *in-situ* test programme within the more difficult strata has proved increasingly popular. Two concepts may guide the engineer in the use of *in-situ* soil tests. He may either envisage the test as an unconventional test on a soil element which happens to remain in the ground, and which will offer him a soil stiffness modulus followed by a shear strength parameter ϕ' or c_u, or he may think of the test as of a small model foundation whose results can be scaled in some way to predict directly the behaviour of the full-scale construction. The tests themselves usually require that an instrumented probe of some sort is introduced into the ground and is then activated in some way which applies a measured force to shear the soil around or beneath the probe. Figure 10.21 depicts the general nature of a representative group of tests. Schmertmann (1975) offers an excellent review of the performance of these and other *in-situ* tests with regard to their ability to generate meaningful information concerning soil strength.

The SPT, mentioned in section 6.6, consists of hammering a thick-walled tube into the ground beneath a bored hole. After an initial drive of 150 mm the number of blows of a standard hammer which are required to drive a standard tube a further distance of 305 mm is referred to as N, the SPT blow-count. The energy lost by friction in the hammer-drop or resonance in the driving tubes, the testing of borehole debris rather than undisturbed ground, the nonequalisation of the water pressure in the ground with that in the hole, can each completely invalidate the SPT. Even when the test is properly conducted it is now widely recognised that the value of N is a very complex function of both static and dynamic soil parameters, and that the test itself is not a good representation of any full-scale construction or process, other than the driving of certain piles perhaps. As the tube is hammered downwards the soil is forced to 'flow' around and above it. The denser the soil the higher will be the N value owing to

(1) the greater angle of friction $\phi' = \phi'_c + \phi'_v$ where ϕ'_v is roughly proportional to the relative density of the soil, as explained in section 8.6.1

(2) the greater transient suction in the soil due to the sudden shear process, as explained in section 8.5.3, which causes a transient increase in the effective friction

(3) the greater stiffness of the surrounding ground which allows greater lateral stresses to be generated during driving, which then create extra friction on the side of the tube.

Although the relative proportion of these effects will vary from soil to soil, their cumulative effect causes a crude correlation to be obtained between a proper N value and the relative density of a sand. Because of the relationship between ϕ' and relative density this results in a crude correlation between ϕ' and N, such as that presented by Schmertmann which was based on de Mello (1971) and which is shown in figure 10.22. It should be no surprise that the effective overburden pressure also has a positive effect on the N-value of a soil of given ϕ', there being a contribution to driving resistance caused by the generation of extra friction by

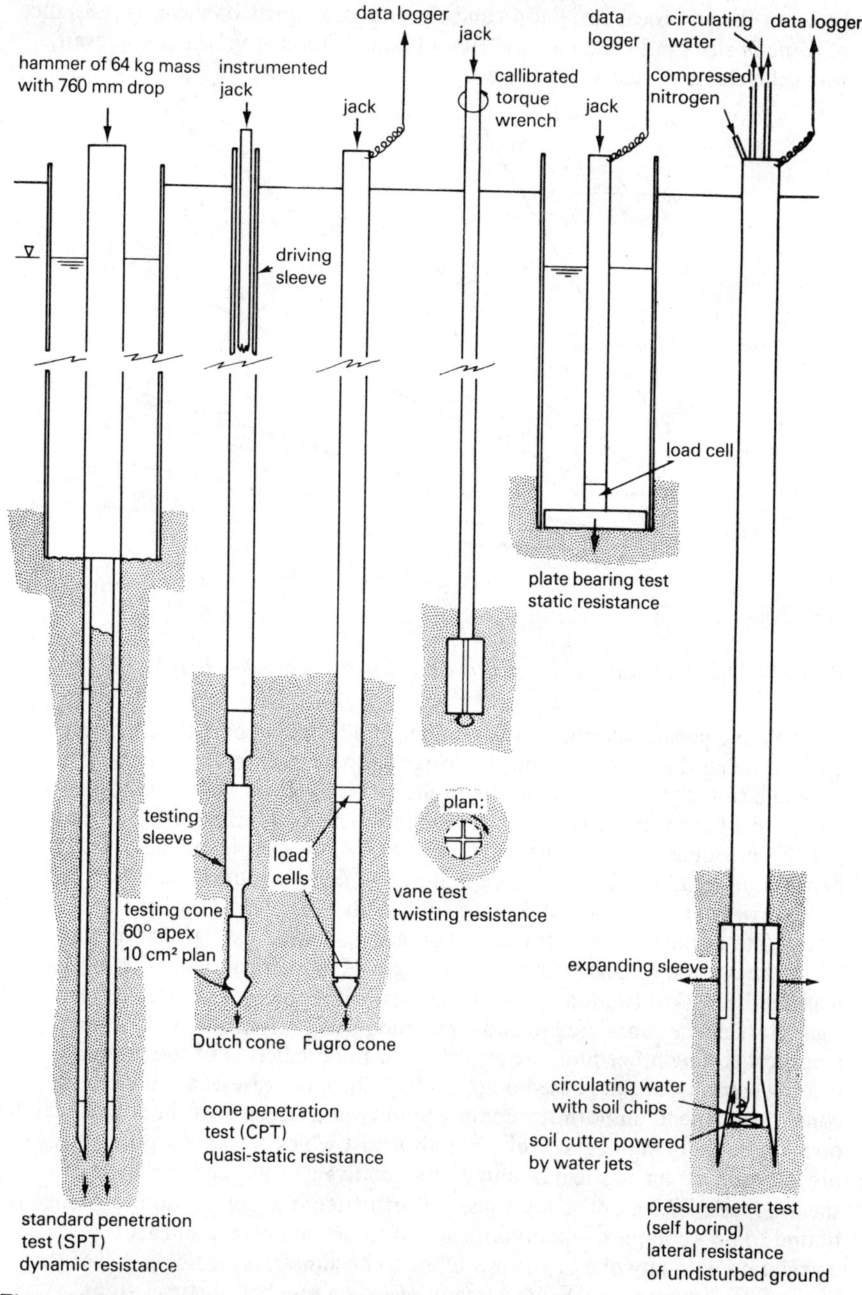

Figure 10.21 *In-situ soil probes (roughly 1/5 scale)*

the *in-situ* stresses. The dependence of N on such a variety of partially independent soil properties makes it likely that variations of ±5° will occur between the empirical correlation and the true peak angle of shearing resistance of some *in-situ* sand which has a different constitution from the norm, even when the test is properly conducted.

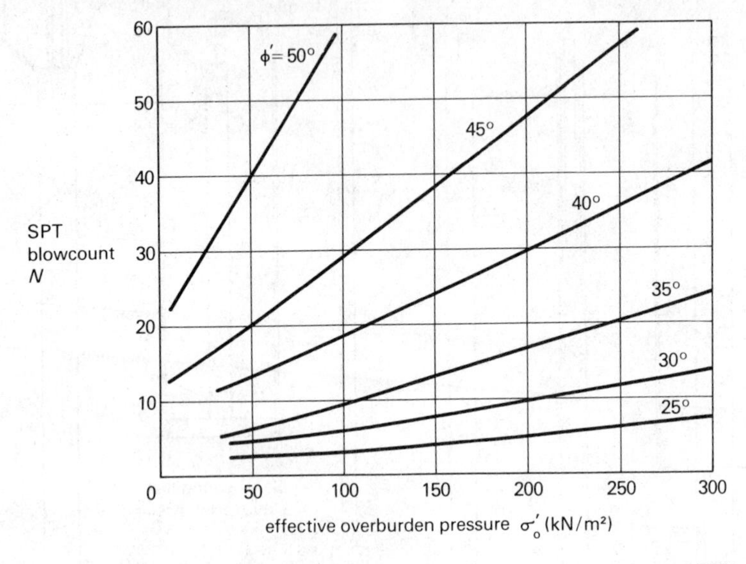

Figure 10.22 *Empirical correlation between N and ϕ' (from Schmertmann, 1975)*

The cone penetration test has a number of advantages over the SPT, the greatest being that it is based on the penetration of rods in the absence of a borehole so that the soil around the probe is less likely to suffer from the sort of gross disturbance that often attends an improperly conducted SPT. Sanglerat (1972) introduces some of the wide range of CPT probes, including those depicted in figure 10.21. The force required to drive one of the standard 60° cones downwards at a rate of 5 mm/s, divided by the plan area of the cone, is referred to as the cone penetration resistance q_c kN/m^2 (or kgf/cm^2 in most literature). If the force developed on a length of cylindrical sleeve is also measured, the skin friction f_s kN/m^2 can also be reported. It should be clear that such a test can be considered to be a direct model of a driven pile, and a number of empirical scaling procedures are available for the prediction of the ultimate bearing capacity of piles based on q_c and f_s. Since the ease of advance of the cone must depend on both the compressibility and strength of the soil, it may be no surprise that empirical correlations also exist which separately purport to predict each of: an elastic modulus E, the relative density, and the angle of shearing resistance ϕ', of sandy soils. Unfortunately the penetration resistance is bound to be a complex function of each of these parameters, so that the correlation with any one of them is likely to be almost as mediocre as was that of the SPT. Nevertheless, Schmertmann reviews a number of correlations between q_c and ϕ' which are based on bearing capacity formulae and which offer the designer a little guidance on the peak *in-situ* value of ϕ' in sands and other relatively free-draining media. The uncertainty over transient pore pressures, whether positive or negative, can easily be overcome by ensuring that the CPT is conducted slowly enough to ensure their dissipation. Correlations also exist between q_c and c_u in soils which are supposed to have been tested so quickly that they remained undrained. Typical relationships are

$$q_c = 1.3 N_q \sigma_0' \tag{10.6}$$

and

$$q_c = \sigma_0 + 9c_u \tag{10.7}$$

in which the factor 1.3 is an empirical adjustment to equation 9.61 to account presumably for the circular rather than plane strain geometry, and the factor 9 is an empirical adjustment to the 5.14 of equation 9.54 to account for circular geometry and the strength of the soil around the body of the penetrometer. The overburden pressure $\sigma_0 = \gamma z$ relates to the depth of the probe and the unit weight of the overlying soils. The groundwater pressure must also be estimated if the effective overburden pressure is to be found, $\sigma_0' = \gamma z - u$: the angle of friction ϕ' can then be deduced from N_q once the value of q_c has been determined. Precisely the same correlations might be used with plate bearing tests conducted across the full diameter of a borehole, but the problems of disturbance and the avoidance of debris would be a consideration once again. Plate tests also lend themselves to treatment as small-scale model tests: Parry (1978) considers some of the issues involved. Indeed most *in-situ* testers prefer to predict safe or limiting foundation pressures directly from their probe measurements.

Sanglerat (1972) points out that equations 10.6 and 10.7 are defective when the pressure q_c is so large as to cause substantial volume changes around the tip. Once such a large critical cone pressure has been attained it remains constant with any further increase in depth of the probe, whereas the simple bearing equations would predict an ever-increasing resistance. Although this borderline between the relative importance of shearing strength at shallow depths but of soil stiffness at greater depths causes untold problems for the parameter-minded, it is less significant to the model-minded engineer who simply accepts the measured value of q_c at whatever depth as a starting point for the prediction of safe bearing pressures.

Schmertmann reviews (rather harshly) the vane test which has been widely used in Scandinavia to estimate the peak undrained shear strength of silts and clays. The torque T developed on the vanes is supposed to cause a cylinder of soil, undisturbed by the downward insertion of the thin blades, to rotate relative to the main body about a vertical axis. If shear stresses c_u develop on the side and the ends of the cylinder of diameter D and height H, then

$$T = \pi DH \frac{D}{2} c_u + \frac{2\pi D^2}{4} \frac{D}{3} c_u$$

or

$$c_u = \frac{2T}{\pi D^2 (H + D/3)} \tag{10.8}$$

Although criticisms are sometimes made of the vane's inability to deal precisely with anisotropic, creeping, varved, sensitive or cemented clays, such criticisms might also be applied to most other methods of test. The technique remains valuable in relatively impermeable soft clays which do not contain large stones.

Pressuremeter tests were developed by Ménard in France and consisted originally in the inflation of a close-fitting borehole sleeve in order to estimate the initial stiffness of the ground followed by its limiting strength. The recent development of self-boring pressuremeters has provided ground engineers with their closest approach to undisturbed soil testing. The Cambridge Camkometer depicted in figure 10.21 is a cylindrical probe which is jacked into the ground while the soil entering at its open base is removed by a rotating cutter and flushed to the surface by circulating water. This procedure has been shown to cause negligible disturbance of the surrounding soil. Once in place, compressed nitrogen can be applied to a long cylindrical rubber membrane in the wall of the probe. Transducers can transmit the radial expansion of the membrane, and the pore-pressure response of the soil, as the applied pressure is increased at whatever rate may be required. The development of the Camkometer was described by Wroth and Hughes (1973), and the method of converting the pressuremeter data into values of ϕ' and ν for sands appears in Hughes, Wroth and Windle (1977). Palmer (1972) undertook the task of converting undrained pressuremeter data into the equivalent plane strain configuration. Schmertmann reviews other contemporary instruments and analyses. Although the construction industry will complain of the cost of a sophisticated pressuremeter test manned by graduates and skilled technicians compared with that of a routine borehole test manned by labourers under the inspection of a trained foreman driller, it is the client who ultimately must decide. Clients in the medical sector place the knife in the hands of a surgeon, to whom are paid great respect and the lion's share of the fee. Why should the ground investigator be less well educated or respected than the back-room boys? Why should blood and mud be treated so very differently? It is widely held that a major proportion of the money (claimed and recovered from clients) which was not included in the original contract estimates is due to unforeseen ground conditions: this excess could be 10 to 100 per cent of the original contract sum, whereas a conventional site investigation would account for a fraction of 1 per cent. There is ample scope to believe that more and better site investigation would lead to greater harmony and efficiency during construction, to fewer delays, and almost certainly to net reductions in cost to the client.

10.6.5 Capacity of Foundations of Various Shapes and Depths to withstand Vertical Loads

Simple mechanics presently leads only to the prediction of the maximum bearing capacity for infinitely long strip footings on the surface of a soil bed, as expounded in chapter 9 and in particular in section 9.9. In order to extrapolate to more common geometries many engineers have performed model tests. The interpretation of such tests, or similar tests which an engineer might perform on a particular site, is very difficult. Consider, for example, a footing in homogeneous dry sand: at least five problems occur.

(1) Whereas the analysis of section 9.9 ignored the strength of the soil above the foundation plane, some contribution should in practice be expected: this could be called the depth factor.

(2) Whereas the analysis of section 9.9 ignored the roughness of the base of the footing, some contribution might be expected if the subgrade were to attempt to expand laterally. This was mentioned in section 9.9, and may be referred to as the roughness factor.

(3) Whereas the analysis of section 9.9 concerned an endless footing the soil under a real footing would generate extra friction at its ends: this could be called the shape factor.

(4) Whereas the experiments of Terzaghi on long footings on sand suggested that the effective foundation plane might be considered to be $B/2$ below a footing of width B, as suggested in section 9.9.5, other shapes could have different effective depths: this could be called the subgrade factor.

(5) Whereas the collapse of a long strip footing would clearly call for a plane strain angle of shearing resistance, that of a square footing might be considered to be roughly triaxial in nature. Section 8.6 suggested that although the basic critical state angle ϕ'_c would be similar in plane or triaxial strain, the dilatancy component ϕ'_v is certainly different, being 50 per cent greater in plane than in triaxial strain. If the engineer is attempting to predict failures, therefore, he should be prepared to use rather lower ϕ' values for square or rough footings than he did for long footings. Furthermore, the greatest foundation load might mobilise true peak soil strengths in some places but critical states where strain concentrations had developed: this was discussed in sections 9.1 and 9.2. Furthermore, the dilatancy component is much enhanced at small confining stresses, as discussed in section 8.6. This adds to the variation in ϕ' values in the soil underneath and around a model footing, and creates extra difficulty if results on lightly stressed models are intended to be applied to heavily stressed field constructions. These deviations in dilatancy might be amalgamated and referred to as the peak strength factor.

Not one of the many contributors to the debate has fully resolved the interference of these four factors, and in particular of (5) and (4) and (3). It has, nevertheless, become common to account for some of them individually by multiplying the simple theoretical bearing capacities of equations 9.54 and 9.61 with 9.65, by correction factors. At this point you are well advised to immerse yourself in one of the authoritative texts, written by experienced foundation engineers, such as Tomlinson (1975) or Peck *et al.* (1974). Although empirical knowledge may not be as long lasting as the principles of formal mechanics, it is the clear responsibility of the engineer to extend his craft by the use of whatever knowledge or intuition might be available. Nevertheless, all authorities seem to agree that the methods of section 9.9 are broadly pessimistic and safe predictors of the bearing capacity of soils, when they are linked to pessimistic estimates of the strength of the soil and when potential settlements have been considered separately. An exception, according to Terzaghi, is the empirical subgrade factor (4), the effective depth being reduced from $0.5B$ below the long footing to $0.4B$ below a square and $0.3B$ below a circular footing.

The degree of conservatism may be judged partially from the corrections often applied to the basic bearing formulae for a smooth flexible strip B wide and D

deep, which were developed in section 9.9: for

$$\tau_f = c_u$$
$$\sigma_f - \sigma_0 = N_c c_u$$
$$N_c = 5.14$$
$$\sigma_0 = \gamma D \qquad \text{from (9.54)}$$

and for

$$\tau_f = \sigma' \tan \phi'$$
$$\sigma'_f \approx \sigma'_E N_q \qquad \text{from (9.61 and 9.65)}$$

N_q given in table 9.3 as $K'_p \exp(\pi \tan \phi')$ and

σ'_E effective overburden pressure at depth $E = D + 0.5B$

Typical corrections for a horizontal loaded area B wide x $L(\geqslant B)$ long x D deep are

(1) depth factor

$$N_c \times (1 + 0.2D/B) \text{ up to a limit of } N_c \times 1.5$$

as suggested by Skempton (1951)

$$N_q$$

for the base of deeply embedded piles multiplied by factors of 1.2 in loose sand up to 5 in dense sand, as suggested by Meyerhof (1976).

(2) roughness factor

$$N_c \times 1.1$$
$$N_q \times 1.2$$

equivalent to those suggested by Terzaghi and Peck (1967), but without much justification and now frequently ignored

(3) shape factor

$$N_c \times (1 + 0.2\, B/L)$$

as suggested by Skempton (1951)

$$N_q \times (1 + 0.2\, B/L)$$

as suggested by Lambe and Whitman (1969)

(4) subgrade factor

$$E = D + 0.5B\,(1 - 0.3\, B/L)$$

equivalent to Terzaghi and Peck (1967).

(5) peak strength factor

most engineers using the peak strength of the least dilatant soil likely to occur, as measured in a triaxial compression test under the largest stresses likely to occur.

Consider the collapse of a conventional 2 m square reinforced concrete pad footing to support a single column, shown in figure 10.23, which is to be constructed in a stiff glacial silty clay. Suppose that the evidence of nearby trial

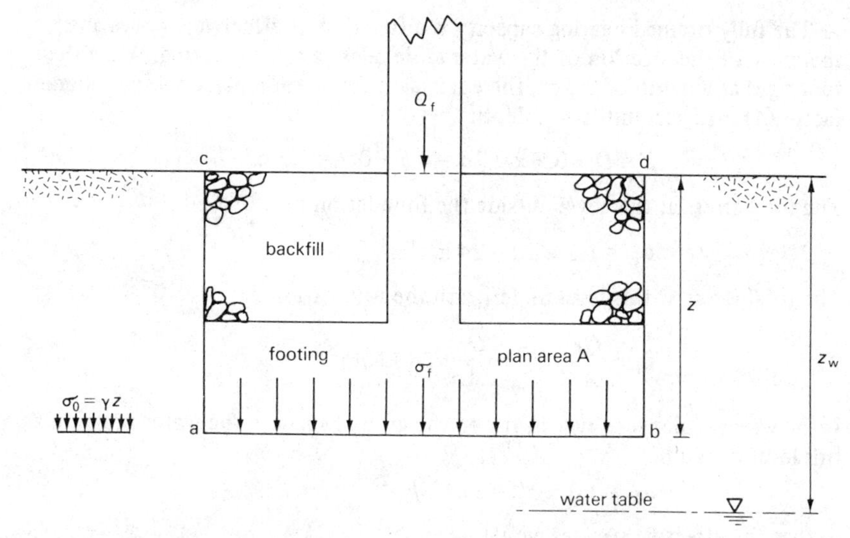

Figure 10.23 *Collapse of a footing*

pits shows the clay to have a unit weight $\gamma = 20\ \text{kN/m}^3$, and typical undrained strengths in the range $c_u = 100$ to $200\ \text{kN/m}^2$, and that very slow shear-box tests on samples cut from blocks reveal that the drained angle of shearing resistance $\phi' = 30°$ at $\sigma' = 400\ \text{kN/m}^2$, increasing to $\phi' = 35°$ at $\sigma' = 50\ \text{kN/m}^2$. The initial estimates of bearing capacity would be based on $c_u = 100\ \text{kN/m}^2$ and $\phi' = 30°$. It will be instructive to consider three possible depths of the foundation $z = 0.5$ m, 1.0 m and 2.0 m, and three possible locations for the highest conceivable future groundwater table $z_w = 0$, z and $(z + 2)$ m. The uncorrected immediate undrained bearing capacity is

$$\sigma_f = \gamma z + N_c c_u$$

But of the stress σ_f delivered across the base ab, some is simply due to the weight of the block abcd while the rest contributes to supporting the structural load Q_f. Since reinforced concrete weighs very little more than soil it is often sufficiently accurate to deduct a stress γz from σ_f to account for the weight of the foundation block, so that

$$Q_f \approx A(\sigma_f - \gamma z) \approx AN_c c_u$$

This makes the uncorrected undrained collapse load

$$Q_f \approx 4 \times 5.14 \times 100 \approx 2056\ \text{kN}$$

at any depth z. If factors (1) and (3) are used, the corrected undrained collapse loads increase very slightly with increased depth of the footing

at $z = 0.5$ m, $Q_f = 2056 \times 1.05 \times 1.2 = 2591$ kN

at $z = 1.0$ m, $Q_f = 2056 \times 1.1 \times 1.2 = 2714$ kN

at $z = 2.0$ m, $Q_f = 2056 \times 1.2 \times 1.2 = 2961$ kN

The fully drained bearing capacity is a function of effective stresses and therefore of the location of the water table relative to the footing. When the footing is at a depth of 0.5 m, the equivalent foundation plane taking subgrade factor (4) into account is at a depth

$$E = D + 0.5 \times 0.7B = 0.5 + 0.7 = 1.2 \text{ m}$$

The total stress at this level outside the foundation is

$$\sigma_E = 1.2 \times 20 = 24 \text{ kN/m}^2$$

The total stress at this level underneath the foundation is

$$\sigma_f \approx \frac{Q_f}{A} + \sigma_E \approx \frac{Q_f}{A} + 24 \text{ kN/m}^2$$

If the water table were ever to rise to the ground surface, the water pressure at this level would be

$$u_E = 1.2 \times 10 = 12 \text{ kN/m}^2$$

so that the effective stresses would become

$$\sigma'_E = 24 - 12 = 12 \text{ kN/m}^2$$

$$\sigma'_f = \frac{Q_f}{A} + 24 - 12 = \frac{Q_f}{A} + 12 \text{ kN/m}^2$$

The bearing capacity equation corrected using factor (3) would then provide that

$$\sigma'_f = \sigma'_E N_q \times 1.2$$

where N_q takes the value 18 from table 9.3 so that

$$\frac{Q_f}{A} + 12 = 12 \times 18 \times 1.2 = 259 \text{ kN/m}^2$$

and

$$Q_f = 247 \times 4 = 989 \text{ kN}$$

If, on the other hand, the water table could only rise to the level of the foundation, the water pressure on the equivalent foundation plane would be

$$u_E = 0.7 \times 10 = 7 \text{ kN/m}^2$$

so that the effective stresses thereon would be

$$\sigma'_E = 24 - 7 = 17 \text{ kN/m}^2$$

$$\sigma'_f = \frac{Q_f}{A} + 17 \text{ kN/m}^2$$

The bearing equation would then stipulate that

$$\frac{Q_f}{A} + 17 = 17 \times 18 \times 1.2 = 367 \text{ kN/m}^2$$

and

$$Q_f = 350 \times 4 = 1401 \text{ kN}$$

If the highest conceivable water table were 2 m below the footing, then the most likely water pressure on the equivalent foundation plane 0.7 m below the footing would be

$$u_E = -1.3 \times 10 = -13 \text{ kN/m}^2$$

when it is remembered that the clayey soil will certainly be able to withstand capillary suction for many metres. The effective stresses on the equivalent foundation plane then become

$$\sigma'_E = 24 - (-13) = 37 \text{ kN/m}^2$$

$$\sigma'_f = \frac{Q_f}{A} + 37 \text{ kN/m}^2$$

so that the collapse condition provides that

$$\frac{Q_f}{A} + 37 = 37 \times 18 \times 1.2 = 799$$

and

$$Q_f = 3049 \text{ kN}$$

You should now apply similar arguments to the cases of $z = 1.0$ m and 2.0 m with the same three groundwater conditions of $z_w = 0, z$ and $(z + 2)$ m, after which you would be able to check the validity of table 10.1, which amalgamates all the estimates of the force Q_f necessary to collapse the footing. It should be obvious from table 10.1 that foundations should be buried deeply enough to protect them against some future rise in groundwater levels. If the engineer

TABLE 10.1
Collapse Load Q_f(kN) in Various Conditions

Depth of footing z m	Undrained	Drained		
		$z_w = 0$	$z_w = z$	$z_w = z + 2$ m
0.5	2591	989	1401	3049
1.0	2714	1401	2225	3873
2.0	2961	2225	3873	5521

wished to prevent the safety of the foundation considered above from deteriorating with time, he could take positive measures to drain the land to a depth of 2 m at least and place the footing at 1 m. Such a measure would be a typical outcome from a careful reading of CP 2004 Foundations, although the placing of the footing at a depth of 1 m would there be regarded as a protection against seasonal volume changes.

Before the designer chose a safe load for the 1 m deep foundation, he would scrutinise table 10.1 rather carefully and supplement it with further considerations. In section 6.8 we deduced that first yield under the edges of a foundation on undrained clay might occur at roughly one-half the collapse load. If the designer was aware that even elastic settlement calculations based on initial tangent moduli might later force him to alter his plans, he would certainly wish to prevent extra settlement due to yielding and would therefore apply a factor of at least 2 to the undrained collapse load, to obtain a safe load of 1350 kN. In addition, equation 6.31 represented a reduced N_q value at which frictional yield would occur under the edges of the footing: at $\phi' = 30°$ this value of N_q at first yield was only 5 instead of the 18 appropriate to collapse. If the designer had interpolated between 2225 and 3873 to obtain 3050 kN as the drained collapse load with a water table 1 m below the footing, he might then apply a reduction factor of 5/18 to achieve 847 kN. This, coincidentally, would have been the order of load adopted by a lay-reader of CP 2004 who would find from tables 1 and 2 therein that the presumed bearing value of clay is $2c_u$, corresponding to 200 kN/m^2 over 4 m^2. Other systems of thought would arrive at the same answer. Terzaghi and Peck (1967) and most other foundation engineers recommend a safety factor of 3 against bearing failure, which would have offered a safe load of 900 kN. If, on the other hand, you had followed the injunction on safety in section 9.9.4 and chosen a definitely safe angle of shearing resistance $\phi' = 20°$, you would have then been limited to a value of 6 for N_q with identical results. There is more than one way to skin a cat: some of them also work with dogs.

Now consider the vertical collapse load of a vertical 2.26 m diameter bored pile which was to be constructed in the same sort of clay in a location where the undrained strength rose from 40 kN/m^2 at the surface to 100 kN/m^2 at 10 m, rising more slowly for a further 25 m to sandstone. Such a pile would be in the upper range of diameters reported in the excellent technological review by Weltman and Little (1977) of types of bearing piles. I have chosen the cross-sectional area of the pile, 4 m^2, to coincide with that of the pad in the previous example. Suppose that in this location the groundwater table was at ground level, and that it was required to calculate the collapse load and deduce a safe bearing load if the pile were 10 m deep. First calculate the immediate undrained bearing strength of the base, using factors (1) and (3)

$$Q_b = 5.14 \times 1.5 \times 1.2 \times 100 \times 4 = 3701 \text{ kN}$$

Now calculate the long-term bearing strength. The effective foundation plane using the subgrade factor (3) is at a depth of 10.7 m where the effective vertical stress is roughly

$$\sigma'_E = 10.7 \times 20 - 10.7 \times 10 = 107 \text{ kN/m}^2$$

The effective bearing pressure on the equivalent base which would cause collapse would then be

$$\sigma_f' = 107 \times 18 \times 1.2 \times 1.2 = 2773 \text{ kN/m}^2$$

if a depth factor of 1.2 were employed. This would correspond to 2770 kN/m² of effective stress on the base of the pile which would be aided by 100 kN/m² of water pressure. The fully-drained bearing strength would then be

$$Q_b = 2870 \times 4 = 11\,480 \text{ kN}$$

Before the collapse load at the surface can be calculated it is necessary to consider the weight of the pile and the possible effect of friction on its side, as exemplified in figure 10.24. If the pile were to fail immediately after construction

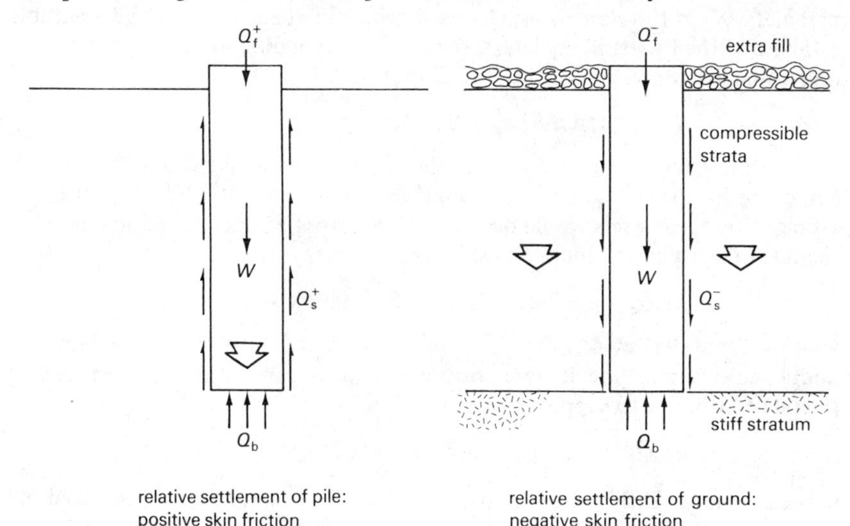

Figure 10.24 *Skin friction on piles: the extremes*

it would necessarily tend to drag down the surrounding soil, which would therefore tend to support it with 'positive' skin friction. Foundation engineers normally apply a reduction factor α to the undrained strength c_u of the soil over the circumference of the pile, in order to take account of the softening which is possible while the pile is being constructed. If the clay had been allowed to swell freely in the presence of water, whether from permeable layers or from the wet concrete, its immediate strength could drop dramatically: the value of α is therefore best estimated from a pile test on the site concerned (which might make its calculation redundant if the test pile were entirely satisfactory). If, in the present case, the engineer were to use $\alpha = 0.5$ he would go on to calculate the skin friction force

$$Q_s = \pi D L \alpha c_u^{av} = \pi \times 2.26 \times 10 \times 0.5 \times \frac{(40 + 100)}{2} = 2485 \text{ kN}$$

The weight of the pile

$$W = \frac{\pi D^2}{4} L \gamma_c = 4 \times 10 \times 24 = 960 \text{ kN}$$

so that the immediate failure load applied to the pile would be

$$Q_f^+ = Q_b + Q_s - W = 5226 \text{ kN}$$

In the long term, and perhaps even in the short term, an effective friction calculation should be used to calculate the skin friction. The coefficient of friction between smooth concrete and soil is often taken to be tan ϕ'_c for the soil, but concrete placed to set against soil is likely to be able to develop the full angle of friction ϕ' of soil on soil; Tomlinson (1975) offers further guidance. In order to estimate the effective friction it is then necessary to calculate the effective lateral stress on the pile. For piles bored in clay this is usually taken to mean the pre-existing lateral stress, and where no information is available Jaky's parameter $K'_0 = 1 - \sin \phi'$ is usually used, which has been shown to be a sensible estimate of the lowest likely lateral stresses corresponding to the virgin one-dimensional compression of the ground. In this case

$$\tau_s = \tan \phi' K'_0 \sigma'_v \approx 0.29\sigma'_v$$

The average effective vertical stress in the soil around the pile will in this case coincide with the stress half way down the stem, which is 50 kN/m^2, so that the average skin shear stress would be 14.5 kN/m^2. Applied over the perimeter of the pile this produces a long-term skin friction force

$$Q_s = \pi \times 2.26 \times 10 \times 14.5 = 1030 \text{ kN/m}^2$$

which is somewhat smaller than the crudely estimated initial value based on undrained strength. Now it would be possible to conceive of the soil surrounding the pile continuing to support it so that

$$Q_f^+ = Q_b + Q_s - W = 11\,550 \text{ kN}$$

in the long term. If fill were placed on the ground surface, however, the resulting settlement could well cause a reversal of the skin friction as indicated in the second diagram of figure 10.24. With negative skin friction the structural collapse load is reduced to

$$Q_f^- = Q_b - Q_s - W = 9490 \text{ kN}$$

If negative skin friction appears to pose a problem it can be reduced by sleeving a bored pile, or by coating a driven pile with bitumen.

The choice of a possible safe bearing load requires the normal degree of sensitivity to the issues involved. It is common to require a factor of 3 against end-bearing but to accept a pessimistic skin friction force unfactored. This would generate a safe load in the short term of

$$Q_{safe} = \frac{3701}{3} + 2485 - 960 = 2759 \text{ kN}$$

and in the long term with negative skin friction

$$Q_{safe} = \frac{11\,480}{3} - 1030 - 960 = 1837 \text{ kN}$$

It may be sensible, here, to notionally reduce the full theoretical negative skin friction by supposing that the compression of the soil under the pile base would in any case exceed the relative compression of the lower and stronger layers of the clay surrounding the pile due to any likely increase in overburden or reduction in groundwater level. Perhaps 2700 kN would be a sensible stab at the safe load of a pile which was allowed to settle somewhat. This would be treble the safe capacity previously calculated for the shallow footing of the same area which was lucky enough to rest on clay of the same strength.

The calculation of the maximum bearing capacity of piles in sand, frequently driven, is subject to at least as many empirical and technological factors as that of piles in clay. Although it is possible to arrive at bearing pressures which take account of the compaction or dilation of the natural soils during construction, the exponential linkage between N_q and ϕ' frequently makes the task unnecessary since the predicted failure stress for moderate depths in moderately dense sand will usually far exceed those stresses that will cause excessive settlements. So although a good deal of information on the bearing strength of piles can be discovered in papers such as that by Meyerhof (1976), it would probably be more useful to continue your study by reference to more practical reference books such as Tomlinson (1975) or Whitaker (1970).

Skin friction generates a further problem for the designer who wants to support a raft over a very deep sequence of compressible soils by using a large group of closely-spaced piles which rely for their strength on the supposed positive skin friction around them, while developing little base resistance. Not only would the whole raft/pile/soil assembly rise and fall together in the event of regional settlement, but the assembly might also prefer to *fail* as a block. If the perimeter of the whole group of piles was less than the sum of the perimeters of the individual members of the group, then the pile-group may prefer to fail as a single composite 'pile' with friction developed over the perimeter of the *group* and with base resistance developed over the base of the *group*. Once he has discovered a threat of group action, the designer should avoid it by increasing the spacing between piles made narrower and longer.

10.6.6 The Collapse of Foundations under Complex Loads

Foundations are frequently required to carry both horizontal forces and bending moments in addition to the normal vertical load. The impact of a horizontal stress component τ_f on the vertical component σ_f on the base of a shallow footing on the point of collapse was displayed in table 9.2 with respect to the cohesion model, and in table 9.4 with respect to friction within a weightless medium. Meyerhof (1953) gave the results of small model tests normalised against a more complete theoretical treatment, which confirm the very great significance of the lateral component. His friction results are often expressed in terms of a sixth set of factors against the previously described bearing capacities for a strip footing:

(6) load inclination factors

$$N_q \times (1 - \delta/90°)^2$$

$$E = D + 0.5B\,(1 - \delta/\phi')^2$$

which can be used in conjunction with the other five when the resultant force on some horizontal footing base is inclined at δ to the vertical.

The reduction in vertical capacity by a factor of about 2 when the resultant force is inclined at 15° does not detract too much from the economy of a shallow footing. Inclinations steeper than 15° usually call for some more positive response such as the provision of a deeper base so that forces can be developed on the sides, or ground beams to distribute horizontal reactions equally to the various footings or a group of piles some of which rake at the appropriate angle.

A footing base can easily be made to carry constant moments in addition to shear and normal forces. Figure 10.25 depicts a shallow footing with no side

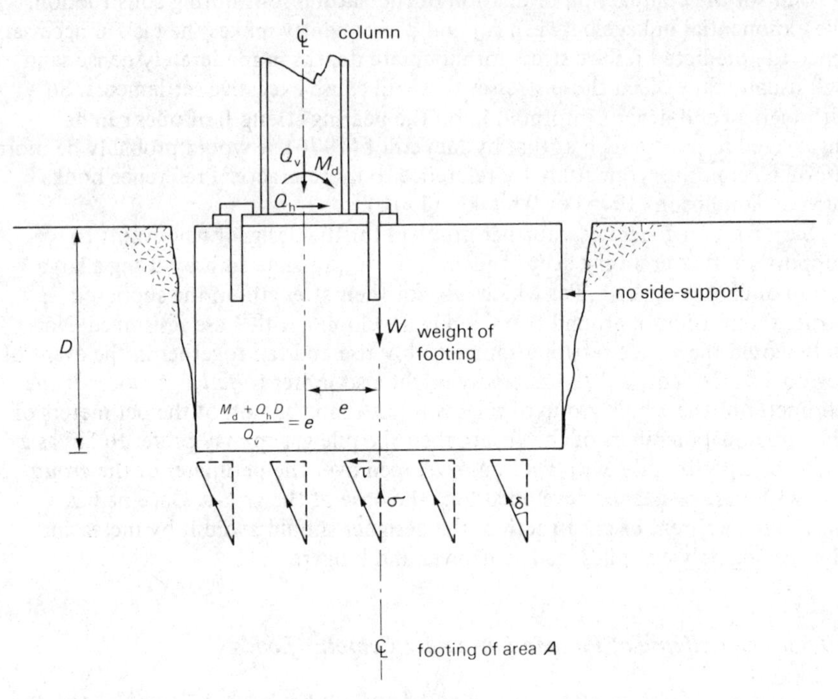

Figure 10.25 *Use of column eccentricity to balance moment of dead loads*

support which is supposed to receive constant 'dead' actions from the structure above of Q_v, Q_h and M_d. The secret of carrying the dead moment M_d is simply to place the centroids of the column and footing at an eccentricity $e_d = (M_d + Q_h D)/Q_v$ in the appropriate direction, as shown. The designer's job is to produce a footing such that credible soil pressures on its boundaries can be in equilibrium with its weight and with the applied load. He has only soil pressures to do the job, and must therefore assume intelligent distributions. In this case it is necessary only to invoke a constant inclined base stress whose lateral component is $\tau = Q_h/A$ and whose vertical component is $\sigma = (Q_v + W)/A$. These were found simply by resolving forces. The trick of placing the column eccentrically over the footing allowed the stresses to be uniform because the moments were balanced. The weight W of the block had no moment about the centroid of the footing.

The structural load Q_v had an anticlockwise moment $Q_v e$, the load Q_h had a clockwise moment $Q_h D$ and the applied moment M_d was also clockwise: the eccentricity e was chosen to balance the moments so that the soil underneath the footing was not aware that any moments were being applied. After choosing the depth D of the footing the designer can calculate

$$\tan \delta = \frac{\tau}{\sigma} = \frac{Q_h}{Q_v + W} = \frac{Q_h}{Q_v + \gamma_c AD}$$

and he can then explore values of A and D in relation to table 9.2 or 9.4 or factor (6) until the stresses τ and σ are found to be safely remote from the failure stresses τ_f and σ_f.

The problem is more difficult if a footing base is required to carry a variable 'live' moment M. It is then inevitable that the soil pressures on the base must continually redistribute themselves so as to attempt to balance the applied moment, while remaining constant in total if the vertical load remains constant. Consider a footing base which receives only a constant vertical force Q_v (including the weight of the footing) at its centroid but an increasing moment M about its centroid. The route to success is to replace the centroidal actions by a single equivalent force Q_v acting at an eccentricity $e = M/Q_v$. It should then be clear that for equilibrium the soil pressures must sum to Q_v and that their resultant must share the line of action of the eccentric equivalent force, as shown in figure 10.26. The effect is to reduce the contact stress under one side of the footing and to increase it under the other. Soil is not thought to be very competent in tension, so a crack is usually invoked if the total stress attempts to become negative at the 'upstream' edge. Neither should the soil at the 'downstream' edge be allowed to exceed its bearing capacity σ_f. Within these constraints the designer must attempt to show that some pressure distribution can in fact exist which would equilibrate the applied loads. Figure 10.26 demonstrates a succession of equivalent forces Q_v with an increasing eccentricity, together with statically admissible base stress distributions, for a typical pad footing with an initial safety factor of 3 against a collapse due solely to Q_v. When the base stress 'upstream' has dropped to half its original value and that 'downstream' has increased accordingly, it is easy to show that the centre of base pressure has become eccentric by $B/12$ so that a moment $M = Q_v B/12$ would be in balance. When the upstream stress has reduced to zero and the downstream stress has doubled it is clear that the centre of pressure under the base is $B/3$ from the downstream side which is $B/6$ from the centroid: a moment $M = Q_v B/6$ is being balanced. Any further moment causes the triangle of pressure to leave a crack 'upstream': this accentuates the increase of pressure at the 'downstream' edge because the area of the pressure diagram must remain constant at Q_v. The yield stress σ_f is soon reached, and the final distribution of stress corresponding to collapse by rotation about P is a rectangular pressure distribution of width $B/3$ and of magnitude σ_f which carries a moment $Q_v B/3$ due to its eccentricity of $B/3$. Most designers respond by limiting the working eccentricity to $B/6$ corresponding to the onset of uplift 'upstream', after checking that the 'downstream' stress could not have reached its limiting value σ_f. If the base is forced to transmit shear at the same time then σ_f must be reduced according to the inclination δ of applied load as previously described. This makes the task of

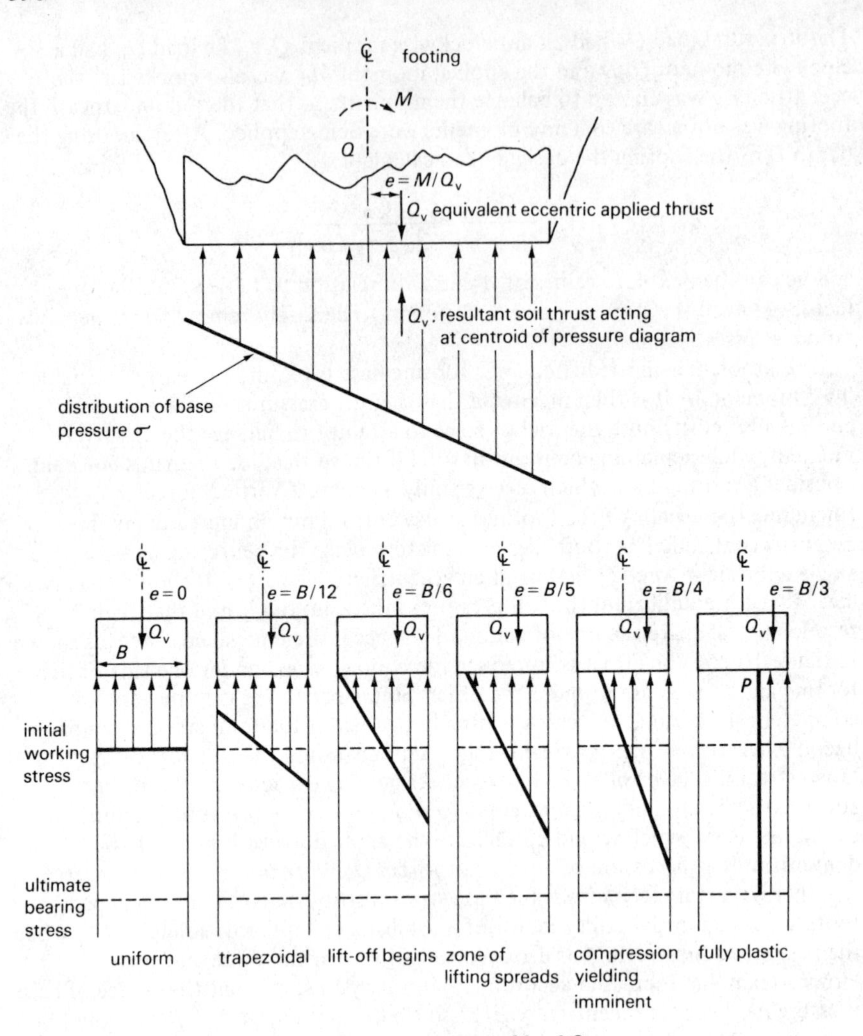

Figure 10.26 *A footing carrying an eccentric vertical load Q_v*

the designer rather difficult since every combination of Q_v, Q_h and M on a footing base must be separately proved to be safe. The most critical condition on the footings of a light shed might either be when Q_v is greatest due to applied loads or when the wind blows when the shed is empty so as to increase Q_h and M while decreasing Q_v due to the relative suction above the roof.

The task of determining the critical loading condition is made easier if an approximate interaction diagram such as figure 10.27 can be used. The figure maps the factor by which the vertical component of the *collapse* load reduces in the presence of lateral loads and moments. For example, if actions $Q_v = 600$ kN, $Q_h = 150$ kN, $M = 200$ kN m were to afflict a 2 m square base, the inclination would be $\delta = 14^\circ$ and the eccentricity ratio would be $(e/B) = 0.17$ so that if these actions were to increase while remaining in proportion they would eventually

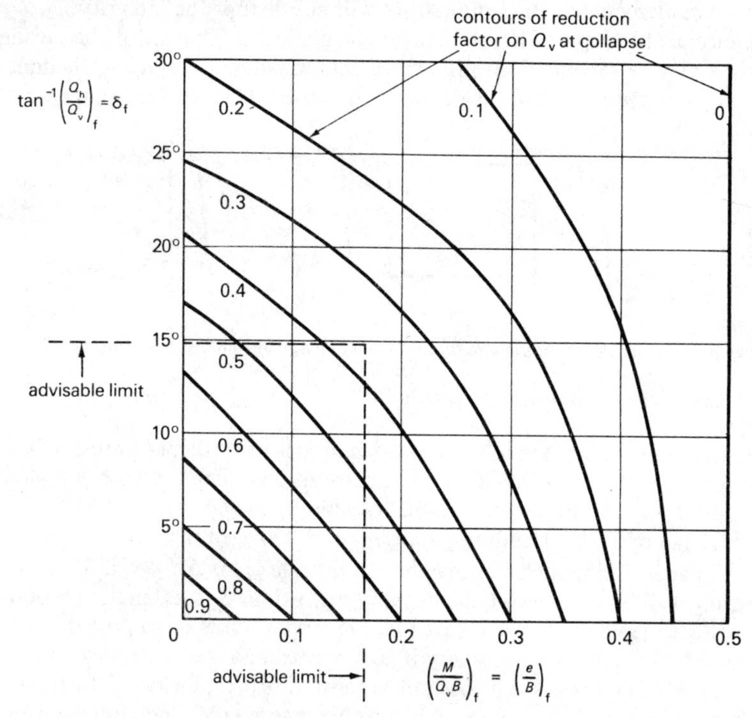

Figure 10.27 *Reduction factors on vertical collapse load Q_V due to inclination δ and eccentricity e*

lead to a collapse at a reduction factor of about 0.37 on the vertical collapse load. To prevent this happening the vertical collapse load in the absence of lateral loads or moments should exceed 600/0.37, which is 1620 kN, by whatever margin is deemed safe. Every load combination generates a just-safe vertical load, and the largest must then be chosen.

A group of piles, some of which may be raking at angles of 1/3 or steeper and which are connected together at the surface by a monolithic reinforced concrete pile cap, often make the most efficient foundations for structures which carry large lateral loads or moments. Although pile groups can be analysed in an oversimplified way by attempting to create a list of pure compressions which happen to satisfy the external loads, the precise interaction of piles and the distribution of normal and shear forces and bending moments which will actually occur within them is still a matter of some speculation.

10.6.7 Avoiding Damage to the Structure

After the designer has tentatively chosen a scheme for the foundations appropriate to their environment, assessed the forces which they might be asked to carry, and proportioned them so that they would not collapse under any possible combination of forces, the designer's next task is to ascertain that the combined settlements of the foundations will not disrupt the integrity, appearance, or function of the structure. The general settlement of the façade of a building may be separated into the three components of average settlement ρ_{av}, tilt ω and distortion Δ/L as depicted in figure 10.28. The tolerance which a

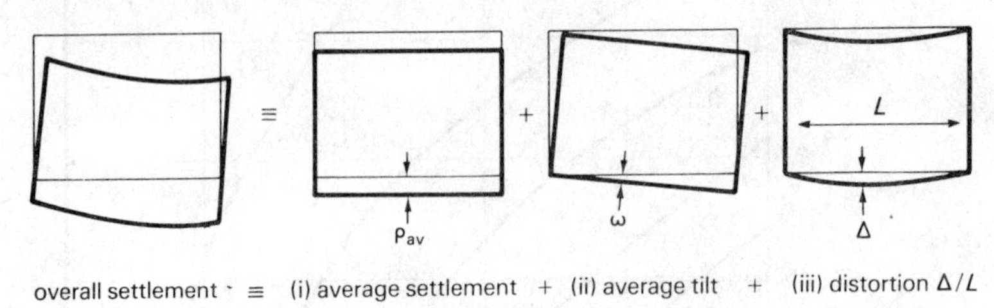

Figure 10.28 *Three components of settlement*

new structure may have of each of the three modes of settlement must be a matter for individual scrutiny. Typically, conventional piped services are likely to fracture if $\rho > 150$ mm, conventional machinery may be disturbed if $\omega > 0.001$ rad or $1/20°$, building frames may be damaged if $\Delta/L > 0.003$, brick wall panels infilling frames may be visibly cracked if $\Delta/L > 0.001$, and loadbearing walls may be visibly cracked by distortions Δ/L as small as 0.0003. Whereas steps may be taken to create flexible service joints or to provide self-levelling beds for sensitive machinery if they are necessary, the damage due to the distortion of architects' finishes in the form of brick, blockwork or plaster is a much more common problem which is difficult to predict and control. The foundation designer should attempt to prevent distortions which would have a greater effect than the inevitable shrinkage due to drying and setting which also causes cracks in cements, mortars and plasters. Burland and Wroth (1974) introduce the concept of a critical gross tensile strain due to the distortion of cemented materials which will just promote cracks of the order of 0.1 mm wide which would then be visible to the owner. According to their experimental data the magnitude of this critical tensile strain is roughly: 0.0005 in reinforced concrete and brickwork with cement mortar, 0.001 in brick panels acting as infill within a frame, and 0.0015 in clinker blocks acting as infill. They go on to show how the greatest tensile strain induced in a wall by the distortion of its foundation can be roughly related to the magnitude of that distortion (Δ/L). Consider, for example, the racking of the wall panels ABDC and CDFE in figure 10.29. Supposing that the panels were framed in by rigid members hinged at the joints, and that the settlement of C relative to A and E caused no shortening of the sides, the Mohr circle of strain drawn in figure 10.29 would show that a simple shear angle of $\Delta/(L/2)$ creates a central circle of radius Δ/L. The maximum tensile strain would occur at 45° to the structural members and be Δ/L in magnitude. The critical tensile strain criterion would then be

$$\left(\frac{\Delta}{L}\right)_{crit} = \epsilon_{crit} \tag{10.9}$$

which would predict just visible cracks at $\Delta/L = 0.001$ in brick infill panels, agreeing quite well with the data that Burland and Wroth collected.

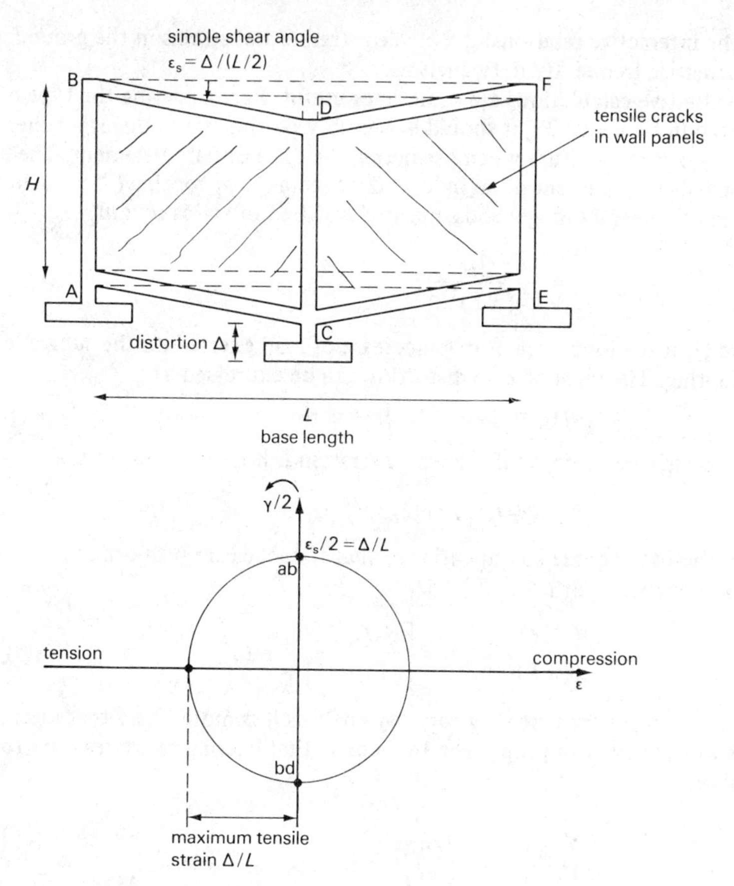

Figure 10.29 *Simplified racking strains due to distortion Δ over base L*

Our capacity to predict the consequences of settlement is not yet very great, but it is greater than our capacity to predict the components of settlement as such: this was a continuing theme during the B.G.S. Conference on the Settlement of Structures (1974) at which Burland and Wroth presented their paper. The problem is one of interaction between the structure and the soil. A light unclad frame or a steel oil tank are effectively perfectly flexible relative to firm soils so that their settlements are independent and do not cause any redistribution of load in the structure. A loadbearing wall or a concrete tank on soft clay could well be effectively rigid, however, settling as a monolith with surprisingly little damage due to distortion because of the redistribution of stress within and beneath the structure. Brick houses on concrete strip footings exemplify the usefulness and complexity of soil–structure interaction, with the walls able to span quite large temporary cavities yet able to creep with the ground over long periods of time so as to come into slightly deformed configurations without being too unsightly or leaky. The satisfactory choice and detailing of foundations to suit building methods presently rests more on experience than calculation, because the constitutive equations of the materials and the interactive relationship between stresses and strains in the ground and the structure frequently defy analysis.

Qualitative calculations are sometimes useful. Reconsidering the type of frame depicted in figure 10.29, it should be sensibly pessimistic to disregard the stiffness of the structure when predicting the differential settlement. The long-term average settlement of an independent rectangular footing ($X \times Y$) on the surface of a deep homogeneous, linear elastic bed of soil is roughly

$$\rho_t = \frac{Q_t}{E'_y(XY)^{1/2}} \qquad \text{(from 6.21)}$$

where Q_t is the long-term average increase of load exerted on the subgrade by the footing. The ideal of zero distortion can be expressed as

$$(\rho_t)_A = (\rho_t)_C = (\rho_t)_E = \text{constant}$$

It can clearly be achieved if the whole stratum is homogeneous so that

$$(E'_y)_A = (E'_y)_C \text{ etc.}$$

and if the footings are so dimensioned that the loads are proportional to their geometric mean width

$$\left[\frac{Q_t}{(XY)^{1/2}}\right]_A = \left[\frac{Q_t}{(XY)^{1/2}}\right]_C \text{ etc.} \qquad (10.10)$$

After completing a bearing capacity analysis it is more likely that interim areas were roughly in proportion to loads so that bearing pressures were roughly equalised.

$$\left(\frac{Q_t}{XY}\right)_A = \left(\frac{Q_t}{XY}\right)_C \text{ etc.} \qquad (10.11)$$

The designer of footings on deep homogeneous beds may then need to increase the areas of the more heavily loaded footings in order to achieve the criterion of equation 10.10, a corollary being that square footings should carry loads in proportion to their width not their area. Gibson (1974), on the other hand, showed that in an incompressible medium in which the stiffness was proportional to depth

$$E = \Lambda z \qquad (10.12)$$

the settlement of any shape of area carrying surcharge q on the surface would be perfectly uniform, and given by

$$\rho = \frac{1.5q}{\Lambda} \qquad (10.13)$$

so that zero distortion would be achieved by leaving bearing pressures constant as in equation 10.11. Many soils display stiffness qualities similar to this, although most building sites would feature a finite modulus at the surface which then increased with depth. Gibson implies that where the stiffness modulus at a depth $z = (XY)^{1/2}$ was more than twenty times the modulus directly under the footing, equation 10.13 will roughly apply and generate a requirement for equal bearing pressures by equation 10.11. If the modulus only doubles over the same interval, homogeneous elasticity with equation 10.10 offers a closer model for zero distortion.

And what are the consequences likely to be if the designer adopts the wrong strategy? Clearly distortion will occur, but the more likely cause of distortion is not that stiffness varies in some wrongly assumed regular pattern, but that the stiffness of the soil varies at random. The greatest danger is associated with having failed to discover the most flexible materials, whether a cavity in filled ground, a peat-filled pond in otherwise competent glacial soils, a lens of clay in a bank of glacial sand, or a pocket of loose wind-blown sand in an otherwise dense sand. Only after a thorough site investigation does it make any sense to proceed to calculate differential settlement. Having conducted such an investigation it might be possible to use such as equation 6.21 in conjunction with the lowest and highest moduli taken from a string of measurements, to infer that zones exhibiting these extreme values may occur under neighbouring columns, and thereby to estimate the damage due to distortion.

For example, suppose that the following determinations had been made for the upper zone of a deep stratum of firm/stiff clay: $c_{u\,min} = 50$ kN/m^2, $c_{u\,av} = 75$ kN/m^2, $c_{u\,max} = 150$ kN/m^2, $E'_{0\,av} = 7500$ kN/m^2, and that the possible distortion of a frame similar to figure 10.29 had to be calculated. Suppose also that the spans were each 10 m so that $L = 20$ m and that the loads on A, C and E would be vertical and constant at 200, 400 and 200 kN respectively and that these forces represent the extra loading of the ground by the new structure. A bearing capacity analysis in this simple loading case would probably have been based on $c_{u\,min}$ and would have come to a safe net bearing pressure of 100 kN/m^2. The interim areas of A, C and E would then have been 2, 4 and 2 m^2 respectively, and in the absence of lateral forces the footings might have been squares of dimension 1.4, 2 and 1.4 m. The designer's next step might be to view the soil as roughly homogeneous and use equation 10.10 to scale up the larger footing until the dimensions were 1.4, 2.8 and 1.4 m. The average settlement from equation 10.13 would then be

$$\rho_{av} = \frac{200/1.4}{E'_y}$$

and by choosing $E'_y \approx 7000$ kN/m^2 after perusing the ratio E'_y/E'_0 in table 6.3, this would generate $\rho_{av} \approx 20$ mm. In estimating distortion, the designer might assume that $E'_y \propto c_u$ so that $E'_{y\,min} \approx 4700$ kN/m^2 and $E'_{y\,max} \approx 14\,000$ kN/m^2. The extreme settlements would then be 30 mm and 15 mm, so that a 15 mm relative settlement of C relative to A and E might be the worst that could be encountered. The distortion Δ/L becomes 0.75×10^{-3}, which is just small enough to make visible cracks in the brick infill panels unlikely. Practitioners often follow Terzaghi and Peck (1967) and assume that

$$\Delta = \frac{3}{4}\rho_{av} \tag{10.14}$$

which takes account of just this magnitude of variation.

A further example of the value of qualitative calculations can be found in the structural design of a strip or raft foundation. Whereas small plain or reinforced concrete pad footings can be designed quite economically and safely by assuming a uniform distribution of pressure under the base, larger areas of foundation demand more thought. Section 6.7 demonstrated that a perfectly flexible foundation carrying a uniform pressure would create a dish-shaped depression in the surface of a deep, homogeneous, elastic bed, the settlement increasing from the edge to the centre. The surcharge pressure exerted on the foundation would then be transmitted directly to the subgrade, with no requirement for structural action within the foundation. If the foundation were perfectly rigid, however, it would be bound to impose a uniform depression on the homogeneous bed, and it would therefore be bound to generate a nonuniform contact pressure distribution with a reduction in the centre and an increase around the edges. Bending moments would then arise in a rigid foundation carrying a uniform surcharge, with a net downward pressure in a central zone being resisted by a net upward pressure around its edges. Borowicka (1936) succeeded, for the special case of circular rafts, in solving the problem of a raft of finite stiffness on the surface of a homogeneous elastic bed, results from which are quoted by Poulos and Davis (1974). These rafts of intermediate stiffness sag a little, but generate smaller bending moments than would a rigid raft. One aspect of the solution was that Borowicka's elastic model generated infinite stresses under the extreme rim of the raft, and it must therefore offer the designer a safe pessimistic view of the degree of arching of the contact stress, and therefore of the bending moments and shear forces that can actually occur.

Recently, Gibson (1974) has pointed out that the likely increase in stiffness with depth in the soil beneath a raft also has a dramatic influence on its pattern of deformation. If the stiffness of an incompressible soil is proportional to depth, the immediate settlement of any foundation representing a constant surcharge pressure q would be monolithic, as expressed by equations 10.12 and 10.13. No structural actions would then be generated within a uniformly pressurised raft. A stiffness distribution in the ground intermediate between constancy and an increase proportional to depth would offer finite bending moments smaller than those generated by Borowicka. Figure 10.30 is adapted from Gibson (1974) and traces the effects of the relative stiffness of a circular raft, the soil in immediate contact with it, and the soil at a depth of one radius, on the maximum bending moment at the centre of the raft caused by an increase of q in the average contact pressure. A rational response to the problem of a raft under uniform pressure might then be to assess the stiffness profile of the soil and firstly determine whether the deformation of the raft would be excessive if it were flexible, probably by applying Newmark's technique with the ground idealised in layers. Only if the distortion $\Delta/2R$ were excessive would it be necessary to ensure that the structure of the raft was capable of reducing it. A large enough raft thickness d should then be chosen to generate a substantial relative raft/soil stiffness, perhaps greater than unity, and to withstand the bending moments which would then arise according to some interaction chart such as figure 10.30. This procedure could well indicate that the thickness of a reinforced concrete raft should be of the order of one-twentieth of its width. Even more severe problems afflict the designer of a raft which must carry variable loads such as those which

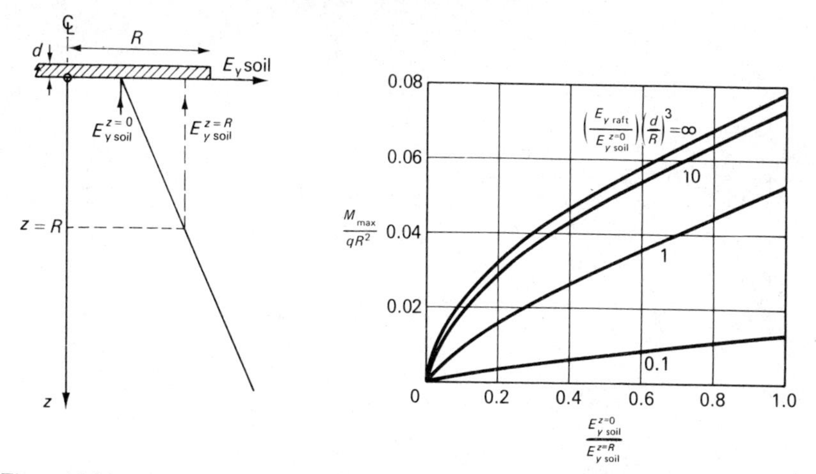

Figure 10.30 *Bending moments in a circular raft ($\mu = 0.3$) on a deep bed of soil ($\mu = 0.5$) whose stiffness increases linearly with depth (from Gibson, 1974)*

might be experienced by a warehouse floor, or which must rest on erratic or variable soil. Most designers presently rely on experience and pessimism in such circumstances and attempt to imagine distributions of load above, or cavities beneath, the raft which will generate the worst conceivable structural reactions at every location.

Nevertheless, the basement raft solution provides the foundation engineer with one of his most attractive remedies to the problem of building on compressible soils, or those soils which might contain soft zones. By constructing a deep basement with stiff monolithic walls and floor his solution approaches that of a flat-bottomed barge. It is necessary to dig out only 10 000 tonnes of soil and replace it by 10 000 tonnes of building in order to guarantee that no *overall* settlement can occur. All that remains is to keep out the soil and the water by virtue of simple pessimistic calculations with regard to structural integrity, and great attention to the details of preventing leakage. The British Code of Practice 102 (1973) and textbooks such as Cedergren (1967), or Tomlinson (1975) would be of some assistance.

10.6.8 Predicting the Magnitude of Settlement

All the calculations which relate to the avoidance of structural damage depend on the ability to predict the settlement of the foundations. Chapter 6 made clear that such a prediction should be based on three areas of knowledge: the stiffness profile of the soil, an elastic solution relating changes of foundation stress to settlement which respects the conditions of the particular foundation and its environment, and a knowledge of the changes in stress which are to be applied.

It is surprisingly difficult to assess that change of stress in the subgrade upon which to base a settlement calculation. Consider, for example, the long-term settlement of a square pad footing of area $B \times B$ to be established at some shallow depth D in a firm to stiff saturated clay of depth H with a groundwater table

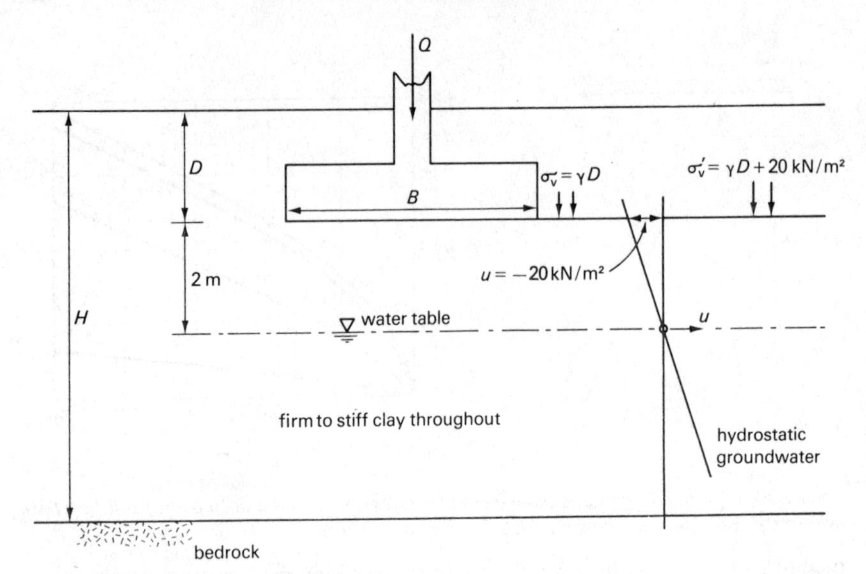

Figure 10.31 *Initial stresses in clay*

presently at a depth $D + 2$ m as shown in figure 10.31. The long-term settlement ought to be based on effective stress analysis, and therefore on the changes in effective stress in the subgrade. If the hole for the footing is excavated and filled with concrete very quickly, the initial effective stress will hardly have been altered by the construction of the foundation. For example, the effective stress immediately under the footing would remain at roughly

$$\sigma'_v = \sigma_v - u = \gamma D - (-2)\gamma_w = \gamma D + 2\gamma_w \tag{10.15}$$

if the difference in density between soil and concrete were ignored. If a constant structural load Q were then applied to the foundation, the long-term effective stress immediately under the footing would be

$$\sigma'_v = Q/B^2 + \gamma D + 2\gamma_w \tag{10.16}$$

if the groundwater table remained 2 m beneath. It is easy to see, in these circumstances, that the long-term settlement could be obtained by using an equation such as 6.21 in conjunction with an effective modulus E'_y measured for the soil under the footing at the appropriate average magnitude of σ'_v, and the extra force Q

$$\rho \approx \frac{Q}{BE'_y} \tag{10.17}$$

These whole elastic solutions take account of the reduction in the induced stress from Q/B^2 to roughly zero at a depth of $2B$, and rely on the engineer being able to infer a roughly unique modulus for the soil in the zone of influence of the footing. But now consider what would be the effect of a lowering by 1 m of the groundwater table, caused by the reduction in infiltration of rainwater after the building was constructed and the surrounding area paved. Not only would the long-term effective stress under the footing be increased to

$$\sigma'_v = Q/B^2 + \gamma D + 3\gamma_w \tag{10.18}$$

but also every element of clay at every depth would also have suffered an increase of $1 \times \gamma_w$ in its effective stress. So in addition to the settlement ρ of equation 10.17, which was of the footing relative to the ground surface, the designer must also consider an extra regional settlement

$$\rho \approx \frac{1 \times \gamma_w H}{E'_0} \tag{10.19}$$

of the whole ground surface due to the 1 m groundwater lowering. The modulus E'_0 must here represent the harmonic mean one-dimensional stiffness $H/\Sigma(\Delta H/E'_0)$ of the whole stratum. This regional settlement would be erratic if the average modulus was geographically variable: only then would those structures spanning the erratic ground suffer extra distortion. Remember the unhappy home-owner of section 1.3.

Now consider the consequences if the excavation for the footing is left open for some time before the concrete is placed. If rainwater is allowed to pond in the hole, it will begin to soak into the clay in order to relieve the local transient pore suction $u = -\gamma D$ generated by the excavation, and the pre-existing pore-suction $u = -2\gamma_w$ due to elevation above the water table. If the ponded water is successful in obliterating all suction in the clay at the bottom of the pit, the effective stress will have decreased to zero and the consistency of that clay will have been reduced to that of a slurry. If the contractor subsequently places the concrete in the pit without cleaning up the base, the ensuing settlement will be so enormous that it is not worth calculating. If on the other hand he scrapes away any clay which has suffered softening, the settlement will remain much as it was in equation 10.17, except that the foundation would be deeper than previously intended. It is clearly important to keep a foundation excavation dry, to place a blinding layer of concrete on the bottom as soon as it has been dug, and to expedite the construction of the footing. Remember that the speed of advance of the transient softening process should be roughly $L = (12C_0 t)^{1/2}$ by the analysis of section 7.3, so that in a typical firm clay with $C_0 = 1$ m²/year the softening front would have travelled 0.2 m in a day of which 0.1 m would be badly softened, 0.5 m in a week of which 0.25 m would be serious, and 1.0 m in a month of which 0.5 m should certainly be removed. Silt or sand lenses below the water table could, of course, reduce the times for softening by an order of magnitude.

As a concluding exercise in appreciation of the difficulty of estimating changes in stress, consider now the problem of predicting the settlement under a load Q which varies. If a footing on clay is to carry a constant load Q_d with an occasional extra live load Q_l which was not long lasting, the maximum settlement during the application of the live load would be of the form

$$\rho = \frac{Q_d}{BE'_y}(1 - \mu'^2) + \frac{Q_l}{BE^u_y}[1 - (\mu^u)^2] \tag{10.20}$$

following the application of equation 6.21 with effective stress parameters on

Q_d and undrained parameters on Q_l. Since it was demonstrated by equation 6.25 that the undrained component of settlement was always at least one-half the fully drained settlement, the application of equation 10.20 is rarely likely to be any more profitable than the much more simple pessimistic assumption that the peak load $Q_{max} = Q_d + Q_l$ could be fully drained, and therefore substituted entirely in the effective stress equation.

The determination of an appropriate stiffness profile for the subgrade faces the predictor of settlement with his most taxing problem. It is firstly necessary for the designer to appoint a resident engineer who will inspect all foundation excavations at regular intervals in order to ensure that the subgrade soils do not deteriorate, or at least that any softened soils are removed immediately prior to concreting. This leaves the problem of discovering the natural stiffness of the undisturbed ground. Since most ways of testing soil disturb it, and because disturbance usually leads to a reduction in stiffness, the foundation engineer frequently uses quite crude testing techniques in the hope that even the pessimistic data he then obtains will predict settlements which his client could tolerate. Occasionally clients will waste thousands of pounds on foundations when a few hundred pounds on high quality tests would have sufficed to prove the merit of some cheaper alternative. Recent appraisals of the measurement and application of soil stiffnesses have been made in the ASCE Conference on In-situ Measurement of Soil Properties (1975) by Wroth and by Mitchell, and in the BGS Conference on the Settlement of Structures (1974) by Sutherland on granular materials, Simons on virgin clays and Butler on heavily overconsolidated clays.

The current practice for sands frequently employs some correlation between the results of the more common *in-situ* tests and the natural stiffness of the soil. For SPT blowcount N, D'Appolonia *et al.* (1970) gave for a virgin compressed sand

$$E'_0 \approx E'_y \approx 22\,000 + 1100N \quad \text{kN/m}^2$$

and for a pre-compressed sand

$$E'_0 \approx E'_y \approx 54\,000 + 1400N \quad \text{kN/m}^2$$

but Mitchell demonstrates an amazing diversity of correlations by other authors, which underlines both the difficulty of achieving a good SPT result and the credibility gap which is inherent in attempting to correlate any single parameter with a measurement such as the SPT that induces a complex multi-parameter soil deformation. For the cone penetration resistance q_c, De Beer (1965) proposed the average correlation

$$E'_0 \approx E'_y \approx 2q_c$$

between the bounds $1.5q_c$ and $3q_c$ for many virgin-compressed sands. For precompressed sands he increased the modulus by the factor λ_0/κ_0 from independent oedometer tests on recompacted samples, being the ratio between the gradients λ_0 of the virgin plastic compression line and κ_0 of the elastic recompression line on an $e/\ln \sigma'$ diagram. The justification for such a correction is presumably that the large stresses around an advancing soil probe inevitably generate plastic compressions, so that the correlation is properly based on λ_0 while the designer truly wants to apply rather small stresses. Equation 6.19

demonstrated that $E'_0 \propto 1/\kappa_0$ for a logarithmically straight elastic recompression line, and could equally well have demonstrated that $E'_0 \propto 1/\lambda_0$ for a logarithmically straight plastic compression line, so that the ratio between recompression modulus and plastic compression modulus would indeed be λ_0/κ_0 at a given vertical effective stress.

De Beer's interpretation of the CPT should have pricked your conscience with regard to the fundamental interpretation of the one-dimensional compression test: now is the time to rediscover section 6.5! If the data of compression and swelling were actually to follow paths such as APRS in figure 6.15, two uncomfortable corollaries must be faced. The first corollary was derived after equation 6.19, and rests with the fact that even elastic soil on a recompression line has the awkward property that its stiffness is roughly proportional to the magnitude of the local effective stress. This not only predicts an almost uniform increase in natural stiffness of such a soil with depth, which should be correctly reflected by SPT or CPT results, but also threatens chaos when the variable stresses around the base of a foundation are considered, since the stiffness of the soil immediately underneath a shallow footing should then increase dramatically as the footing was loaded. How can a CPT take account of the enhancement of soil stiffness under foundations applying various stresses, if this were to occur? The second corollary concerns the approach at P in figure 6.15 of the plastic compression curve along which rather large irrecoverable strains can occur. The preceding correlations for E'_0 against N or q_c in the case of virgin-compressed sand presumably referred to sand in just such a condition. Strictly, therefore, such a soil should not be considered to be elastic at all, nor should the methods of elasticity be applied to it. If it were found to be convenient to choose some gradient of a stress–strain curve such as figure 6.13 at point X along the plastic part of the compression line, to refer to the gradient as E'_0, and then to apply some elastic system of logic to deduce settlements, then so be it: but the results should be treated with even more scepticism than usual! If the engineer feels that his subgrade is nearly in a virgin condition he might prefer to preload it with fill which he can then remove, or compact it, in order to release its potential plastic strains before he builds over it: he would then have protected his client from the catastrophic settlements sometimes caused in loose sands by ground-water lowering, small earthquakes or other vibrations quite separate from the effects of the dead weight of the structure.

Current practice for determining the stiffness of a clay usually depends on the one-dimensional compression test, which is frequently spoiled by the smallness of the samples tested, their sparseness and degree of disturbance, and occasionally by errors in specifying the equilibrium effective stresses between which the compression was observed. Errors due to the disturbance of stiff clays which were *known* to be heavily overconsolidated may perhaps be removed pragmatically by cycling the effective stress repeatedly between its initial and final value, and using the results only when the compression curve has stabilised. Errors due to the disturbance of soft or sensitive clays would simply be catastrophic and must therefore be avoided. Advanced *in-situ* tests such as Wroth's self-boring pressure-meter probably offer the greatest promise to those who cannot avoid applying extra stresses to soft soils in circumstances which make the accurate prediction of settlement important. Those who can avoid creating extra stresses on compressible soils usually do, either by piling through them if they are shallow enough or by constructing within them a fully buoyant structure with a tanked basement if they are deep. Where extra loading is unavoidable it is usual to reduce the significance of the settlement by constructing over a rigid raft and employing flexible service joints to the outside world.

Before proceeding to an important calculation of the displacement of a foundation the designer should first assure himself that the soil surrounding his construction will be able to support the magnitude of the intended stresses without indulging in some form of plastic strain of which his stiffness tests had omitted to warn him. You should recall that section 6.8 dealt with the first approach of a limiting shear stress, $\tau = c_u$ or $\tau = \sigma' \tan \phi'$, along certain planes and within certain zones of initially elastic soil with no initial shear stresses which is made to support a shallow flexible strip footing. From equation 6.28 it follows that if the yield point in some elastic-cohesive soil were at a shear stress $\tau = c_e$, then first yield of the flexible strip would occur at a net pressure of $\sigma_e = \pi c_e$. And by equation 6.31 it follows that if some irrecoverable sliding of an elastic-frictional soil were to occur at an inclination of stress ϕ'_e across any plane, that the first yield of the strip would occur at a net foundation pressure of roughly

$$\sigma_e = \pi \sigma'_0 / (\phi'_e + \cot \phi'_e - \pi/2)$$

where σ'_0 was the effective overburden pressure on the plane of the foundation. These rough solutions for a shallow strip merely set an example and provide a warning. The elastic stress distribution around any foundation is complex, and the various zones of soil are subject to a variety of stress paths. Figure 8.30 demonstrated that the onset of either contractile yield or dilatant yield was a function both of initial stresses and stress paths. Notwithstanding the inherent complexity of the problem, it is unlikely that serious yielding will invalidate an elastic settlement calculation if the stress distribution comprising the superposition of initial and induced stresses can be shown to satisfy three criteria

(1) no soil element under the edge of the foundation shall be in a state of stress in which $q > Mp'$, which is equivalent to saying that on no plane should $\tau > \sigma' \tan \phi'_c$

(2) no soil element under the edge of the foundation shall be in a state of stress in which $q > q_u$, which is equivalent to saying that on no plane should $\tau > c_u$

(3) the soil under the centre of the foundation shall not be subjected to a vertical effective stress which exceeds the precompression demonstrated in one-dimensional compression tests.

What then remains is a known system of foundation stresses which must be carried by a system of soil elements whose stiffness is known. Either the algebraic and numerical solutions gathered together by Poulos and Davis (1974), or the *ad hoc* approach of section 6.7 and figure 6.22 may then be used to generate the settlements required for an analysis of damage.

10.7 Retaining Walls

10.7.1 Function, Type and Modes of Failure

An earth-retaining wall in open country is usually the last resort of a designer who has insufficient space to form natural soil slopes between constructions which are at different levels. New road schemes frequently generate the need for retaining walls, principally where the road must be in a cutting through land which is expensive or restricted, or where a new overpass is required which will be built up on an earth embankment rising towards an abutment wall, which serves the dual purpose of retaining the earth bank and supporting the new bridge deck over the existing carriageway. Basement walls in buildings serve the analogous function of retaining the external earth and transmitting the structural forces down to the foundation slab. Retaining walls may also be subject to counter-balancing water pressures, such as when they are used to train rivers, or to create docks, wharves and navigation locks. Finally, the walls of vessels to retain water, sewage, flour, coal or other materials, whether above or below external ground level, must clearly be considered part of the family. I shall concentrate on rather simple walls retaining earth, in order to reveal some principles.

Figure 10.32 demonstrates the wide diversity of structural forms which can retain earth, and indicates with arrows the more significant resultant forces exerted on the structural components, which increase in flexibility from type (i) to type (vii). The designer's initial preference will depend on at least three criteria

(1) the need to choose an economical system which will often imply that the soil behind the wall should be replaced by a well-drained, compacted, granular fill: this would render type (iii) unhelpful

(2) the need to reduce to within tolerable limits the settlement of the retained soil relative to the retaining wall, which is important for bridge abutments to prevent a step appearing as the high level road passes on to the bridge deck: this tends to push the designer either towards reducing all settlements such as by a costly type (ii) wall with granular backfill, or towards a much cheaper type (vii) structure which would all settle together

(3) the need never to remove support from the adjoining ground which leads to a type (iii) wall; in sensitive city-centre sites, the wall need not be driven but can be constructed as an *in-situ* reinforced concrete diaphragm by the use of a slurry trench of the same depth and thickness; such a wall could either be anchored, or propped by ground-floor and basement slabs as the excavation proceeded.

Each of these forms of construction should be checked against three types of failure. The first mode of failure is as an inclusion in a general landslide, whether triggered by the weight of the new wall and its backfill, or purely coincidental, but seated in the natural ground such as that depicted in figure 10.33. The methods of slope stability analysis described in chapter 9, and particularly that of slip circles with slices, should be used in conjunction with pessimistic undrained and drained soil strengths appropriate to the case.

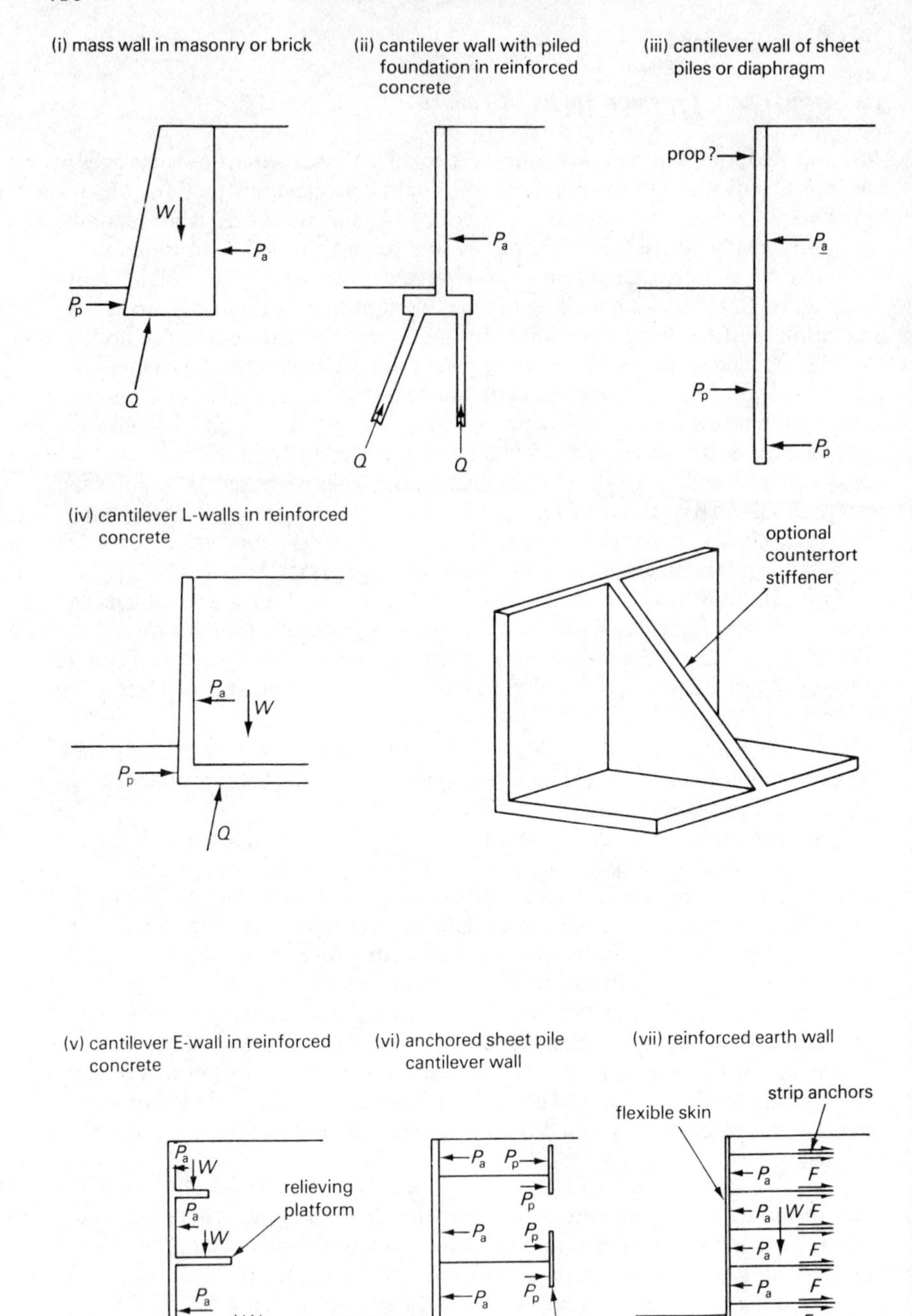

Figure 10.32 *Types of earth retaining wall*

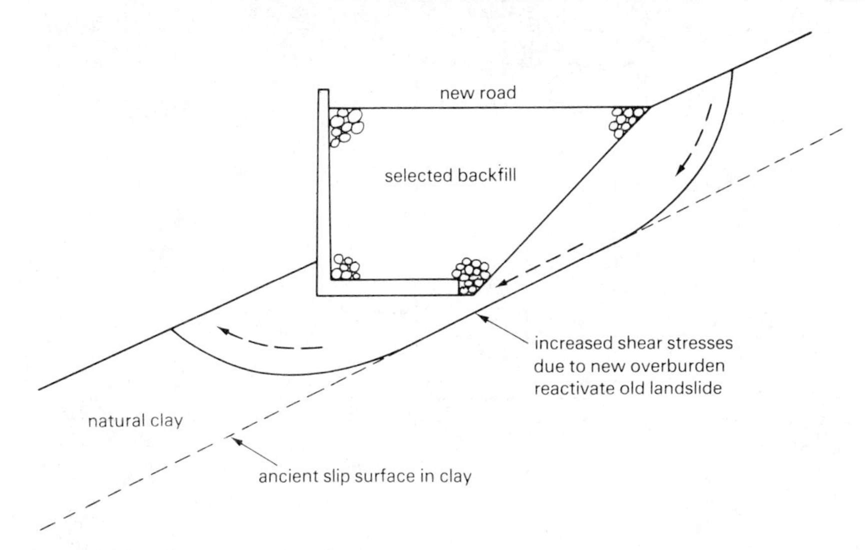

Figure 10.33 *Inclusion of a wall in a deep-seated landslide*

The second mode is a global failure of the whole soil construction acting as a rigid embedded monolith. Collapse can only occur when there is a simultaneous approach to limiting stresses all around the construction with active pressures behind the wall, passive pressures in front of the wall, and limiting foundation stresses beneath the wall. This can be prevented by merging the relevant limiting wall thrusts taken from chapter 9, with the weight of the construction, in order to calculate the magnitude, direction and location of the foundation thrust, and altering the proportions of the wall until there is no chance of exceeding the limiting foundation thrusts discussed in section 10.6. Figure 10.34 depicts in (i) the rotation of an excessively narrow mass wall with active and passive zones on its sides and a complex foundation failure with the transmission of an eccentric inclined thrust which is the resultant of the other forces, and shows in (ii) the sliding of an excessively narrow cantilever L-wall due to the generation of an

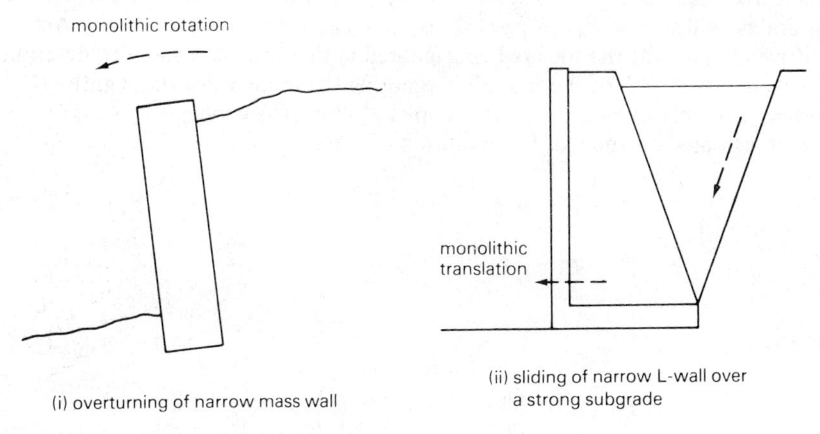

Figure 10.34 *Modes of global failure*

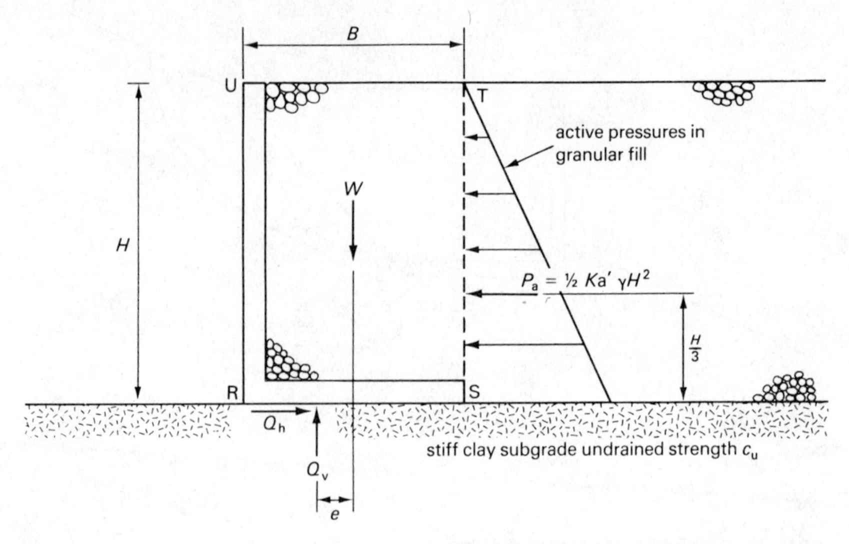

Figure 10.35 *Forces per unit length acting on a monolith comprising an L-wall with infill which is on the point of collapsing*

active wedge of backfill which can drive the wall forward by creating a foundation failure due simply to sliding.

The production of a stable free-body diagram for a typical soil-structure monolith is undertaken below with respect to the choice of the length B of the base of the long L-wall of height H originally depicted in figures 10.32 (iv) and 10.34 (ii) and redrawn in figure 10.35. By erecting a vertical plane ST through the potential active wedge the stability of the monolith RSTU, which contains the wall, can be examined. The first step is to employ some pessimistic angle of shearing resistance ϕ' for the fill in order to calculate its active earth pressure coefficient K'_a. Horizontal active stresses can then be inferred on the buried plane TS which, in the absence of pore pressures due to the supposed existence of drains, will sum to $P_a = K'_a \gamma H^2/2$, as shown acting at a height of $H/3$. Any passive pressures at the toe are being ignored in this case. It remains to determine the weight $W = \gamma HB$ of the monolith, ignoring the presence of the slightly heavier concrete in order to create simpler algebra. The concepts of section 10.6.6 can now be applied. By resolution of forces

$$Q_v = W = \gamma HB$$

$$Q_h = P_a = \frac{K'_a \gamma H^2}{2}$$

so that

$$\delta = \tan^{-1}\frac{Q_h}{Q_v} = \tan^{-1}\left(\frac{K'_a H}{2B}\right)$$

By moments, the eccentricity

$$e = \frac{M}{Q_v} = \frac{1}{2}\frac{K'_a\gamma H^2}{\gamma HB}\frac{H}{3} = \frac{1}{6}\frac{K'_a H^2}{B}$$

If the designer were to impose the restriction $e \not> B/6$, he must then satisfy

$$\frac{B}{H} \not< (K'_a)^{1/2}, \text{ typically} \not< 0.5$$

The restriction $\delta \not> 20°$, taking into account both the reduction in bearing capacity caused by inclined stresses and the possible angle of friction between the concrete base and the underlying clay, would imply

$$\frac{B}{H} \not< 1.4K'_a, \text{ typically} \not< 0.34$$

which will usually be less severe a condition. If the eccentricity condition $B/H = (K_a)^{1/2}$ is used to determine the inclination, the designer will obtain

$$\delta = \tan^{-1}((K'_a)^{1/2}/2)$$

which has a sensible typical value of 14°. He must then additionally compute the bearing capacity of the foundation, perhaps using figure 10.27 to create an approximate reduction factor of 0.39 so that the immediate safety factor could be written

$$F = 0.39 \times \frac{5.14c_u}{\gamma H} \approx \frac{2c_u}{\gamma H}$$

If it were required to achieve a safety factor of 2.5 it would then be essential to check that

$$c_u \not< 1.25\gamma H$$

failing which the designer can opt to improve the bearing capacity of the base by making it longer in order to reduce e/B and δ, or to avoid the undrained strength problem by ensuring that the subgrade reacted in a drained fashion, or to avoid the whole problem by employing a piled foundation. If the designer decided to persevere with the original scheme he would certainly continue to peruse the bearing capacity and settlement problems along the lines of the previous section.

If the designer can show that even pessimistic active thrusts in such situations do not create a limiting state of stress beneath and in front of the wall, he may describe the proposal as 'safe'. This does not preclude movement, of course, and his final treatment of the wall as a rigid monolith must follow the rationale of the previous section in determining whether the client is likely to find the movement of the construction acceptable. Not only might some overall settlement, translation and rotation be aesthetically offensive, but also the haphazard movement of one section of wall relative to another might be more so. The designer must not forget that a retaining wall is three dimensional, and that although it might be considered relatively rigid in the plane of its cross section it is rather unlikely to be other than flexible along its length. As with any foundation, differential settlement along the axis may cause unsightly cracking of the face of the structure, and this must be mitigated by the provision either of regular joints which allow relative motion, or at least of deep vertical grooves in the face which can act as orderly crack generators. A bold, knobbly pattern on the face can not only hide a great deal of deformation but will also weather more pleasantly than would an attempt at a smooth face. As with all structures, the movements caused by changes in the pressure of the groundwater may be significant, and good drainage is an essential concern of the wall designer.

The third fundamental mode which must be checked is of some local failure due, for example, to overstressing or corrosion. Figure 10.36 demonstrates the local failure of a reinforced concrete wall in (i) and a soil anchor in (ii). The hinge at the base of the cantilever wall in case (i) could be due to the generation of extra lateral stress on the wall such as by the failure of drains to prevent ponding of storm water in the gravel backfill or by the unforeseen application of a heavy vehicle on top of the fill, or to the corrosion of the reinforcement if it were insufficiently covered by concrete. The yield of the anchor in case (ii) could likewise be due to extra lateral stress on the wall, or to the corrosion of the cable, or to the softening by inward drainage of previously stiff cohesive soil in the passive zone in front of the anchor. It is therefore necessary to be realistically pessimistic about drainage, surcharge due to vehicles, and corrosion, and to make suitably stringent provisions. It is frequently permissible to analyse the likelihood of some local failure by inferring pessimistic active stresses behind the wall, such as by choosing a low value of $\phi' = \phi'_c$ in a sand which ought to possess a good deal of extra dilatant strength ϕ'_v if its compaction was up to standard. Occasionally, and most dangerously, the structure may be able to fail locally without mobilising the full strength of the soil: the designer must then attempt to estimate the lateral soil stresses at failure which would, of course, exceed the active stresses. Section 6.10 demonstrated that this occurred when a wall was stiffer in lateral motion than its retained soil, and figure 6.31 showed that a

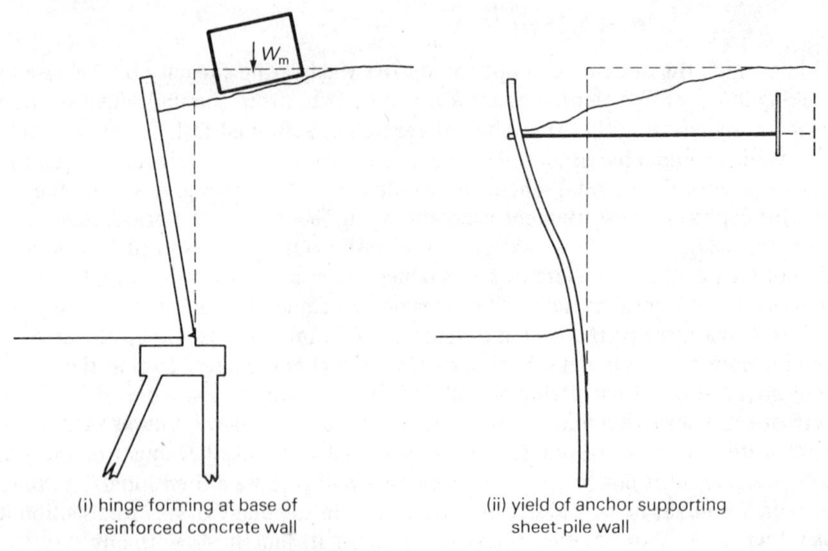

Figure 10.36 *Modes of local failure*

criterion $E_sH^3/EI > 15$ might be appropriate to distinguish clayey fills which should be stiff enough to avoid the problem. Section 6.10 went on to show that conventional reinforced concrete cantilever walls were likely to be safely flexible with respect to moderate heights of granular fill or stiff clayey fill. Excessive late lateral stresses due to an insufficient soil-structure stiffness ratio are likely to be confined to the following situations, therefore

(1) an attempt to retain more than 15 m of well-compacted granular material, or more than 6 m of well compacted stiff clay, with a conventional reinforced concrete wall

(2) the designer choosing too low a structural stiffness in order to test the ratio: pessimism here flows from the highest structural stiffness that may occur

(3) the designer providing lateral supports for the cantilever wall, such as the counterfort webs sketched in figure 10.32 case (iv). The wing walls on either side of a bridge abutment create just such a stiffened box construction. When a bridge deck is used to prop a cantilevered abutment wall at its top, a similar result may be obtained.

The response of the designer might then be either to adopt a more flexible alternative scheme such as reinforced earth, or to use some one-dimensional compression parameter K_0 instead of the active earth pressure parameter K'_a. As discussed in section 6.10 an appropriate coefficient for saturated clay against a rigid wall would be $K_0 = 1$, while engineers frequently and empirically use $K'_0 = 1 - \sin \phi'$ for the equivalent effective coefficient of well-drained granular soils. In a marginal case the designer might like to use some active-type earth pressure coefficient K'_a but based on some angle ϕ'_e at which inelastic strains would have hardly begun to offer any reduction in stiffness: $\phi'_e \approx \phi'_c$ may then be sufficiently accurate in dilatant granular materials, if it was not felt to be worth while to perform the appropriate compression test such as that idealised in figure 8.6.

10.7.2 Pore-water Pressures, and their Effects on Lateral Stresses

Each of the modes of failure discussed above are very sensitive to the pore-water pressures in the ground around and beneath the wall. Section 9.6.4 showed that the active thrust of submerged granular fill could be more than double that of the same fill when dry. It follows that good drainage is central to the safe and economical design of retaining walls. If a cheap clean gravel backfill is available, it is only necessary to provide regular weepholes towards the base of the wall and to attempt to prevent storm water from percolating in at the crest. Otherwise it is necessary to detail a blanket drain such as those depicted in figure 10.37 to reduce or eliminate water pressures in the backfill. If it is possible to lay out the drain beneath the fill as shown in (i), a storm severe enough to cause flooding can only generate a vertical downward percolation, which was shown in figure 4.16 to offer zero pore-water pressures at every point. If it is only possible to construct a drain against the back face of the wall as shown in (ii), the resulting flownet generates small but significant water pressures amounting to average heads of roughly

$$h \approx 0.2\theta H$$

on planes inclined at angle θ to the vertical wall of height H. The method of trial wedges described in section 9.4.4 can then be used to determine the thrust on the wall: with $\gamma = 20$ kN/m^3, $\gamma_w = 10$ kN/m^3, $\phi' = 35°$ and $\delta = 25°$ it is easy to show that the influence of the water pressures is to increase the lateral thrust by roughly 35 per cent. Of course both of these solutions depend on the drains being perfectly efficient at conducting the water away without generating losses of head, and Cedergren (1967) demonstrates that the ideal drainage material would possess a permeability at least thirty times greater than the fill it was supposed to be draining, so that its characteristic grain sizes should be at least five times greater. If the designer has proportioned the wall on the basis of zero pore-water pressures in the soil, any serious blockage in the drainage system would be likely to double the shear forces and overturning moments, which in turn would reduce the bearing capacity of the foundation soil; one or other of these effects would be likely to generate a collapse. For this reason the designer must take care to filter out fine soil particles from the fill before they enter and clog the drain: this may be achieved by either a protective layer of medium sized well-graded soil or a suitable sheet of nonrotting fabric. These considerations would be all that was necessary in the case of a granular soil.

If a designer has to deal with clayey soils he inherits certain opportunities and difficulties in addition to those which would have applied had the soils been coarser. These extra considerations chiefly involve pore-water pressures which may be either helpfully negative or dangerously positive, although they are usually dealt with in terms of a corresponding cohesion or undrained strength which may be either helpfully larger or dangerously smaller than the simple drained strength. Although figure 10.37 would represent the steady state of seepage of ponded groundwater through a retained mass of soil of whatever pore size, the designer of a wall to retain clayey soil would have additional transient states of pressure to concern him. The soil may initially possess larger pore pressures which dissipate only slowly and which may allow the wall to be pushed over as it is being constructed: such soil would be looser than its eventual critical state, and a cohesion analysis after the fashion of 9.4.6 or 9.6.5 would demonstrate that larger active forces would be available to push over the wall while smaller passive forces would be available to prevent failure. Alternatively, the soil may initially possess smaller pore pressures than those of the ponded flownet, which make the construction temporarily stronger so that a cohesion analysis would demonstrate smaller overturning forces and larger restraining forces than had the friction analysis based on long-term pore pressures. Unfortunately, as I demonstrate below, the relaxation of pore suction may cause extra lateral pressures due to swelling.

If a clayey soil has been chosen as fill to be compacted behind a retaining wall then it is likely to be relatively strong and dense, and to be positively drained at its base. If the designer felt that he could eliminate all ingress of water to the clay fill, he might allow himself the luxury of postulating negative long-term hydrostatic pore pressures in the clay fill above a lowered water table, following figure 3.4. If in addition the short-term pore pressures were very negative,

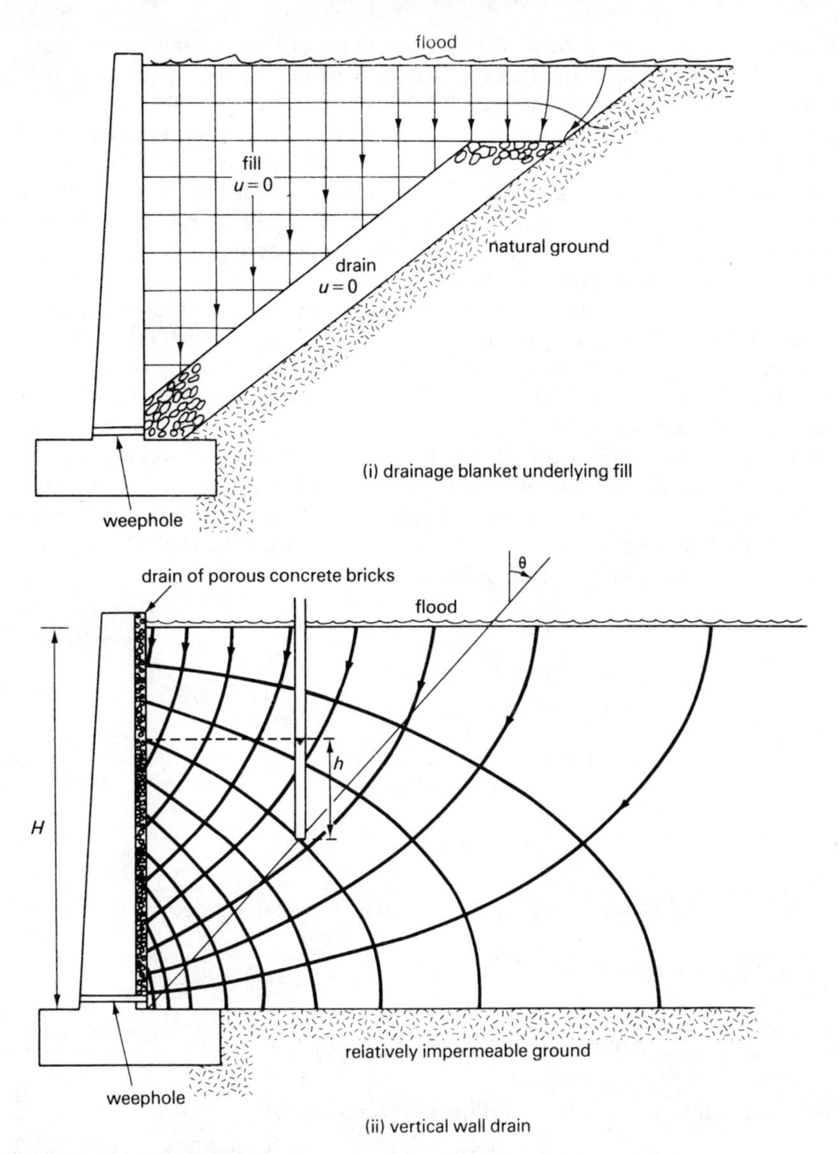

Figure 10.37 *Drainage behind retaining walls*

demonstrated perhaps by a cohesion analysis which generated small or negative active earth pressures, the designer might be tempted to base his retaining wall design on one or other of these negative pore-water pressure systems. Any relaxation of the pore suctions would then tend to lead towards a collapse: such a contingency might be due to the presence of ponded water in a badly constructed surface drain, or to the migration of water vapour through a rather unsaturated fill which was compacted too dry. In such cases the pessimistic designer will perform calculations identical to those he would use for a sand fill, based on the effective angle of shearing resistance of the clay and an assumption of zero pore-water pressure: he need then only guarantee the longevity of the base drain and weepholes if his wall is sufficiently flexible to allow the generation of active earth pressures. 'Loose' clay soils, on the other hand, can be recognised by their relatively large undrained active pressures following equation 9.41, which will provide the critical criterion for the design of the wall. Only if the soil could be replaced by imported granular fill, or be drained while the wall were constructed, or if the wall could be temporarily propped, could the dangerous 'undrained' condition be circumvented.

A flexible wall retaining clayey fill of sufficient relative stiffness should not be overstressed immediately after construction. If the long-term lateral stresses conform to his assumption $\sigma_h = \sigma'_h = K'_a \gamma z$ the designer should also have been successful in preventing a long-term collapse of the structure. But consider the transition of an element of dense clay from an initial state of suction $u = -s$ to the state of long-term downward percolation with $u = 0$ which was the designer's assumed condition. Will the soil come into an effective active condition, or will its tendency to swell cause extra lateral pressures? Suppose that the suction was initially strong enough to prevent the clay from stressing the wall, although it was just in contact with it. Then initially

$$(\sigma_v)_i = \gamma z$$

$$(\sigma'_v)_i = \gamma z + s$$

$$(\sigma_h)_i = 0$$

$$(\sigma'_h)_i = s$$

And finally

$$(\sigma_v)_f = (\sigma'_v)_f = \gamma z$$

$$(\sigma_h)_f = (\sigma'_h)_f = \Delta\sigma \quad \text{which is to be determined}$$

The average effective stress will then reduce by

$$\Delta p' = \frac{(\gamma z + s) + s - \gamma z - \Delta\sigma}{2} = s - \frac{\Delta\sigma}{2}$$

This must cause a volumetric swelling of magnitude

$$\Delta\epsilon_v = \frac{(s - \Delta\sigma/2)}{E'_v} \tag{10.21}$$

Figure 10.38 illustrates how volumetric expansion can be crudely added to the simple shear mode of retaining wall deformation discussed in section 6.10 with respect to figures 6.29 and 6.30, in order to generate a composite mechanism. The volumetric expansion of the triangle of fill is assumed to be uniform so that

$$\Delta\epsilon_v = \frac{\text{increase in volume}}{\text{original volume}} = \frac{\alpha_v H\sqrt{2}}{H/\sqrt{2}} = 2\alpha_v$$

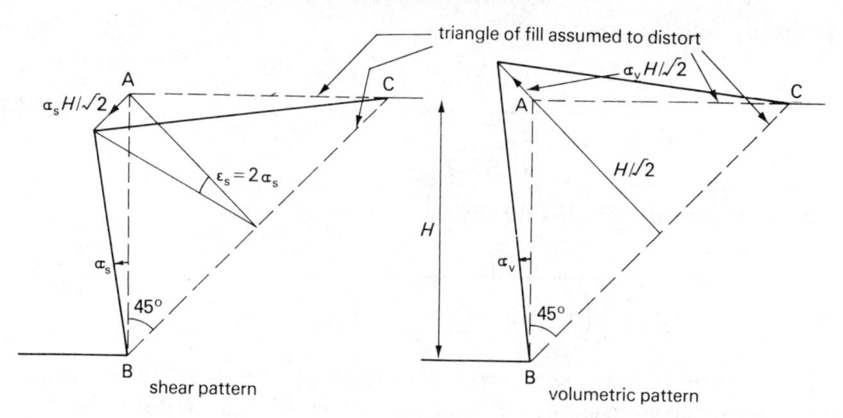

Figure 10.38 *Patterns of strain in a mass of soil retained behind a wall AB*

Now the total wall rotation is

$$\alpha = \alpha_s + \alpha_v$$

$$\alpha = \frac{\epsilon_s}{2} + \frac{\Delta\epsilon_v}{2}$$

The shear strain on 45° lines generated by a simultaneous wall rotation α and volumetric expansion $\Delta\epsilon_v$ is therefore

$$\Delta\epsilon_s = 2\alpha - \Delta\epsilon_v \tag{10.22}$$

It is now necessary to substitute for α and $\Delta\epsilon_v$ in the case of the relief of suction. It is easy to show that the postulated uniform increase $\Delta\sigma$ in the lateral stress on the cantilever wall produces a rotation at mid-height

$$\alpha = 0.145\Delta\sigma\frac{H^3}{(EI)} \tag{10.23}$$

This, and equation 10.21, can be substituted into equation 10.22 in order to achieve

$$\Delta\epsilon_s = 0.29\Delta\sigma\frac{H^3}{(EI)} - \frac{(s - \Delta\sigma/2)}{E'_v}$$

But if the soil has remained elastic

$$\Delta\tau_s = E_s\Delta\epsilon_s$$

$$\Delta\tau_s = 0.29\Delta\sigma\frac{E_s H^3}{(EI)} - \frac{E_s}{E'_v}(s - \Delta\sigma/2)$$

But the increase in shear stress was inferred at the very outset

$$\Delta\tau_s = \frac{\Delta\sigma_v - \Delta\sigma_h}{2} = -\frac{\Delta\sigma}{2}$$

so that by collecting terms

$$\frac{\Delta\sigma}{s} = \frac{2\dfrac{E_s}{E'_v}}{0.58\dfrac{E_s H^3}{(EI)} + \dfrac{E_s}{E'_v} + 1} \tag{10.24}$$

A typical value of E_s/E'_v is 1.0 at which

$$\frac{\Delta\sigma}{s} = \frac{1}{\left(1 + 0.29\dfrac{E_s H^3}{EI}\right)} \tag{10.25}$$

which happens to bear a very close resemblance indeed to the previous elastic earth pressure coefficient K_e derived in equation 6.42. This confirms the importance to the designer of achieving flexible walls and of preventing the ingress of water into clay. If the designer had achieved $K_e \approx 0.25$ then the following stress changes might be expected at 5 m depth in a fill which was subjected to percolation after being in a strong state of suction

	Initial stresses	Final stresses
σ_v	100	100
σ_h	0	50
u	−200	0
σ'_v	300	100
σ'_h	200	50

This makes clear that even with a relatively flexible wall, the consequences of allowing water into a clay fill in suction can be severe. Only if very flexible forms of construction can be used ($K_e < 0.1$), or if water can be kept out, can clays with strong suctions be used economically for fill near sensitive structures. Of course, such materials may be hosed as they are placed so that large suctions are eliminated: care must then be taken to ensure that the watering is not excessive! The ideal pore-water pressure to achieve in newly completed clayey ground is that equilibrium value which will persist for all time.

10.7.3 Lateral Pressures due to Compaction Machines and Other Traffic

Observations have shown that the extra lateral pressures caused by surcharge loads on compacted fill can be large in comparison with those caused simply by the self-weight of the soil, and particularly so in shallow zones of fill where the effects of surcharge are greatest and the weight of soil overburden least. These extra stresses usually remain when the surcharge load has departed. The reason for this is clear. Consider a well-designed flexible wall which would have enjoyed an elastic earth-pressure coefficient of 0.05 had not the limiting strength of the soil expired at an active pressure coefficient of 0.20. If a uniform surcharge pressure q is applied to the completed fill, it will probably find the soil in an active condition and will generate a uniform lateral pressure increment of $0.2q$. If the surcharge is now removed, the soil will be able to recover elastically and the corresponding reduction in lateral pressure will be only $0.05q$. Three-quarters of the original increment will remain. Rehnman and Broms (1972) describe the extra lateral pressures which may be inferred after compaction by various weights of machine. The effects of such machines will be localised rather than uniform, and will build up as successive layers are compacted. Nevertheless, a designer who adopts some pessimistic equivalent surcharge relating to the weight of a machine spread over its gross area, and who applies such a surcharge to the completed cross section of fill, has taken a major step in overcoming this difficult problem. The magnitude of this extra surcharge may certainly lie between 10 and 60 kN/m^2, and it can therefore be likened roughly to an extra 0.5 to 3 m of overburden increasing the equivalent height of the construction. There is no doubt that traffic corresponding to the heavy end of this range should be avoided if possible, since it is bound to have a dominant effect on every aspect of the design of the wall.

10.7.4 Ingenious Structural Systems of Greater Flexibility

Most of the difficulties encountered in the design of retaining walls are generated by the flexibility problem. The designer decides that the wall should not translate or rotate very much: this creates earth pressures larger than the active values which the designer hoped to use: this in turn forces the designer to adopt a stiffer and heavier structure, which will probably need to be piled in order to resist the overturning stresses. Some engineers have attempted to sever this vicious circle by designing simple flexible retaining structures behind which active earth pressures may always be assumed. The stiffness of a cantilever wall is caused by the necessity to prevent bending moments from breaking it. Unfortunately the bending resistance of a concrete wall is proportional to the square of its thickness while its stiffness is proportional to the cube: if bending moments are increased by a factor 4 the thickness must be doubled and the stiffness must be increased by a factor 8. If walls are to be designed to be flexible, the moments in their face must be reduced to a minimum.

Tsagareli (1967) describes the function of continuous relieving platforms rigidly connected to the back face of a cantilever wall as shown in figure 10.32(v). He points out that two distinct mechanisms reduce the outward bending moments in the stem of the wall, as demonstrated in figure 10.39. Firstly, the weight of soil overlying a platform causes an inward bending moment at the joint with the stem: this causes a step reduction in moment in the face which then begins to increase again below the shelf. Secondly, the vertical stresses below the platform are much reduced since the weight of the overlying soil is transmitted through the structure: if σ_v is reduced then so likewise is σ_h, so that the curvature of the bending moment diagram beneath a shelf is reduced. Of course, the length of the base of the wall must be such as to withstand monolithic sliding or rotation.

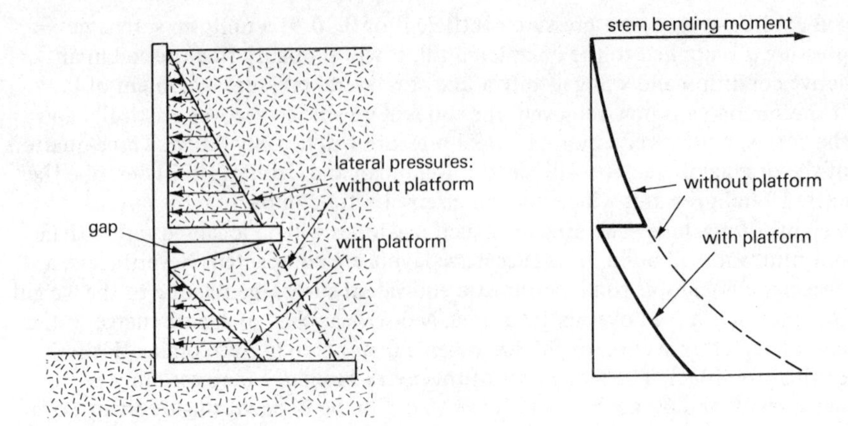

Figure 10.39 *Tsagareli's relieving platform*

Vidal (1969) carried the reduction in bending moments in the stem even further in his design of 'reinforced earth' depicted in figure 10.32(vii) in which loosely articulated wall panels are held in place by thin strip anchors. Schlosser and Long (1974) propose that these constructions can be treated at the global level as equivalent to mass walls, and at the local level as individual anchored wall units. Consider the design of a small wall panel S_v high and S_h wide which is retained by a single strip anchor attached to its centroid at depth z and which is freely embedded in dry sand. Suppose that the length of each of the anchors is B, as shown in figure 10.40. A control volume B wide and z deep may be chosen

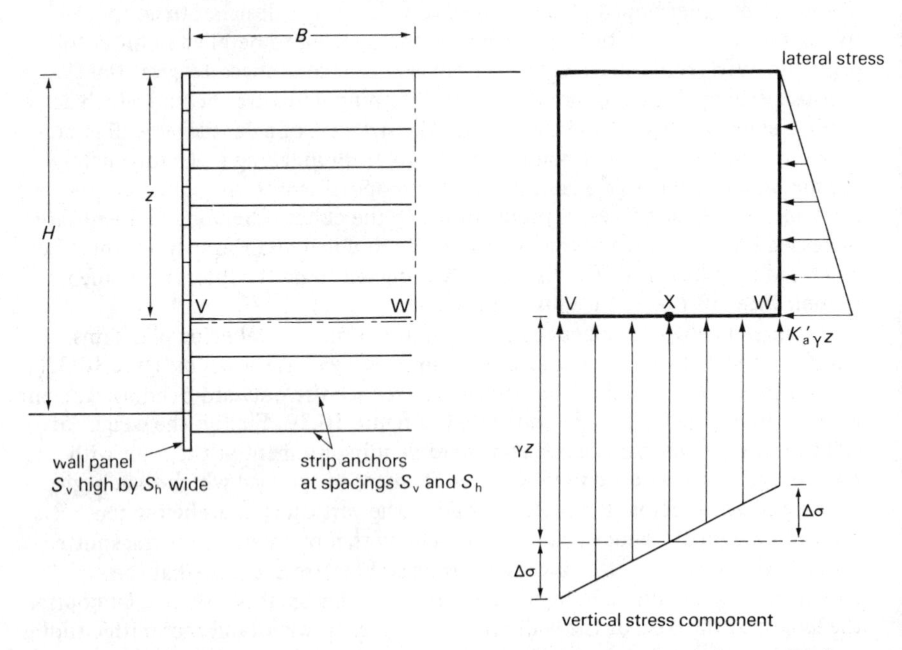

Figure 10.40 *Trapezoidal distribution of stress normal to a horizontal plane at depth z*

which should be in equilibrium. If horizontal pressures $K'_a\gamma z$ exist on its buried side, a moment $K'_a\gamma z^3/6$ must exist about the mid-point X of the base. Following figure 10.26 this could be counterbalanced by a trapezoidal distribution of vertical stress on the base: an extra vertical stress $\Delta\sigma$ acting at V, with a corresponding reduction $\Delta\sigma$ at W, causes a restraining moment $\Delta\sigma B^2/6$. For rotational equilibrium about X therefore

$$\frac{\Delta\sigma B^2}{6} = \frac{K'_a\gamma z^3}{6}$$

$$\Delta\sigma = \frac{K'_a\gamma z^3}{B^2}$$

The vertical stress at any depth immediately behind the wall panels would then be

$$\sigma_v = \gamma z + \Delta\sigma$$

$$\sigma_v = \gamma z(1 + K'_a z^2/B^2)$$

If the wall panels are so slender that they would buckle under the action of vertical thrust, they may be designated frictionless. Using Rankine's active stresses behind the wall panels, therefore, and with $\sigma = \sigma'$ since $u = 0$

$$\sigma_h = K'_a\gamma z(1 + K'_a z^2/B^2)$$

The area of wall tied back by a single anchor is $S_v S_h$, so that the tension in the anchor at its joint must be

$$T = S_v S_h K'_a\gamma z(1 + K'_a z^2/B^2)$$

Bolton *et al.* (1977) support the judgement of Schlosser and Long (1974) that the strip anchor may now be designed very simply. Suppose that the anchor is rectangular, being A_h wide, A_v deep and B long and that it has a coefficient of friction μ against the sand and ultimate tensile stress σ_{ult}. It appears to be sufficiently accurate to say that its tensile strength is

$$P_T = A_h A_v \sigma_{ult}$$

so that tensile rupture is avoided if $P_T > T$. Rather more surprising, it also appears to be sufficiently accurate for widely spaced strips to say that the ultimate sliding resistance of the strip anchor is

$$P_F = (\mu\gamma z)\, 2A_h B$$

with the shear stress $\mu\gamma z$ being fully developed on both top and bottom surfaces of the strip. The sliding of the anchor may then be avoided if $P_F > T$. Figure 10.41 summarises the stresses which may be inferred for the purpose of designing the structural components of a 'reinforced earth' wall: the lateral stresses on the panel causing a peak tensile force in the anchor at its front joint which then dissipates to zero at its free end by virtue of the friction between itself and the surrounding soil.

The source of the undoubted economy of 'reinforced earth' can be seen to lie in the method by which the inevitable lateral forces are absorbed. Instead of

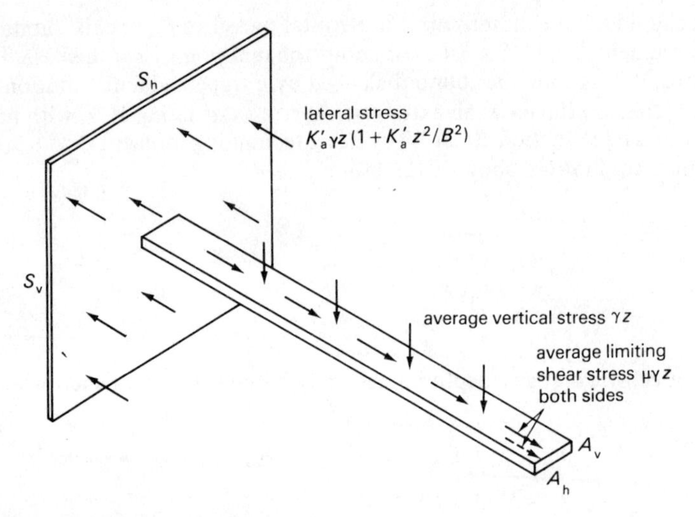

Figure 10.41 *Panel/anchor assembly*

allowing them to be carried down to the foundation soil as bending stresses in the front wall they are carried a similar distance into the retained mass itself as tensile stresses in the anchors. Tension is always a more economical source of support than bending. Of course the small wall panels of dimension S may tend to bend due to the moment of the lateral stresses. These are, however, only of the order $K'_a\gamma zS^2/8$ per unit length in comparison with $K'_a\gamma z^3/6$ per unit length in a cantilever wall: if $S \approx H/10$ then the bending moments in the facing panels never exceed one-hundredth of the base moment in a cantilever wall. Indeed Vidal's early designs featured semi-elliptical facing units with a horizontal axis so that the lateral soil pressures created only tensile face stresses, just as wind creates membrane tension in the sails of a yacht. A great variety of flexible facing systems may be used without altering the basic design of strip anchors at spacing S_v and S_h.

The source of most anxiety about 'reinforced earth' concerns the potential deterioration of the buried anchors. Whereas the steel reinforcement in a cantilever wall ought to remain safely covered by impermeable and inert concrete, the heavily galvanised mild steel anchors which are commonly used in such constructions are exposed to whatever chemical or biological predators are accidentally washed into their vicinity through the sandy fill. The source of most interest in 'reinforced earth' is its capacity to withstand soil deformations without distress, and therefore its ability to stand on compressible soils without the need for a piled foundation.

10.8 Modelling

Civil engineers are characteristically employed in design, decision-making and mutual communication. These tasks are as complex as any in which a human being can engage, and demand intellectual skills which are quite distinct from those that often govern those more menial tasks called examinations which are invented by teachers. This, perhaps, excuses the frequently heard claim amongst engineers that the 'subjects' which they were taught at school and university were of little use to them when they came into practice. Information, certainly, is best given very shortly *after* it was required and skills are best learnt when they can be practised: if engineering were solely about information and skills it would therefore follow that a craft apprenticeship was what was required rather than an education divorced from the real working environment. By this view schools, universities, teachers and books would be regarded as potentially inefficient or irrelevant. But what of design, decision-making and communication? Surely one purpose of tertiary (and indeed secondary and primary) education was to encourage just such habits of mind as sensitivity, critical appraisal, creative thinking and clear exposition which would lead to improvements in these very activities? Whether or not graduate engineers believe that their university education was vital for them, they probably agree that the more beneficial aspects were concerned with learning to think, learning to learn, learning to decide, and learning to speak and write. They might also remark that such opportunities for intellectual development were as oases in a desert of indigestible facts and routines. And so here comes the crunch question. Can you now hope to think positively and usefully about some ground engineering problem which you are about to face, or will you be plunged totally, hopelessly and irrecoverably into chaos?

When a problem arises it will not come knocking on the door with a label stuck to its forehead saying 'I'm a seepage problem' or 'I'm supposed to be a duplicate foundation to the one you built last week'. First of all it will not come knocking: you will have to go out there and perceive it amidst a welter of irrelevant and confusing events that happen to be obscuring it. Secondly it will not be labelled: you will have to attempt to diagnose it, make some measurements of it, analyse it, predict it, revise your diagnosis in the light of the appalling errors of your prediction, and finally attach such labels to it as may be intelligible to your colleagues who are going to have to help you deal with it. Thirdly it will not be a replica of any previous problem: time does not go round twice. These observations are as true of life in general as of soil mechanics or civil engineering, of course, and so it is not surprising that various philosophers, psychologists, and teachers have developed systems of thought and action appropriate to problem-solving, scientific enquiry, perception and communication. If you are interested, you will find some of the roots of my own approach to decision-making in the Penguin book *Inquiring Man* by D. Bannister and Fay Fransella (1971), to teaching and learning in the Pelican book *How Children Learn* by John Holt (1972), to thinking in the Cape book *The Use of Lateral Thinking* by Edward de Bono (1967), and to design in the Corgi book *Zen and the Art of Motorcycle Maintenance* by Robert M. Pirsig (1974). All this is of less importance than that you develop your own flexible style of constructive thinking, and it is for this reason that I have stressed the notion of mental models. A model should clearly 'work' in some direct repeatable and easily recognisable fashion and yet it is clearly not the reality that it is attempting to reflect. Its ambiguity makes the notion of a model an ideal instrument for engineers. Internally it must be logical and consistent like a tool, but like a tool the model must also have a well-defined boundary so that it can be picked up and put down independently of the system that the engineer is attempting to design. Some students treat their mental world as though it was meant to be a monolithic sculpture in which every idea was fixed relative to every other. This is no use to the busy engineer: he has the real world to deal with, which will appear to be an organic and fast-moving kaleidoscope of some complexity. The most successful way to make progress in such a world may be to travel light, with a mind containing a few robust and well-chosen tools, a readiness to use them skilfully and a deep respect for the authority of subsequent events which will prove whether the tools were appropriate and the proper skills practised.

How, then, is it possible to use models creatively? Figure 10.42 is an attempt

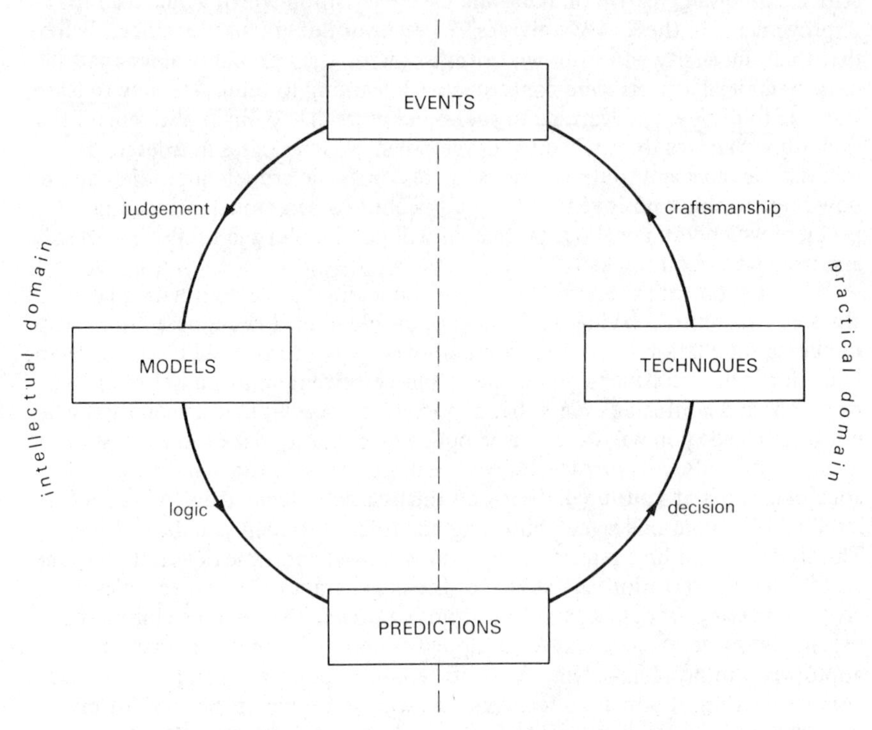

Figure 10.42 *The creativity cycle*

to display a cycle of intellectual and physical activity which results in the creation of an 'event'. Perhaps the cycle begins in the realm of past events and starts with an act of judgement as to which of those events is relevent to the desired creation, and which mental models may be relevant to those events. The second step must then be to recover those relevant models from within whichever book or mental recess they reside, and to apply logic and reason to the task of making predictions relevant to the creation. In the third step, decisions must be taken based on those predictions, leading to a choice of techniques. Finally the techniques must be skilfully applied in order to bring the creation into being as a new event. You will see that there is some strong polar symmetry in the diagram, predictions being the intellectual ghost of future physical events, the introspective judgement of relevance being the counterpart to decisions that turn into reality, mental models being analogous to a store of physical techniques, and the formal deductive logic which turns out predictions being the twin of the craftsmanship that skilfully turns out real articles. Engineering education is concentrated unhappily and unnecessarily in the lower left-hand corner of figure 10.42 where logic leads to predictions – such as the stress at A or the deflection at B. In the first chapter of this textbook, however, I attempted to demonstrate the whole creativity cycle. In the central part I hoped to cover that part of the cycle from models to predictions, with an occasional excursion into a decision. In the final chapter I have intended to widen the scope by sweeping from judgements through to techniques. It might appear, therefore, that it is your relative ignorance of craft skills, and above all your ignorance of events themselves, which will be mainly responsible for any stunting of your creative ability. And yet these will shortly be amply corrected as you gain practical experience. If you are to fail in your attempt to reach the heights of the engineering profession it will probably be due to your inability to take good decisions once the calculations have been made, and crucially your inability to judge what was significant and relevant in the first place. These purely creative acts lie between the inner world of the mind and that outer world which it is attempting to change, and they resist most forms of enquiry. Perhaps you had better get into those books on thinking, after all!

Of course, anything as complex as a civil engineering construction is not completed in one wave of concerted action, and this allows the whole process to be fine-tuned to produce optimum performance. A typical spiral of thought and action relevant to a large project is modelled in figure 10.43. Peck (1969) followed Terzaghi in pointing out that some observation of relevant 'events' is necessary for the creation of an economical solution, and that sometimes such events are nonexistent before the project begins in the sense that no-one may have ever attempted to build on any site similar to the one in question, and site investigation and testing techniques may not be able to precreate all the relevant aspects of behaviour. For large projects it may then be worth creating a special prototype event, such as the loading of a specimen foundation or the construction of a trial embankment, so that the designer can see some good events to help clinch his opinions regarding the relevance of various mental models and the detailed logic that will then flow from them. To make such a trial worth while, the designer must have options to choose between: if there is only one fixed technique available that can create the required solution, the decision to use it was presumably inevitable and any previous intellectual activity was therefore redundant.

The creation of prototype events in the field has an analogue in the laboratory: small-scale physical models can be made and observed. Of course, some creativity cycle must be followed before the physical model can be produced. The engineer must judge what attributes of the full-scale construction are most important to him; he must then choose mental models of behaviour which are supposed to reflect these attributes, and deduce logically what must be measured, observed and achieved if his physical model is to be representative and useful. Then he must decide how to build it and test it, and after using

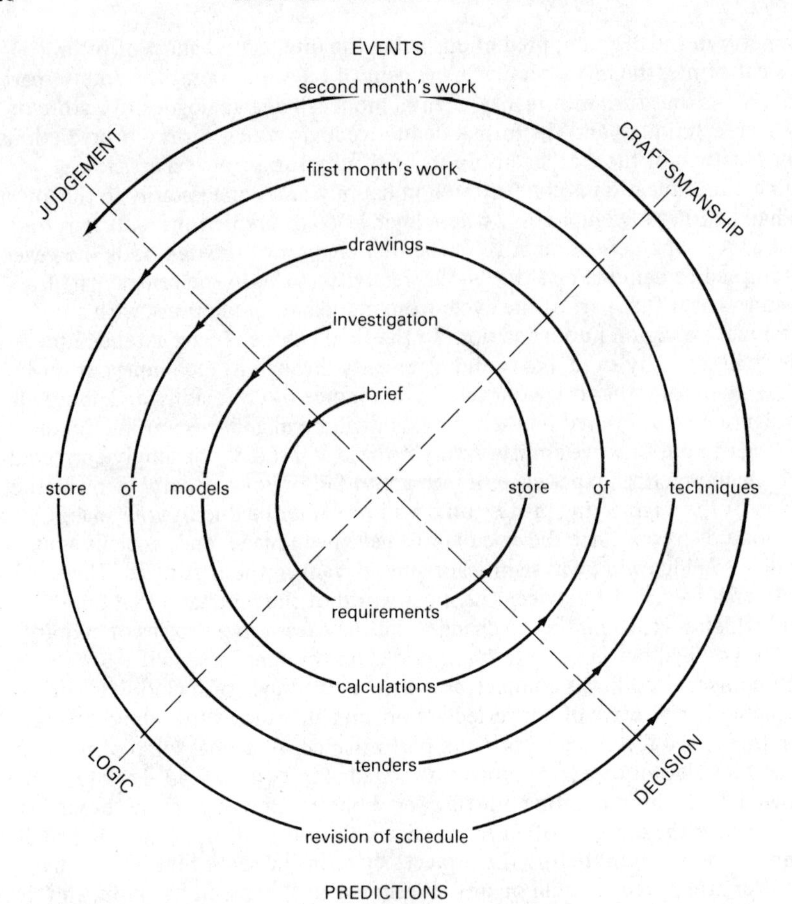

Figure 10.43 *A construction project viewed as a creativity spiral*

whatever practical skills may be appropriate he will then have succeeded in creating a worth-while event in the laboratory. Most soil tests attempt to replicate just one mode of behaviour of one element of the field construction. For example, the oedometer test creates a model of a zone of one-dimensional compression which might represent one soil element beneath a full-scale foundation: in detail, that element which was at the corresponding void ratio over the corresponding range of effective vertical stress. Because the laboratory test has not featured the geometry and range of soil properties and stresses of the full-scale foundation, the engineer will need to invoke a mechanical model such as that of section 6.7 for the full-scale event, and a corresponding model such as that of section 6.5 for the laboratory test, so that he can predict full-scale performance using an over-arching model such as that of section 6.4 to interrelate the others. If the *real* foundation problem were going to be one of collapse, however, this laboratory test would have nothing relevant to say. And if any of the three mechanical models were in error or irrelevant, the predictions would be in error or irrelevant. Consider a foundation on sloping soil, or near an excavation, or with an eccentric distribution of soils in the subgrade, or which was not horizontal, or which was to suffer vibrating loads, or which was to be loaded slowly at first and then faster later so that a period of partial consolidation was to be followed by an undrained compression process: will it be possible to extend the small-strain model of chapter 6, the large-strain models of chapter 8, and the collapse models of chapter 9 to include any of these effects?

One very powerful technique for studying the more complex soil construction problems is that of centrifugal testing. Reconsider the opening lines of the last paragraph. The most important attributes of full-scale constructions are presumably

(1) the shape of the landform
(2) the shape, disposition and properties of each zone of soil
(3) the effective stresses and water pressures at every point, caused by gravity acting on the water, the soil and any structures which are to be built.

Pokrovsky in the USSR in the 1920s hit on the ideal method of replicating each of these quite accurately in the laboratory. First conduct a detailed site investigation and recover very large undisturbed blocks of soil from the various strata which may be present. Create a three-dimensional mental model of the disposition of these soil types, return to the laboratory, and sculpt a $1/N$ scale model landform from the soil that has been recovered, making sure that the boundaries of the model container are as far as possible from the zone of interest, with the walls smooth and the floor rough in an attempt to create similarity. Now centrifuge the model landform at N times Earth's gravity, injecting and draining water in such locations that the field groundwater regime is replicated as closely as possible. When the landform has settled down, and in particular when the groundwater regime has settled down and transient flow is completed, the engineer must create a model event, perhaps powering a jack to initiate the loading of a foundation while he closely observes the results. What then ensues in terms of stresses, strains or cracks should be a close analogue of the same sort of event in the field. Figure 10.44 demonstrates that the key to the attempted replication of events is the replication of stresses between the field-scale construction and the centrifugal model in equilibrium. From this equality follows equality of angles of friction and dilation, stiffness, changes in void ratio, permeabilities, etc. The errors in stress due to the nonuniformity of the acceleration field in the centrifuge (curvature, different accelerations at different radii, and the effect of Earth's gravitation) can be kept to below 10 per cent in a large enough machine. Access to the detailed application of centrifugal modelling can be gained through Hird *et al.* (1978). The unique advantage of having seepage, friction, cohesion, elasticity, yield, and transient flow all taking place together in its proper three-dimensional environment is that in most instances the modeller can treat his TV picture of the experiment as though it were the real thing: no further mental models are usually required. Only when the full-scale version of a satisfactory centrifugal model embankment collapses due to an unforeseen soft spot in some part of the full-scale subgrade that was not sampled, and which did not therefore feature in his model, will the

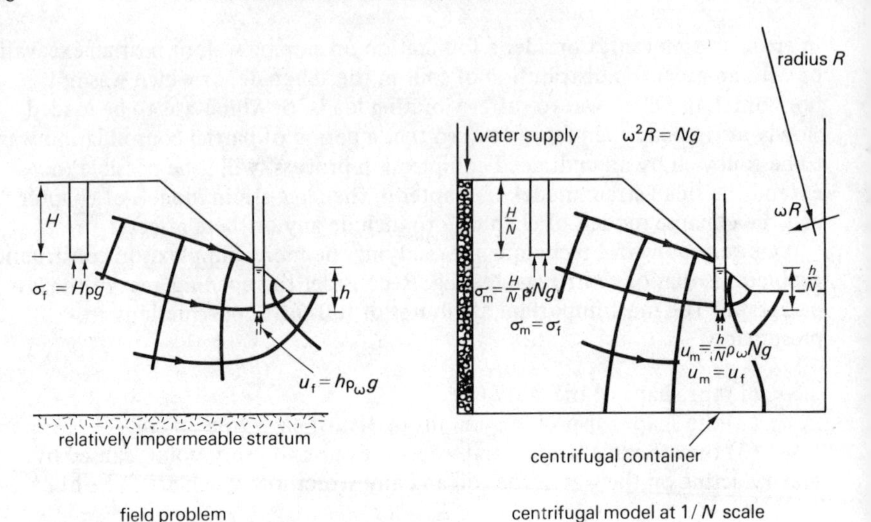

Figure 10.44 *Replication of stresses by centrifuging a model comprising identical soils*

modeller be reminded with a jolt that his physical model was only as good as his mental model and only as representative as the soil with which he built it.

Concentrate therefore on the application of judgement with regard to which ideas to pursue, and on the taking of decisions which account for the inability of mere ideas to replicate reality. The mechanics of soil represents the intellectual meat in this rather challenging sandwich.

References

Amerasinghe, S. F., and Parry, R. H. G., Anisotropy in heavily overconsolidated kaolin. *J. Geotech. Div. Am. Soc. civ. Engrs,* 101 (1975) 1277–93

Arthur, J. R. F., Dunstan, T., Al-Ani, Q. A. J. L., and Assadi, A., Plastic deformation and failure in granular media, *Géotechnique*, 27 (1977) 53–74

ASCE Conference on In-situ Measurement of Soil Properties, North Carolina State University, June 1975

Atkinson, J. H., and Potts, D. M., Stability of a shallow circular tunnel in cohesionless soil, *Géotechnique*, 27 (1977) 203–15

BGS Conference on the Settlement of Structures, Cambridge University, April 1974 (Pentech, London, 1974)

Bishop, A. W., The use of the slip circle in the stability analysis of slopes, *Géotechnique*, 5 (1955) 7–17

Bishop, A. W., Shear strength parameters for undisturbed and remoulded soil specimens, *Stress-Strain Behaviour of Soils, Proc. Roscoe Memorial Symp., Cambridge, March 1971,* (ed. R. H. G. Parry) (Foulis, Yeovil, 1972) 3–58

Bishop, A. W., and Henkel, D. J., *The Measurement of Soil Properties in the Triaxial Test*, 2nd ed. (Arnold, London, 1962)

Bishop, A. W., Green, G. E., Garga, V. K., Andresen, A., and Brown, J. D., A new ring shear apparatus and its application to the measurement of residual strength, *Géotechnique*, 21 (1971) 273–328

Bjerrum, L., Geotechnical properties of Norwegian marine clays, *Géotechnique*, 4 (1954) 49–69

Blyth, F. G. H., and de Freitas, M. H. *A Geology for Engineers* (Arnold, London, 1974)

Bolton, M. D., Choudhury, S. P., and Pang, P. L. R., Modelling reinforced earth, *Proc. Symp. Reinforced Earth, Transport and Road Research Laboratory and Heriot-Watt University, Edinburgh, September, 1977*

Borowicka, H., Influence of rigidity of a circular foundation slab on the distribution of pressures over the contact surface, *Proc. 1st Int. Conf. Soil Mech., Cambridge Mass.*, 2, (1936) 144–9

Bowden, F. P., and Tabor, D., *The Friction and Lubrication of Solids*, 2 vols (Oxford University Press, 1964)

Bransby, P. L., and Milligan, G. W. E., Soil deformations near cantilever sheet pile walls, *Géotechnique,* 25 (1975) 175–95

BS 1377: 1975 Methods of test for soil for civil engineering purposes

Burland, J. B., and Wroth, C. P., Settlement of buildings and associated damage, *BGS Conference on the Settlement of Structures, Cambridge University, April 1974* (Pentech, London, 1974) 611–54

Burland, J. B., Longworth, T. I., and Moore, J. F. A., A study of ground movement and progressive failure caused by a deep excavation in Oxford clay, *Géotechnique*, 27 (1977) 557–91
Cedergren, H. R., *Seepage, Drainage, and Flow Nets* (Wiley, New York, 1967)
Chandler, R. J., Shallow slab slide in the Lias clay at Uppingham, *Géotechnique*, 20 (1970) 253–60
Chandler, R. J., Lias clay: the long term stability of cutting slopes, *Géotechnique*, 24 (1974) 21–38
Chandler, R. J., and Skempton, A. W., The design of permanent cutting slopes in stiff fissured clays, *Géotechnique*, 24 (1974) 457–66.
Cornforth, D. H., Prediction of drained strength of sands from relative density measurements, Am. Soc. Test. Mat., Special Tech. Publ. No. 523. (1973) 281–303
CP 102: 1973 Protection of Buildings against Water from the Ground
CP 2003: 1959 Earthworks
CP 2004: 1972 Foundations
D'Appolonia, D. J., D'Appolonia, E., and Brisette, R. F., Discussion on Settlement of Spread Footings on Sand, *J. Soil Mech. and Found Div., Am. Soc. Civ. Engrs*, 96, (1970) SM2, 754–61
De Beer, E., Bearing capacity and settlement of shallow foundations on sand, *Proc. Symp. Bearing Capacity and Settlement of Foundations, Duke University, 1965*, 15–33
De Josselin de Jong, G., Rowe's stress-dilatancy relation based on friction, *Géotechnique*, 26 (1976) 527–34
de Mello, V., The standard penetration test – a state of the art report, *4th Pan Am. Conf. on Soil Mech. and Found. Engng, Puerto Rico, 1971*, 1, 1–86
Gibson, R. E., The analytical method in soil mechanics, *Geotéchnique*, 24 (1974) 115–40
Gibson, R. E., and Henkel, D. J., Influence of duration of tests at constant rate of strain on measured 'drained' strength, *Géotechnique*, 4 (1954) 6–15
Gould, J. P., A study of shear failure in certain tertiary marine sediments, *Res. Conf. on Shear Strength of Cohesive Soils, Am. Soc. Civ. Engrs, Boulder, 1960*, 615–41
Harr, E., *Groundwater and Seepage* (McGraw-Hill, New York, 1962)
Henkel, D. J., The relationships between the strength, pore-water pressure and volume-change characteristics of saturated clays, *Géotechnique*, 9 (1959) 119–3
Heyman, J., *Coulomb's Memoir on Statics: an essay in the history of civil engineering* (Cambridge University Press, 1972)
Hird, C. C., Marsland, A., and Schofield, A. N., The development of centrifugal models to study the influence of uplift pressures on the stability of a flood bank, *Géotechnique*, 28 (1978) 85–106
Hirschfeld, R. C., and Poulos, S. J., *Embankment Dam Engineering* (Wiley, New York, 1973)
Hughes, J. M. O., Wroth, C. P., and Windle, D., Pressuremeter tests in sands, *Géotechnique*, 27 (1977) 455–77
Hyde, A. F. L., and Brown, S. F., The plastic deformation of a silty clay under creep and repeated loading, *Géotechnique*, 26 (1976) 173–84
Lambe, T. W., and Whitman, R. V., *Soil Mechanics* (Wiley, New York, 1969)

Marsland, A., The shear strength of stiff fissured clays, *Stress-Strain Behaviour of Soils, Proc. Roscoe Memorial Symp., Cambridge, March 1971*, (ed. R. H. G. Parry) (Foulis, Yeovil, 1972) 59–68

Meyerhof, G. G., The bearing capacity of foundations under eccentric and inclined loads, *Proc. 3rd Int. Conf. Soil Mech. and Found. Engng, Zurich, 1953*, 1, 440–5

Meyerhof, G. G., Bearing capacity and settlement of pile foundations, *Geotech. Div. Am. Soc. Civ. Engrs*, 102, (1976) 197–228

Milligan, G. W. E., and Bransby, P. L., Combined active and passive rotational failure of retaining wall in sand, *Géotechnique*, 26 (1976) 473–94

Palladino, D. J., and Peck, R. B., Slope failures in an overconsolidated clay, Seattle, Washington, *Géotechnique*, 22 (1972) 563–95

Palmer, A. C., Undrained expansion of a cylindrical cavity in clay: a simple interpretation of the pressuremeter test, *Géotechnique*, 22, (1972) 451–7

Parry, R. H. G., Triaxial compression and extension tests on remoulded saturated clay, *Géotechnique*, 10, (1960) 166–80

Parry, R. H. G., and Amerasinghe, S. F., Components of deformation in clays, *Proc. Symp. on Plasticity and Soil Mech. Cambridge, September 1973* (ed. A. C. Palmer) 108–26

Parry, R. H. G., Estimating foundation settlements in sand from plate bearing tests, *Géotechnique*, 28 (1978) 107–18

Peck, R. B., Advantages and limitations of the observational method in applied soil mechanics, *Géotechnique*, 19 (1969) 171–87

Peck, R. B., Hanson, W. E., and Thornburn, T. H., *Foundation Engineering* (Wiley, New York, 1974)

Ponce, V. M., and Bell, J. M., Shear strength of sand at extremely low pressure, *J. Soil Mech. and Found. Engng Div., Am. Soc. Civ. Engrs*, 97, (1971) 625–38

Poulos, H. G., and Davis, E. H., *Elastic Solutions for Soil and Rock Mechanics* (Wiley, New York, 1974)

Rehnman, S. E., and Broms, B. B., Lateral pressures on basement wall: results from full-scale tests, *Proc. 5th Eur. Conf. Soil Mech. and Found. Engng, Madrid, 1972*, 189–97

Roscoe, K. H., and Burland, J. B., On the generalised stress-strain behaviour of 'wet' clay, in *Engineering Plasticity*, ed. J. Heyman and F. A. Leckie (Cambridge University Press, 1968) 535–609

Rowe, P. W., Cantilever sheet piling in cohesionless soil, *Engineering* (1951) 316–19

Rowe, P. W., A theoretical and experimental analysis of sheet pile walls, *Proc. Inst. Civ. Engrs*, 4 (1955) 32–69

Rowe, P. W., The stress-dilatancy relation for static equilibrium of an assembly of particles in contact, *Proc. R. Soc., A*, 269 (1962) 500–27

Rowe, P. W., The relation between the shear strength of sands in triaxial compression, plane strain and direct shear, *Géotechnique*, 19 (1969) 75–86

Rowe, P. W., The relevance of soil fabric to site investigation practice, *Géotechnique*, 22 (1972) 195–300

Rowe, P. W., Oates, D. B., and Skermer, N. A., The stress-dilatancy performance of two clays. *Laboratory shear testing of soils,* Am. Soc. Test. Mat., Special Tech. Publ. No. 361, (1964) 134–43

Sanglerat, G., The penetrometer and soil exploration (Elsevier, Amsterdam, 1972)
Sangrey, D. A., Naturally cemented sensitive soils, *Géotechnique*, 22 (1972) 139–52
Schlosser, F., and Long, N. T., Recent results in French research on reinforced earth, *J. Const. Div. Am. Soc. Civ. Engrs*, 100 (1974) 223–37
Schmertmann, J., Measurement of in-situ shear strength, *Proc. Conf. on in-situ measurement of soil properties, Am. Soc. Civ. Engrs., North Carolina State University, 1975*, 2, 57–138
Schofield, A. N., and Wroth, C. P., *Critical State Soil Mechanics* (McGraw-Hill, New York, 1968)
Skempton, A. W., *The Bearing Capacity of Clays* (Building Research Congress, London, 1951)
Skempton, A. W., Long term stability of clay slopes, *Géotechnique*, 14, (1964) 77–101
Skempton, A. W., First time slides in overconsolidated clays, *Géotechnique*, 20 (1970) 320–4
Taylor, D. W., *Fundamentals of Soil Mechanics* (Wiley, New York, 1948)
Terzaghi, K., *Theoretical Soil Mechanics* (Wiley, New York, 1943)
Terzaghi, K., and Peck, R. B., *Soil Mechanics in Engineering Practice* (Wiley, New York, 1967)
Timoshenko, S. P., and Goodier, J. N., *Theory of Elasticity* (McGraw-Hill, New York, 1951)
Tomlinson, M. J., *Foundation Design and Construction*, 3rd ed. (Pitman, London, 1975)
Tsagareli, Z. V., New methods of lightweight wall construction, Stroiizdat, Moscow, 1967 (in Russian).
Vidal, H., Principle of reinforced earth, *Highway Research Record* (U.S. National Research Council) No. 282, (1969) 1–16
Weltman, A. J., and Little, J. A., A review of bearing pile types, Dept. of Environment and CIRIA Piling Development Group, Report PG 1 (CIRIA, London, 1977)
Whitaker, T., *The Design of Piled Foundations* (Pergamon, Oxford, 1970)
Wroth, C. P., and Bassett, R. H., A stress-strain relationship for the shearing behaviour of a sand, *Géotechnique*, 15, (1965) 32–56
Wroth, C. P. and Hughes, J. M. O., An instrument for the in-situ measurement of the properties of soft clays, *Proc. 8th Int. Conf. Soil Mech. and Found. Engng, Moscow, 1973*, 1.2, 487–94

Index